Integrated Analytical Systems

Series editor
Radislav A. Potyrailo
Niskayuna, USA

This comprehensive and interdisciplinary series offers the most recent advances in all key aspects of development and applications of modern instrumentation for chemical and biological analysis on the microscale.

These key aspects will include (1) innovations in sample introduction through micro- and nano-fluidic designs, (2) new types and methods of fabrication of physical transducers and ion detectors, (3) materials for sensors that became available due to the breakthroughs in combinatorial materials science and nanotechnology, and (4) innovative data processing and mining methodologies that provide dramatically reduced rates of false alarms.

Clearly, a true multidisciplinary effort is required to meet objectives for a system with previously unavailable capabilities. This cross-discipline fertilization is driven by the expanding need for chemical and biological detection and monitoring and leads to the creation of instruments with new capabilities for new demanding applications. Indeed, instruments with more sensitivity are required today to analyze ultra-trace levels of environmental pollutants, pathogens in water, and low vapor pressure energetic materials in air. Sensor devices with faster response times are desired to monitor transient in-vivo events and bedside patients. More selective instruments are wanted to analyze specific proteins in vitro and analyze ambient urban or battlefield air. For these and many other applications, new features of modern microanalytical instrumentation are urgently needed. This book series is a primary source of both fundamental and practical information on both the current state of the art and future directions for microanalytical instrumentation technologies. This book series is addressed to the rapidly growing number of active practitioners and developers and those who are interested in starting research in this direction, directors of industrial and government research centers, laboratory supervisors and managers, students and lecturers.

More information about this series at http://www.springer.com/series/7427

Jin-Ming Lin
Editor

Cell Analysis on Microfluidics

Springer

Editor
Jin-Ming Lin
Department of Chemistry
Tsinghua University
Beijing
China

ISSN 2196-4475 ISSN 2196-4483 (electronic)
Integrated Analytical Systems
ISBN 978-981-10-5393-1 ISBN 978-981-10-5394-8 (eBook)
https://doi.org/ 10.1007/978-981-10-5394-8

Library of Congress Control Number: 2017952915

Printed on acid-free paper

This Springer imprint is published by Springer Nature
The registered company is Springer Nature Singapore Pte Ltd.
The registered company address is: 152 Beach Road, #21-01/04 Gateway East, Singapore 189721, Singapore

Preface

It is an old Chinese saying that "proper tools for proper works." To crack the mystery inside cells, special tool is demanded. At this moment, microfluidic chip represents an efficient and capable tool for cell analysis and biological study, and its rapid development is prompted by the great demands and wishes to one more step further the cellular biology research. The beginning of cellular biology studies started from the employment of petri dish as cell culture container. And until now, after repeated optimization and modification, dishes are still utilized in every biochemical laboratory. Typically, cell population at millions or tens of millions scale are cultured, stimulated, and harvested in one dish, and through these operations, researchers can carry out drug screening test and cellular components' extraction from different cell batches. Depending on experiments consuming large cell population, results can be highly accurate and reliable, but also averaged. Imaging that cells of different phases in mitotic cycle and of different physiological situations and metabolism activities are equally averaged in signal calculation from one dish, individual behavior is completely wiped off. If we are going to look closer to cells, to reveal changes inside cell body and concern more about cell migration, division and apoptosis as well as cell–cell interaction by direct contact, a handy tool with cell manipulation and analysis capability can be indispensible. Under this circumstance, microfluidics was brought to the center of stage.

Unique advantages of microfluidics include less sample consumption, miniaturization and integration and function-oriented design. The invention of first microfluidic device can be dated back to 1970s, but not until 1990s did academic society start paying attention to this tiny chip. The bottleneck for its large-scale utilization is the expensive infrastructure and high cost to produce single piece of chip. After the successful fabrication of PDMS chip by soft lithography reported in late 1990s, PDMS-based microfluidic chip has spread to almost every laboratory inside university and academic institute. The most significant features of microfluidic chip are its microscale channel and flexible structure design. The dimension of microfluidic channel is on the same level as single cell, and therefore by flowing cell suspension inside microchannel and harnessing hydrodynamics in microregion, various manipulations such as cell transportation, isolation, sorting, and lysis are realized. And different microfluidic techniques are invented according to different physics used. Besides, to realize cell components analysis, several types of detector can be coupled with microchip to implement online or offline analysis, such as noninvasive optical and electrochemical detection or invasive mass and electrophoresis.

As the continuous development of microfluidics and broadening of its application, more attention has been paid to employing microfluidic system to build up in vitro cell culture model or organ mimicry. Normally, complex cell culture model involves co-culture of multiple cell types, embedding cells into biocompatible hydrogels and adjusting interactions between cells and matrixes. Many works have been reported which made breakthroughs in improving bio-functionality of cell culture model, but fully recapitulation of organ function is still far away. In 13 chapters of this book, researchers and experts with deep insights in different fields present a detailed review and discussion over the design, formatting, application, and development of microfluidic chip related to cell biology research. Several techniques are being paid special attention, such as 3D cell culture, microfluidic droplet technique, and microfluidic chip-mass spectrometry interfaces.

The first part of this book concentrates on the history of invention and evolution of microfluidic chip and also the prospect of its future development. Chapter 1 (by Luyao Lin) summarizes from different aspects such as the choice of supporting materials, fabrication methods, functional units of microfluidic chips and also gives a brief discussion about its recent development. For chip fabrication, multiple 3D printing techniques that have been widely used in academic researches as well as industrial intention are introduced and compared to explore their further potential in chip formatting. Chapter 2 by Ziyi He highlights the advances in realizing cell culture, cell manipulation, cell stimulation, and cell analysis on chip. The highly flexible chip design permits implementation and integration of different functional units on the same chip, and makes microchip a powerful processing platform.

The second part of this book targets on several specific applications of microfluidic platform in cell biology researches. Chapter 3 by Qiushui Chen introduces two schematics to realize cell isolation, that is, physical and affinity-based cell isolation. Applications such as CTCs' capture and recognition and stem cells' purification are demonstrated. Chapter 4 by Linglu Yi depicts the critical factors influencing cell culture behaviors such as culture matrix and physical/chemical gradients. Nondestructive observation methods like optical and electrochemical detection are reviewed. In Chap. 5, Jinxin Dou introduces the application of microchip in studying cell migrating behavior and deciphering the influencing factors like chemotaxis, nutrients, and matrix rigidity. Ruizhi Ning of Chap. 6 presents the evolution of matrix materials for cell culture and highlights some novel biocompatible hydrogels with notable functions in establishing 3D cell culture environment and rehabilitating cell functionality. In Chap. 7, Junming Wang introduces the basic principle, different types of nozzle generators and compatible detectors of microfluidic droplet technology. Normally, optical observation through fluorescence labeling or simply bright field is the most convenient detecting method to determine the existence of target cells or biomarkers inside droplets, and this noninvasive method allows continuous observation without interfering cell biological process. Other detectors such as mass spectrometry and electrochemical device are also included in this chapter. In Chap. 8, another unique advantage of microfluidics for cell analysis is well exemplified by Qiushi Huang. The content in this chapter involves the generation of single cells as well as the analysis of cellular components. Single cell sorting and lysis on microfluidics through controlling of hydrodynamics as well as other physical principles are described in detail. Chapter 9 by Ling Lin discusses the scheme of "From sample to data" by microchip online preparation and pretreatment with coupled mass detection, and varieties of chip-mass interfaces to achieve high throughput analysis are introduced.

The third part focuses on the construction of complex cell culture model on microfluidics and the on-chip analysis of cell metabolites as well as other cellular components. Chapter 10 by Jing Wu enumerates the latest design and realization of various chip functions about cell manipulation and analysis by taking advantage of physics phenomena such as electronic and magnetic field and surface acoustic wave, and their applications in genetic and protein analysis are introduced. In Chap. 11, Mingsha Jie describes the construction of "organ on chip" system for drug screening and evaluation. Duplication of real organ on chip includes not only the incorporation of multiple cell types, but also the implantation of complex cell–cell interactions. Chapter 12 by Xuexia Lin introduces the on-chip analysis methods of cell metabolites, among which self-designed fluorescent probes are specially emphasized. Aptamers with high degree of structure design freedom have been widely used for biomarker probes to achieve on-site and continuous fluorescence analysis, and therefore are an ideal candidate for microchip-based detection. In Chap. 13, Lin Zhou gives an overview about microorganism culture on chip and protocols of bacteria-related testing. There is no essential difference between culturing bacteria and eucaryote cells, but the rapid proliferation of bacteria makes it hardly a companion in direct co-culture with mammal cells.

Microfluidic platform has provided a powerful tool for microvolume sample analysis and cellular biology study, but there are still challenges waiting to be overcome. Highly remarked in novel device development, few of the microchip prototypes have been commercially transformed into established instruments, which may be due to its lack of competence comparing to well-validated conventional devices. But in recent years, this embarrassment has been alleviated and we have witnessed more and more large apparatus and equipment applying microchips as their core functional units. Some scholars may also question the accuracy and reliability of microchip in biological applications. It has to be admitted that cell-based microchip model is yet capable to replicate animal or clinical testing in drug screening or to verify a potential signal pathway among living organism. But as far as we consider, the structural mimicking on microchip is actually the first step on the road to build up more complex and functional in vitro organ models, and in some researches, on-chip method has made cell–cell interactions more detectable. Although chapters in this book only provide a brief review of microfluidics and limited collection of its applications on cell analysis, plenty of knowledge is included about almost every element to establish a microfluidic platform. And we hope this book can be helpful to those ardent researchers and students who wish to know more, explore more, and achieve more in the fields of microchips and cellular biology. Also, I should gratefully thank the authors for their hard working and contributions to this book.

Beijing, China
June 2017

Jin-Ming Lin

Contents

Chapter 1
Design and Preparation of Microfluidics Device

Luyao Lin and Jin-Ming Lin

Abstract A rapid development has been witnessed since the birth of this miniaturized analytical equipment about forty years ago. And microfluidics, a discipline studying the manipulation and analysis of micro volume sample is also attracting the attention of academics with the annually increasing publications. At present, microfluidic device has become a powerful platform with diverse functions implicated in the fields of cellular biology, environmental study, food safety monitoring and micro synthesis, and has crossed the barrier of academic study into the daily life of normal customers. In this chapter, we will focus on several aspects concerning the design and fabrication of microfluidic devices, such as the development of device material, different methods of chip fabrication and functional units to realize purpose of intention. And in the end of this part, a brief future perspective is given to evaluate the potential applications of microfluidic device, especially with other portal devices like smart phone.

Keywords Microfluidics design · Chip fabrication · Functional units integration

1.1 Introduction

Microfluidic devices are termed as micro total analysis system (μTAS) or laboratory on a chip (LOC), with the first reported work dating back to 1975. A gas chromatography at micro scale was reported then [1]. However, the academics did not respond actively to the new-born of this epoch-marking device firstly, but rather turned indifference to it with no follow up studies in a long term. Not until 1990s did the microfluidics device start to receive broad attention. The rapid development of integrated circuit processing technique and computer industrial, along with the urgent demand of bio-molecules separation devices redirected analytical devices to

L. Lin · J.-M. Lin (✉)
Department of Chemistry, Tsinghua University, Beijing 100084,
People's Republic of China
e-mail: jmlin@mail.tsinghua.edu.cn

J.-M. Lin (ed.), *Cell Analysis on Microfluidics*, Integrated Analytical Systems,
https://doi.org/10.1007/978-981-10-5394-8_1

a path of miniaturization. Manz and coworkers designed an open-channel liquid chromatography based on microfabricated silicon chip in 1990 [2]. It is noteworthy that Manz is also the proposer of μTAS concept [3]. In this work, the authors displayed integrated functions of sample pretreatment, separation and detection on one chip. In the last 20 years, researches concerning novel microfabrication strategies [4], development of new chip material [5] and application of microchip to detection [6, 7] have been increasingly studied and published. And microfluidic chip has become a multi-functional platform for various targets analysis (Fig. 1.1a) [8] and a powerful tool for biomedical applications (Fig. 1.1b) [9].

The advantages of applying microfluidic devices to analytical testing include: less consumption of reagents and samples, rapid reaction in microscale chamber,

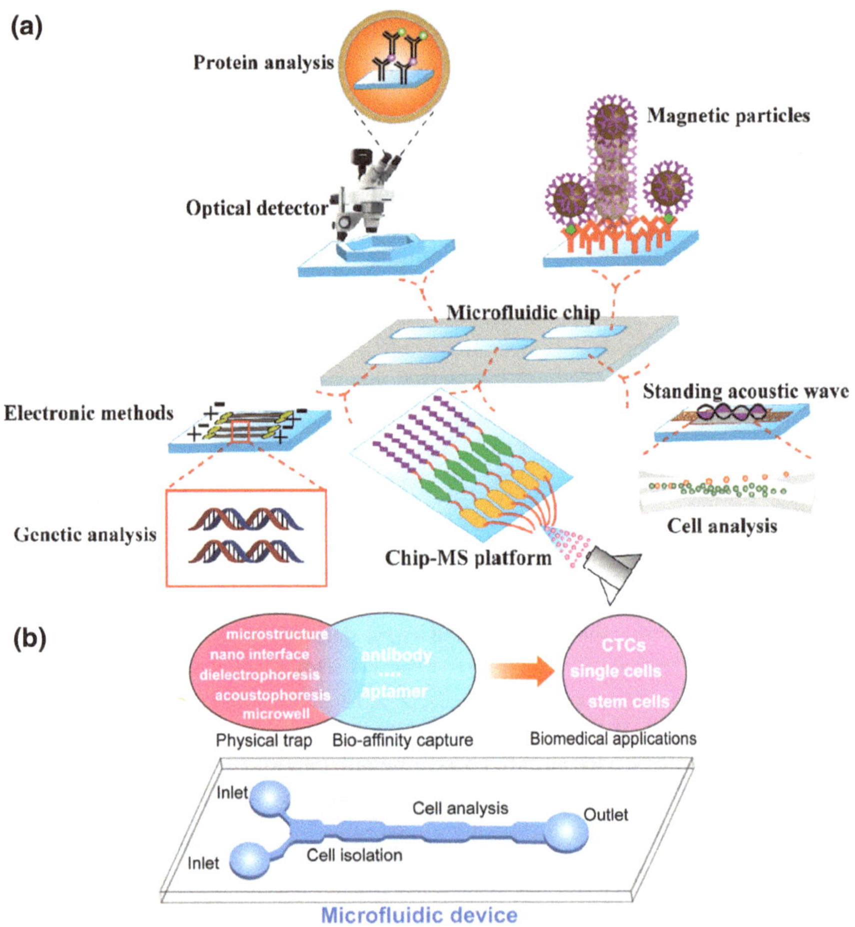

Fig. 1.1 Illustration of lab-on-chip concept. **a** Multifunctional microfluidic platform(Reprinted with permission from Ref. [8]). **b** Microfluidic analysis process for biomedical applications (Reprinted with permission from Ref. [9])

high integration allowing for high-throughput parallel test, miniaturization for good portability, diverse chip designs for various functioning intentions, and precise control of cell culturing environment in cell in vitro models. The processing technique of microfluidic chip originated from the semi-conductor industrial, and was further developed by micro-electromechanical systems fabrication. The wide utilization of microfluidic chip in research fields is benefiting from its analytical ability and manipulation of microscale objects. The fluids dynamics in micro scale is distinct from macro world (Fig. 1.2a) [7]. A variant as Re is used to interpret the characteristics of flow in micro regions. The connection of Re with fluid properties as well as channel dimensions is illustrated in following equation:

$$\mathrm{Re} = \frac{\rho v d}{\mu}$$

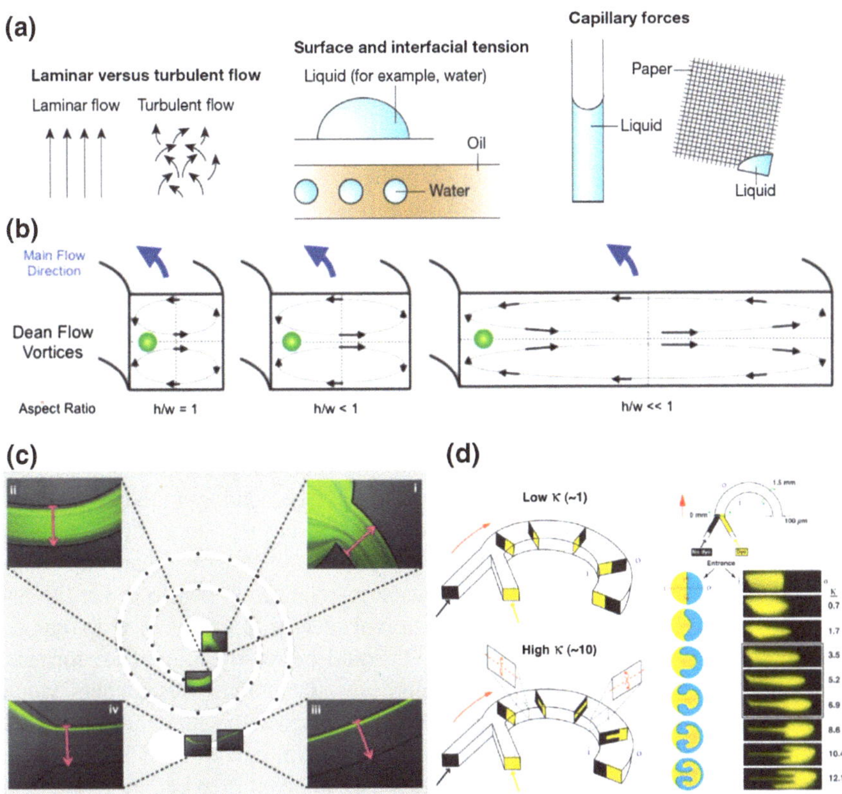

Fig. 1.2 Schematic of flow dynamics in microfluidic region. **a** Laminar and turbulent flow, surface tension and capillary forces (Reprinted with permission from Ref. [7]). **b** Generation of Dean Flow in curved micro channel. **c** Flow focusing in curved channel induced by Dean Flow (Fig. 1.2**b** and **c** were reprinted with permission from Ref. [12]). **d** Dean flow enhances flow mixing (Reprinted with permission from Ref. [13])

where ρ represents fluid density (kg/m^3), v stands for average linear velocity (m/s), d for the dimension of channel and μ for fluid viscosity. When Re is larger than 2000, the main characteristic of flow is turbulent with vortexes happening. For cases when Re is smaller than 1, the flow is steady and laminar. In micro channels, Re is basically among the section belonging to laminar flow due to the miniaturized channel dimensions, thus making the flow highly predictable. Lack of convection leads to slow fluid mixing by diffusion. Another difference in microfluidic region is that gravity no longer dominates the object motion as it does in macro world. Surface interfacial tension and capillary forces become two major influencing factors of objects behaviors [10, 11]. Besides, there are Dean flow induced inertia forces (Fig. 1.2b) [12] which can be used for particle focusing (Fig. 1.2c) [12] and enhancing mixing process (Fig. 1.2d) [13].

One of the big future developments for microfluidic chip is to design and prepare function oriented chip prototypes with optimized and facile fabrication steps. And generally, the basic rule for chip design rests in taking the maximum advantage of microscale fluid physics to realize analytical test as well as objects manipulation. Therefore various functional units were invented and integrated into microchip along with its development, such as surface tension valve [14], concentration gradient generator [15, 16] and droplet nozzle[17–19]. To match up the micro analytical system, popular detection methods like fluorescent [20], electrochemical [21, 22] and MS [23, 24] detectors as well as other biosensors [25, 26] were combined to the microchip. In material selection, the choice for chip fabrication evolved from inorganic material of silicon and glass to elastomer, thermoset and thermoplastic, and again further to the paper based and other functional hybrid materials. At present, the microfluidic device has become one of the most popular research tools in analytical chemistry and bio-analytical fields with frequent applications in environmental pollutants analysis [27, 28], food safety [29, 30], disease diagnosis [31, 32] and construction of cell in vitro culture model [33–35].

However, it has to be admitted that the commercialization of microfluidic devices is not successful despite of its broad application in academics and research institutes [36]. Reason for this mediocrity is that comparing to other maturely established detection methods, microfluidic devices are still short of some killer applications, which makes researchers prefer conventional assays than microchips [37]. On the other hand, the microfluidics commercialization is also limited by the difficulties in mass production. The realization of specific functions on chip requires specifically designed chip structure, and it would be overly difficult to format a universal chip model for all kinds of purposes. Commonly researchers would design and prepare homemade chips themselves, especially after the widely utilization of PDMS as a replacement for glass and silicon in chip fabrication and the acceptance of facial and cheap soft lithography as standard preparation methods. In 2013, the market scale of microfluidic devices was about 1.6 billion dollars, with compound annual growth rate of 18–29% to a predicted size of 3.6–5.7 billion dollars in 2018 [38]. The rapid development of this industrial is largely depending on the fast expansion of genome sequencing [39, 40] and in vitro diagnosis applications.

In this chapter, we will have a detailed discussion about the development of microfluidic devices materials and fabrication methods. The design of functional units as well as the field's prospect will also be covered.

1.2 Development of Microfluidic Chip Material

At the primitive stage of microfluidic chip, glass and silicon were the basic materials for chip fabrication. With the invention of new materials and improvement of processing technique, a flood of materials were applied in microchip preparation(Fig. 1.3a) [41]. Currently, the chip materials can be roughly divided into three categories as polymers (Fig. 1.3b) [42], inorganic(Fig. 1.3c) [43] and paper based materials (Fig. 1.3d) [44]. And the material properties are among the basic considering factors for material selection, such as air tightness, biocompatibility, conductivity, transparency and solvent tolerance. For example, glass and silicon are with good stability and highly tolerant to solvents, while the polymer materials have poor resist to solvents, and become partly dissolved or even disintegrate among organic solvents. Besides, the choice of material should also meet the requirement from chip miniaturization and integration. In laboratory environment, the major concern of researchers is how to realize rapid prototype of newly designed chip model and conduct performance testing, but for commercial intention lowing the fabrication cost of product and improving the reliability during usage would be more important. Next, we will have a closer look of the material development history.

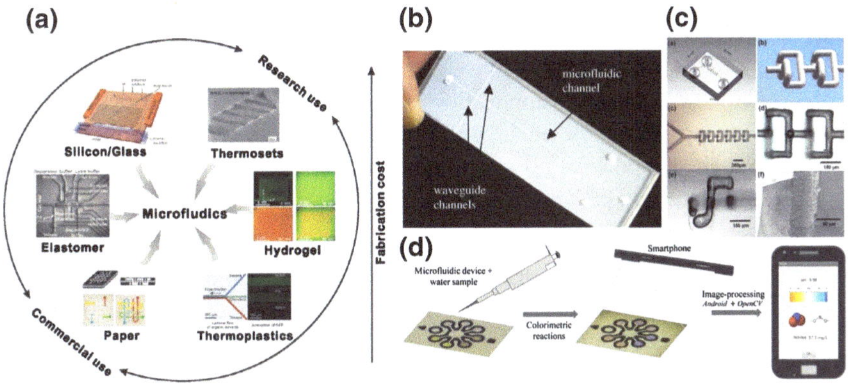

Fig. 1.3 Material choice of microfluidic device preparation. **a** Examples of microfluidic chip material (Reprinted with permission from Ref. [41]). **b** PDMS based micro chip (Reprinted with permission from Ref. [42]). **c** Glass microfluidic channel by laser direct writing method (Reprinted with permission from Ref. [43]). **d** Paper based multiple targets analysis chip (Reprinted with permission from Ref. [44])

1.2.1 Inorganic Material in Chip Preparation

The earliest type of microfluidic device such as miniaturized gas [1] or liquid chromatography [2] and capillary electrophoresis devices was prepared by glass and silicon, under the processing technique inherited from semiconductor and micro-electronics industrials [45]. The advantages of glass and silicon chip are fine chemical stability in high temperature, good strength and thermal conductivity and processability for nano-resolution surface patterns [46, 47].

Silicon is the first material which has been used in microfluidics devices, but was once being replaced by glass and polymers. With excellent thermal conductivity, the use of silicon in some devices is intended to achieve a uniform distribution of temperature [48]. The modification of silicon is easy through the reaction between silylating reagent and silicon hydroxyl exposed on surface [49]. Silicon material has been widely applied in the manufacture of complex 3D structures such as micro reaction chamber [50, 51], cell culture unit [52–54], electrospray nozzle [55] and solvent extraction device [56]. However, the weakness of silicon chip is also significant. Due to lack of elasticity, it is hard to produce flow control elements like pump and valve from silicon. The opacity of material makes optical detection almost incompatible with silicon chip. But still the weakness can be circumvented by comprising hybrid microchip with transparent and elastic material. It is safe to conclude that because of the very appearance of hybrid material chip, the application of silicon in microfuidics is once again expanded.

As contrast, glass is highly transparent with low fluorescence background, which allows for combination with real-time monitoring methods. The thermal conductivity of glass is relatively poor and has been utilized in generation of temperature gradients on device. Fluid flow in glass channel is witnessed with rare non-specific absorption and the modification of glass through silicon hydroxyl reaction is also easy. The air tightness of glass makes it unsuitable for long term cell culture. Glass is well-known for its role as channel material in capillary and chip electrophoresis [57]. Benefiting from high material electroosmotic mobility, fluid in glass capillary is driven by strong electroosmotic flow without the need for external pumping device and being rapidly mixed at good reproducibility. In addition, glass is widely applied in PCR equipment [58] and gas chromatography [59].

Currently, the main processing techniques of glass and silicon material include:

(a) Standard photolithography [60]. Firstly, photoresist is spin coated to form a thin layer over the surface of glass or silicon, and then exposed to UV radiation against predesigned mask. Next, the patterns transferred from photo mask is developed by developer solution to reveal bare glass or silicon surface for further etching step. The exposed surface is etched by hydrofluoric acid solution for a concave channel. After removal of the photoresist, channels are finally sealed to a plain surface.
(b) Surface micromachining. As shown in Fig. 1.4a, the fabrication involves steps of patterning sacrificial layer on material surface, casting of amorphous material over the sacrificial layer to form structural layer and etching of sacrificial layer

to achieve the corresponding channels [61]. Comparing to other processing methods, surface micromachining is more capable in fabricating multi-layer devices (Fig. 1.4b) and nano channel due to the minimum thickness of sacrificial layer to a few nano meters [62, 63]. Because multiple steps are required, this machining method would take a long time to finish. Figure 1.4c presents micro featured devices fabricated by surface machining [64].

(c) Buried-channel techniques. In this method, a tunnel is carved through anisotropic deep reactive ion etching (DRIE) in vertical direction. Then the tunnel walls are blocked by chemical vapor deposition (CVD) only leaving a reaction site for further etching to form a cylinder-shaped channel [65–67]. This method is particularly suitable for preparation of high aspect radio channel.
(d) Bulk micromachining. Bulk micromachining is a common micro fabrication method usually employed in preparing micro or nano scale channels [68, 69]. Formation of channel is realized due to the different etching rates for different lattice planes of silicon substrate in electrochemical corrosion. The channel depth is regulated by controlling the etching time, but the accuracy is often unsatisfactory. Another etching method with better accuracy is to dope boron into selected regions, where after doping slower etching rate is witnessed. However the doping process might cause stress in structure and lead to easy break.
(e) Laser direct writing (LDW). Laser direct writing is a method of fabrication employing laser beam to carve channels directly specially for photosensitive glass material [70–72].
(f) Electrochemical micromachining. Electrochemical machining was firstly used in aerospace industrial half century ago, and when it was optimized and adapted for fabrication at micro scale region, a feasible method for micro structure building was born. Electrochemical micromachining utilizes cathode and anode to simultaneously and electrically depositing and electrolyzing material at the

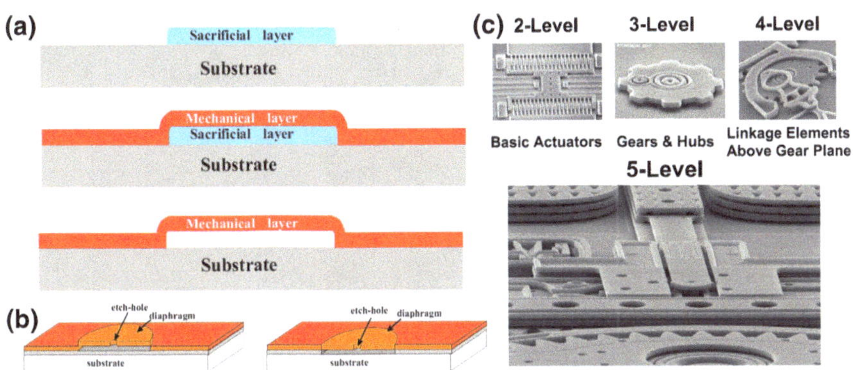

Fig. 1.4 Schematic of surface micromachining. **a** Three steps of surface machining. **b** Illustrations of surface machined devices (**a** and **b** were reprinted with permission from Ref. [61]). **c** SEM photos of surface machined devices (Reprinted with permission from Ref. [64])

same time, and was suitable for processing silicon (Fig. 1.5a, b) [73] and metallic (Fig. 1.5c) [74] material. Combined with computer controlled processor or scanning tunneling microscopy (Fig. 1.5d) [75], high-resolution and three dimensional patterns can be made on material surface with micro scales or even lower dimension (Fig. 1.5e) [76].

Generally the processing steps for glass and silicon based microchip include etching, bonding and connecting to external equipment, and the etching resolution determines the smallest structure that could be achieved on glass and silicon. For glass, wet etching by hydrofluoric acid has been frequently used and resulted in arc shape of the channel side walls [77]. Other etching methods such as plasma etching would be slow and hard to control. For silicon, method of reactive ion beam etching is well established and has been applied in monocrystalline silicon processing to form channels with vertical side walls [78, 79]. Judging from processability of glass and silicon, the disadvantages of their application in chip preparation are high material cost, tedious fabrication steps, employment of dangerous chemical reagents, high standards of bonding condition for chip seal such as high temperature and pressure and ultra-clean environment. Moreover, functional structures like microvalve are hard to be integrated into glass or silicon microchips, which would require additional use of PDMS to assemble hybrid chip. There are other inorganic

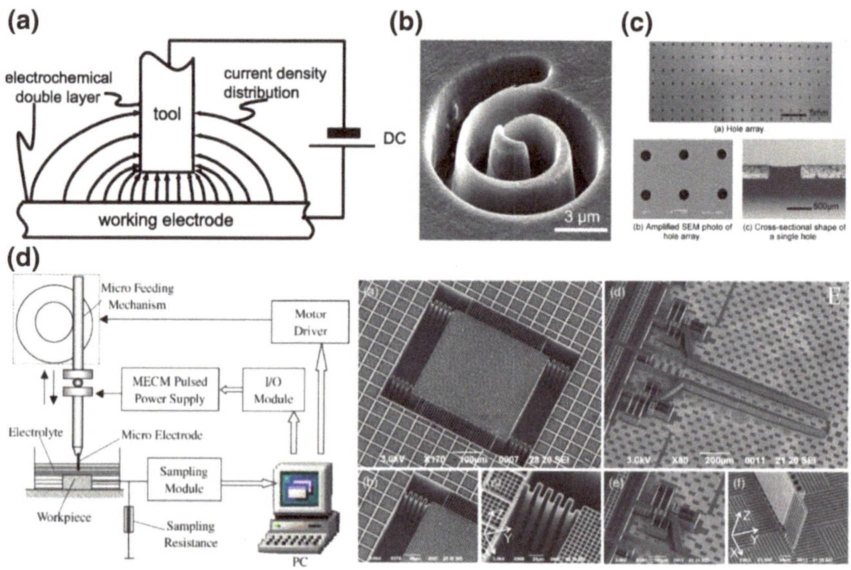

Fig. 1.5 Schematic of Electrochemical micromachining. **a** Principle diagram of electrochemical micromachining **b** Spiral indentation fabricated by electrochemical micromachining (Fig. 1.5**a** and **b** were reprinted with permission from Ref. [73]). **c** Through hole array fabricated by electrochemical micromachining (Reprinted with permission from Ref. [74]). **d** Set up of electrochemical micromachining device (Reprinted with permission from Ref. [75]). **e** Electrochemical micromachining of silicon substrate. (Reprinted with permission from Ref. [76])

materials that have been reported in the preparation of microchip, such as low-temperature co-fired ceramic (LTCC) [80, 81], but we will not have a detailed discussion here.

1.2.2 Silicon Elastomer

Polymer materials have been frequently applied to microfluidic devices since 2000s [82, 83]. Comparing to inorganic material, polymers are finely processable and castable, cheap in cost and amendable for mass production. The molecular structure of polymer is variable, and by facile modification its chemical properties can be easily regulated. Polymers applied in chip preparation can be roughly classified into three types, silicon elastomer, thermoset and thermoplastic material.

In the last decade, PDMS has become the primary choice for microfluidics device fabrication [84, 85]. Elastomer allows for rapid prototype of microchip model and assembling with glass substrate for a transparent device. The elasticity makes possible the integration of micro pump and valve into one chip [86–89]. Through placing PDMS layer by layer, complex chip structures can be realized.

Elastomer casting technique was invented by Bell group in 1970s [90], and firstly applied to microfluidic devices in 1980s [91]. After Whitesides reported the PDMS chip fabrication through soft lithography in late 1990s [92], soft lithography as well as PDMS has been broadly accepted by researchers and become a primary chip preparation method. In soft lithography, liquid elastomer is polymerized at a mild temperature condition (40–70 °C), which lowers the infrastructure requirement for laboratory use. The resolution of mold-casting method can reach nano scale, and owing to the small surface tension of PDMS, casted chip device can be easily peeled from photoresist template with good integrity [93, 94]. By simple physical attachment, PDMS channels can be reversibly sealed to another PDMS, glass or other surface. With the assistance of thin layer PDMS as adhering agent or plasma surface treatment, PDMA chip can be irreversibly sealed to another PDMS or glass substrate. Comparing to air tight materials such as glass, silicon, PMMA and PC, PDMS is permeable to gas and permits exchange of oxygen and carbon dioxide in cell culture model [95]. Also the hydrophobic surface of PDMS makes it preferable for cell adhesion [96].

Advantages mentioned above as well as cheap material cost account for the broad application of PDMS in cell biology researches such as cell screening [97], culture [98, 99] and biochemical analysis [100]. And the ability of microfluidics to process samples in nano or pico liter enables the building of single cell analysis platform by PDMS microchips with integration of functional units like micropump and microvalve[101].

However the limits of PDMS are also apparent. Hydrophobic surface of PDMS increases the chances of non-specific absorption of drug components and biomolecules [102]. Although surface treatment by plasma or corona discharge will temporarily reduce this unwanted effect, the treatment often does not last long

[103]. From the view of molecular structure, PDMS is a porous bulk with Si-O chains as skeleton and alkyl groups covering all over it. Therefore, PDMS is extremely vulnerable to organic solvent, and the application of PDMS is only for aqueous system. The air permeability benefits gas exchange but also causes the volatilization of water through micro channel and alternates the solution concentration. Besides, when fabricating channels of high aspect ratio, the flexibility of PDMS material would be more likely to induce channel collapse. The operation condition of PDMS chip has to be mild in prevention of unrecoverable deformation. And due to the existing uncertainties between different baths that may influence product quality, PDMS has been excluded from commercialization.

Substitutions of PDMS are applied to chip manufacture to overcome some disadvantages. For example, perfluoro polymers were introduced for great inertness and anti-fouling properties [104], and hydrogels were used in 3D cell culture matrix and concentration gradient generator[105, 106].

1.2.3 Thermoset and Thermoplastic Materials

Thermoset material was employed in soft lithography as photo resist before it was assumed as device building material. When being heated or radiated, molecules inside thermoset will cross link to form rigid structure, which can't be reshaped once accomplished. Normally, thermoset materials are transparent, stable at high temperature and tolerant of most solvents. Microfluidic chip can be prepared by sole material of thermoset via proper bonding technique [107, 108]. Thermoset materials are proper candidates for constructing complex microstructures through photo initiated polymerization, and the high rigidity permits implement of extreme aspect-ratio channels. However, the high rigidity also makes it impossible to integrate membrane based microvalve on chip. Common thermoset materials include thermoset polyester (TPE) and epoxy plastics.

Different from thermoset polymers, thermoplastics can be shaped and reshaped repeatedly at high temperature. The raw material of thermoplastics is at solid phase when purchased, and when being heated to its glass phase transition temperature, the material is significantly softer and suitable for next fabrication process. Thermoplastics is widely used in modern industrial, typical products of thermoplastics include PMMA, PC, PS, PET, PVC and perfluorinated compounds. Application of Teflon PFA and FEP to the fabrication of microfluidics devices has been reported [109, 110]. Teflon is well resistant to the harmful impact from solvents and chemical reagents, and flexible with good transparency, which can be used for microvalve integration on chip. Generally, thermoplastics have a better tolerance to solvents than PDMS, but are still vulnerable to organic solvents such as ketone and alkanes. The air tightness limits long term cell culture on thermoplastics. Through thermomolding, mass production of thermoplastics based devices can be realized at low cost, which is favored by the commercial companies. However, utilization of thermoplastics in laboratory is limited by the infrastructures of expensive metal or silicon template. Whitesides and Xia developed a method of transferring patterns from photo resist template to thermoplastics by using PDMS as intermediary material [111]. Restricted by the mild operating condition of PDMS, this method is feasible only with materials of transition temperature lower than 150 °C.

Thermal bonding and adhesive agent assisted sealing are two major packaging methods for thermoplastics, and the operating condition is much milder comparing to glass material. Surface modification of thermoplastics can endure a long time by covalent reaction. For example, the surface hydrophilicity of thermoplastics can last for years after plasma treatment [112].

1.2.4 Hydrogel Material

The skeleton of bulk hydrogel is formed by entangling hydrophilic polymer chains in three dimensions, and there are water molecules occupying nearly 99% of the material space [113]. The porous structures inside hydrogel permit diffusion of small molecules and bio-molecules. Owing to the similarity of hydrogel to extracellular matrix in properties such as permeability and mechanical strength, hydrogels are frequently applied in cell encapsulation and in vitro culture model. However, long-term culture scale is still limited by the effective diffusion distance of nutrients and oxygens, and requires more precisely controlled culturing conditions. The application of hydrogels as supporting material for vascular mimicry model was reported [114], and enhancement of mass transportation was observed [115].

Though hydrogels are generally biocompatible, there are different affinities to cell adhering of different gel materials. For example, animal derived gels such as matrigel and collagen contain factors enhancing cell proliferation, while plant derived gels like alginate and agarose and other synthesized polymers like PEG and polyacrylamide are lack of adhering points for cells. The deficiency of less cell adhering affinity can be overcome by simple grafting modification.

Hydrogel is a low-density soft material, and therefore not ideal for high-resolution manufacturing. The minimum texture in hydrogel is currently at micro scale. Major processing techniques for hydrogel based fabrication include LDW and 3D printing. Due to the mild gelating condition of hydrogel, the only requirement of employed mold template is insolubility in aqueous system. But real problem of gel based devices is the difficulty of sealing. Hydrogels are widely applied in the fields of 3D cell culture and tissue engineering[116, 117].

1.2.5 Paper Based Microfluidic Devices

Microfluidic paper based analytical devices (μPADs) is an emerging field in rapid development (Fig. 1.6a–f) [118–123]. Whitesides group reported the first paper microchip in 2007 [124], while the prototype of its design can be dated back to

1940s, when Muller and Clegg proposed a method of pH testing by filter paper strips [125]. Muller patterned micro channels on filter paper through paraffin coating and observed plant pigment diffusion and separation over the channel after sample addition of different pigments mix. This very micro channel can be regarded as the embryo of μPADs. Till now, the applications of paper based devices have expanded to disease diagnosis [126], environmental monitoring [127, 128] and food safety control [129]. Rapid development of paper based devices is based on their unique advantages: cheap and easily obtained materials with flexibility (Fig. 1.6g) [130] and biocompatibility, facile disposal by burning or burying, porous structures and large specific surface area of paper for analytes filtering, transportation and separation, convenient surface modification and patterning by wax coating or covalent reaction, inherent white background for colorimetric detection. Besides, when operating paper based analytical devices, liquids are passively driven by capillary effect without necessity of connecting to external pump, which makes it a proper candidate for portable device fabrication. Functional reagents can be preserved on paper with post bake after addition.

At present, the major directions of equipment development lay on the facile and convenient chip preparation methods and novel chip based test for important biomarkers. Common processing techniques for paper based microchips include lithography [131], inkjet modification [132], plasma treatment [133], direct cutting, patterned wax coating [134], flexographic printing [135], silk-screen printing [136] and laser processing [137]. And the principles are basically the same, which include

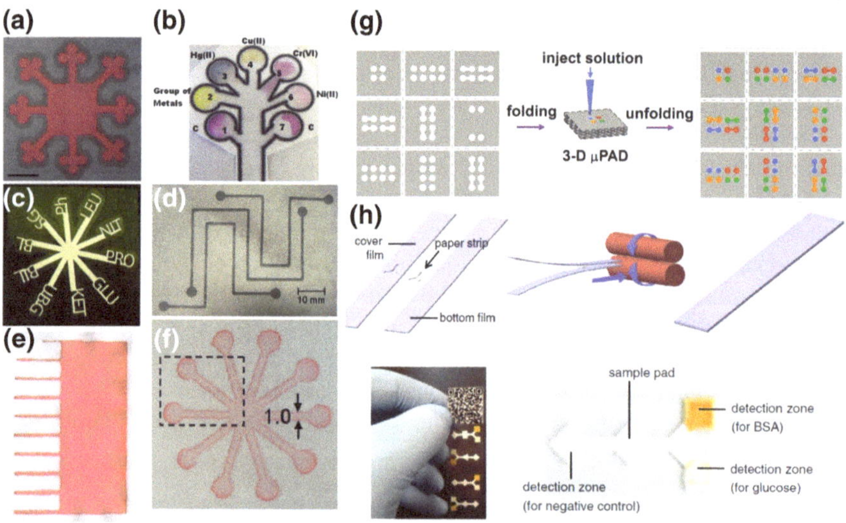

Fig. 1.6 Paper based analytical device. **a–f** Examples of paper based analytical device. (Reprinted with permission from Ref. [118–123]). **g** Origami inspired folding paper device (Reprinted with permission from Ref. [130]). **h** A ready-to use analytical paper strip (Reprinted with permission from Ref. [139])

the formation of hydrophobic/hydrophilic contrast area and restricting liquid flows in predefined microchannels. Normally, the width of micro paper channel is about several hundred or thousand micrometers, and due to the hydrophilicity of paper surface, hydrophobic reagent is employed in paper modification. According to the different cost of surface modification, hydrophobic reagents can be classified into three categories: expensive ones as SU-8 photo resist (modifying 100 cm^2 requires 0.1 dollars), cheaper ones as wax (modifying 100 cm^2 requires 0.01 dollars) and the cheapest ones like alkyl ketene dimer (modifying 100 cm^2 requires 0.00001 dollars). And modification principles can also be differentiated as physical block of surface porous structures, deposition of hydrophobic layer and chemical regulation of surface property. Modification by chemical reactions can be hardly removed by solvent dissolving, while physical deposition on surface is vulnerable to organic solvents, which is utilized in channel formation on paper based devices.

The processing techniques for surface modification include hydrophobic coating in predesigned patterns such as wax printing and ink jet printing, and overall coating followed by selective removal of coating material through lithography, microfluidics methods and region confined plasma treatment. Currently, the most frequently used methods are ink jet printing of AKD reagents and wax coating. Common advantages shared by these two techniques are low in cost, easy to control and short fabrication periods.

Detection methods accommodated with paper based microfluidic chips include colorimetric, electrochemical, chemiluminescence and electrochemiluminescence detections. Among all these methods, colorimeter has been often applied to qualitative and semi-quantitative testing due to its direct observation by raw eyes of the reaction products under enzymes catalyzed systems [138]. And because of no requirement of external equipment, colorimetric methods have also become well commercialized to ready-to-use products (Fig. 1.6h) [139]. Electrochemical methods have high sensitivity which is capable for quantitative detection of substances as low amount as nM level. Besides, the application of electrochemical detection can also effectively avoid the interference from other colored matters in detection matrix [140]. Studying from the published works, the major detection methods employed in paper based devices are colorimetric and electrochemical, while in few works chemiluminescenceand electrochemiluminescence were applied.

Future development directions for paper based devices lay on the studies of inventing cheap, easily-handled and portal devices to help improve diseases diagnosis, especially in the developed regions which suffer from the lack of infrastructures and trained medical workers. Also development of quantitative detecting papers of multiple targets on the basis of current qualitative and semi-quantitative test papers is in avid demand. At present, the functional test papers can be roughly divided into two types: on demand devices and ready to use devices. Different reacting reagents can be added to the former ones according to different target substances, while for the latter ones, specific reagents for single target test have been placed into the paper in advance and simple sample addition can lead to detection results.

To reinforce the detection capability of paper based devices, integration of different functional units is needed, such as sample separation channel and reaction chambers. The development of 3D paper chip has allowed the transportation of solution not only inside 2D plane, but also in vertical directions [141]. The simplest method for 3D paper chip fabrication is physical folding and overlaying of 2D paper chip. Comparing to 2D chips, the appearance of 3D chips has largely increase the device detection throughput, where each layer can be designed to implement analysis for one specific substance. Of course, one popular application for paper chip is as supportive material for cell biology related researches. For example, paper chip has been applied as 3D cell culture matrix for in vitro model construction [142].

As it is said, every coin has two sides. The limited factors for paper based devices including:

(a) The diffusion and carryover of sample inside paper would significantly decrease the valid concentration of target for detection. Also open channels on paper would enhance water volatilization and harm sample transportation. Normally, amount of sample reaching detection area is no more than 50% of the sample addition.
(b) For fluids with low surface tensions (bio-sample containing surfactant, for example), hydrophobic modification on surface is not enough to confine sample solution inside predesigned channels. In this situation, sample channeling is not effective by controlling surface hydrophilicity/hydrophobicity.
(c) Frequently used in conventional methods, microfluidic hydrodynamics such as laminar and micro droplet technology can't be accommodated to paper devices. Micro structures like micro valves are unable to be integrated into paper chip.
(d) Detection limit of colorimetric method is not low enough for trace amount pollutants in drinking water, which is usually in ppb or ppt level.

1.2.6 Hybrid Material Chip

When single material is unable to fulfill all the technique requirements to realize device function, coupling different types of materials in one microchip should be considered. We term the chip which extensively employs two or more types of materials the hybrid chip. Design of hybrid chip should following the principle that combined materials can't antagonize each other, but serve as complements to improve over all properties. One typical example of the hybrid chip includes glass as substrate and PDMS as channel and microvalve formatting material in one microchip [143–145]. There are also other examples such as glass channel and electrode substrate [146], and microchannel embedded with hydro gel to eliminate bulk flow while preserving diffusion.

1.3 Chip Fabrication

We have had a brief introduction about the chip materials and different fabrication methods that vary according to material as mentioned above. Next, we shall supplement some frequently used chip fabricating methods to make it more complete.

1.3.1 Soft Lithography

After the elastomer became the major material for microfluidic devices preparation, its fabrication methods have also been developed since. Currently the basic method for PDMS chip preparation is soft lithography. Soft lithography was invented in the late 1990s by Whiteside's group [92], and its principle originated from standard lithography which is employed for glass and silicon chip preparation. In standard lithography, glass or silicon surface is covered with a layer of photo resist, which is selectively exposed to UV radiation and developed. The bare surface without photo resist covering is vulnerable to chemical etching, and thus ready for channel formation. For soft lithography, elastomer is mold-casted to a photo resist template standing on silicon base. After elastomer solidification under hours' heating, the whole PDMS bulk is peeled from the template and sealed to a glass substrate. Main processing steps of soft lithography include: UV mask design, photo resist spin-coating on silicon wafer, exposure to UV and development, mold-casting and sealing. With the advantages of low infrastructure cost and simple operation, soft lithography has been widely applied in laboratories of universities and institutes to achieve fast prototype of chip model. However, the disadvantages are multiple processing steps. From preliminary design to real products, it often requires several hours to days to cover a single fabrication period. If the mold chip couldn't pass performance testing, preparation would have to restart over from design, which makes a great liability on time and labor. Limited to the lithography principle, some three-dimensional structures are unable to be formed. In the aspect of resolution, no matter the specific lithography processing steps, the final resolution is always limited by diffraction effect. Slight differences may be observed by using different masking technologies. For example, resolution of contact photolithography is about 0.5 μm but 0.3 μm for projection photolithography [4]. Generally, microchip prepared by lithography can meet the requirement of micro volume bio-sample analysis, but when more precisely controlled structure feature is needed, electron beam etching or nano imprint lithography may be necessary. Both of these methods can lower down resolution to below 100 nm, but with heavy cost of equipment and time.

1.3.2 Fabrication of Extended-Nano Channels

The scale of extended-nano channel (10–100 nm) lies between the conventional nanotechnology (1–10 nm) and microtechnology (>1 μm). In the past, researches concerning fluid properties at extended-nano scale were limited by the lack of appropriate tools. Extended-nano channel is too large for bottom-up synthesis or assembling from atoms or molecules of nanotechnology, and too precise for micromachining of microtechnology. Therefore, specialized methods are required to study channels at extended-nano scale. Focused ion beam milling and electron beam lithography have become alternatives to traditional photo lithography and applied to nano channel formation(Fig. 1.7a) [147]. These methods are able to realize precise nanomachining in selected regions, but unsuitable for large scale and high throughput fabrication. Focused ion beam milling mainly uses gallium ion beam to directly carve 2D nano channels on silicon or fused silica. Normally, the formed extended-nano channel is embed and connected with surrounding microchannels to constitute complete flow system. The microchannels are intended for solution supply.

In electron beam lithography, electron resist and conductive polymer are spin coated to quartz wafer respectively. Nano patterns are created by electron beam drawing without the requirement of photo mask. Then with the coupled plasma and chemical vapors treatment, extended-nano channels are achieved on substrate. And chip sealing is accomplished by thermal bonding with or without external pressure. In addition, laser direct writing can also be applied for fabrication of extended-nano channels(Fig. 1.7b) [148].

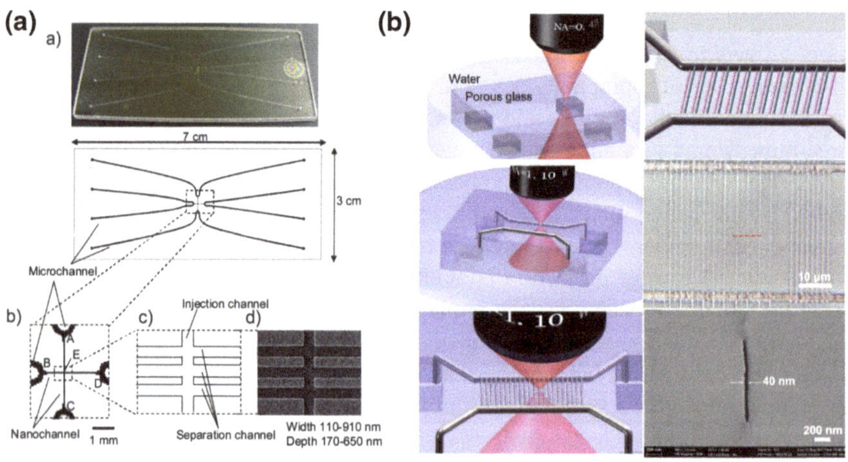

Fig. 1.7 Extended-nano channel fluidics. **a** Extended-nano channel produced by focused ion beam milling and connected to micro channels (Reprinted with permission from Ref. [147]). **b** Extended-nano channel bunch fabricated by direct laser writing (Reprinted with permission from Ref. [148])

1.3.3 3D Printing Technology

3D printing technology firstly appeared in the 1980s [149], but subject to expensive equipment, limited printing material and few choices of printing nozzle, the application of 3D printing technology in microfluidics was only restricted to chip prototype and exploring fabrication methods. Thirsty years since then, as the development of technology and decrease of equipment and material cost, 3D printing has not only been widely employed to the researches of various disciplines in laboratory, but also come to the daily life of ordinary customers. Customer grade 3D printing device has now been successfully commercialized with price at 1000 dollars [150].

3D printing is a highly precise manufacturing method, enabling facile and easy preparation of complex 3D structures. Comparing to conventional microfluidic device fabrication methods like lithography and micromachining, 3D printing is simpler in operation, enables full automation and on-step fabrication of microchip. 3D printing spares the design of UV mask, and testing performances can be directly used as reference for design optimization, thus increasing the speed of experience based modifying. At present, the choice of materials for 3D printing is broadly extended, and materials from polymers to various bio substances such as cells [151] can be printed.

3D printing is also called additive manufacturing (ASTM F2792), that is, device manufacture by continuous addition of new material to substrate. Commonly, steps of 3D printing include:

(a) Building 3D model of target object on CAD or other engineering drawing software. (b) Transferring the model data to 3D printing devices where 3D model is sliced to be successive 2D layers. The final amount of 2D slices is determined by the resolution of 3D printing devices, and normally printing resolution can be smaller than 10 μm.

(c) Layer by layer 3D printing of material to rebuild the 3D features of model objects. (d) In construction of some structures, post-processing is needed to remove the supporting frames or other exogenous materials. 3D printing devices are equipped with precise positioning and delivering modules, which can build structures of any shape in 3D space theoretically, like curved surface, slope and irregular shape that can't be prepared by other methods.

According to different printing material and polymerization methods, principles of 3D printing can be further classified as stereolithography (SLA) (Fig. 1.8a, b), fused deposition method (FDM) (Fig. 1.8c), electron beam melting (EBM) and bio-printers, the first two of which have been well exemplified [152].

1.3.3.1 Stereolithography

Stereolithography was the principle of both the first proposed 3D printing device and the first batch of commercialized products of 3D printing, and still the most

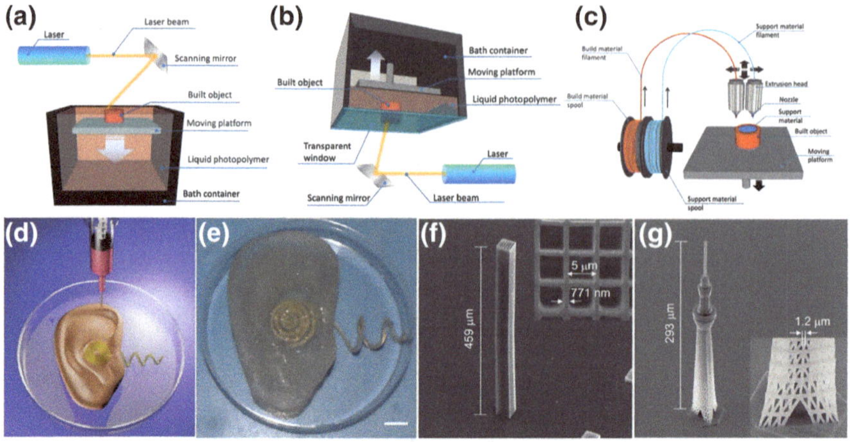

Fig. 1.8 Schematic illustration of **a** Bath configuration of stereolithography, **b** Layer configuration ofstereolithographyand **c** Fused deposition modeling. (Figure 1.8**a–c** were reprinted with permission from Ref. [152]). **d** Computer model design of human ear. **e** 3D printed ear. (Figure 1.8**d** and **e** were reprinted with permission from Ref. [157]). **f** and **g** 3D printed structures (Fig. 1.8**f** and **g** were printed with permission from Ref. [158])

widely used and successful models. In 1988, Chuck Hull invented the first 3D printing device by using UV laser to initiate liquid resin polymerization and localized solidification. The vertical dimension of object was expanded by the lifting of carrier platform. During printing, fresh resin was painted to the section and then shaped by laser beam. And above procedures were repeated until the recapitulation of 3D objects. The vertical resolution of stereolithography can reach sub micrometers or tens of micrometers, while the planar resolution is determined by the size of laser beam spot. Currently, single layer of resin being painted is 25 μm in thickness with minimum laser beam shifting distance of 10 μm. In addition to laser beam, digital light processing projector and LED can serve as light sources.

The advantages of stereolithography are precise patterns on 2D plane, modest consumption of liquid material, low cost in printing and ability to produce customer grade 3D printing devices. However, limited by printing procedures, only one epoxy or acrylate resin can be applied during one printing operation. And requirement of post-processing may be another disadvantage of stereolithography. Besides, the printing process may be influenced by uncontrollable non-linear factors such as resin polymerization around the vicinity of laser spot and temperature shift induced polymerization. Based on the mechanism of stereolithography, researchers developed multijet or polyjet technique to print multiple photo sensitive inks for UV induced polymerization [153, 154]. Multijet technique has equivalent or even better resolution comparing to conventional stereolithography, and employment of multiple nozzles enables switch of materials during printing. The major defect of this method is expensive hardware cost, which corresponds to 10 or 100 times of the cost of customer grade printing devices. Besides, other derivations based on stereolithography include digital micromirror device-based projection printing

(DMD-PP) and two photon polymerization (2PP). Both of these two methods have better printing resolution than stereolithography, but the time required by processing is prolonged.

1.3.3.2 Fused Deposition Method

In fused deposition method, melted thermoplastic material of several millimeters diameter is squeezed out from a tiny nozzle of hundreds micrometers wide and added to the printing structure [155]. Common printing material include polylactic acid,polycarbonate and acrylonitrile butadiene styrene (ABS), and through mixing stainless steel particles, wood chips, ceramic powder, carbon nanotubes and grapheme into thermoplastics, hybrid materials can be obtained. After squeezing, printing material is piled up shoulder by shoulder in the shape of cylinders, therefore resulting in larger roughness than other 3D printing methods. Certainly, fused deposition method has its own unique advantages, including compatibility with various cheap and easily obtained materials, low cost in equipment, alternative printing of several materials by integration of multiple nozzles and cost performance allowing for development of customer grade devices. Similar with fused deposition method, direct ink writing (DIW) uses viscoelastic ink with regulated rheological properties and pneumatic pump controlled injection to directly paint material to substrate [156]. For DIW, the minimum diameter of nozzle lumen can be 1 μm, and normally used materials are ceramics, polymers and metal nano particles.

When assessing the performance of a 3D printing device, major concerned parameters are resolution, printing speed, material compatibility and manufacture size. And there is huge space for further development of current printing devices in these aspects. As for hardware, cost of printing devices is mainly ranging from thousands or tens of thousands dollars and it is necessary to further cut down the cost of equipment. For printing material, the choice is largely limited by the processing procedure, and functional properties of material like biocompatibility and transparency become subordinate factors. In the future, 3D printing technique should provide more options when it comes to material choices and strategies for multiple materials alternative printing (Fig. 1.8d–g) [157, 158]. Finally, the resolution of 3D printing needs to be improved comparing with soft lithography [159]. And it is a large challenge to avoid significant prolonged time period when increasing printing resolution. One feasible manner is to combine high resolution printing and low resolution printing into one device to balance the time cost and requirement of printing resolution [160]. Of course, the incoming problem is largely increased cost of equipment to implement two printing system in one machine.

1.4 Integration of Microfluidics Functional Units

The miniaturized features of microfluidic chip make possible not only the manipulation of micro objects and analysis of micro volume sample, but also the integration of different functional units into one chip. Functional units on chip can realize sample pretreatment, flow control, cell culture and sensing detection, which greatly extend the detection capability of device. Next, we shall have some introduction about the common functional units in microfluidic chips.

1.4.1 Flow Manipulation: Micropump, Microvalve and Mixer

In microfluidic devices, flow control can be divided into two types according to the underlain principles. The first type of control relies on the physical effects that are significant in micro scale such as surface tension and capillary effect, and it is called passive manipulation. The second type depends on external forces like pump to control flow, and is called active manipulation. Common driving forces for flow pump include electricity [161], magnetofluid power [162] and electroosmotic flow [163]. Besides, flow driven by centrifuge force from device rotation is also subject to active manipulation. Pump has been regarded as the pivot for flow system, and when building microfluidic chip based analytical platform in laboratory, external pump is often essential. The advantages of connecting a pump are obvious: precise and steady control of flow and realization of complex programmed pump behaviors. But in development of mobile device, external pump would harm the device portability, and extending conductor pipe would cause dead volume and result in sample waste. There are publications reporting integration of micropump on chip, and according to different driving manners, passive or active pump is called (Fig. 1.9) [164]. Principle of passive micropump is the capillary effect, while the active pump has variety of driving forces, such as electroosmotic, electrochemical, pneumatic peristaltic, acoustic, electrostatic and magnetic power. Electroosmotic driving provides steady flow rate, and through switching the polarity of electrode, flow direction can be easily reversed. Owing to mechanical-motion-free feature of electroosmotic unit, this type of pump can be simply integrated into micro chip [165]. Electrochemical pump relies on the generating bubbles from water electrolysis to drive flow forwards [166], and pneumatic pump depends on the pressure induced deformation of membrane to control the flow in lower channels [167].

Pump provides driving forces for fluid flowing, while valve is in charge of the on-off of flow circuit. Microvalve on chip can also be divided into two types as passive and active valve. Passive valves refer to those independent of mechanical opening and closing, the typical examples of which include cantilever valve, diaphragm valve and diffuser valve. Active valves are oppositely dependent of mechanical forces for the valve switching, and may be based on the principles of

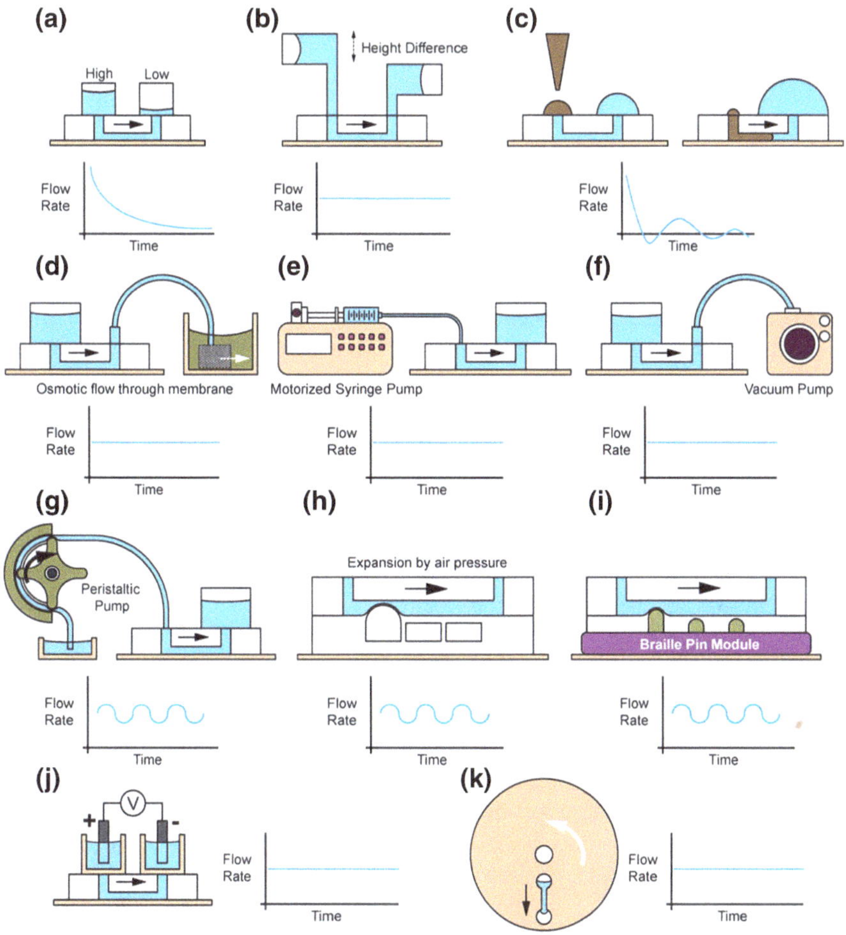

Fig. 1.9 Schematics of different type of pumps for microfluidic devices. **a**, **b** Gravity driven pump. **c** Surface tension droplet pump. **d** Osmosis based pump. **e** Syringe pump. **f** Vacuum pump. **g** Peristaltic pump. **h** Pneumatic peristaltic pump. **i** Braille pin peristaltic pump. **j** Electrokinetic pump. **k** Centrifugal pump. (Reprinted with permission from Ref. [164])

piezoelectric ceramics, thermal pneumatic power, electrostatic forces and electromagnetic interaction. Currently, pneumatic membrane microvalve prepared by PDMS is the most widely used integrated valve in microchip [168].

In the field of microfluidics, the tiny scale of channel as well as the laminar makes solution mixing a very slow process basically through free diffusion. And this is an unfavorable fact for mixing, but has been ingeniously applied for the generation of concentration gradients. Methods to enhance flow mixing include producing convection and shortening diffusion distance. For example, protruding geometries were designed inside micro channel to effectively induce turbulent flow and enhance solution mixing, which is easy to realize on microchip [169]. Besides, external forces have also been employed to increase the turbulent component of fluid flow.

1.4.2 Concentration Gradient Generator

Biomolecules regulate the behaviors of cells such as proliferation and differentiation among inside-body environment. There are researches indicating the close relation between the concentration gradient of cellular factors and cancer invasion [170], and this chemotaxis of cancer cells has been employed to develop anti-cancer medicine. Conventional drug concentration screening experiments are conducted on multi-well plate, and as the growing maturity and broader application of microfluidic technique on biological researches, methods of concentration gradients based on microfluidics have been rapidly developed. Advantages of concentration gradients on chip include wider gradient ranges, more precise partition of concentration on space distribution, maintained gradient persistency and compatibility with real time monitoring method of cell behaviors. Generally, the strength of chip based concentration gradients reflects its powerful control of the cell culturing environment. And until now, concentration gradients have played important parts in the researches of wound healing [171], inflammation response [172] and cancer metastasis [173] as well as chemical synthesis of drug. According to its time dependent stability, concentration gradient can be classified as dynamic and static. Major methods for the concentration gradient realization on chip include:

(a) Semi-permeable membrane diffusion based concentration gradient. As mentioned above, fluid mixing through diffusion inside a tiny channel is very slow, which provides chance for concentration gradient creation (Fig. 1.10a) [174]. To preserve diffusion based concentration gradient, the priority is to eliminate the convection flow through methods such as setting up semi-permeable membrane to separate two flows [175]. Another method is to deploy narrow connecting channels between different chambers (Fig. 1.10b) [176]. However, normally the gradient can't last long.

(b) Convection based concentration gradient generator. Convection is actually also applied to create concentration gradient. Typically, convection requires rather higher flow rate and precise flow control, which may cause harm to cell biology due to the large shear force. On the other hand, running flow is critical for the medicine transportation and removal of metabolic waste. Therefore the flow rate can't be too small. According to the equation of shear force,

$$\tau(\mathrm{cell}) = 6\mathrm{Q}\mu/\mathrm{h}^2\mathrm{w}$$

shear force is related to flow volume rate, fluid viscosity and channel dimension (height and width). Increasing the channel height and lowering flow linear velocity can help reduce shear force subject to cell and alleviate cell damage. Another feature of convection based concentration gradient is that the

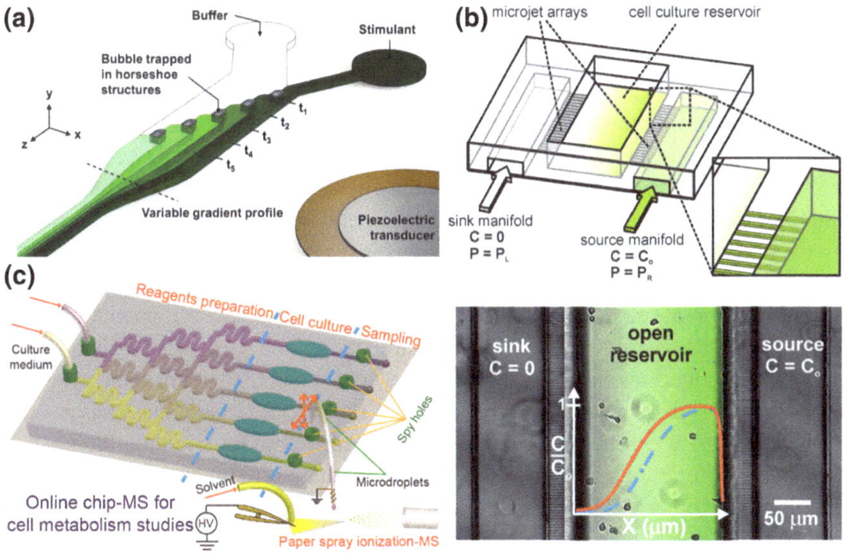

Fig. 1.10 Concentration gradient generator. **a** Concentration gradient produced in a Y channel with multiple barrier units (Reprinted with permission from Ref. [174]). **b** Drug concentration gradient controlled by narrow connecting channels (Reprinted with permission from Ref. [176]). **c** Tree-shape concentration gradient generator. (Reprinted with permission from Ref. [15])

formation of gradient is always perpendicular to flow direction, for example the well-known "Christmas Tree" model (Fig. 1.10C) [15]. Christmas tree shaped gradient generator was proposed by Jeon in 2000 [177], and by regulating flow rates and concentrations of different flows, concentration gradients of special and temporal variety can be created. Convection based gradients can be preserved for a long time, and similar structures include Y-shape channel.

(c) Hydrodynamic focusing device. Through hydrodynamics related method, externally added drugs or cell factors are confined in a predesigned area [178].

(d) Concentration gradients on micro droplets. Droplet technology is a unique product born along with the development of microlfuidics and has been applied in micro-volume reaction, single cell analysis and material synthesis. Regulation of concentration in the dispersion phase can achieve a gradient distribution over different droplets [179]. The combination of droplet technology with concentration gradient method avoids the external interference and provides a stable culturing environment for cells. Besides, the application of droplet technology greatly improves the analysis throughput, which can be further applied in drug screening.

1.4.3 Cell Culture Chamber

Integration of all kinds of functional units on chip makes microfluidic chip a platform for various research purposes rather than a specific analytical device. In biology related researches, the functions of microchip is extended from pure analysis of cell lysis and metabolites to construction of cell in vitro culture model, studying intercellular interaction and single cell analysis. And one critical step involved in the realization of above functions is the cell online culturing. Compared to conventional cell culture methods, advantages of using microfluidics rest with the precise control of microenvironment and targeting smaller cell population even single cell as research subject. Cell culture experiments in petri dish normally ignore the distinctions between different cell individuals, but rather take averaged responses from mass amount of cells as final results. In some situations, research concerning single cell behavior bears larger significance, for example the study of stem cell proliferation and differentiation under certain circumstances [180]. And at this point microfluidic chip has become one ideal candidate from not many available choices for single cell research. Series of derivative technologies concerning cell manipulation have been developed such as 3D hydrogel encapsulated culture and microwell capture array, which we will discuss in the later chapters.

1.4.4 Integrated Biosensors

When establishing microfluidic chip based analytical platform in laboratory, optional external detectors include mass spectrometer, fluorescence monitor and electrochemical detector. However, for analysis of complex biosamples, common detecting methods usually fail to achieve high specificity and sensitivity. And to solve this problem, specific biosensors are required and integrated into microchip to improve analysis efficiency (Fig. 1.11a, b) [26, 181]. On chip biosensors can be categorized as enzyme catalyzed reacting biosensor, immune biosensor, optical detector and DNA detecting array. Integrated biosensors can effectively reduce the consumption of sample and reagents, shorten operation time, lower detection cost and help development of portal devices. We shall have a brief introduction of on chip biosensors in the following content.

1.4.4.1 Enzyme Catalyzed Biosensor

For enzyme biosensor, enzymes with bioactivity are fixed inside channel or to the electrode surface (Fig. 1.11c) [182]. When passing through detection area, the analytes are catalyzed by enzymes, transferred to color changes or electrochemical signals (Fig. 1.11d) [183] and detected. At present, enzyme biosensors are widely applied in the detection of blood glucose, urea and cholesterol, among which glucose assay is the most successful commercialized product. The first glucose

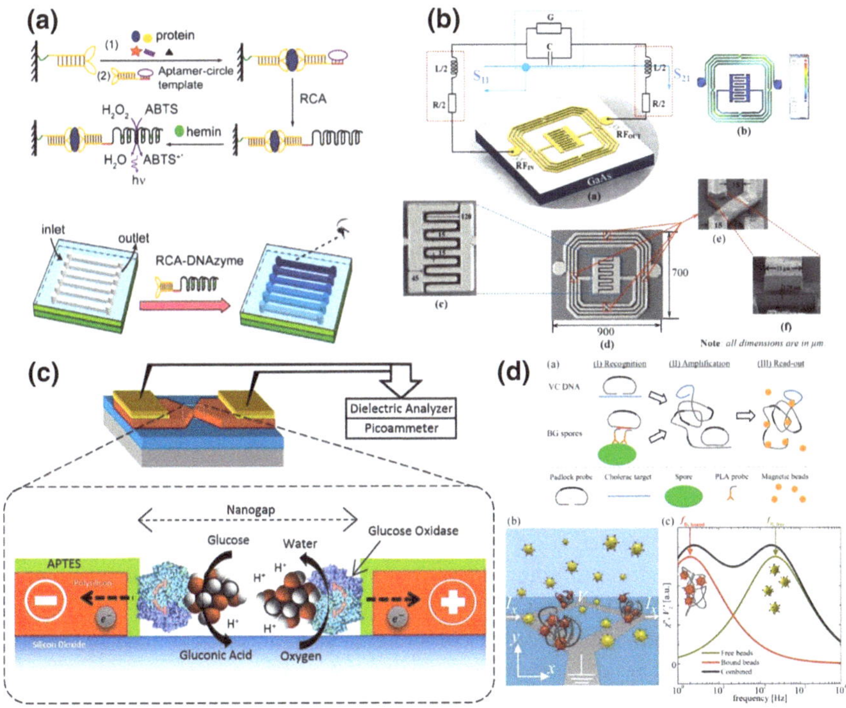

Fig. 1.11 On chip integrated biosensor. **a** Biosensor based on aptamer recognition. (Reprinted with permission from Ref. [26]). **b** Circuit design of an electrode based sensing chip (Reprinted with permission from Ref. [181]). **c** A dielectric analyzer for glucose detection (Reprinted with permission from Ref. [182]). **d** DNA recognition and analysis on bio-sensing chip (Reprinted with permission from Ref. [183])

biosensor was invented in 1962 [184], and decades since, there had been many optimized versions while with little alternation of the basic principle of electrochemical detection by enzyme modified electrodes. Urea level in blood is also an important indicator of many diseases like renal failure, nephritis, urinary obstruction and gastrointestinal bleeding. For urea detection, urease and glutamic acid dehydrogenase are employed.

1.4.4.2 Immune Sensor

Immune sensor is a detection method based on antigen-antibody recognition and turns the existence of analytes into fluctuation of resistance, electric current, electric capacity and refraction index. Specificity of immune sensor is high with good reproducibility. But the limits are high cost of monoclonal antibodies and tedious detection cycles. Immune sensors are mainly used in on chip detection of pathogen, bacteria, and viruses.

1.4.5 Microfluidic Centrifuge Device

A special type of device with the shape of laser Discs is invented with the development of microfluidics and called microfluidic centrifuge device (Fig. 1.12a–d) [185–188]. During operation, the high speed rotation of device generates centrifuge force to drive sample flow into treatment chamber, without the need of connecting to external pump. History of centrifuge device can be traced back to 1960s, when ORNL micro analytical device was employed in clinical to transport sample solution by centrifuge driving. But the idea of functional integration and miniaturization hadn't come to stage back then. After decades' development, microfluidic centrifuge device has become a very important tool for sample treatment and analysis, especially in the fields of clinical detection, immune diagnosis, protein analysis, cellular sample handling and biomarker detection.

As mentioned above, microfluidic device spares the trouble of connecting to an external pump and allows for set up of enclosed system. Due to the removal of external conductor system and accompanying dead volume, waste of sample is avoided. Centrifuge device is capable of processing sample in the volume of nano liter to milliliter, and the driving of flow in centrifuge device is continuous due to the rotation movement of device. Other unique functionalities reflected by centrifuge device include density based components separation, automatic processing

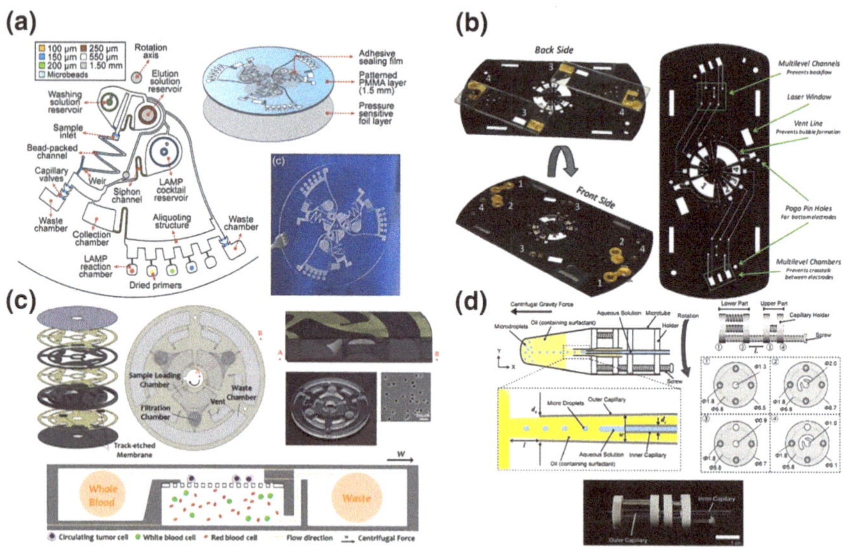

Fig. 1.12 Examples of centrifuge device. **a** Schematic of colorimetric foodborne pathogen detectionon centrifugal device (Reprinted with permission from Ref. [185]). **b** Centrifugal device for electrophoretic separation of DNA (Reprinted with permission from Ref. [186]). **c** Centrifugal device for tumor cells separation from blood sample (Reprinted with permission from Ref. [187]). **d** Centrifugal device for droplet production (Reprinted with permission from Ref. [188])

procedures such as sample inflow and eviction under different rotation speeds and miniaturization and integration. Generally, the working procedures of microfluidic centrifuge device are sample addition, reagents storage and release, liquid transportation, valve switch, mixing and detection. Direction of liquid transportation is normally from center of the device to outer margin due to the centrifuge force driving. Therefore, the sample inlet is also placed in the center of the device. Detectors compatible with centrifuge device are mainly optical and electrochemical. Although numerous companies have participated in the development of microfluidic centrifuge devices, few commercialized products have come into the market. The reasons for this weak commercial transformation are in two aspects. In one hand, employment of centrifuge movement saves part of the trouble in developing portal device by removing external pump, but also results in extra requirement for external power supply. Therefore, centrifuge device doesn't actually improve portability. On the other hand, flow control by centrifuge force is apparently not as precise or stable as flow pump does, and special liquid transporting manner causes the chip design some problems.

1.5 Development and Outlooks

Going through the brief history of microfluidics development, we have witnessed a path of favored diversity. In the choice of chip materials, there were different times when inorganic and polymers taking the dominant positions. In specific formats, there are ready-to-use test papers and analytical platforms with multiple functionalities. Actually when speaking to the future development of microfluidics, it is hard to cover all kinds of devices by a single standard. Each microfluidic device has its own development according to the different functionalities. Next, we will focus on several emerging fields of microfluidics which might have important applications in the future.

1.5.1 Point of Care (POC)

Analytical methods of series of biomarkers have been well established in laboratory, which require extensive work as well as professionally trained personnel. The common procedures include sampling, detection and results analysis. To improve the convenience of biochemical test and shorten detection time, point of care detection methods based on microfluidics technology have been developed. The major detection targets of POC methods are small molecules (blood glucose and urea), cells (CD4 + T-cell counter for AIDS diagnosis) and nucleic acids. The purpose of POC diagnosis is to make preliminary screening test closer to patients and may be used in community medical center, clinics, emergency center or at home.

Although microfluidic methods receive mass attention in academic circles and funding supports from government and private capitals, current examples of commercialization of POC device are still few. The major defects of POC device are the compromised sensitivity and quantification ability in analysis of complex biosamples comparing to laboratory large equipment. The development of POC device largely depends on the discovery of new disease biomarkers, especially those existing in the peripheral body fluids. For certain diseases such as cancer, diabetes, AIDS, tuberculosis and Alzheimer's disease, positive results have been obtained in biomarkers discovery, but there is a lack of unified standard and theoretical basis for biomarker based disease diagnosis. At present, commercialized POC devices can be classified into two types, and one is the lateral flow test that uses test paper or semi-permeable membrane strip for protein detection. The scale of global market for in vitro diagnosis has exceeded 40 billion dollars and 1 billion among is occupied by test paper business. Normal working procedures for test paper are sample addition, solution flowing through areas with labeling reagents and capture array, and finally formation of a colored line by labeled and captured biomarkers which can be told by raw eye observation. Typical examples are pregnancy test and flu test. In areas suffering from lack of infrastructure, test paper is also used for HIV diagnosis. Lateral flow based test is capable for target analysis at μM or mM level. Lower amount of analytes requires specific amplification strategies for detection, which may increase the complexity of device structure. In the last decade, plenty of funds and manpower have been invested to improve test paper performance, but little payback is obtained.

Another type of POC device is the electronic detecting instruments represented by glucose meter. Commercialization of glucose meter is an unrivaled success, which can be hardly duplicated by other products. There are unique favorable factors for the market expansion of glucose meter. Firstly the content of blood glucose in human body reaches mM level and exceeds the lower detection limit of electrochemical methods. Secondly, frequency of glucose meter usage can be several times a day for patients of diabetes, which provides large demands for the consumptive test papers.

There are problems needing to be solved in the future development of POC device.

First question is how to integrate multi-step sample treatment and analysis into one device and realize full automation of processing from samples to diagnosis results. Procedures of analysis include sample collection, pretreatment, analyte reaction, signal transformation and report. Due to the vast disparities of technologies involved in different steps, it is hard to meet all requirements by one scheme.

Other questions are how to simplify operation of device and make untrained personnel capable of operating, and how to improve test reliability.

1.5.2 Implantable Device

First implantable device was invented in 1958, and since then numerous companies and researchers have dedicated themselves to the development and improvement of this field (Fig. 1.13a–d) [189–192]. The continuously emerging demand for implantable device is related with the rise of aging population in major developed countries. According to the research from Freedonia group, the market scale of implantable device in America is increased with a rate of 7.7% and reached 52 billion dollars in 2005 [193], and the largest share of market is taken by implantable rectification equipment such as artificial joints and bones. Other types of implantable device include cardiovascular related apparatus (cardiac pacemaker and intravascular stent), nerve stimulator and drug delivery carrier. Though many implantable devices of medical intention are aimed at aging population, more and more newly emerged products are targeting at the body shaping market of young people. Medical implantable device that is totally or partly prepared by micromachining has been well commercialized and applied in treatment to different diseases. For example, acceleration sensor is equipped to cardiac pacemaker and defibrillator to sense body location and movement. Electronic instruments of this type require

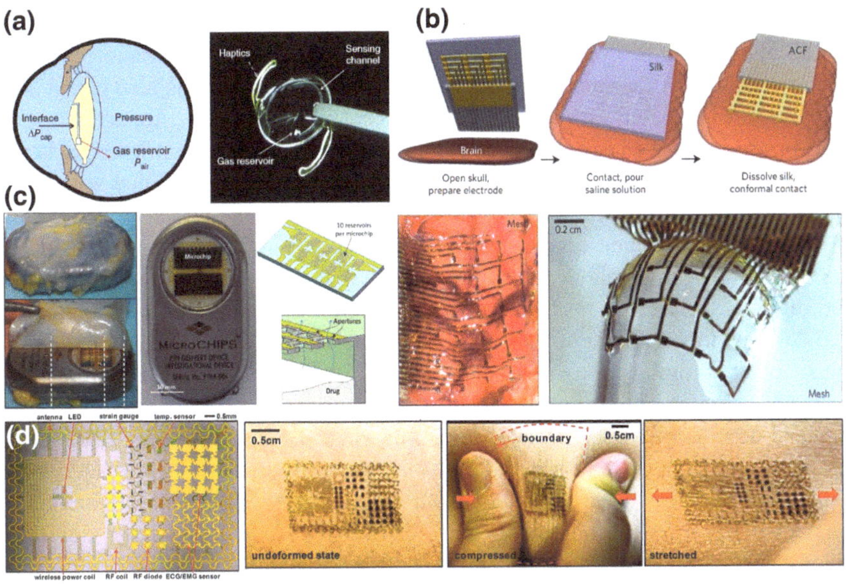

Fig. 1.13 Implantable microchip device. **a** An implantable microchip for intraocular pressure monitoring. (Reprinted with permission from Ref. [189]). **b** Ultrathin conformal bio-integrated electronics for brain monitoring (Reprinted with permission from Ref. [190]). **c** Wirelessly controlled drug delivery microchip (Reprinted with permission from Ref. [191]). **d** Multifunctional electronics with physical properties matched to the epidermis (Reprinted with permission from Ref. [192])

built-in power source, which occupies large space and may do harm to patient's health if leakage happened. Recently developed wireless charging system can avoid the problems encountered by built-in power design and extend the validity period of device. But the disadvantage is requirement of external supporting equipment.

Material choice of implantable device is strictly limited by several standards. Firstly, the candidate must be biocompatible which causes no toxic, allergic, inflammatory or immune reaction of the body. Secondly, stable must the material maintain in sterilization treatment such as heating, chemical reagents washing, radiation and high pressure. These treatments can't harm the normal functionality of device. Finally, the chosen material should possess corresponding electronic or mechanical properties to realize full device functions.

Future development of implantable device rests with the integration of monitoring and treatment in one device to achieve a closed loop from diagnosis to therapy according to the personalized situation of patients. Besides, through online information interchange, doctors at remote terminals can receive the monitoring data and give medical intervention in time.

1.5.3 Smart Mobile Device

Another direction for microfluidic device development is to combine microchip portal detection with the data analysis, management and transportation of smart phone and establish a mobile analytical platform. This concept is also called MS^2 (Mobile sensing based on microfluidics and smartphone) [194]. Actually, various built-in sensors as well as external detectors have been used to acquire health data of user with smart phone APP. The combination of microfluidic device with smart phone greatly expands the data acquisition ability of mobile phone (Fig. 1.14a–c) [195–197]. At the same time, features of immediate information processing and transportation of smart phone makes detection results no longer restricted to temporal and regional factors and instant feedback available. Common characters shared by two devices such as miniaturized structures, portability and simple operation enable the successful combination and establishment of mobile analytical platform. Smart mobile sensing can realize remote in-field monitoring, help improve family health care and provide medical aids to resource limited areas. Its application has involved fields of food quality monitor, regular health examination and disease diagnosis.

As a pioneer of this discipline, Martinez and coworkers from Harvard firstly connected paper based microchip to a smart phone and applied the device to glucose and protein detection in artificial urine sample at 2008 [198]. Then in 2009, Lu and colleagues designed an immune test device assembled by PDMS chip and smart phone [199]. The combination of microfluidics with mobile phone not only provides a simple solution for portable detection, but also makes remote medical diagnosis a reality.

However, the biggest limit of applying mobile sensing device to regular physical examination now is that mobile device doesn't reflect an irreplaceable detecting

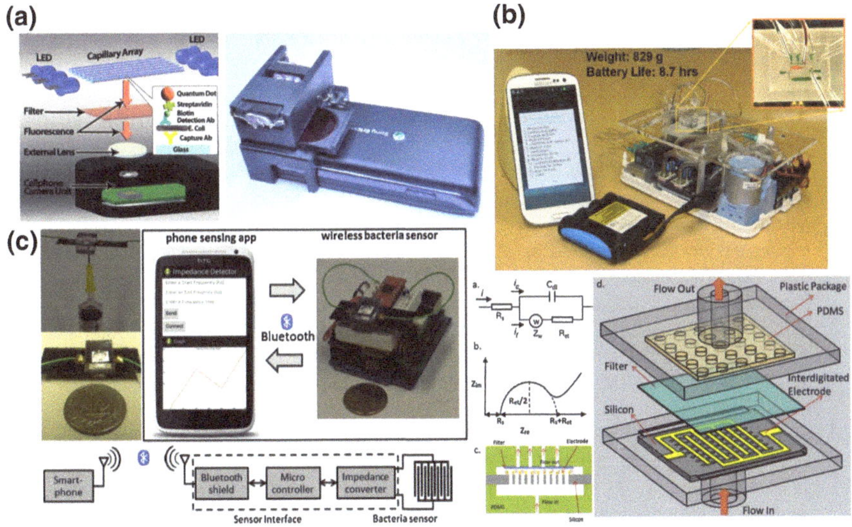

Fig. 1.14 Smart phone combined microchip device. **a** Quantum dots based assay for E. coli detection on cell phone. (Reprinted with permission from Ref. [195]). **b** Smartphone controlled handheld microfluidic liquid handling system (Reprinted with permission from Ref. [196]). **c** Smartphone based portable bacteria pre-concentrating microfluidic sensor (Reprinted with permission from Ref. [197])

feature, and medical workers prefer to use conventional methods. Moreover, the sampling of MS^2 equipment requires additional pipette or flow pump, which largely affects the portability and easy operation. In the future, the human-device interface needs further optimization by integrating sample inlet directly on device, such as installing blood taking needle or swab. Certain protection must be installed to prevent sample contamination in a complicated working environment. And owing to poor telecommunication and infrastructures in some regions, detection results must be stored in device for a certain period. This limitation can be overcome by increasing battery capacity and endurance.

Only after mass production and wide acceptance by consumers, can a value of technology be reflected. Smart mobile sensing is an emerging territory of microfluidics and we will wait and see what a role it can play in future.

References

1. Terry SC, Jerman JH, Angell JB (1979) Gas-chromatographic air analyzer fabricated on a silicon-wafer. IEEE T Electron Dev 26(12):1880–1886. doi:10.1109/T-Ed.1979.19791
2. Manz A, Miyahara Y, Miura J, Watanabe Y, Miyagi H, Sato K (1990) Design of an open-tubular column liquid chromatograph using silicon chip technology. Sensor Actuat B-Chem 1(1–6):249–255. doi:10.1016/0925-4005(90)80210-Q

3. Manz A, Graber N, Widmer HM (1990) Miniaturized total chemical-analysis systems—a novel concept for chemical sensing. Sensor Actuat B-Chem 1(1–6):244–248. doi:10.1016/0925-4005(90)80209-I
4. Iliescu C, Taylor H, Avram M, Miao JM, Franssila S (2012) A practical guide for the fabrication of microfluidic devices using glass and silicon. Biomicrofluidics 6(1):016505. doi:10.1063/1.3689939
5. Nge PN, Rogers CI, Woolley AT (2013) Advances in microfluidic materials, functions, integration, and applications. Chem Rev 113(4):2550–2583. doi:10.1021/cr300337x
6. Beebe DJ, Mensing GA, Walker GM (2002) Physics and applications of microfluidics in biology. Annu Rev Biomed Eng 4:261–286. doi:10.1146/annurev.bioeng.4.112601.125916
7. Sackmann EK, Fulton AL, Beebe DJ (2014) The present and future role of microfluidics in biomedical research. Nature 507(7491):181–189. doi:10.1038/nature13118
8. Wu J, He ZY, Chen QS, Lin JM (2016) Biochemical analysis on microfluidic chips. Trac-Trend Anal Chem 80:213–231. doi:10.1016/j.trac.2016.03.013
9. Wu J, Chen Q, Lin JM (2017) Microfluidic technologies in cell isolation and analysis for biomedical applications. Analyst 142(3):421–441. doi:10.1039/c6an01939k
10. Squires TM, Quake SR (2005) Microfluidics: Fluid physics at the nanoliter scale. Rev Mod Phys 77(3):977–1026. doi:10.1103/RevModPhys.77.977
11. Janasek D, Franzke J, Manz A (2006) Scaling and the design of miniaturized chemical-analysis systems. Nature 442(7101):374–380. doi:10.1038/nature05059
12. Martel JM, Toner M (2012) Inertial focusing dynamics in spiral microchannels. Phys Fluids 24(3):032001. doi:10.1063/1.3681228
13. Sudarsan AP, Ugaz VM (2006) Multivortex micromixing. P Natl Acad Sci USA 103 (19):7228–7233. doi:10.1073/pnas.0507976103
14. Mao SF, Zhang J, Li HF, Lin JM (2013) Strategy for signaling molecule detection by using an integrated microfluidic device coupled with mass spectrometry to study cell-to-cell communication. Anal Chem 85(2):868–876. doi:10.1021/ac303164b
15. Liu W, Lin JM (2016) Online monitoring of lactate efflux by multi-channel microfluidic chip-mass spectrometry for rapid drug evaluation. Acs Sensors 1(4):344–347. doi:10.1021/acssensors.5b00221
16. Liu W, Chen QS, Lin XX, Lin JM (2015) Online multi-channel microfluidic chip-mass spectrometry and its application for quantifying noncovalent protein-protein interactions. Analyst 140(5):1551–1554. doi:10.1039/c4an02370f
17. Wang NJ, Mao SF, Liu W, Wu J, Li HF, Lin JM (2014) Online monodisperse droplets based liquid-liquid extraction on a continuously flowing system by using microfluidic devices. Rsc Adv 4(23):11919–11926. doi:10.1039/c4ra00984c
18. Zhang J, Chen FM, He ZY, Ma Y, Uchiyama K, Lin JM (2016) A novel approach for precisely controlled multiple cell patterning in microfluidic chips by inkjet printing and the detection of drug metabolism and diffusion. Analyst 141(10):2940–2947. doi:10.1039/c6an00395h
19. Chen QS, Utech S, Chen D, Prodanovic R, Lin JM, Weitz DA (2016) Controlled assembly of heterotypic cells in a core-shell scaffold: organ in a droplet. Lab Chip 16(8):1346–1349. doi:10.1039/c6lc00231e
20. Lin XX, Leung KH, Lin L, Lin LY, Lin S, Leung CH, Ma DL, Lin JM (2016) Determination of cell metabolite VEGF(165) and dynamic analysis of protein-DNA interactions by combination of microfluidic technique and luminescent switch-on probe. Biosens Bioelectron 79:41–47. doi:10.1016/j.bios.2015.11.089
21. Fantuzzi A, Capria E, Mak LH, Dodhia VR, Sadeghi SJ, Collins S, Somers G, Hug E, Gilardi G (2010) An electrochemical microfluidic platform for human P450 drug metabolism profiling. Anal Chem 82(24):10222–10227. doi:10.1021/ac102480k
22. Zimmerman WB (2011) Electrochemical microfluidics. Chem Eng Sci 66(7):1412–1425. doi:10.1016/j.ces.2010.03.057

23. Wu J, Jie MS, Dong XL, Qi HB, Lin JM (2016) Multi-channel cell co-culture for drug development based on glass microfluidic chip-mass spectrometry coupled platform. Rapid Commun Mass Sp 30:80–86. doi:10.1002/rcm.7643
24. Jie MS, Li HF, Lin LY, Zhang J, Lin JM (2016) Integrated microfluidic system for cell co-culture and simulation of drug metabolism. Rsc Adv 6(59):54564–54572. doi:10.1039/c6ra10407j
25. Lin XX, Chen QS, Liu W, Zhang J, Wang SQ, Lin ZX, Lin JM (2015) Oxygen-induced cell migration and on-line monitoring biomarkers modulation of cervical cancers on a microfluidic system. Sci Rep 5:0001. doi:10.1038/Srep09643
26. Lin XX, Chen QS, Liu W, Li HF, Lin JM (2014) A portable microchip for ultrasensitive and high-throughput assay of thrombin by rolling circle amplification and hemin/G-quadruplex system. Biosens Bioelectron 56:71–76. doi:10.1016/j.bios.2013.12.061
27. Lefevre F, Chalifour A, Yu L, Chodavarapu V, Juneau P, Izquierdo R (2012) Algal fluorescence sensor integrated into a microfluidic chip for water pollutant detection. Lab Chip 12(4):787–793. doi:10.1039/c2lc20998e
28. Lafleur JP, Senkbeil S, Jensen TG, Kutter JP (2012) Gold nanoparticle-based optical microfluidic sensors for analysis of environmental pollutants. Lab Chip 12(22):4651–4656. doi:10.1039/c2lc40543a
29. Yoon JY, Kim B (2012) Lab-on-a-chip pathogen sensors for food safety. Sensors-Basel 12(8):10713–10741. doi:10.3390/s120810713
30. Atalay YT, Vermeir S, Witters D, Vergauwe N, Verbruggen B, Verboven P, Nicolai BM, Lammertyn J (2011) Microfluidic analytical systems for food analysis. Trends Food Sci Tech 22(7):386–404. doi:10.1016/j.tifs.2011.05.001
31. Govindarajan AV, Ramachandran S, Vigil GD, Yager P, Bohringer KF (2012) A low cost point-of-care viscous sample preparation device for molecular diagnosis in the developing world; an example of microfluidic origami. Lab Chip 12(1):174–181. doi:10.1039/c1lc20622b
32. Kim MS, Kim T, Kong SY, Kwon S, Bae CY, Choi J, Kim CH, Lee ES, Park JK (2010) Breast cancer diagnosis using a microfluidic multiplexed immunohistochemistry platform. PLoS ONE 5(5):e10441. doi:10.1371/journal.pone.0010441
33. Ziółkowska K, Kwapiszewski R, Brzózka Z (2011) Microfluidic devices as tools for mimicking the in vivo environment. New J Chem 35(5):979–990. doi:10.1039/c0nj00709a
34. Sung JH, Shuler ML (2012) Microtechnology for mimicking in vivo tissue environment. Ann Biomed Eng 40(6):1289–1300. doi:10.1007/s10439-011-0491-2
35. Huh D, Matthews BD, Mammoto A, Montoya-Zavala M, Hsin HY, Ingber DE (2010) Reconstituting organ-level lung functions on a chip. Science 328(5986):1662–1668. doi:10.1126/science.1188302
36. Webb DP, Knauf B, Liu CQ, Hutt D, Conway P (2009) Productionisation issues for commercialisation of microfluidic based devices. Sensor Rev 29(4):349–354. doi:10.1108/02602280910986601
37. Blow N (2007) Microfluidics: in search of a killer application. Nat Methods 4(8):665–668. doi:10.1038/nmeth0807-665
38. Volpatti LR, Yetisen AK (2014) Commercialization of microfluidic devices. Trends Biotechnol 32(7):347–350. doi:10.1016/j.tibtech.2014.04.010
39. Jang JS, Simon VA, Feddersen RM, Rakhshan F, Schultz DA, Zschunke MA, Lingle WL, Kolbert CP, Jen J (2011) Quantitative miRNA expression analysis using fluidigm microfluidics dynamic arrays. BMC Genom 12:144. doi:10.1186/1471-2164-12-144
40. Moonsamy PV, Williams T, Bonella P, Holcomb CL, Hoglund BN, Hillman G, Goodridge D, Turenchalk GS, Blake LA, Daigle DA, Simen BB, Hamilton A, May AP, Erlich HA (2013) High throughput HLA genotyping using 454 sequencing and the Fluidigm Access Array (TM) system for simplified amplicon library preparation. Tissue Antigens 81(3):141–149. doi:10.1111/tan.12071
41. Ren KN, Zhou JH, Wu HK (2013) Materials for microfluidic chip fabrication. Accounts Chem Res 46(11):2396–2406. doi:10.1021/ar300314s

42. Fleger M, Neyer A (2006) PDMS microfluidic chip with integrated waveguides for optical detection. Microelectron Eng 83(4–9):1291–1293. doi:10.1016/j.mee.2006.01.086
43. Liao Y, Song JX, Li E, Luo Y, Shen YL, Chen DP, Cheng Y, Xu ZZ, Sugioka K, Midorikawa K (2012) Rapid prototyping of three-dimensional microfluidic mixers in glass by femtosecond laser direct writing. Lab Chip 12(4):746–749. doi:10.1039/c2lc21015k
44. Lopez-Ruiz N, Curto VF, Erenas MM, Benito-Lopez F, Diamond D, Palma AJ, Capitan-Vallvey LF (2014) Smartphone-based simultaneous pH and nitrite colorimetric determination for paper microfluidic devices. Anal Chem 86(19):9554–9562. doi:10.1021/ac5019205
45. Abgrall P, Gue AM (2007) Lab-on-chip technologies: making a microfluidic network and coupling it into a complete microsystem—a review. J Micromech Microeng 17(5):R15–R49. doi:10.1088/0960-1317/17/5/R01
46. Fu JP, Mao P, Han J (2008) Artificial molecular sieves and filters: a new paradigm for biomolecule separation. Trends Biotechnol 26(6):311–320. doi:10.1016/j.tibtech.2008.02.009
47. Reidy S, Lambertus G, Reece J, Sacks R (2006) High-performance, static-coated silicon microfabricated columns for gas chromatography. Anal Chem 78(8):2623–2630. doi:10.1021/ac051846u
48. Pipper J, Inoue M, Ng LFP, Neuzil P, Zhang Y, Novak L (2007) Catching bird flu in a droplet. Nat Med 13(10):1259–1263. doi:10.1038/nm1634
49. Wang ZH, Jin G (2004) Silicon surface modification with a mixed silanes layer to immobilize proteins for biosensor with imaging ellipsometry. Colloid Surface B 34(3):173–177. doi:10.1016/j.colurfb.2003.12.012
50. Northrup MA, Benett B, Hadley D, Landre P, Lehew S, Richards J, Stratton P (1998) A miniature analytical instrument for nucleic acids based on micromachined silicon reaction chambers. Anal Chem 70(5):918–922. doi:10.1021/Ac970486a
51. Wilding P, Shoffner MA, Kricka LJ (1994) Pcr in a silicon microstructure. Clin Chem 40(9):1815–1818
52. Torisawa Y, Shiku H, Yasukawa T, Nishizawa M, Matsue T (2005) Multi-channel 3-D cell culture device integrated on a silicon chip for anticancer drug sensitivity test. Biomaterials 26(14):2165–2172. doi:10.1016/j.biomaterials.2004.05.028
53. Torisawa YS, Kaya T, Takii Y, Oyamatsu D, Nishizawa M, Matsue T (2003) Scanning electrochemical microscopy-based drug sensitivity test for a cell culture integrated in silicon microstructures. Anal Chem 75(9):2154–2158. doi:10.1021/ac026317u
54. Park TH, Shuler ML (2003) Integration of cell culture and microfabrication technology. Biotechnol Progr 19(2):243–253. doi:10.1021/bp020143k
55. Sainiemi L, Nissila T, Kostiainen R, Ketola RA, Franssila S (2011) A microfabricated silicon platform with 60 microfluidic chips for rapid mass spectrometric analysis. Lab Chip 11(17):3011–3014. doi:10.1039/c1lc20275h
56. Kralj JG, Sahoo HR, Jensen KF (2007) Integrated continuous microfluidic liquid-liquid extraction. Lab Chip 7(2):256–263. doi:10.1039/b610888a
57. Harrison DJ, Fluri K, Seiler K, Fan ZH, Effenhauser CS, Manz A (1993) Micromachining a miniaturized capillary electrophoresis-based chemical-analysis system on a chip. Science 261(5123):895–897. doi:10.1126/science.261.5123.895
58. Easley CJ, Humphrey JAC, Landers JP (2007) Thermal isolation of microchip reaction chambers for rapid non-contact DNA amplification. J Micromech Microeng 17(9):1758–1766. doi:10.1088/0960-1317/17/9/002
59. Lewis AC, Hamilton JF, Rhodes CN, Halliday J, Bartle KD, Homewood P, Grenfell RJP, Goody B, Harling AM, Brewer P, Vargha G, Milton MJT (2010) Microfabricated planar glass gas chromatography with photoionization detection. J Chromatogr 1217(5):768–774. doi:10.1016/j.chroma.2009.12.009
60. Lin CH, Lee GB, Lin YH, Chang GL (2001) A fast prototyping process for fabrication of microfluidic systems on soda-lime glass. J Micromech Microeng 11(6):726–732. doi:10.1088/0960-1317/11/6/316

61. French PJ, Sarro PM (1998) Surface versus bulk micromachining: the contest for suitable applications. J Micromech Microeng 8(2):45–53. doi:10.1088/0960-1317/8/2/002
62. James CD, Okandan M, Mani SS, Galambos PC, Shul R (2006) Monolithic surface micromachined fluidic devices for dielectrophoretic preconcentration and routing of particles. J Micromech Microeng 16(10):1909–1918. doi:10.1088/0960-1317/16/10/001
63. Abgrall P, Nguyen NT (2008) Nanofluidic devices and their applications. Anal Chem 80 (7):2326–2341. doi:10.1021/ac702296u
64. Sniegowski JJ, de Boer MP (2000) IC-compatible polysilicon surface micromachining. Annu Rev Mater Sci 30:299–333. doi:10.1146/annurev.matsci.30.1.299
65. de Boer MJ, Tjerkstra RW, Berenschot JW, Jansen HV, Burger CJ, Gardeniers JGE, Elwenspoek M, van den Berg A (2000) Micromachining of buried micro channels in silicon. J Microelectromech S 9(1):94–103. doi:10.1109/84.825783
66. Paik SJ, Byun A, Lim JM, Park Y, Lee A, Chung S, Chang JK, Chun K, Cho DD (2004) In-plane single-crystal-silicon microneedles for minimally invasive microfluid systems. Sensor Actuat a-Phys 114(2–3):276–284. doi:10.1016/j.sna.2003.12.029
67. Dijkstra M, de Boer MJ, Berenschot JW, Lammerink TS, Wiegerink RJ, Elwenspoek M (2007) A versatile surface channel concept for microfluidic applications. J Micromech Microeng 17(10):1971. doi:10.1088/0960-1317/17/10/007
68. Kovacs GT, Maluf NI, Petersen KE (1998) Bulk micromachining of silicon. P IEEE 86 (8):1536–1551. doi:10.1109/5.704259
69. Lee S, Park S, Cho D (1999) The surface/bulk micromachining (SBM) process: A new method for fabricating released MEMS in single crystal silicon. J Microelectromech S 8 (4):409–416. doi:10.1109/84.809055
70. Zhang YL, Chen QD, Xia H, Sun HB (2010) Designable 3D nanofabrication by femtosecond laser direct writing. Nano Today 5(5):435–448. doi:10.1016/j.nantod.2010.08.007
71. Cheng JY, Yen MH, Wei CW, Chuang YC, Young TH (2005) Crack-free direct-writing on glass using a low-power UV laser in the manufacture of a microfluidic chip. J Micromech Microeng 15(6):1147–1156. doi:10.1088/0960-1317/15/6/005
72. Hanada Y, Sugioka K, Kawano H, Ishikawa IS, Miyawaki A, Midorikawa K (2008) Nano-aquarium for dynamic observation of living cells fabricated by femtosecond laser direct writing of photostructurable glass. Biomed Microdevices 10(3):403–410. doi:10.1007/s10544-007-9149-0
73. Kock M, Kirchner V, Schuster R (2003) Electrochemical micromachining with ultrashort voltage pulses-a versatile method with lithographical precision. Electrochim Acta 48(20–22):3213–3219. doi:10.1016/S0013-4686(03)00374-8
74. Zhu D, Qu NS, Li HS, Zeng YB, Li DL, Qian SQ (2009) Electrochemical micromachining of microstructures of micro hole and dimple array. Cirp Ann-Manuf Techn 58(1):177–180. doi:10.1016/j.cirp.2009.03.004
75. Li Y, Zheng YF, Yang G, Peng LQ (2003) Localized electrochemical micromachining with gap control. Sensor Actuat a-Phys 108(1–3):144–148. doi:10.1016/S0924-4247(03)00371-6
76. Bassu M, Surdo S, Strambini LM, Barillaro G (2012) Electrochemical micromachining as an enabling technology for advanced silicon microstructuring. Adv Funct Mater 22(6):1222–1228. doi:10.1002/adfm.201102124
77. Iliescu C, Chen BT, Miao J (2008) On the wet etching of Pyrex glass. Sensor Actuat a-Phys 143(1):154–161. doi:10.1016/j.sna.2007.11.022
78. Reyntjens S, Puers R (2001) A review of focused ion beam applications in microsystem technology. J Micromech Microeng 11(4):287–300. doi:10.1088/0960-1317/11/4/301
79. Tseng AA (2004) Recent developments in micromilling using focused ion beam technology. J Micromech Microeng 14(4):R15–R34. doi:10.1088/0960-1317/14/4/Ro1
80. Golonka LJ, Zawada T, Radojewski J, Roguszczak H, Stefanow M (2006) LTCC microfluidic system. Int J Appl Ceram Tec 3(2):150–156. doi:10.1111/j.1744-7402.2006.02072.x

81. Vasudev A, Kaushik A, Jones K, Bhansali S (2013) Prospects of low temperature co-fired ceramic (LTCC) based microfluidic systems for point-of-care biosensing and environmental sensing. Microfluid Nanofluid 14(3–4):683–702. doi:10.1007/s10404-012-1087-3
82. Sollier E, Murray C, Maoddi P, Di Carlo D (2011) Rapid prototyping polymers for microfluidic devices and high pressure injections. Lab Chip 11(22):3752–3765. doi:10.1039/c1lc20514e
83. Becker H, Gartner C (2008) Polymer microfabrication technologies for microfluidic systems. Anal Bioanal Chem 390(1):89–111. doi:10.1007/s00216-007-1692-2
84. McDonald JC, Whitesides GM (2002) Poly(dimethylsiloxane) as a material for fabricating microfluidic devices. Accounts Chem Res 35(7):491–499. doi:10.1021/ar010110q
85. Stroock AD, Whitesides GM (2003) Controlling flows in microchannels with patterned surface charge and topography. Accounts Chem Res 36(8):597–604. doi:10.1021/ar0202870
86. Cha KJ, Kim DS (2011) A portable pressure pump for microfluidic lab-on-a-chip systems using a porous polydimethylsiloxane (PDMS) sponge. Biomed Microdevices 13(5):877–883. doi:10.1007/s10544-011-9557-z
87. Eddings MA, Gale BK (2006) A PDMS-based gas permeation pump for on-chip fluid handling in microfluidic devices. J Micromech Microeng 16(11):2396–2402. doi:10.1088/0960-1317/16/11/021
88. Zhang WH, Lin SC, Wang CM, Hu J, Li C, Zhuang ZX, Zhou YL, Mathies RA, Yang CYJ (2009) PMMA/PDMS valves and pumps for disposable microfluidics. Lab Chip 9 (21):3088–3094. doi:10.1039/b907254c
89. Lee SW, Kim DJ, Ahn Y, Chai YG (2006) Simple structured polydimethylsiloxane microvalve actuated by external air pressure. P I Mech Eng C-J Mec 220(8):1283–1288. doi:10.1243/09544062JMES177
90. Aumiller GD, Chandross EA, Tomlinson WJ, Weber HP (1974) Submicrometer resolution replication of relief patterns for integrated optics. J Appl Phys 45(10):4557–4562
91. Masuda S, Washizu M, Nanba T (1989) Novel method of cell-fusion in field constriction area in fluid integrated-circuit. IEEE T Ind Appl 25(4):732–737. doi:10.1109/28.31255
92. Duffy DC, McDonald JC, Schueller OJA, Whitesides GM (1998) Rapid prototyping of microfluidic systems in poly(dimethylsiloxane). Anal Chem 70(23):4974–4984. doi:10.1021/Ac980656z
93. Zhao XM, Xia YN, Whitesides GM (1997) Soft lithographic methods for nano-fabrication. J Mater Chem 7(7):1069–1074. doi:10.1039/A700145b
94. Qin D, Xia YN, Whitesides GM (2010) Soft lithography for micro- and nanoscale patterning. Nat Protoc 5(3):491–502. doi:10.1038/nprot.2009.234
95. Merkel TC, Bondar VI, Nagai K, Freeman BD, Pinnau I (2000) Gas sorption, diffusion, and permeation in poly(dimethylsiloxane). J Polym Sci Pol Phys 38(3):415–434. doi:10.1002/(Sici)1099-0488(20000201)38:3<415:Aid-Polb8>3.0.Co;2-Z
96. Fuard D, Tzvetkova-Chevolleau T, Decossas S, Tracqui P, Schiavone P (2008) Optimization of poly-di-methyl-siloxane (PDMS) substrates for studying cellular adhesion and motility. Microelectron Eng 85(5–6):1289–1293. doi:10.1016/j.mee.2008.02.004
97. Clausell-Tormos J, Lieber D, Baret JC, El-Harrak A, Miller OJ, Frenz L, Blouwolff J, Humphry KJ, Koster S, Duan H, Holtze C, Weitz DA, Griffiths AD (2008) Merten CA (2008) Droplet-based microfluidic platforms for the encapsulation and screening of mammalian cells and multicellular organisms (vol 15, pg 427. Chem Biol 15(8):875. doi:10.1016/j.chembiol.2008.08.004
98. Leclerc E, Sakai Y, Fujii T (2003) Cell culture in 3-dimensional microfluidic structure of PDMS (polydimethylsiloxane). Biomed Microdevices 5(2):109–114. doi:10.1023/A:1024583026925
99. Di Carlo D, Wu LY, Lee LP (2006) Dynamic single cell culture array. Lab Chip 6(11):1445–1449. doi:10.1039/b605937f

100. Fan R, Vermesh O, Srivastava A, Yen BKH, Qin LD, Ahmad H, Kwong GA, Liu CC, Gould J, Hood L, Heath JR (2008) Integrated barcode chips for rapid, multiplexed analysis of proteins in microliter quantities of blood. Nat Biotechnol 26(12):1373–1378. doi:10.1038/nbt.1507
101. Wang DJ, Bodovitz S (2010) Single cell analysis: the new frontier in 'omics'. Trends Biotechnol 28(6):281–290. doi:10.1016/j.tibtech.2010.03.002
102. Liu V, Song YA, Han JY (2010) Capillary-valve-based fabrication of ion-selective membrane junction for electrokinetic sample preconcentration in PDMS chip. Lab Chip 10 (11):1485–1490. doi:10.1039/b923214a
103. Zhou JW, Ellis AV, Voelcker NH (2010) Recent developments in PDMS surface modification for microfluidic devices. Electrophoresis 31(1):2–16. doi:10.1002/elps.200900475
104. Begolo S, Colas G, Viovy JL, Malaquin L (2011) New family of fluorinated polymer chips for droplet and organic solvent microfluidics. Lab Chip 11(3):508–512. doi:10.1039/c0lc00356e
105. Lin LY, Jie MS, Chen FM, Zhang J, He ZY, Lin JM (2016) Efficient cell capture in an agarose-PDMS hybrid chip for shaped 2D culture under temozolomide stimulation. Rsc Adv 6(79):75215–75222. doi:10.1039/c6ra15734c
106. Eng G, Lee BW, Parsa H, Chin CD, Schneider J, Linkov G, Sia SK, Vunjak-Novakovic G (2013) Assembly of complex cell microenvironments using geometrically docked hydrogel shapes. P Natl Acad Sci USA 110(12):4551–4556. doi:10.1073/pnas.1300569110
107. Sticker D, Rothbauer M, Lechner S, Hehenberger MT, Ertl P (2015) Multi-layered, membrane-integrated microfluidics based on replica molding of a thiol-ene epoxy thermoset for organ-on-a-chip applications. Lab Chip 15(24):4542–4554. doi:10.1039/c5lc01028d
108. Fiorini GS, Lorenz RM, Kuo JS, Chiu DT (2004) Rapid prototyping of thermoset polyester microfluidic devices. Anal Chem 76(16):4697–4704. doi:10.1021/ac0498922
109. Ren KN, Dai W, Zhou JH, Su J, Wu HK (2011) Whole-Teflon microfluidic chips. P Natl Acad Sci USA 108(20):8162–8166. doi:10.1073/pnas.1100356108
110. Grover WH, von Muhlen MG, Manalis SR (2008) Teflon films for chemically-inert microfluidic valves and pumps. Lab Chip 8(6):913–918. doi:10.1039/b800600h
111. Xia YN, Whitesides GM (1998) Soft lithography. Angew Chem Int Edit 37(5):550–575. doi:10.1002/(Sici)1521-3773(19980316)37:5<550:Aid-Anie550>3.0.Co;2-G
112. Alves P, Coelho JFJ, Haack J, Rota A, Bruinink A, Gil MH (2009) Surface modification and characterization of thermoplastic polyurethane. Eur Polym J 45(5):1412–1419. doi:10.1016/j.eurpolymj.2009.02.011
113. Ahmed EM (2015) Hydrogel: preparation, characterization, and applications: a review. J Adv Res 6(2):105–121. doi:10.1016/j.jare.2013.07.006
114. Tibbitt MW, Anseth KS (2009) Hydrogels as extracellular matrix mimics for 3D cell culture. Biotechnol Bioeng 103(4):655–663. doi:10.1002/bit.22361
115. Breslin S, O'Driscoll L (2013) Three-dimensional cell culture: the missing link in drug discovery. Drug Discov Today 18(5–6):240–249. doi:10.1016/j.drudis.2012.10.003
116. Jayawarna V, Ali M, Jowitt TA, Miller AE, Saiani A, Gough JE, Ulijn RV (2006) Nanostructured hydrogels for three-dimensional cell culture through self-assembly of fluorenylmethoxycarbonyl-dipeptides. Adv Mater 18(5):611–614. doi:10.1002/adma.200501522
117. Huh D, Hamilton GA, Ingber DE (2011) From 3D cell culture to organs-on-chips. Trends Cell Biol 21(12):745–754. doi:10.1016/j.tcb.2011.09.005
118. Zhang Y, Zhou CB, Nie JF, Le SW, Qin Q, Liu F, Li YP, Li JP (2014) Equipment-free quantitative measurement for microfluidic paper-based analytical devices fabricated using the principles of movable-type printing. Anal Chem 86(4):2005–2012. doi:10.1021/ac403026c
119. Hossain SMZ, Brennan JD (2011) beta-galactosidase-based colorimetric paper sensor for determination of heavy metals. Anal Chem 83(22):8772–8778. doi:10.1021/ac202290d

120. Fenton EM, Mascarenas MR, Lopez GP, Sibbett SS (2009) Multiplex lateral-flow test strips fabricated by two-dimensional shaping. Acs Appl Mater Inter 1(1):124–129. doi:10.1021/am800043z
121. Li X, Tian JF, Garnier G, Shen W (2010) Fabrication of paper-based microfluidic sensors by printing. Colloid Surface B 76(2):564–570. doi:10.1016/j.colsurfb.2009.12.023
122. Dungchai W, Chailapakul O, Henry CS (2011) A low-cost, simple, and rapid fabrication method for paper-based microfluidics using wax screen-printing. Analyst 136(1):77–82. doi:10.1039/c0an00406e
123. Olkkonen J, Lehtinen K, Erho T (2010) Flexographically printed fluidic structures in paper. Anal Chem 82(24):10246–10250. doi:10.1021/ac1027066
124. Martinez AW, Phillips ST, Butte MJ, Whitesides GM (2007) Patterned paper as a platform for inexpensive, low-volume, portable bioassays. Angew Chem Int Edit 46(8):1318–1320. doi:10.1002/anie.200603817
125. Muller RH, Clegg DL (1949) Automatic paper chromatography. Anal Chem 21(9):1123–1125. doi:10.1021/Ac60033a032
126. Martinez AW, Phillips ST, Whitesides GM, Carrilho E (2010) Diagnostics for the developing world: microfluidic paper-based analytical devices. Anal Chem 82(1):3–10. doi:10.1021/ac9013989
127. Martinez AW, Phillips ST, Whitesides GM (2008) Three-dimensional microfluidic devices fabricated in layered paper and tape. P Natl Acad Sci USA 105(50):19606–19611. doi:10.1073/pnas.0810903105
128. Jokerst JC, Emory JM, Henry CS (2012) Advances in microfluidics for environmental analysis. Analyst 137(1):24–34. doi:10.1039/c1an15368d
129. Zhang YL, Zuo P, Ye BC (2015) A low-cost and simple paper-based microfluidic device for simultaneous multiplex determination of different types of chemical contaminants in food. Biosens Bioelectron 68:14–19. doi:10.1016/j.bios.2014.12.042
130. Liu H, Crooks RM (2011) Three-dimensional paper microfluidic devices assembled using the principles of origami. J Am Chem Soc 133(44):17564–17566. doi:10.1021/ja2071779
131. Klasner SA, Price AK, Hoeman KW, Wilson RS, Bell KJ, Culbertson CT (2010) Paper-based microfluidic devices for analysis of clinically relevant analytes present in urine and saliva. Anal Bioanal Chem 397(5):1821–1829. doi:10.1007/s00216-010-3718-4
132. Delaney JL, Hogan CF, Tian JF, Shen W (2011) Electrogenerated chemiluminescence detection in paper-based microfluidic sensors. Anal Chem 83(4):1300–1306. doi:10.1021/ac102392t
133. Li X, Tian JF, Nguyen T, Shen W (2008) Paper-based microfluidic devices by plasma treatment. Anal Chem 80(23):9131–9134. doi:10.1021/ac801729t
134. Carrilho E, Martinez AW, Whitesides GM (2009) Understanding wax printing: asimple micropatterning process for paper-based microfluidics. Anal Chem 81(16):7091–7095. doi:10.1021/ac901071p
135. Maattanen A, Fors D, Wang S, Valtakari D, Ihalainen P, Peltonen J (2011) Paper-based planar reaction arrays for printed diagnostics. Sensor Actuat B-Chem 160(1):1404–1412. doi:10.1016/j.snb.2011.09.086
136. Siegel AC, Phillips ST, Dickey MD, Lu NS, Suo ZG, Whitesides GM (2010) Foldable printed circuit boards on paper substrates. Adv Funct Mater 20(1):28–35. doi:10.1002/adfm.200901363
137. Chitnis G, Ding ZW, Chang CL, Savran CA, Ziaie B (2011) Laser-treated hydrophobic paper: an inexpensive microfluidic platform. Lab Chip 11(6):1161–1165. doi:10.1039/c0lc00512f
138. Dungchai W, Chailapakul O, Henry CS (2010) Use of multiple colorimetric indicators for paper-based microfluidic devices. Anal Chim Acta 674(2):227–233. doi:10.1016/j.aca.2010.06.019
139. Cassano CL, Fan ZH (2013) Laminated paper-based analytical devices (LPAD): fabrication, characterization, and assays. Microfluid Nanofluid 15(2):173–181. doi:10.1007/s10404-013-1140-x

140. Nie ZH, Nijhuis CA, Gong JL, Chen X, Kumachev A, Martinez AW, Narovlyansky M, Whitesides GM (2010) Electrochemical sensing in paper-based microfluidic devices. Lab Chip 10(4):477–483. doi:10.1039/b917150a
141. Zang DJ, Ge L, Yan M, Song XR, Yu JH (2012) Electrochemical immunoassay on a 3D microfluidic paper-based device. Chem Commun 48(39):4683–4685. doi:10.1039/c2cc16958d
142. Chen QS, He ZY, Liu W, Lin XX, Wu J, Li HF, Lin JM (2015) Engineering cell-compatible paper chips for cell culturing, drug screening, and mass spectrometric sensing. Adv Healthc Mater 4(15):2291–2296. doi:10.1002/adhm.201500383
143. Mu X, Liang QL, Hu P, Ren KN, Wang YM, Luo GA (2009) Laminar flow used as "liquid etch mask" in wet chemical etching to generate glass microstructures with an improved aspect ratio. Lab Chip 9(14):1994–1996. doi:10.1039/b904769g
144. Blazej RG, Kumaresan P, Mathies RA (2006) Microfabricated bioprocessor for integrated nanoliter-scale Sanger DNA sequencing. P Natl Acad Sci USA 103(19):7240–7245. doi:10.1073/pnas.0602476103
145. Zhou JH, Ren KN, Zheng YZ, Su J, Zhao YH, Ryan D, Wu HK (2010) Fabrication of a microfluidic Ag/AgCl reference electrode and its application for portable and disposable electrochemical microchips. Electrophoresis 31(18):3083–3089. doi:10.1002/elps.201000113
146. Xu J, Wu D, Ip JY, Midorikawa K, Sugioka K (2015) Vertical sidewall electrodes monolithically integrated into 3D glass microfluidic chips using water-assisted femtosecond-laser fabrication for in situ control of electrotaxis. Rsc Adv 5(31):24072–24080. doi:10.1039/c5ra00256g
147. Kato M, Inaba M, Tsukahara T, Mawatari K, Hibara A, Kitamori T (2010) Femto liquid chromatography with attoliter sample separation in the extended nanospace channel. Anal Chem 82(2):543–547. doi:10.1021/ac9017605
148. Liao Y, Cheng Y, Liu CN, Song JX, He F, Shen YL, Chen DP, Xu ZZ, Fan ZC, Wei XB, Sugioka K, Midorikawa K (2013) Direct laser writing of sub-50 nm nanofluidic channels buried in glass for three-dimensional micro-nanofluidic integration. Lab Chip 13(8):1626–1631. doi:10.1039/c3lc41171k
149. Ho CMB, Ng SH, Li KHH, Yoon YJ (2015) 3D printed microfluidics for biological applications. Lab Chip 15(18):3627–3637. doi:10.1039/c5lc00685f
150. Bishop GW, Satterwhite-Warden JE, Kadimisetty K, Rusling JF (2016) 3D-printed bioanalytical devices. Nanotechnology 27(28):0001. doi:10.1088/0957-4484/27/28/284002
151. Mironov V, Boland T, Trusk T, Forgacs G, Markwald RR (2003) Organ printing: computer-aided jet-based 3D tissue engineering. Trends Biotechnol 21(4):157–161. doi:10.1016/S0167-7799(03)00033-7
152. Ambrosi A, Pumera M (2016) 3D-printing technologies for electrochemical applications. Chem Soc Rev 45(10):2740–2755. doi:10.1039/c5cs00714c
153. Gross BC, Erkal JL, Lockwood SY, Chen CP, Spence DM (2014) Evaluation of 3D printing and its potential impact on biotechnology and the chemical sciences. Anal Chem 86(7):3240–3253. doi:10.1021/ac403397r
154. O'Neill PF, Ben Azouz A, Vazquez M, Liu J, Marczak S, Slouka Z, Chang HC, Diamond D, Brabazon D (2014) Advances in three-dimensional rapid prototyping of microfluidic devices for biological applications. Biomicrofluidics 8(5):0001. doi:10.1063/1.4898632
155. McDonald JC, Chabinyc ML, Metallo SJ, Anderson JR, Stroock AD, Whitesides GM (2002) Prototyping of microfluidic devices in poly(dimethylsiloxane) using solid-object printing. Anal Chem 74(7):1537–1545. doi:10.1021/ac010938q
156. Lewis JA (2006) Direct ink writing of 3D functional materials. Adv Funct Mater 16(17):2193–2204. doi:10.1002/adfm.200600434
157. Mannoor MS, Jiang ZW, James T, Kong YL, Malatesta KA, Soboyejo WO, Verma N, Gracias DH, McAlpine MC (2013) 3D printed bionic ears. Nano Lett 13(6):2634–2639. doi:10.1021/nl4007744

158. Obata K, El-Tamer A, Koch L, Hinze U, Chichkov BN (2013) High-aspect 3D two-photon polymerization structuring with widened objective working range (WOW-2PP). Light-Sci Appl 2:0001. doi:10.1038/lsa.2013.72
159. Au AK, Lee W, Folch A (2014) Mail-order microfluidics: evaluation of stereolithography for the production of microfluidic devices. Lab Chip 14(7):1294–1301. doi:10.1039/c3lc51360b
160. Hengsbach S, Lantada AD (2014) Rapid prototyping of multi-scale biomedical microdevices by combining additive manufacturing technologies. Biomed Microdevices 16(4):617–627. doi:10.1007/s10544-014-9864-2
161. Nguyen NT, Huang XY, Chuan TK (2002) MEMS-micropumps: a review. J Fluid Eng-T Asme 124(2):384–392. doi:10.1115/1.1459075
162. Duwairi H, Abdullah M (2007) Thermal and flow analysis of a magneto-hydrodynamic micropump. Microsyst Technol 13(1):33–39. doi:10.1007/s00542-006-0258-0
163. Brask A, Kutter JP, Bruus H (2005) Long-term stable electroosmotic pump with ion exchange membranes. Lab Chip 5(7):730–738. doi:10.1039/b503626g
164. Byun CK, Abi-Samra K, Cho YK, Takayama S (2014) Pumps for microfluidic cell culture. Electrophoresis 35(2–3):245–257. doi:10.1002/elps.201300205
165. Zeng SL, Chen CH, Mikkelsen JC, Santiago JG (2001) Fabrication and characterization of electroosmotic micropumps. Sensor Actuat B-Chem 79(2–3):107–114. doi:10.1016/S0925-4005(01)00855-3
166. Sheybani R, Cobo A, Meng E (2015) Wireless programmable electrochemical drug delivery micropump with fully integrated electrochemical dosing sensors. Biomed Microdevices 17 (4):0001. doi:10.1007/s10544-015-9980-7
167. Clime L, Brassard D, Geissler M, Veres T (2015) Active pneumatic control of centrifugal microfluidic flows for lab-on-a-chip applications. Lab Chip 15(11):2400–2411. doi:10.1039/c4lc01490a
168. Kim J, Stockton AM, Jensen EC, Mathies RA (2016) Pneumatically actuated microvalve circuits for programmable automation of chemical and biochemical analysis. Lab Chip 16 (5):812–819. doi:10.1039/c5lc01397f
169. Sahu PK, Golia A, Sen AK (2013) Investigations into mixing of fluids in microchannels with lateral obstructions. Microsyst Technol 19(4):493–501. doi:10.1007/s00542-012-1617-7
170. DeLong SA, Moon JJ, West JL (2005) Covalently immobilized gradients of bFGF on hydrogel scaffolds for directed cell migration. Biomaterials 26(16):3227–3234. doi:10.1016/j.biomaterials.2004.09.021
171. Wang F (2009) The signaling mechanisms underlying cell polarity and chemotaxis. Csh Perspect Biol 1(4):0001. doi:10.1101/cshperspect.a002980
172. Schmidt S, Friedl P (2010) Interstitial cell migration: integrin-dependent and alternative adhesion mechanisms. Cell Tissue Res 339(1):83–92. doi:10.1007/s00441-009-0892-9
173. Chiang HC, Wang YS, Chou CH, Liao AT, Chu RM, Lin CS (2012) Overexpression of chemokine ligand 7 is associated with the progression of canine transmissible venereal tumor. Bmc Vet Res 8:0001. doi:10.1186/1746-6148-8-216
174. Ahmed D, Chan CY, Lin SCS, Muddana HS, Nama N, Benkovic SJ, Huang TJ (2013) Tunable, pulsatile chemical gradient generation via acoustically driven oscillating bubbles. Lab Chip 13(3):328–331. doi:10.1039/c2lc40923b
175. Diao JP, Young L, Kim S, Fogarty EA, Heilman SM, Zhou P, Shuler ML, Wu MM, DeLisa MP (2006) A three-channel microfluidic device for generating static linear gradients and its application to the quantitative analysis of bacterial chemotaxis. Lab Chip 6(3):381–388. doi:10.1039/b511958h
176. Keenan TM, Frevert CW, Wu A, Wong V, Folch A (2010) A new method for studying gradient-induced neutrophil desensitization based on an open microfluidic chamber. Lab Chip 10(1):116–122. doi:10.1039/b913494h
177. Jeon NL, Dertinger SKW, Chiu DT, Choi IS, Stroock AD, Whitesides GM (2000) Generation of solution and surface gradients using microfluidic systems. Langmuir 16 (22):8311–8316. doi:10.1021/La000600b

178. Liu Y, Butler WB, Pappas D (2012) Spatially selective reagent delivery into cancer cells using a two-layer microfluidic culture system. Anal Chim Acta 743:125–130. doi:10.1016/j.aca.2012.06.054
179. Sun M, Bithi SS, Vanapalli SA (2011) Microfluidic static droplet arrays with tuneable gradients in material composition. Lab Chip 11(23):3949–3952. doi:10.1039/c1lc20709a
180. Brammer KS, Choi C, Frandsen CJ, Oh S, Jin S (2011) Hydrophobic nanopillars initiate mesenchymal stem cell aggregation and osteo-differentiation. Acta Biomater 7(2):683–690. doi:10.1016/j.actbio.2010.09.022
181. Kim NY, Adhikari KK, Dhakal R, Chuluunbaatar Z, Wang C, Kim ES (2015) Rapid, sensitive, and reusable detection of glucose by a robust radiofrequency integrated passive device biosensor chip. Sci Rep-Uk 5:0001. doi:10.1038/Srep07807
182. Balakrishnan SR, Hashim U, Letchumanan GR, Kashif M, Ruslinda AR, Liu WW, Veeradasan P, Prasad RH, Foo KL, Poopalan P (2014) Development of highly sensitive polysilicon nanogap with APTES/GOx based lab-on-chip biosensor to determine low levels of salivary glucose. Sensor Actuat a-Phys 220:101–111. doi:10.1016/j.sna.2014.09.027
183. Osterberg FW, Rizzi G, Donolato M, Bejhed RS, Mezger A, Stromberg M, Nilsson M, Stromme M, Svedlindh P, Hansen MF (2014) On-chip detection of rolling circle amplified DNA molecules from Bacillus globigii spores and Vibrio cholerae. Small 10(14):2877–2882. doi:10.1002/smll.201303325
184. Clark LC, Lyons C (1962) Electrode systems for continuous monitoring in cardiovascular surgery. Ann N Y Acad Sci 102(1):29. doi:10.1111/j.1749-6632.1962.tb13623.x
185. Oh SJ, Park BH, Choi G, Seo JH, Jung JH, Choi JS, Kim DH, Seo TS (2016) Fully automated and colorimetric foodborne pathogen detection on an integrated centrifugal microfluidic device. Lab Chip 16(10):1917–1926. doi:10.1039/c6lc00326e
186. Thompson BL, Birch C, Nelson DA, Li JY, DuVall JA, Le Roux D, Tsuei AC, Mills DL, Root BE, Landers JP (2016) A centrifugal microfluidic device with integrated gold leaf electrodes for the electrophoretic separation of DNA. Lab Chip 16(23):4569–4580. doi:10.1039/c6lc00953k
187. Lee A, Park J, Lim M, Sunkara V, Kim SY, Kim GH, Kim MH, Cho YK (2014) All-in-one centrifugal microfluidic device for size-selective circulating tumor cell isolation with high purity. Anal Chem 86(22):11349–11356. doi:10.1021/ac5035049
188. Yamashita H, Morita M, Sugiura H, Fujiwara K, Onoe H, Takinoue M (2015) Generation of monodisperse cell-sized microdroplets using a centrifuge-based axisymmetric co-flowing microfluidic device. J Biosci Bioeng 119(4):492–495. doi:10.1016/j.jbiosc.2014.09.018
189. Araci IE, Su BL, Quake SR, Mandel Y (2014) An implantable microfluidic device for self-monitoring of intraocular pressure. Nat Med 20(9):1074–1078. doi:10.1038/nm.3621
190. Kim DH, Viventi J, Amsden JJ, Xiao JL, Vigeland L, Kim YS, Blanco JA, Panilaitis B, Frechette ES, Contreras D, Kaplan DL, Omenetto FG, Huang YG, Hwang KC, Zakin MR, Litt B, Rogers JA (2010) Dissolvable films of silk fibroin for ultrathin conformal bio-integrated electronics. Nat Mater 9(6):511–517. doi:10.1038/NMAT2745
191. Farra R, Sheppard NF, McCabe L, Neer RM, Anderson JM, Santini JT, Cima MJ, Langer R (2012) First-in-human testing of a wirelessly controlled drug delivery microchip. Sci Transl Med 4(122):0001. doi:10.1126/scitranslmed.3003276
192. Kim DH, Lu NS, Ma R, Kim YS, Kim RH, Wang SD, Wu J, Won SM, Tao H, Islam A, Yu KJ, Kim TI, Chowdhury R, Ying M, Xu LZ, Li M, Chung HJ, Keum H, McCormick M, Liu P, Zhang YW, Omenetto FG, Huang YG, Coleman T, Rogers JA (2011) Epidermal electronics. Science 333(6044):838–843. doi:10.1126/science.1206157
193. Meng E, Sheybani R (2014) Insight: implantable medical devices. Lab Chip 14(17):3233–3240. doi:10.1039/c4lc00127c
194. Yang K, Peretz-Soroka H, Liu Y, Lin F (2016) Novel developments in mobile sensing based on the integration of microfluidic devices and smartphones. Lab Chip 16(6):943–958. doi:10.1039/c5lc01524c
195. Zhu HY, Sikora U, Ozcan A (2012) Quantum dot enabled detection of Escherichia coli using a cell-phone. Analyst 137(11):2541–2544. doi:10.1039/c2an35071h

196. Li BC, Li L, Guan A, Dong Q, Ruan KC, Hu RG, Li ZY (2014) A smartphone controlled handheld microfluidic liquid handling system. Lab Chip 14(20):4085–4092. doi:10.1039/c4lc00227j
197. Jiang J, Wang XH, Chao R, Ren YK, Hu CP, Xu ZD, Liu GL (2014) Smartphone based portable bacteria pre-concentrating microfluidic sensor and impedance sensing system. Sensor Actuat B-Chem 193:653–659. doi:10.1016/j.snb.2013.11.103
198. Martinez AW, Phillips ST, Carrilho E, Thomas SW, Sindi H, Whitesides GM (2008) Simple telemedicine for developing regions: camera phones and paper-based microfluidic devices for real-time, off-site diagnosis. Anal Chem 80(10):3699–3707. doi:10.1021/ac800112r
199. Lu Y, Shi WW, Qin JH, Lin BC (2009) Low cost, portable detection of gold nanoparticle-labeled microfluidic immunoassay with camera cell phone. Electrophoresis 30(4):579–582. doi:10.1002/elps.200800586

Chapter 2
Recent Development of Cell Analysis on Microfludics

Ziyi He and Jin-Ming Lin

Abstract Cells are basic structural and functional units of living organisms. Understanding the composition, structure and function of cells, and exploring cellular activities, are quite important for the cognition of phenomena and rules of life. Microfluidics, combined with advanced molecular, imaging and bioinformatics techniques, constitute a robust 'toolbox' and revolutionize the way for cell biology researches. In microfluidic systems, small amounts of fluids are manipulated using precisely designed channels with dimensions at micrometer level. Various chemical and biological processes can be transferred and integrated in a small single device, achieving multiple chemical and biological functions. Microfluidic technology displays a number of unique merits over conventional approaches, and has been extensively applied to various fields of cell research. In this chapter, we will review the recent developments and outstanding achievements of microfluidic technology in cell researches. Based on the cell study procedure, the main content is divided into four parts: cell culture, cell manipulation, cell stimulation and cell analysis. This review will also discuss the challenges and directions of microfluidic-based cell analysis, providing important references and ideas for the development of biological and medical researches and applications.

Keywords Microfluidics · Cell analysis · Cell culture · Cell manipulation
Cell stimulation

2.1 Introduction

Cells are basic structural and functional units of living organisms. Understanding the composition, structure and function of cells, and exploring cellular activities, are quite important for the cognition of activity rules and phenomena of life. Cells have small size, low amount of contents and diverse species. And in organisms, cells are

---

Z. He · J.-M. Lin (✉)
Department of Chemistry, Tsinghua University, Beijing 100084, People's Republic of China
e-mail: jmlin@mail.tsinghua.edu.cn

J.-M. Lin (ed.), *Cell Analysis on Microfluidics*, Integrated Analytical Systems,
https://doi.org/10.1007/978-981-10-5394-8_2

located in complex microenvironments and subjected to multiple cues that vary in time and space, including temperature, oxygen concentration, gradients of cytokines and signaling molecules, mechanical forces, and interactions with extracellular matrix (ECM) and other cells [1–3]. The microenvironments are significant for maintaining cellular functions [4, 5]. Therefore, it is highly important and desirable to develop advanced technologies that enable precise cell manipulation, physiologically relevant microenvironment simulation, as well as sensitive, selective, high-throughput and reliable cell analysis.

Since the 1990s, microfluidic technology has experienced explosive developments, and gradually become an important tool for cell research. In microfluidic systems, small (10^{-9}–10^{-18} L) amounts of fluids are manipulated using precisely designed channels with dimensions of tens to hundreds micrometres [6, 7]. Various chemical and biological processes, such as synthesis, reaction, separation, detection, and cell culture, cell isolation, cell lysis and cell analysis, can be integrated in a single device to achieve multiple chemical and biological functions [8–10]. Microfluidic technology displays a number of unique merits over conventional approaches, which promote the applications in cell researches [11–13]. The dimensions of microfluidic channels are comparable to the sizes of cells, thus facilitating precise cell manipulation. Microchannels reduce sample consumption, avoid sample dilution, and allow rapid mass and heat transfer [14]. These features contribute to highly sensitive cell analysis. And due to the size effect, microfluidic fluids at low flow rate are laminar, which enables precise fluid control [15], and can be applied to partitioned channel modification [16], concentration gradient establishment [17] and regional cell stimulation [18]. A variety of complex microstructures can be designed and fabricated on microfluidic devices, allowing better control of multiple biological cues and more physiologically relevant mimic of cellular microenvironments. The ability of integration is a prominent advantage of microfluidic technology. Microfluidics not only enable the assembly of multiple cell analysis units on a single device, but also can integrate with diverse cell detecting techniques (such as optical and electrical detection, mass spectrometry), thus facilitating comprehensive cell researches [19–21]. For example, combining microvalves, micropumps and microchambers, an automatic single-cell analysis pipeline was established on a microfluidic device, which incorporated cell culture, precise cell stimulation, live-cell microscopy, computerized cell tracking, on-chip staining of key proteins and subsequent retrieval of cells for high-throughput gene expression analysis [22]. Using droplet-based microfluidics [23, 24] or microarrays [25, 26], high throughput cell analysis can be achieved to improve research efficiency and shorten analysis time. Owing to these advantages, recently microfluidic technology has been applied to various fields of cell researches, such as cell culture [27, 28], cell sorting [29], cancer research [30], stem cell research [31], clinic diagnosis [32, 33], drug screening [34, 35] and tissue engineering [36].

In this chapter, we will introduce the recent developments and outstanding achievements of microfluidic technology in cell researches. Based on the cell study procedure, the main content is divided into four parts: cell culture, cell manipulation, cell stimulation and cell analysis. This chapter will also discuss the challenges and directions of microfluidics in cell research, providing important reference and ideas for the development of biological and medical research and application.

2.2 Cell Culture

Cell culture in vitro is the cornerstone of cell biology research. In living organisms, cells are located in complex microenvironments and subjected to multiple cues, including physicochemical properties such as temperature, oxygen concentration, pH, osmotic pressure, stimulation factors such as gradients of cytokines and signaling molecules, mechanical forces, and interactions with ECM and other cells [37]. Compared to conventional cell culture methods using petri dishes or culture plates, microfluidic devices have many unique advantages [38]. Micro-sized channels are comparable to in vivo cellular microenvironment, and enable precise regulation of cell number, cell density and spatial location; controllable fluids allow cell culture under flow condition and precise cell stimulation; using microstructures or hydrogels, three dimensional (3D) cell culture and cell co-culture can be achieved, which maintain cell-cell, cell-ECM interactions and are more physiologically relevant; microfluidic technology also facilitates parallel and automated cell culture, improving throughput and reproducibility of cell researches [27, 39].

In microfluidic cell study, how to develop more physiologically relevant cell culture models and make subsequent biological and medical researches more realistic and reliable, is an important research topic. We will review recent progresses of cell culture on microfluidics in this section, according to three parts: 3D cell culture, cell co-culture and tissues/organs-on-chips.

2.2.1 3D Cell Culture

The majority of microfluidic cell-culture systems are 2D cell culture, in which cells are grown as a monolayer on a flat substrate surface (e.g., glass or plastic). Although it is simple to handle, the 2D systems have certain limitations in mimicking the in vivo cellular microenvironment, and lack diffusion-limited distribution of soluble factors and cell-cell, cell-ECM interactions [40]. The transition from 2D to 3D cell culture is an important step for better mimicking the in vivo microenvironment [41, 42]. Owing to the complex microstructures and well-controlled parameters, microfluidics provide a versatile platform for 3D cell culture, which offer more physiologically relevant cellular morphology and phenotype, and promote metabolic activity and cellular functionality [43, 44].

The most common strategy for microfluidic 3D cell culture is to embed cells in 3D hydrogel scaffolds, such as matrigel, collagen, agarose and synthetic hydrogels [45]. Hydrogels enable cell-ECM interactions and permit diffusive permeability of oxygen, nutrients and metabolites to encapsulated cells [46]. Sung et al. developed a microfluidic platform to examine the influence of 2D and 3D culture of human mammary fibroblasts (HMFs) on the invasive transition of breast cancer cells (MCF-DCIS) [47]. MCF-DCIS encapsulated in the mixture of collagen and matrigel were co-cultured with HMFs either embed in 3D matrix or grown on 2D surface. Results demonstrated that HMFs cultured in 3D secreted more paracrine signaling molecules and intensified the promotion of the invasive progression through the HGF/c-Met interaction. Combined with high-throughput droplet-based microfluidics, a large number of monodisperse 3D liver model were formed by controlled assembly of hepatocytes and fibroblasts in core–shell hydrogel scaffolds [48]. These droplets were able to be cultured for long periods of time and showed enhanced liver-specific functions. Advances in hydrogel materials promote the development of 3D cell culture with better in vivo relevance. Donald et al. developed an injectable, interconnected microporous gel scaffold assembled from monodisperse microgels generated by microfluidic droplet technique (Fig. 2.1a) [49]. In vitro, cells incorporated during scaffold formation proliferated and formed extensive 3D networks within 48 h. In vivo, the scaffold facilitated cell migration that resulted in rapid cutaneous-tissue regeneration and tissue-structure formation within five days. 3D cell culture can also be realized by gel-free microfluidic systems, such as hanging-drop network [50, 51], micro-well array [52] and dielectrophoretic patterning [53].

2.2.2 *Cell Co-culture*

Cell-cell interactions are very important for the development and function of multicellular organisms. They allow cells to communicate with each other, respond to signals in microenvironment and regulate basic cellular functions such as survival, apoptosis, migration, proliferation, and differentiation [54, 55]. Cell-cell interactions occur through multiple mechanisms, including direct cell contact, diffusion of soluble factors, electrical signal transmission and transduction of mechanical cues through ECM [56]. Efforts have been made to investigate the interaction mechanisms and their roles in physiological homeostasis and disease states. Microfluidics, as a flexible and reliable technique which is capable of co-culturing multiple cell types in precisely defined positions and delivering biochemical and biophysical stimuli in a spatiotemporal controllable manner, has been extensively developed and widely applied to cell-cell interaction study [57, 58].

Different intercellular interaction mechanisms have been investigated in various microfluidic cell co-culture systems, which can be divided into two categories: contact and non-contact approaches [59]. The non-contact microchips utilize barriers such as hydrogels [60, 61], semi-permeable membranes [62], porous films [63], pressure-controlled valves [64] and narrow channels [65] to separately culture cell in different regions. These microfluidic systems eliminate the influence of direct cell contact and can be used for the study of paracrine signaling and endocrine signaling. Contact approaches, such as microfluidic droplet co-culture [66],

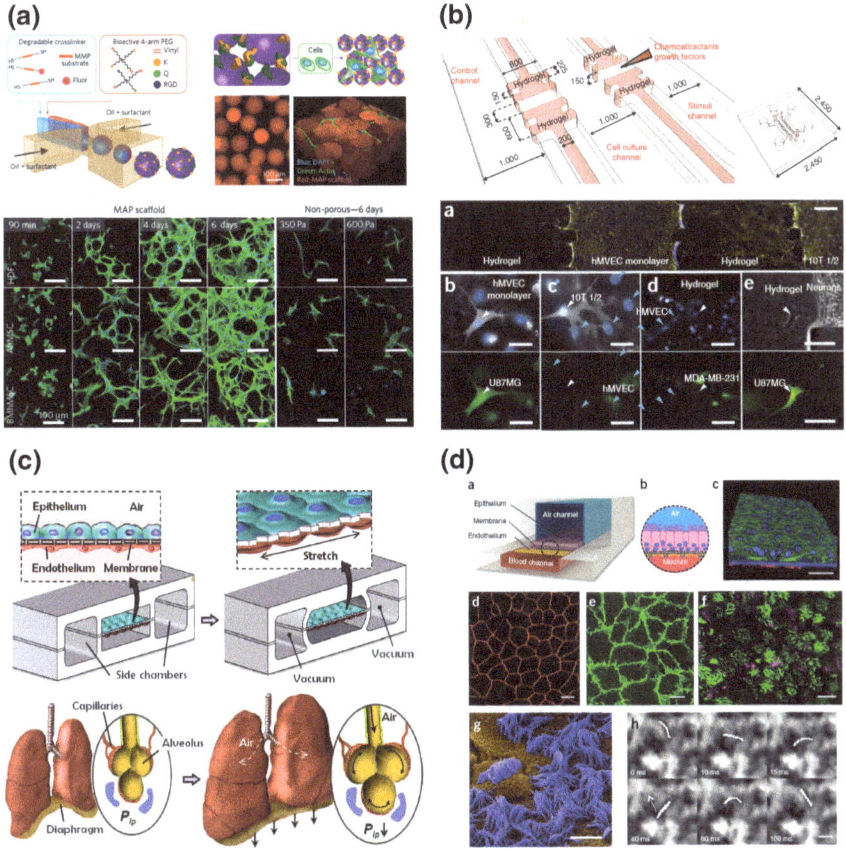

Fig. 2.1 Cell culture on microfluidic chips. **a** 3D cellular network formation on microporous gel scaffolds; **b** Cell migration into scaffolds under co-culture conditions in a microfluidic platform; **c** Human breathing lung-on-a-chip microdevice; **d** The human small airway-on-a-chip. (Reprinted with permission from Ref. [49, 72, 93, 94])

microcontact printing [67] and stencil-based methods [68, 69], culture cells in direct contact and are applicable to all interaction mechanisms, especially gap junction signaling and juxtacrine signaling. Selective cell adhesion on specific functionalized substrates was an alternative method for defined cell co-culture [70].

Cells in microfluidic systems can be co-cultured in 2D, 3D or single cell manner. Combining micropatterned surfaces with microfluidic channels, neurons and astrocytes were cultured in defined locations and communicated with each other through grooves [71]. Pharmacological agents were delivered through microchannels. This device analyzed neuron–astrocyte interactions under both healthy and pathophysiological conditions, and found that calcium dynamics in astrocytes could be modulated by the interactions with neurons. Shin et al. [72] developed a hydrogel-incorporating microfluidic cell culture assay, which facilitated the interaction of cells in 3D ECM scaffolds. It could be used in various applications, including angiogenesis and cancer metastasis, and provide new insights into how biochemical and biophysical factors regulate interactions between populations of different cell types (Fig. 2.1b). Dura et al. [73] described a microfluidic platform that achieved high-throughput single-cell pairing of lymphocytes with a defined contact time, enabling pairwise-correlated multiparametric profiling of lymphocyte interactions over hundreds of pairs in a single experiment. This platform was applied to characterize early activation dynamics of CD8 T cells and investigated the extent of heterogeneity in T-cell activation and the correlation of multiple readouts.

2.2.3 Tissues/Organs-on-Chips

Owing to the progress in 3D cell culture and cell co-culture, as well as the precisely controlled fluid flow and mechanical force, tissue/organ-on-chip microsystems have been extensively developed on microfluidic devices. In tissues/organs-on-chips, living cells are culture in well-organized microchambers, with well-defined physiological factors and mechanical cues, which can not only reconstitute multicellular architectures and microenvironment of living human tissues or organs, but also recapitulate their physiological functions and responses. These systems are more human relevant and cost-effective than animal models, and the optical transparency of microdevices enable the direct real-time imaging and analysis of cellular activities [74, 75]. Tissues/organs-on-chips have great potential to study basic mechanisms of organ physiology and disease such as cancers, and are highly applicable for preclinical drug discovery [76, 77].

Various tissue/organ-on-chip models have been developed, such as blood vessels [78, 79], brain [80], liver [81, 82], lung [83, 84], kidney [85, 86], heart [87, 88], bone [89] and muscle [90]. Microfluidic chip is quite suitable for the development of blood vessel-on-a-chip, because of its fluid perfusion ability which facilitates blood mimic and introduces shear stresses, and the ease of gradient generation for angiogenesis assays [44]. Zheng and co-workers [91] engineered living microvascular networks in type I collagen on a microfluidic scaffold. With long-term (one to two weeks) culture, the microvasculature-on-a-chip emerged appropriate endothelial morphology and barrier functions. This platform could be used to investigate angiogenic remodeling, interactions between endothelial cells and perivascular cells, and interactions between blood components and endothelium with flow. Reconstituting the organ-level functions is the major goal of organs-on-chips study. Ingber's group developed a human 'breathing' lung-on-a-chip which reproduced key structural, functional, and mechanical properties of the human alveolar-capillary interface (Fig. 2.1c) [92, 93]. This bioinspired microdevice demonstrated complex organ-level responses to bacteria and inflammatory cytokines introduced into the alveolar space, and could be used to investigate the role of mechanical breathing motions in lung disease. In follow-up study, this group reconstituted a small airway-on-a-chip containing a differentiated, mucociliary bronchiolar epithelium and an underlying microvascular endothelium that experienced fluid flow (Fig. 2.1d) [94]. This platform achieved greater robustness and fidelity in modeling of pulmonary diseases and recapitulation of lung inflammatory responses in vitro. As complex heterogeneous diseases, cancer in vitro models have also been intensively studied [95, 96]. Alexandre et al. reported a tumor-on-a-chip system where incorporation of tumor-like spheroids into a microfluidic channel permitted real-time analysis of nanoparticle (NP) accumulation at physiological flow conditions [97]. Taking advantage of the integration feature of microfluidics, 'human-on-a-chip' models which interconnect different organ-on-chip compartments through microfluidic circulatory systems have been investigated currently. It can provide more complete physiological biomimicry and become an important direction in further microfluidic study [75, 98].

2.3 Cell Manipulation

Cell manipulation plays an important role in basic cell biology study, drug screening, disease diagnosis and therapy. Because of the unique advantages, microfluidics was an excellent practical technique which provides incomparable possibilities to manipulate cells in an automated, reproducible, fast and efficient way. Various microfluidic techniques have been developed to manipulate cell precisely for diverse biological researches, such as microstructures, integrated valves and pumps, droplet encapsulation, electrokinetic operations, affinity-based surface patterning and free flow manipulation [99, 100]. Different techniques can be integrated to improve performance and functionalities within a single chip. In the next three sections, we mainly focus on the microstructures, electrokinetic operations, and free flow manipulation, and provide some recent examples. Droplet-based microfluidics will be summarized in Chap. 7.

2.3.1 Microstructures

Microstructures, such as microwells, microbarriers and microtraps, can be precisely designed and applied to cell capture, pairing, patterning and subsequent cell culture as well as other biological study [101–104]. This approach is high-throughput, high-efficient and ease of operation, which has been extensively used in both multicellular and single-cell systems. Chung et al. [105] developed a microfluidic embryo-trap array that could rapidly order and vertically orient hundreds of embryos, and this platform was used to quantitatively analyze multiple morphogen gradients in the dorsoventral patterning system (Fig. 2.2a). Sarioglu et al. [106] introduced a Cluster-Chip, which contained a series of triangular pillars to differentiate CTC clusters from single cells in blood. It was used to isolate CTC clusters from unprocessed patient blood samples with high sensitivity, allowing for

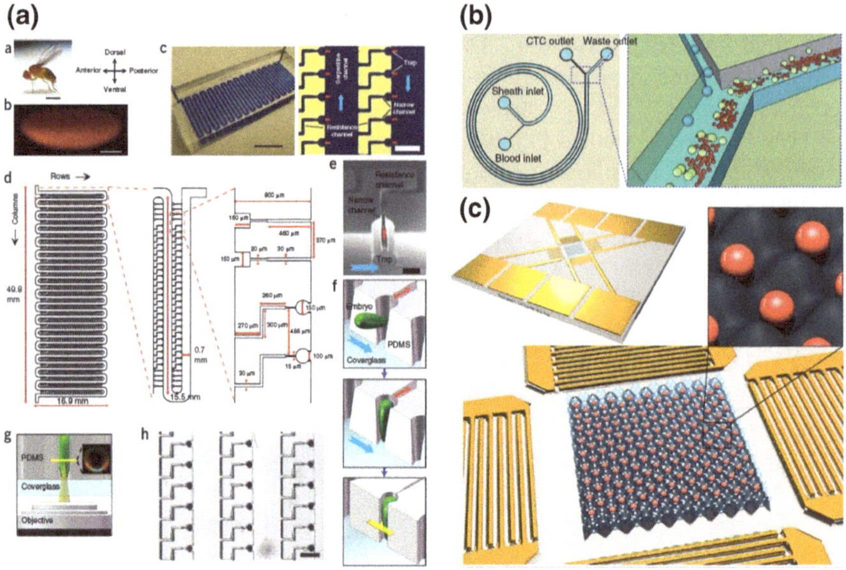

Fig. 2.2 Cell manipulation on microfluidic chips. **a** Microfluidic embryo-trap array for high-throughput arraying of vertically oriented Drosophila embryos; **b** Size-based cell isolation in a spiral microfluidic device; **c** Two-dimensional single-cell patterning driven by surface acoustic waves in a microfluidic device. (Reprinted with permission from Ref. [105, 115, 117])

downstream molecular and functional assays. Integrating the microwell array with microfluidic valve and pump systems, Lecault et al. [107] developed a longer-term mammalian cell culture platform which was able to immobilize nonadherent cells during automated medium exchange and recover the cells for subsequent analysis. This platform was then applied to high-throughput investigation of hematopoietic stem cell proliferation at the single-cell level.

2.3.2 *Free-Flow Manipulation*

The free flow cell manipulation can be divided into passive and active strategies [108]. Passive strategies use rationally designed microfluidic structures to control cell positions, such as pinched flow [109] and deterministic lateral displacement [110]; Active strategies use actuators to manipulate cells based on their electrical, magnetic and mechanical properties, such as dielectrophoresis [111], magnetophoresis [112], acoustophoresis [113] and optical tweezers [114]. These techniques can also be integrated to improve the performance [108]. Warkiani et al. [115] developed a label-free spiral microfluidic device to allow size-based cell isolation by taking advantage of dean migration and inertial focusing in curvilinear microchannels

(Fig. 2.2b). Karabacak et al. [116] presented a CTC-iChip using deterministic lateral displacement, inertial focusing and magnetophoresis to isolate rare circulating tumor cells (CTCs) from blood samples. This device achieved an average of 3.8-log depletion of white blood cells at a rate of 8 ml whole blood/h and a cancer cell yield of 97 ± 2.7%. Collins et al. [117] utilized surface acoustic waves at high frequency to create a 2D acoustic force field with an inter-nodal spacing of the same order as the cell dimensions (Fig. 2.2c). This device was applied to the patterning of multiple spatially separated single cells with one cell per acoustic well.

2.3.3 *Electrokinetic Operations*

Owing to the feasibility of integrating microelectrodes in microfluidic chips, electrokinetic forces stemming from the electric field have been widely applied to microfluidic cell manipulation [118]. Electrokinetic manipulations include several categories: electrophoresis [119], dielectrophoresis [120], electroosmosis [121], electroporation [122], electrofusion [123] and electric cell lysis, and these techniques have all been realized on microfluidic chips.

Dielectrophoresis (DEP) refers to the movement of cells caused by a force, generated due to polarization differences between cells and the medium under a non-uniform electric field. It has been used to move, separate and position cells [124]. Tsutsui et al. developed a quick and active method based on positive DEP traps to pattern embryonic stem cells on PEG hydrogels [53]. Mazutis et al. integrated microdroplet generation and DEP sorting on one microfluidic platform for the high-throughput analysis and sorting of single cells (Fig. 2.3a) [125]. Compartmentalization of single cells in droplets enabled the analysis of proteins secreted by cells, and fluorescence-activated droplet sorting by electrophoretic force enabled target cell enrichment.

Exposing a cell to a strong electric field pulses results in electroporation—the formation of nanoscale aqueous pores in the cell membrane. These permeable structures provide a pathway for diffusive transport of molecules which are physiologically membrane impermeable [126]. Electroporation on microfluidics alleviates heat effect, allows real-time monitoring of cellular response and enables single-cell manipulation [127]. These devices have been applied to cell transfection. Garcia et al. introduced a rapid microfluidic assay to determine the critical electric field threshold required for inducing bacterial electroporation [128]. Qu et al. utilized a droplet electroporation microfluidic platform for nuclear transformation of microalgae, which showed a remarkably higher transformation efficiency than bulk phase electroporation [129]. Kang et al. presented a microfluidic device that coupled long-term cell culture on the device and repeated temporal transfection by localized electroporation (Fig. 2.3b) [130]. This platform enabled on-chip differentiation of neural stem cells and transfection of postmitotic neurons with a green fluorescent protein plasmid.

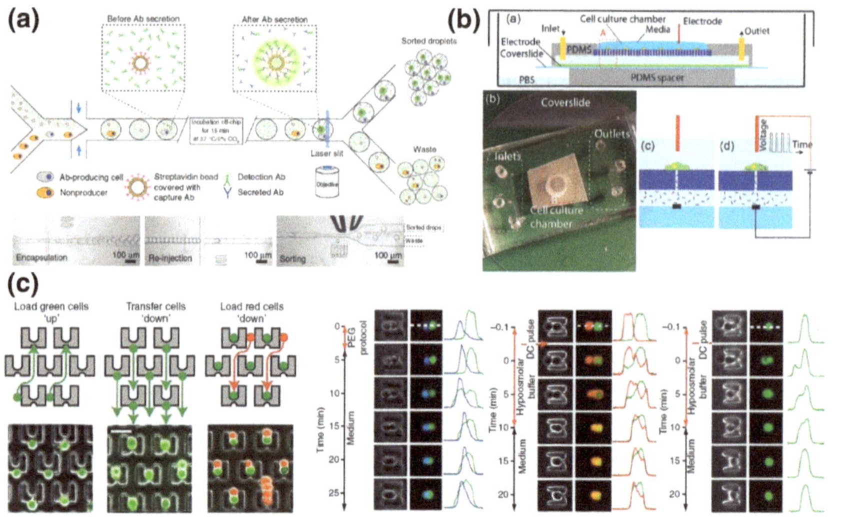

Fig. 2.3 Cell manipulation by electrokinetic operations on microfluidic chips. **a** Dielectrophoresis sorting of microdroplets containing single cells; **b** Microfluidic device for localized electroporation of postmitotic neurons; **c** Microfluidic control of cell pairing and fusion. (Reprinted with permission from Ref. [125, 130, 133])

Cell fusion is an important method to achieve nucleus transfer, hybridoma and epigenetic reprograming of somatic cells [131]. Due to the ability of precisely controlling cell positions, electrofusion in microfluidic devices shows remarkable advantages [132]. Skelleyet al. [133] presented a microfluidic device containing a dense array of weir-based passive hydrodynamic cell traps, which could immobilize and pair thousands of cells at once (Fig. 2.3c). The device is compatible with both chemical and electrical fusion protocols, with better performance of electrical fusion. 50% properly paired and fused cells were achieved over the entire device, fivefold greater than the commercial electrofusion chamber. This platform was successfully applied to the reprogramming in hybrids between mouse embryonic stem cells and mouse embryonic fibroblasts.

2.4 Cell Stimulation

In cellular microenvironment, cells are subject to multiple cues that vary in time and space, including physical conditions such as temperature, oxygen, pH, gradients of cytokines and secreted proteins from neighboring cells, and mechanical forces. Investigating cellular responses to multiple stimulations will facilitate better understanding of biological pathways, cell-fate decisions and tissue functionalities. Microfluidics is a robust technology that enables controlled perturbation of the cellular environment spatiotemporally in vitro. By precise flow control and well-defined microstructures, it is easy to build concentration gradients and mechanical conditions in microfluidic devices. In this section, we will discuss the recent developments of cell stimulation studies in microfluidics, from three aspects: flow control, gradient generation and mechanical stimuli.

2.4.1 Flow Control

Microfluidic devices facilitate precise flow control, owing to the unique features of flow at micrometer length scale and the feasible integration of valves and pumps. The microfluidic flows are always laminar, allowing for highly predictable fluid dynamics and molecular diffusion kinetics. Laminar flow can route different fluid to specific region in a well-ordered manner, which can alter liquid-phase environment over distances and be applied to controlled cell stimulation in a high spatial and temporal resolution [15]. Lucchetta et al. used microfluidic laminar flow to create temperature differences by flowing two converging aqueous streams around an embryo, each at a controlled temperature (Fig. 2.4a) [18]. This platform was applied to differentially control the rate of development in the anterior and posterior halves of the embryo. Similar microfluidic chips were used to deliver small molecules to selected subcellular microdomains [134], for the study of mitochondrial movement [135] or neurite injury [136].

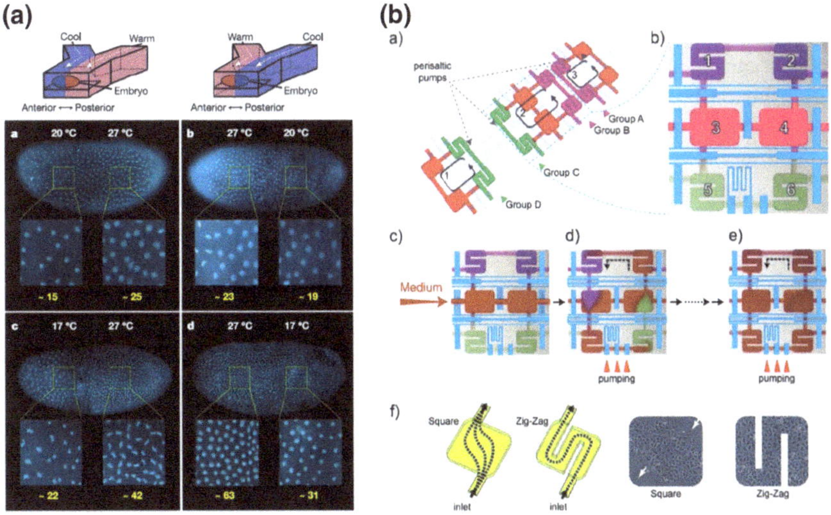

Fig. 2.4 Cell stimulation by flow control. **a** Temperature different stimulation of *Drosophila* embryo by microfluidic laminar flow; **b** Controlled cell stimulation by integrated valves or pumps to study anthrax toxin induced cell lethality. (Reprinted with permission from Ref. [18, 140])

Integrated valves or pumps, which are feasible to be incorporated into elastomeric microfluidic devices, lead to flexible, automated and high-throughput flow control with unparalleled temporal and spatial precision [137–139]. Diverse microfluidic designs with valves and pumps have been developed for various biological researches, including the precisely controlled cell stimulation. For example, Shen et al. reported a microfluidic device that integrated with pneumatic valves and peristaltic pumps to control fluid exchange parallelly and pump conditioned mediums towards the cells that exposed to toxin (Fig. 2.4b) [140]. This platform was used to investigate the roles of Dickkopf-1 in cell susceptibility to anthrax toxin. Taylor et al. developed a microfluidic platform that combined programmable on-chip mixing and perfusion with high-throughput image acquisition and processing [141]. Single-cell network responses under hundreds of combined genetic perturbations and time-varying stimulant sequences were investigated on this platform for dynamic analysis of MAPK signaling.

2.4.2 Gradient Generation

In cellular microenvironment, cells are always exposed to concentration gradients of biochemical signals such as growth factors, hormones and chemokines, which regulate many biological processes including cell differentiation, cell migration, immune responses, angiogenesis and cancer metastasis [142]. Thanks to the ability of accurate and precise flow control, it is convenient to establish concentration gradients in microfluidic systems to mimic the stimulations in microenvironment and study cellular behaviors [143]. Recently, a number of microfluidics-based gradient devices have been developed and used for different cellular studies. These devices can be categorized into two groups: flow-based gradient generators and free-diffusion-based gradient generators [144].

Flow-based gradient generators utilize laminar flows in microfluidic channels and can be divided into two design mechanisms. In Y-junction, T-junction or flow splitter microfluidic devices, streams of fluids composed of different chemical species or concentrations are brought together where the solutes diffuse across the interface as they flow down the microchannel, thus concentration gradients perpendicular to the flow direction can be established [145]. Lin et al. developed a "Y" type microfluidic device to generate concentration gradients of chemokine CCL19 and CXCL12 for T cell chemotaxis investigation [146]. In the other design mechanism, which refers to "Christmas tree", solutions of different concentrations are introduced from the inlet, and then repeatedly split, mixed in serpentine channel regions and recombined to produce multiple streams of mixed solutions having different proportions of input solutions. These streams can be brought together into a single wide channel to generate a gradient across the channel or introduced into separate channels for parallel cell stimulation with defined concentration gradients [145]. Jeon et al. used "Christmas tree" device to generate linear or complex gradients of interleukin-8 in a single channel to investigate neutrophil chemotaxis

[17]. Similar devices were also utilized to study bacterial chemotaxis [147], to investigate neural stem cell differentiation [148], and to generate gradients of substrate-bound laminin to orient axonal specification of neurons [149]. Our group [150] developed an integrated microfluidic device for high-throughput drug screening with an online mass spectrometry analysis (Fig. 2.5a). "Christmas tree" mixer network was used to generate drug gradient, and cells in the culture chambers were stimulated with different drug concentrations separately. Drug absorption and cytotoxicity were then characterized on this platform.

In free-diffusion-based gradient generators, a gradient across a given area is established as the molecules diffuse from high concentration "source" to low concentration "sink" [151]. Porous materials such as hydrogels and semi-permeable membranes are often used to form concentration gradients between sources and sinks [152]. For example, Haessler et al. developed a microfluidic device that allowed rapid establishment of stable gradients in 3D matrices to show that dendritic cells chemotaxis in 3D could respond to CCR7 ligand gradients (Fig. 2.5b) [153]. Nguyen et al. designed blood vessels-on-a-chip by lining endothelial cells in a cylindrical channel encapsulated within a 3D collagen hydrogel [154]. Emanating from a parallel source channel, gradients of angiogenic factors were established in hydrogel and used to stimulate endothelial cells and recapitulate the angiogenic sprouting in vitro. Apart from porous gels, gradient can also be formed across a

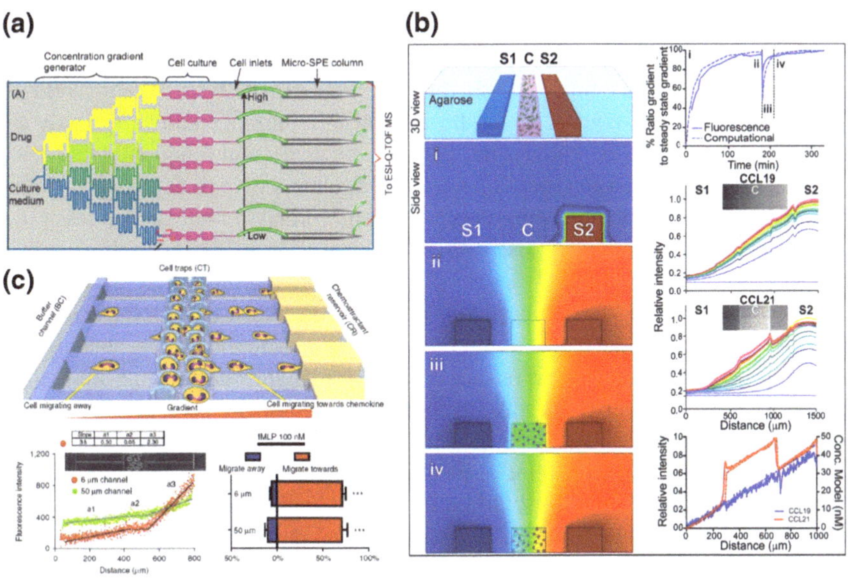

Fig. 2.5 Gradient generation on microfluidic chips for cell stimulation. **a** Drug gradient generated by "Christmas tree" mixer network; **b** Establishment and characterization of matrix bound and soluble protein gradients in a 3D hydrogel microfluidic device; **c** Gradient formed across the interconnecting channels between the source and sink for single-cell migration study. (Reprinted with permission from Ref. [150, 153, 156])

region that connects to the source and sink with microchannels. The interconnecting channels have low height, narrow width and long length, thus creating a high fluidic resistance while minimizing convective flow [152]. Chabaud et al. presented a microfluidic device composed of two fluidic chambers connected by migration microchannels [155]. Perpendicular drug gradients along migration channels were established and applied to investigate the migration and antigen capture processes of dendritic cells. Boneschansker et al. employed two large-scale arrays of microchannels to connect the central main channel with two side channels (Fig. 2.5c) [156]. Cell traps were integrated in the central channel to load precise numbers of leukocytes. Chemokine gradients were developed in the main channel by filling one side channel with chemokines and the other with buffer. This device facilitated the quantification of leukocyte migration patterns at single-cell level.

2.4.3 Mechanical Stimuli

Apart from physical and biochemical signals, cells in vivo are also subject to multiple mechanical cues in microenvironment, including shear stress, interstitial flow, substrate strain, confinement, compression and matrix stiffness. These mechanical processes are important for cell growth, migration, differentiation, apoptosis, and dysfunctional mechanotransduction can lead to numerous diseases [157, 158].

Microfluidics offers an excellent strategy for the study of cellular responses to mechanical stimuli. As blood flows through a vessel, it exerts shear stress on endothelial cells. The control of fluidic flow in microfluidics allows to study the impacts of shear stress on cellular morphology, behavior and functions [159, 160]. Sundd et al. integrated a microfluidic device with quantitative dynamic footprinting microscopy to study the mechanisms of neutrophil rolling at high shear stress [161, 162]. They indicated that step-wise peeling of “slings” at the front of rolling cells is responsible for the rolling of neutrophils. Miura et al. fabricated a multilayer microfluidic device to study the microvilli formation in placental transfer process (Fig. 2.6a) [163]. Abroad range of fluid shear stress was applied to placental barrier cells, and cellular responses were monitored. Results showed that the fluid shear stress serves as a trigger for microvilli formation in human placental trophoblastic cells, and the molecular mechanisms were also intensively investigated.

Substrate strain is another crucial mechanical force that can manipulate cellular alignment and tissue functions [164]. By incorporating flexible substrates into microfluidic platforms, devices have been developed to study the effect of mechanical stretch on the cells cultured on the deformable substrate. Hsieh et al. developed microfluidic chip which consisted of a concentric circular hydrogel pattern and a flexible PDMS membrane (Fig. 2.6b) [165]. A range of gradient static strains on cells can be generated by compressing the cell-laden hydrogels with the membrane, and cells were elongated with the increase of strain. In previously introduced example, by applying and releasing vacuum in two larger, lateral

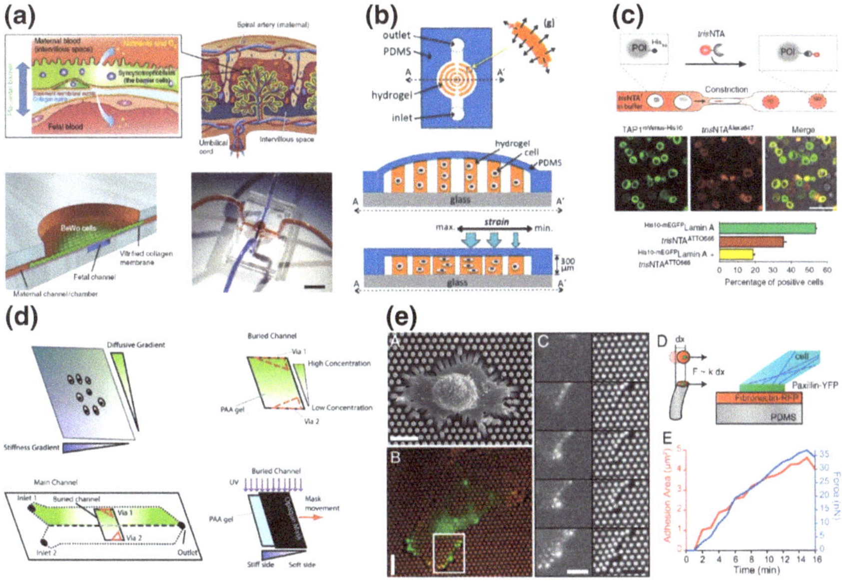

Fig. 2.6 Mechanical cell stimulation on microfluidic chips. **a** Microfluidic platform to study impacts of fluid shear stress on placental barrier cells; **b** Gradient static strain stimulation in a microfluidic chip for 3D cellular alignment; **c** Compression-induced cell deformation for living cell labeling; **d** Generation of stable orthogonal gradients of chemical concentration and substrate stiffness in a microfluidic device; **e** Measurement of mechanical forces generated by cells using micrometre-sized elastomer post arrays. (Reprinted with permission from Ref. [163, 165, 166, 170, 176])

microchambers, PDMS membrane with the adherent tissue layers were stretched and recoiled [93]. This "breathing" human lung-on-a-chip could replicate the dynamic mechanical distortion of the alveolar-capillary interface caused by breathing movements.

Compressive stress on cells was also realized in microfluidic devices. Kollmannsperger et al. utilized the compression-induced cell deformation for rapid and efficient transfer of trisNTA probe into living cells (Fig. 2.6c) [166]. Cells were compressed through micrometre constrictions in a microfluidic device, causing the formation of transient holes in the plasma membrane. Si et al. developed an air-driven microfluidic device to apply a compressive force on Escherichia coli cells [167]. With compression, cells no longer retained their rod-like shapes but grew and divided with a flat pancake-like geometry.

The mechanical stiffness of extracellular matrix has proven to be a crucial regulator for cell growth, differentiation, movement and functions [168]. Microfluidics offer versatile platforms for the study of cellular responses to matrix stiffness. Sundararaghavan et al. designed and developed a microfluidic device to investigate the neurites growth under stiffness gradients [169]. An "H"-shaped, source–sink network was utilized to generate a gradient of genipin (a collagen crosslinker) in type I collagen solution. After self-assembly, hydrogel with a gradient of mechanical stiffness, linearly ranging from 57 to 797 Pa, was formed in the device. Chick dorsal root ganglia was introduced into the gradient and results shown that the neurites extended preferentially down the stiffness gradient. García et al. presented a robust microfluidic device that generated a stable, linear and diffusive chemical gradient over a biocompatible hydrogel with a well-defined stiffness gradient (Fig. 2.6d) [170]. Matrix stiffness was regulated by UV exposure level of the polyacrylamide hydrogel. This device was then used to study the cell scattering in response to perpendicular gradients of hepatocyte growth factor and substrate stiffness.

Except for the mimicking of cellular mechanical environments, microfluidic techniques can also be used to measure mechanical forces generated by cells [171]. Christopher S. Chen group has developed micrometre-sized elastomer post arrays to manipulate and measure mechanical interactions between cells and their underlying substrates. Cells attached to, spread across, and deflected multiple posts. And the deflections of the posts directly reported the subcellular distribution of traction forces. Values of the forces could be measure according to the geometry change. By controlling cell adhesion on these micromechanical sensors, it showed that cell morphology regulated the magnitude of traction force generated by cells. Unspread cells did not contract in response to stimulation by serum or lysophosphatidic acid, whereas spread cells did [172]. Similar micro-fabricated device has also been utilized for mapping mechanical forces during epithelial cell migration [173]. Besides, by alternating the post heights or pillar diameters, the stiffness of substrates could be regulated, and their effects on cell morphology, focal adhesions, cytoskeletal contractility and stem cell differentiation were then investigated [174, 175]. The mechanosensing mechanism for cellular adaptation to substrate stiffness was also intensively studied (Fig. 2.6e) [176].

2.5 Cell Analysis

In native cellular microenvironment, cells are surrounded with multiple biophysical and biochemical cues and also respond to the various stimuli. Thus analyzing cells, including the cell morphology, cellular contents, cell signaling and cell secretion, is crucial for intensive understanding of biological processes and mechanisms. Owing to the unique merits, including flexible design, low sample consumption, high throughput, ease of integration and automation, microfluidic technology has been regarded as a robust and promising tool for cell analysis [11, 31, 177]. Almost every analytical tool available in a conventional biology lab has an equivalent microfabricated counterpart on microfluidic devices. And the microfluidic chips are also feasible to integrate with diverse analytical instruments, such as microscopy, electronic operation, mass spectrometry and nuclear magnetic resonance, to analyze cells for various purposes [21]. In this section, we will introduce the recent developments of cell analysis on microfluidics, both in sample preparation and analytical systems.

2.5.1 Sample Preparation

2.5.1.1 Cell Sorting

Isolating and sorting cells from complex, heterogeneous cell mixtures is a critical preparatory step in many biological and medical assays, enabling the enrichment of cell samples into well-defined populations or the isolation of rare cells such as circulating tumor cells (CTCs) and hematopoietic stem cells (HSCs) from much larger population of background cells [178]. It enhances the efficiency in biological researches and diagnosis, and facilitates the understanding of accurate underlying biochemical information of specific cell types in a mixture. Recent advances in microfluidics promote high-throughput cell sorting, and this has led to various novel diagnostic and therapeutic applications that are difficult to implement using conventional technologies [179]. Microfluidic cell sorting techniques are either based on the inherent physical properties of cells, such as cell size, morphology, electrical properties and cell-fluid interactions, or on account of the differences after affinity labeling [180, 181]. And depending on the utilization of external forces, microfluidic cell sorting can also be classified as passive and active strategies. Passive sorting relies on the channel geometry (pillar and weir structures, microfilters), hydrodynamic forces (pinched flow fractionation, hydrodynamic filtration, inertial separation) and surface modification (affinity-based separation) for functionality. While active sorting applies external forces such as electric, magnetic, acoustic and optical forces for cell separation [182]. Detailed examples are listed in Table 2.1 [183].

Affinity ligands for cell surface markers can be used either to provide a force for separation, such as in cell affinity separation and capture, or as labels in fluorescence activated cell sorting (FACS) and magnetic activated cell sorting (MACS). It improves the selectivity and are quite suitable for cell types that are physically similar to the background cell populations. In cell affinity separation, affinity ligands such as antibodies, aptamers and proteins are modified on chip surface. Cells with specific markers can be selectively captured in channel through the ligand-receptor interactions, while other cell types are passed through the device. The large surface to volume ratio of microfluidic channels significantly increases the possibility of cell-to-surface interactions and leads to a better isolation performance [180, 184]. The geometry of the channel can affect the cell flow and cell capture efficiency. Stott et al. demonstrated a herringbone-chip (HB-chip) that allowed passive mixing of blood cells through the generation of microvortices to significantly increase the number of interactions between target CTCs and the antibody-coated chip surface (Fig. 2.7a) [185]. The HB-chip was used for CTC separations, identifying CTCs in 93% of patients with metastatic disease. The low shear flow properties enabled the isolation of previously unappreciated micro-clusters of CTCs, facilitating the investigation of the association between expression of mesenchymal markers and CTC clusters [186]. Chen et al. developed a microfluidic chip with a microwell array that was encoded with cell-recognizable

Table 2.1 List of various active and passive cell separation methods. (Reprinted with permission from Ref. [183])

	Method	Mechanism	Separation markers	Flow rate/throughput
Active	Flow cytometry (FACS)	Fluorescence	Fluorescence labels	100 s^{-1}
	Magnetic	Homogeneous/inhomogeneous magnetic field	Size, magnetic susceptibility	10,000 s^{-1}
	Dielectrophoresis	Inhomogeneous electric acid	Size, polarizability	10,000 s^{-1}
	Electrophoresis	Homogeneous electric acid	Size, charge density	0.1–0.2 mm s^{-1}
	Optical	Optical force	Size, refractive index, polarizability	1500 min^{-1}
	Acoustic	Ultrasonic standing waves	Size, density, compressibility	0.1–0.4 mL min^{-1}
Passive	Pillar and weir structures	Laminar flow	Size, deformability	1000 μm s^{-1}
	Pinched flow fractionation (PFF)	Hydrodynamic force (parabolic velocity profile)	Size	~4000 min^{-1} 20 μL h^{-1}
	Hydrodynamic filtration	Hydrodynamic force	Size	20 μL min^{-1} >100,000 min^{-1}
	Inertial	Shear-induced and wall-induced lift	Size, shape	~10^6 min^{-1}
	Surface affinity	Specific binding to surface markers	Surface biomarkers	1–2 mL h^{-1}
	Biomimetic	Hydrodynamic force/Fahraeus effect	Size, deformability	10 μL h^{-1}
	Hydrophobic filtration	Pressure field gradient	Size	4×10^3 s^{-1}

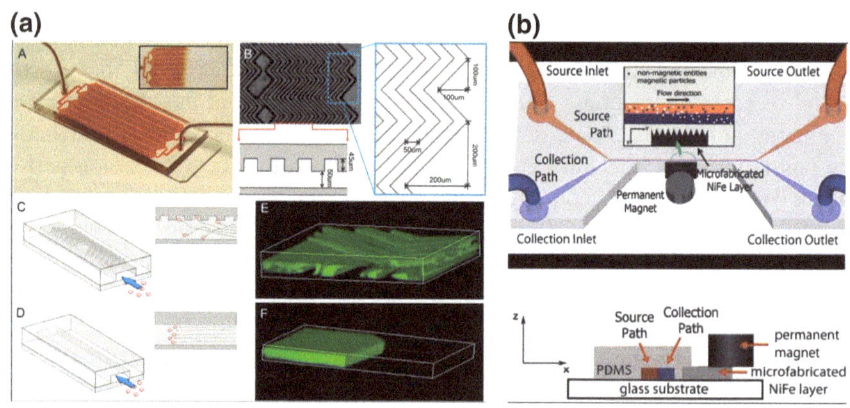

Fig. 2.7 Cell sorting on microfluidic chips. **a** Isolation of CTCs using a microvortex-generating herringbone chip; **b** Lateral magnetophoresis for cell separation. (Reprinted with permission from Ref. [185, 198])

aptamer [187]. Single tumor cells were isolated with 88.2% single-cell occupancy, and various cellular carboxylesterases were studied by time-course measurements of cellular fluorescence kinetics at individual-cell level. Recovery of captured cells can be achieved by using specific enzymes for antibodies or aptamers digestion or photocleavable linkers for ligand immobilization.

In FACS, cells are labeled by antibodies conjugated to fluorophores and selected according to their fluorescence signals. Traditional FACS has been thoroughly developed and widely applied in biological and pharmaceutical researches and industries [188]. More recently, FACS has been implemented in microfluidic devices, achieving high separation purity, high efficiency and low sample consumption. Microfluidic FACS generally operates by ordering cells in flow streams for: (i) serial interrogation by laser light, (ii) real-time classification, and (iii) rapid, command-driven sorting [29]. Sorting process can be driven by electrokinetic mechanisms (including electrophoresis, dielectrophoresis and electroosmotic flow), acoustophoresis, optical manipulations and mechanical forces [189–194]. However, the throughput of microfluidic FACS is not high enough to compete with the commercial, large-scale instruments systems. Thus further efforts should be made to improve the throughput, for example by using a pulsed laser triggered sorting to reduce the switching time [195] or by building a parallel array of microfluidic sorting units for simultaneous operation.

In MACS, cells are first labeled by magnetic beads coated with affinity ligands, and a strong magnetic field is used to isolate the magnetic beads that are attached to the desired cell population. The immunomagnetic-based separation is highly specific, biocompatible and high-throughput. Enrichment of up to 10^{11} cells in less than 30 min has been reported [184]. Many macroscale MACS separations have been adapted in microfluidic systems, which have the advantages of low sample costs, fluid flow condition and parallel operations [29, 196]. Among these approaches, continuous separation by means of magnetophoresis has been well-studied and applied to cell separation. For example, Hoshino et al. developed an immunomagnetic microchip for CTC detection [197]. As the blood sample flowed through the microchannel closely above arrayed magnets, cancer cells labeled with magnetic nanoparticles were captured and separated from blood flow. CTCs with low cancer cell to blood cell ratios (about $1{:}10^7$ to 10^9) were detected, at a fast screening speed (10 mL/h). Xia et al. developed a lateral magnetophoresis for cell separation (Fig. 2.7b) [198]. This method used a horizontal magnetic field to drag labeled target cells from the sample flow into the buffer flow. Living E. coli bacteria bound to magnetic nanoparticles were efficiently removed from flowing solutions containing densities of red blood cells similar to that found in blood. Similar design was also utilized in CTC separation, which isolates about 90% of spiked CTCs in human peripheral blood with a purity of 97%. The overall isolation procedure was completed within 15 min for 200 μL of blood [199].

2.5.1.2 Cell Lysis

Cell lysis is an essential step for the analysis of cellular contents such as proteins and nucleic acids. Microfluidic cell lysis has several advantages over conventional approaches. The unique geometries and precise dimensions allow for finely tuned mechanical or chemical cell perturbation. Micrometer length scale minimizes lysate dilution. And laminar flow limits the convective transport of lysate. These properties facilitate the increase of analyzing sensitivity [200, 201]. Microfluidic cell lysis approaches can be categorized into four major groups: mechanical lysis, thermal lysis, chemical lysis and electrical lysis [202]. In this section, we will discuss these four cell lysis approaches and provide some recent examples.

Mechanical cell lysis tears or punctures cell membranes by mechanical forces, which include shear stress, collision with sharp features, friction forces and compressive stress. In this way, cell structures are disrupted and intracellular components are released [201]. Yun et al. presented a handheld mechanical cell lysis chip with ultra-sharp nano-blade arrays (Fig. 2.8a) [203]. Cells bumped into the blades were easily ruptured by these ultra-sharp nanostructures. This chip can be directly connected to a commercial syringe, and the protein concentration obtained by this chip is quantitatively comparable to the conventional chemical lysis method. Kim et al. developed a microfluidic CD (Compact Disc) platform for cell lysis [204]. Cells were mixed with granular particles, and the solution was placed into an annular channel on the chip. When the disc was rotated at high speeds around a horizontal axis, cells were broken up by the frictions and collisions between the cells and particles. Mammalian cells, bacteria and yeast cells could all be effectively lysed and the lysis efficiency relative to a conventional lysis protocol was approximately 65%. Combining a magnetically actuated bead-beating system with this CD chip could further improve the lysis efficiency [205]. These platforms were utilized in nucleic acid extraction from clinical samples. Mechanical lysis can relatively minimize the protein damage and avoid detergent interferences. However, it requires additional instrumentation or operation for activation, and the cell debris produced in mechanical lysis may hinder subsequent extraction.

In electrical cell lysis, cells are exposed to strong electric fields. The membranes are destabilized and pores are formed. As the osmotic pressure between the cytosol and the surrounding media becomes unbalanced, cells swell, rupture and eventually lyse. The electric field can be tuned for rapid cell lysis without denaturing target biomolecules. And considering difference between trans cell-membrane potential and trans organelle-membrane potentials, appropriate electric field strength and exposure time can selectively rupture the cell membrane while leave organelles intact [2, 201]. Electrical cell lysis is well suitable for microfluidics applications. Mellors et al. developed a microfluidic device for automated real-time analysis of individual cells using capillary electrophoresis (CE) and electrospray ionization mass spectrometry (ESI-MS) [206]. Cell lysis occured at a channel intersection using a combination of rapid buffer exchange and an increase in electric field strength (4 kV). The cell lysis rate is 0.2 cells per second. Jokilaakso et al. reported a microfluidic device which positioned individual cells on silicon nanowire biological field effect transistors by

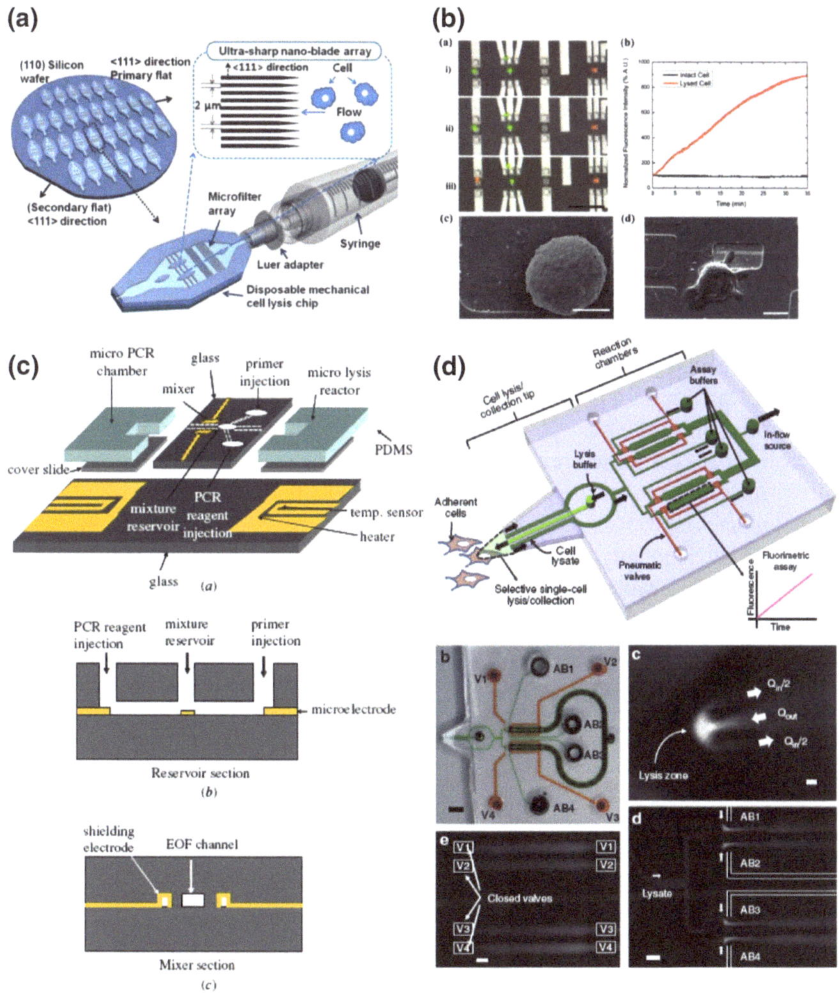

Fig. 2.8 Cell lysis on microfluidic chips. **a** Mechanical lysis; **b** Electrical lysis; **c** Thermal lysis; **d** Chemical lysis. (Reprinted with permission from Ref. [203, 207–209])

manipulating magnetic beads using external magnetic fields (Fig. 2.8b) [207]. Ultra-rapid cell lysis was subsequently performed by applying 600–900 mVpp at 10 MHz for as little as 2 ms across the transistor channel and the bulk substrate. This system was used to study the single cell variation within a population.

Thermal cell lysis utilizes high temperature to denature the proteins within cell membranes, thus irreparably damaging the cells and releasing the cytoplasmic contents. This method is commonly used in nucleic acid preparation, combining with PCR-based assays. Thermal lysis can be performed by ohmic heating, induction heating or heat generated by irradiating nanoparticles. For example,

Lee et al. presented an automated microfluidic chip capable of performing thermal cell lysis, electrokinetic sample/reagent transportation and mixing, and DNA amplification (PCR) (Fig. 2.8c) [208]. Two sets of micro-heaters and micro temperature sensors were integrated in cell lysis reactor and PCR chamber, to regulate and monitor the temperature. Cells could be lysed within 2 min at a constant temperature of 95 °C and PCR amplification of a 273 bp Streptococcus pneumoniae were demonstrated. Thermal lysis is a simple method and has been well-established for nucleic acid analysis. However, the thermal damage of proteins restricts its application in immunoassays.

Chemical cell lysis uses lytic agents to break down the cell wall and/or membrane. There are various chemical reagents which are chosen based on the cell types and target molecules. Detergents such as sodium dodecyl sulfate (SDS) and Triton X-100 disrupt cell membranes by solubilizing membrane proteins and lipids and creating pores. Ammonium chloride can only lyse the erythrocytes. For bacterial lysis, enzymatic degradation step such as lysozyme treatment is essential to destroy the cell wall. Chaotropic salts such as guanidinium thiocyanate and guanidinium chloride lyse cell membranes by disrupting protein intermolecular forces. They are most commonly used in nucleic acid preparations [200]. Chemical cell lysis is easily incorporated in microfluidic chips. Sarkar et al. presented a microfluidic probe that chemically lysed single adherent cells from standard tissue culture using commercial lysis buffer and captured the contents to perform single-cell biochemical assays (Fig. 2.8d) [209]. This device was applied to measure kinase and housekeeping protein activities from single human hepatocellular carcinoma cells in adherent culture. Chemical cell lysis is simple to implement and needs only mixing for activation. However, the chemical reagents in lysis buffer may interfere subsequent extraction and detection, for example when target molecules are analyzed by mass spectrometry. Thus additional separation steps will be required to remove the regents.

2.5.1.3 Sample Separation

Preconcentration and separation of biomolecules (nucleic acids, proteins, metabolites etc.) from complex cell lysate are quite important for sensitive detection and successful downstream processing. Microfluidics offer unique advantages for sample separation compared to conventional bench-top methods [32]. Microscale channels reduce sample and reagent consumption. The ease of integration on microfluidic devices enables the coupling of multiple separation techniques, which improves the separation efficiency. The separation step can also be integrated with lysis and analysis steps in an automatic manner, and this can minimize manual sample handling and preserve sample integrity, thereby improving the accuracy and reproducibility [210].

Many separation techniques can be operated in microfluidic devices, including chromatography, electrokinetic separation, solid-phase extraction (SPE), liquid-liquid extraction and filtration [211]. Each method has its own merits and applicative analytes. These separation techniques can also be integrated in one device to develop multidimensional separations works, which increase the capacity, throughput and efficiency.

Microfluidic electrokinetic separation, such as microchip capillary electrophoresis (MCE), gel electrophoresis, electrochromatography, isoelectric focusing (IEF) and isotachophoresis (ITP), are more commonly used for biomolecule fractionation and enrichment [212–214]. In microchip electrophoresis, charged molecules are separated based on their electrophoretic mobility under an electric field. Liu et al. developed a portable lab-on-a-chip system comprising a PCR reactor and a 7-cm-long separation channel for capillary electrophoretic analysis (Fig. 2.9a) [215]. This device was utilized in real-time forensic short tandem repeat (STR) analysis. Alleles which differed by 1 bp could be separated, and the CE separation process was completed in 8 min. Lin et al. presented a multiplex protein assay based on tunable aptamer by MCE [216]. Different lengths of aptamers could modulate the electrophoretic mobility of proteins, allowing the proteins to be effectively separated. This method was applied to analyze the PDGF-BB and $VEGF_{165}$ from cell secretions. ITP uses a heterogeneous buffer system consisting of high-mobility ions in the leading electrolyte (LE) and low-mobility ions in the terminating electrolyte (TE); an applied electric potential separates the ionic species based on their electrophoretic mobility, thus focusing the analytes at the LE/TE interface [217]. Schoch et al. demonstrated an on-chip ITP for the simultaneous extraction, isolation, preconcentration and quantitation of small RNAs from cell lysate [218]. It separated small RNAs from precursor miRNAs in less than 3 min, and the minimal cell number for small RNA extraction and detection is 900 (from a 5 mL sample volume).

In microchip chromatography system, stationary phase plays a critical role in molecules separation. There are a variety of stationary phase materials, including chromatography resins, monoliths and nanowires. In addition, surface

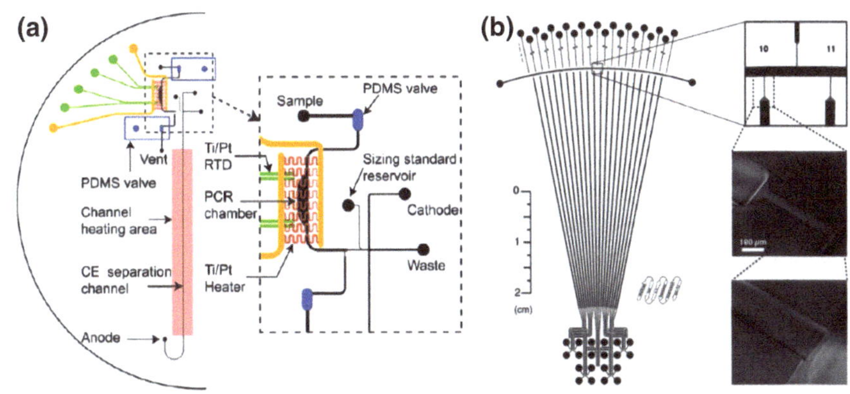

Fig. 2.9 Sample seperation on microfluidic chips. **a** A portable lab-on-a-chip system comprising a PCR reactor and a 7-cm-long separation channel for capillary electrophoretic analysis; **b** Two-dimensional differential gel electrophoretic separations (IEF-CE). (Reprinted with permission from Ref. [215, 231])

functionalized microchannels (open channel system) and "pillar structured" microchannels are also studied as stationary phases [219, 220]. Several companies have launched nanoLC systems with integrated liquid chromatography (LC) microchips for chip-based LC-MS analysis. The cHiPLC-nanoflex system from Eksigent Technologies, Trizaic UPLC nanoTile system from Waters and Chip Cube from Agilent are currently commercially available and have been used in various fields of bioanalytical applications including proteomics, glycomics, and clinical and pharmaceutical studies [221]. Chen et al. coupled the cHiPLCnanoflex system to an Orbitrap spectrometer [222]. After multistep sample preparation, peptide mixtures were separated in cHiPLC-nanoflex equipped with a 15 cm × 75 cm C18 column. high peptide coverage for 12 γ-aminobutyric acid type A receptor ($GABA_A$ receptor) subunits was obtained from 2 pmol of affinity-purified $GABA_A$ receptors from rat brain neocortex.

Microchip solid-phase extraction is also a widely used method for sample pre-concentration and clean-up. In microchip SPE, sample is retained on a solid phase, allowing the matrix to be rinsed away, and then the retained analytes was eluted for analysis [211]. Various solid phases, including packed-bead columns [223], monolith columns [224] and affinity columns [225] have been incorporated in microfluidic devices. Our group has developed a series of microfluidic systems which integrated drug injection, cell culture chambers and micro-SPE column in one device and coupled to mass spectrometry. The micro-SPE was used for desalination and extraction of cell metabolites. These devices have been applied to study drug metabolism in cell co-culture systems [150, 226–228]. Hagan et al. utilized chitosan-coated silica as a solid phase for RNA purification in a microfluidic device [229]. The effectiveness of the chitosan phase was demonstrated by the successful purification of RNA from cancer cells, with 3.5-fold greater extraction efficiencies than that purified by the simple silica phase.

Multidimensional separation is a significant advantage for microfluidic separation techniques. Various multidimensional microfluidic systems have been developed, such as micellar electrokinetic chromatography (MEKC)-CE, IEF-CE, ITP-CE, IEF-free-flow electrophoresis (FFE) [210, 230]. Emrich et al. developed a microfluidic separation system to perform two-dimensional differential gel electrophoretic separations of complex protein mixtures (Fig. 2.9b) [231]. This two-layer borosilicate glass microdevice consisted of a single 3.75 cm long channel for IEF, which was sampled in parallel by 20 channels effecting a second-dimension separation by native CE. The connection between the orthogonal separation systems was much shallower, narrower channels, which could prevented media leakage between the two dimensions and enabled facile loading of discontinuous gel systems in each dimension. Reproducible separations of both purified proteins and complex protein mixtures produced by E. coli were performed with minimal run-to-run variation.

2.5.2 *Cell Analysis*

Microfluidic devices are feasible to integrate with diverse analytical techniques, including optical detector, electrical analysis, mass spectrometry and nuclear magnetic resonance, to perform both qualitative and quantitative cell analysis for various purposes [21]. Optical detection, such as absorbance, fluorescence, infrared (IR) and surface plasmon resonance (SPR), are the most widely used techniques in microfluidics for cell analysis [232–234]. Owing to the transparency of microfluidic chips, various types of microscopy can be integrated to imaging the cell morphology, structure, movements, as well as specifically labeled cellular contents. Optical detector such as absorbance, laser-induce fluorescence (LIF), chemiluminescence (CL) can also be coupled with microfluidic separation channels, for the detection and quantification of fractionated biomolecules [235]. Electrical analysis, such as amperometry, electrochemical impedance spectroscopy (EIS), or patch-clamp, can be incorporated in microdevice to monitor cell secretion, cell morphology and migration, cell monolayer permeability, cell membrane electrophysiologic activity, and so on [236–239]. Mass spectrometry, as a powerful analytical technique, is commonly coupled with microfluidics for the analysis of cellular contents (DNA, protein, glycan etc.) and metabolites [240–242]. Integrating with these analytical techniques, microfluidic systems enable rapid, sensitive, reproducible and high-throughput cell analysis, which promotes the development of both basic biological researches and clinic diagnosis and therapy [243, 244]. In this section, we will introduce cell analysis using microfluidic devices integrated with diverse analytical techniques, according to the analysis objects. Some recent examples are also reviewed here.

2.5.2.1 Cell Morphology and Movement

In cellular microenvironment, cells are surrounded with multiple physical and biochemical cues. Cell morphology and movement are the most intuitionistic parameters that can reflect cellular responses to the diverse stimuli. By integrating biomimetic cell culture systems with various types of microscopy or electrical techniques, microfluidics offers a robust platform for real-time monitoring of cell morphology alteration and cell movements.

Fluorescence imaging are the most commonly used technique for cell observation. Cells cultured in microchannels are labeled by fluorescent dyes or proteins and visualized under fluorescence microscopy. Giobbe et al. described a multistage microfluidics-based approach for the differentiation of human pluripotent stem cells [245]. Cell morphology changed during factor stimulation and cell differentiation were observed by immunofluorescent staining of F-actin. And with the increase of acetaminophen concentration, morphology of human embryonic stem cell was disrupted, especially at 25 mM with almost complete loss of cell structure and function. Cell movement can also be altered under external stimuli, which is a

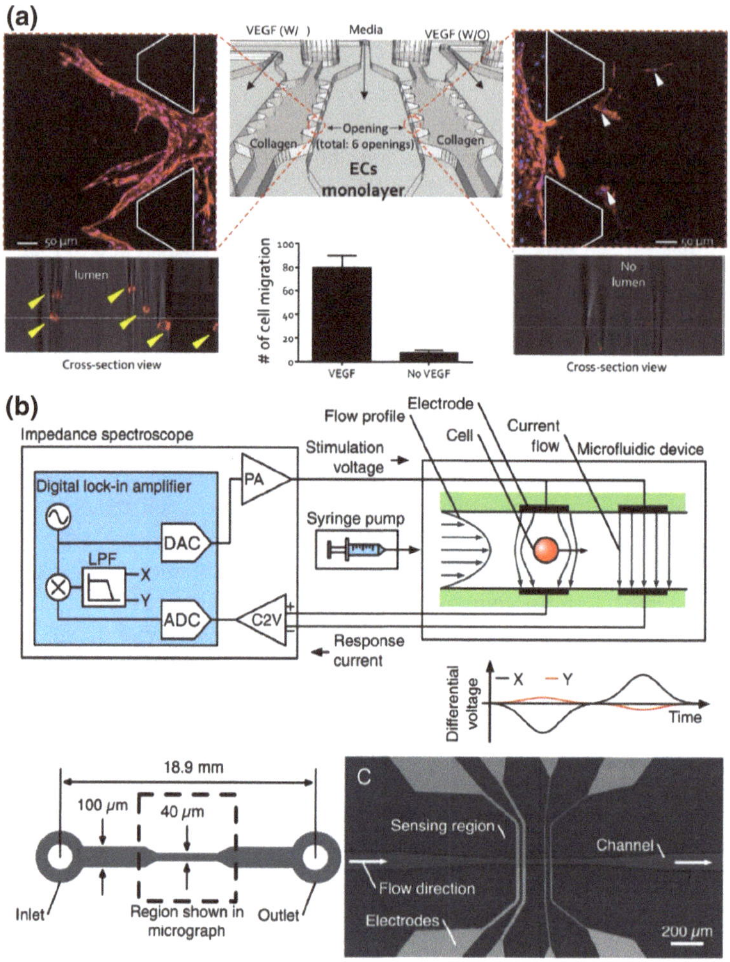

Fig. 2.10 Cell morphology and movement analysis on microfluidic chips. **a** Microfluidic cell angiogenesis assay combining with fluorescence imaging; **b** Microfluidic impedance cytometer for subcellular morphology characterization. (Reprinted with permission from Ref. [249, 250])

pivotal step in angiogenesis, cancer invasion and metastasis [246–248]. Kim et al. proposed a microfluidic platform that could monitor and quantify cellular behaviors, including morphological changes, cell migration and formation of angiogenic sprouts, under the treatment of anti-angiogenic drug (Fig. 2.10a) [249]. Results showed that cells rapidly and actively migrated from the endothelial channel into the 3D hydrogel scaffolds toward the VEGF-supplemented media channel, and this process could be inhibited by anti-angiogenic drug bortezomib.

Electrical techniques can also be incorporated in microfluidic devices for cell morphology analysis. Haandbæk et al. demonstrated a microfluidic single cell impedance cytometer capable of dielectric characterization of single cells at frequencies up to 500 MHz (Fig. 2.10b) [250]. The increased frequency range allowed for characterization of subcellular morphology, such as vacuoles and cell nuclei, in addition to the properties detectable at lower frequencies. This device could discriminate wild-type yeast cells from those with a mutant, which differed in size and distribution of vacuoles in the intracellular fluid.

2.5.2.2 Genetic Analysis

Genetic analysis is one of the most extensively developed field in microfluidic cell analysis. Since the amount of the nucleic acids extracted from cells is relatively small, nucleic acid amplification processes play a critical role in sensitive detection and quantification [251]. Polymerase chain reaction (PCR) is the most widely used non-isothermal amplification technique, which performs thermal cycling to amplify a few copies of DNA across several orders of magnitude, generating thousands to millions of copies of a particular DNA sequence. Quantification of RNA can be achieved by performing reverse-transcription PCR (RT-PCR). There are also various isothermal amplification techniques, such as strand displacement amplification (SDA), nucleic acid sequence based amplification (NASBA), rolling circle amplification (RCA), and loop mediated isothermal amplification (LAMP) [252]. Both isothermal and non-isothermal amplification have been successfully performed in microfluidic devices, with advantages including faster reaction times, low sample consumption, precise temperature distribution and the ease of integrating with separation techniques [253, 254]. Zhang et al. developed a targeted RNA sequencing method that coupled microfluidics-based multiplex PCR with deep sequencing [255]. This system could uniformly and simultaneously amplify up to 960 loci in 48 samples on a single microfluidic chip independent of their gene expression levels. Resulting PCR amplicons were barcoded for each sample and then subjected to deep sequencing to obtain high coverage allowing accurate measurement of allelic ratios. Fang et al. demonstrated a portable microchip-based LAMP platform, which performed rapid DNA release, exponential signal amplification and naked-eye result read-out in single or multiplex format [256]. This device was successfully used for point-of-care identification of bacteria.

Microfluidic techniques have been applied to various areas of genetic analysis, including sequencing, gene expression analysis [257], pathogen detection [258], and forensic STR typing [259]. High-throughput single-cell transcriptome sequencing (RNA-Seq) offers an unbiased approach for understanding the extent, basis and function of gene expression variation between seemingly identical cells. Currently, several microfluidic-based single-cell RNA-Seq platforms have been developed and applied to study transcriptional heterogeneity of cancer [260, 261], immune [262] and stem cells [263]. Streetsv et al. [264] presented a microfluidic-based system to perform single-cell whole-transcriptome sequencing (Fig. 2.11a). Single cells were captured and lysed in a microfluidic device, where mRNAs with poly(A) tails were reverse-transcribed into cDNA. Double-stranded cDNA was then collected and sequenced using a next generation sequencing platform. This technique enabled the identification and quantification of biological variation in a population of mouse embryonic stem cells at the single-cell level. Shalek et al. used a commercially available microfluidic device (C1 single-cell Auto Prep System, Fluidigm) and a transposase-based library preparation strategy to help prepare over 1700 single-cell RNA-seq libraries along time courses of bone-marrow derived dendritic cells responding to different stimuli [265]. Transcriptome-wide changes in single-cell expression variation across a variety of conditions was tracked and how both antiviral and inflammatory response modules in dendritic cells were controlled by positive and negative intercellular paracrine signaling were illustrated.

Chromatin immunoprecipitation coupled with deep sequencing (ChIP-seq) has also been successfully operated on microfluidics for the examining of in vivo genome-wide chromatin modifications (Fig. 2.11b) [266]. Magnetic beads coated

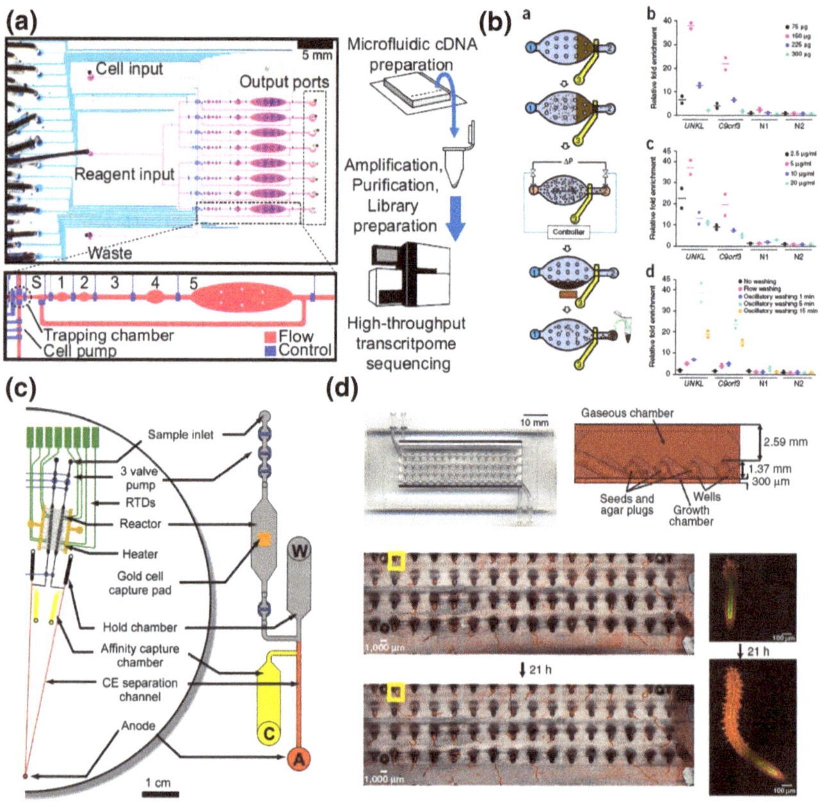

Fig. 2.11 Genetic analysis on microfluidic chips. **a** High-throughput single-cell transcriptome sequencing microfluidic system; **b** Microfluidic ChIP-seq; **c** An integrated microdevice for the analysis of gene expression in single cells; **d** A microfluidic platform for high-throughput live imaging of gene expression. (Reprinted with permission from Ref. [264, 266, 268, 269])

with a ChIP antibody were flowed into the microfluidic chamber and form a packed bed. It facilitated rapid and high-efficiency adsorption of target chromatin fragments, and the nonspecific adsorption was removed by effective oscillatory washing. These target chromatin samples were then purified and sequenced. Using this technology, many new enhancers and super enhancers in hematopoietic stem and progenitor cells were uncovered, suggesting that enhancer activity was highly dynamic during early hematopoiesis.

Even genetically identical cells with seemingly identical cell histories and environmental conditions can have significant differences in gene expression levels, due largely to the alteration of mRNA production by random fluctuations or complex molecular switches. Thus quantitative analysis of gene expression at single-cell level is important for the understanding of basic biological mechanism and disease onset and progression [267]. Toriello et al. developed an integrated microdevice for the analysis of gene expression in single cells (Fig. 2.11c) [268]. This device consisted of integrated nanoliter metering pumps, a 200 nL RT-PCR reactor with a single-cell capture pad, and an affinity capture matrix that was coupled to a microchip capillary electrophoresis separation channel for product purification and analysis. It was used to measure siRNA knockdown of the GAPDH gene in individual Jurkat cells and indicated the presence of 2 distinct populations of cells with moderate or complete silencing. Busch et al. developed a high-throughput microfluidic device in which 64 *Arabidopsis thaliana* seedlings could be grown and their roots were imaged by confocal microscopy over several days without manual intervention (Fig. 2.11d) [269]. This device was used to quantify expression patterns of 12 reporter genes in roots growing in different conditions and identified several cases of transient or heterogeneous expression.

2.5.2.3 Protein Analysis

Proteins are one of basic component of cells, which perform and regulate various cellular functions. Owing to the low abundance and high complexity, the development of sensitive and reliable protein analysis techniques are highly desirable. Microfluidics offer rapid, sensitive, reproducible and high-throughput platforms for protein analysis. Various aspects, including protein species, amounts, activity, as well as protein interaction with other biomolecules, can be analyzed using microfluidic devices, with tremendous advantages over conventional methods [270]. In this section, we will review the main microfluidic-based protein analysis methods, with some recently developed examples.

Cellular staining assays are commonly used methods which are easy to be applied in microfluidic devices for protein analysis. Proteins in cells are specifically labeled by tags or fluorescent antibodies, and their locations and expressions can be imaged using microscopies. Sun et al. reported a microfluidic image cytometry which was capable of quantitative, single-cell proteomic analysis (Fig. 2.12a) [271]. Simultaneous measurement of four critical signaling proteins (EGFR, PTEN, phospho-Akt, and phospho-S6) within the oncogenic signaling pathway in human

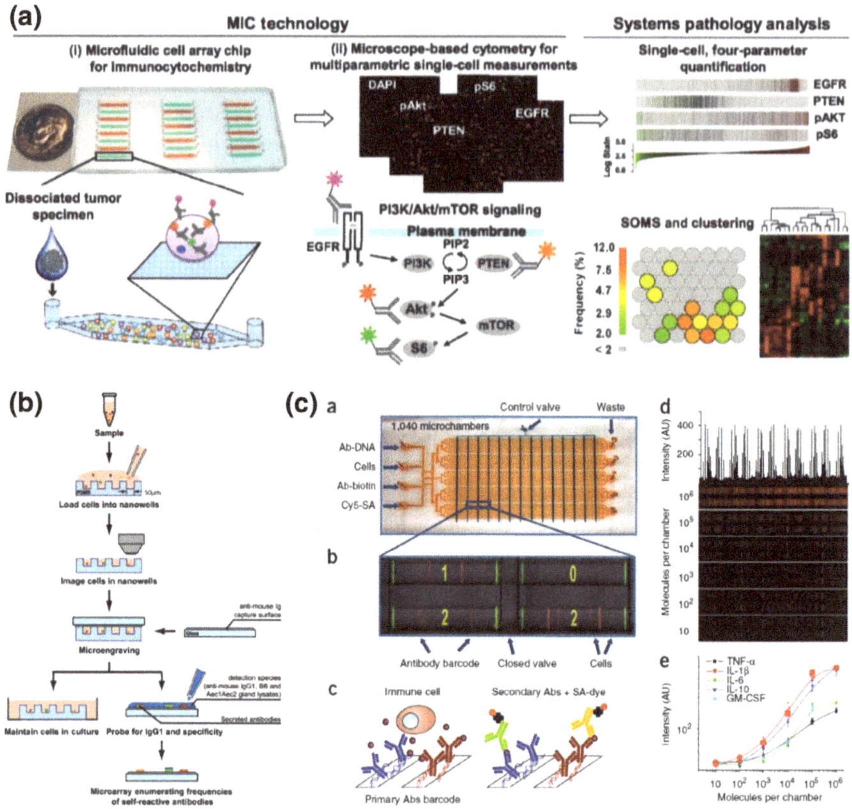

Fig. 2.12 Protein analysis on microfluidic chips. **a** Microfluidic image cytometry for quantitative, single-cell proteomic analysis; **b** Microengraved single-cell proteomics chip; **c** Microfluidic barcode chip for proteomic analysis. (Reprinted with permission from Ref. [271, 272, 275])

brain tumor biopsies was performed by multicolor analysis of fluorescent antibody-labeled cytoplasmic proteins. This platform required much smaller amounts of reagents ($\sim$ 2 μL/channel) and samples (< 3000 cells) than that for flow cytometry, and results were in good correlation with the clinical immunohistochemistry method. In previously mentioned example, native proteins were genetically encoded with oligohistidine sequence (His_{6-10}), which could interacted with a fluorescent multivalent chelator head trisNTA, leading to high-affinity protein labelling in living cells [166]. And the transfer of trisNTA probe into cells was facilitated by microfluidic cell squeezing. Multiplexed protein labeling as well as super-resolution (nanometer precision) tracing of target proteins in live mammalian cells were achieved using this technique.

Another important microfluidic protein analysis technique is surface-based immunoassay. Proteins are specifically captured by affinity ligands modified on microchannel or microbead surface, and sandwich immunoassays are then performed. This method enables high level of multiplexing and quantitation, and intracellular, membrane, and secreted proteins can all be analyzed from the same single cell [270]. Love's group developed a microengraved single-cell proteomics chip which employed an array of microfabricated subnanoliter wells to isolate and culture single cells (Fig. 2.12b) [272]. This microwell array was capped with an antibody-coated microengraved substrate, and proteins secreted by single cells were captured and then characterized using sandwich ELISAs. This devices were used to analyze the antibody production behavior of B cells collected from both a healthy mouse and a mouse model with autoimmune disorder. Microfluidic barcode chip developed by Heath group was also a versatile and robust tool for proteomic analysis (Fig. 2.12c). Single cells, or defined numbers of cells, were isolated within microchambers, each of which contained a full barcode array. Each barcode stripe was initially patterned with a unique ssDNA oligomer, and the barcode was converted into an antibody array using a cocktail of antibodies labeled with complementary ssDNA oligomers (DNA-encoded antibody library, DEAL) [273], just prior to running an assay. This platform could be used to capture and detect secreted proteins from living cells, or proteins, nucleic acids and metabolites from lysed cells, using ELISA or fluoroimmunoassy (FIA) coupled with a standard array scanner [274]. This technique was quantitative, sensitive, highly multiplexed and high-throughput, and has been applied to study single-cell proteomics, cell signaling and cell-cell interaction [275, 276].

Recently, protein immunoblotting assay has been operated on microfluidic devices, overcomes several limitations associated with conventional immunoblotting, including multiple steps requiring manual intervention, low throughput and substantial consumption of reagents [277, 278]. Polyacrylamide (PA) gels in glass microfluidic devices were regional photo-patterned and served as a platform for rapid and automated protein immunoblotting. All assay stages were programmably controlled by a high-voltage power supply and monitored by an epifluorescence microscope equipped with a charge-coupled device camera. Microfluidic single-cell western blotting (scWestern) has also been developed by the same group (Fig. 2.13a) [279, 280]. PA gels were photo-patterned to form a microwell array, in which single cells were settled and lysed in situ. Gel electrophoresis was then performed, and separated proteins were immobilized by photoinitiated blotting and detected by antibody probing. This scWestern method enabled multiplexed analysis of 11 protein targets per single cell with detection thresholds of $< 30{,}000$ molecules. It was applied to monitor single-cell differentiation of rat neural stem cells and their responses to mitogen stimulation.

Protein activity, and the interactions between protein and other biomolecules, can also been analyzed on microfluidics. In previously mentioned example, Sarkar et al. presented a microfluidic probe that lysed single adherent cells from standard tissue culture and captured the contents to perform single-cell assays [209]. Kinase and housekeeping protein activities were measured simultaneously or separately by mixing them with assay reagents in nanolitre scale integrated chambers, and imaging over time for fluorescence. This approach was used to elucidate the biological heterogeneity in Akt kinase activity levels among cells under insulin

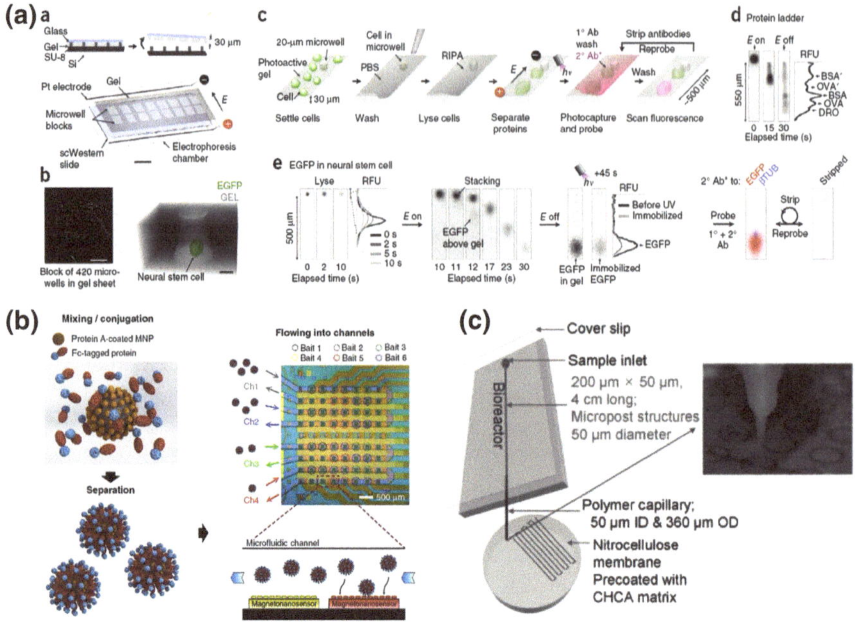

Fig. 2.13 Protein analysis on microfluidic chips. **a** Microfluidic single-cell western blotting (scWestern); **b** A magneto-nanosensor microfluidic platform for protein–protein interaction study; **c** A microfluidic device integrated with MALDI-TOF MS for protein analysis. (Reprinted with permission from Ref. [279, 281, 286])

stimulation. Lee et al. presented a magneto-nanosensor platform integrated with a microfluidic chip that allowed measurement of low-affinity protein–protein interactions (Fig. 2.13b) [281]. Prey proteins were pre-conjugated to magnetic nanoparticles (MNPs), and then introduced into microchannels and specifically interacted with magneto-nanosensors modified with bait proteins. The magneto-nanosensors produced signals proportional to the number of bound complexes, and real-time binding curves were measured and kinetic parameters were calculated. This platform was used to characterize the binding affinities of the PD-1—PD-L1/PD-L2 co-inhibitory receptor system, and discovered an unexpected interaction between the two known PD-1 ligands, PD-L1 and PD-L2.

Coupled with mass spectrometry (MS), microfluidic technique offers a robust platform for proteomic analysis [282, 283]. A commercially available HPLC-chip/MS system has been developed by Agilent Technologies, which uses a multilayer polyimide chip consisting of preparation channels and an integrated electrospray tip [284]. Trypsin-digested proteins are introduced into the HPLC-chip, concentrated on a small volume (40 nL) enrichment column and separated in the analytical nanochannel. Eluting compounds are directly sprayed into the MS instrument with the electrospray tip. This system was combined with strong cation exchange chromatography and applied to the analysis of the nucleolar proteome, resulting in the identification of more than 200 proteins which corresponded to 2024 unique tryptic peptides. Microfluidic devices can also be integrated with matrix-assisted laser desorption and ionization mass spectrometry (MALDI-MS), in either off-line or on-line manner [285]. Lee et al. constructed an automated proteolytic digestion bioreactor and continuous deposition system in a plastic microfluidic device for off-line interfacing to MALDI-TOF MS (Fig. 2.13c) [286]. Proteins were digested by trypsin immobilized on an array of micropost structures in bioreactor, and the obtained peptides were continuously deposited onto a MALDI plate modified with nitrocellulose solution containing a matrix by a 50 mm capillary tube attached to the end of the bioreactor. Various proteins were analyzed in this platform with good sequence coverage.

2.5.2.4 Metabolite Analysis

The intracellular levels and spatial localizations of metabolites reflect the state of a cell and its relationship to its surrounding environment [287]. Microfluidic device is an ideal platform for cellular metabolite profiling both in physiological environment and under drug treatment, owing to the ability of integrating cell culture, stimulation, metabolite enrichment and detection on a single chip coupled with various analytical instruments [288].

Among diverse analytical techniques, MS is the most powerful and promising tool for cell metabolite analysis, because of its broad detection range, high sensitivity, high mass resolution, rapid operation, and the ability for multiplexed analysis. Microfluidic devices can integrate with different types of MS, including ESI-MS, MALDI-MS and paper-spray ionization MS [289]. Recently, our group has developed a series of chip-MS platforms for cell metabolites analysis. Functional sections including cell medium/drug introducing channels, cell culture chambers, micro-SPE columns were integrated in one microchip, and coupled to ESI-MS via capillaries [150, 226–228, 290–292]. Chen et al. developed a stable isotope labeling assisted microfluidic chip electrospray ionization mass spectrometry (SIL-chip—ESI-MS) platform for qualitative and quantitative analysis of cell metabolism (Fig. 2.14a) [293]. MCF-7 cells were cultivated in vitro and exposed in anticancer agent (genistein) for cell-based drug assay. A dual-isotopic labeling was presented for effective qualitative analysis of multiplex metabolites. Three coeluting pairs of isotopomers could be easily recognized and identified. This chip-MS technique can also be extended to study cell-cell communication. Zhuang et al. [294] designed a membrane integrated microfluidic device to achieve the co-culture of PC12 cells and 293 cells to study nephrocyte-neurocyte interaction. The neuro-like differentiated PC12 cells induced by mNGF released neurotransmitter acetylcholine, which stimulated the 293 cells and led to the secretion of hormone epinephrine. Secreted epinephrine could be detected by ESI-MS, providing a chemical insight into the understanding of cellular interaction.

Optical detecting technique can also be integrated with microfluidic device for cell metabolite analysis. Wang et al. presented a flexible high-throughput approach

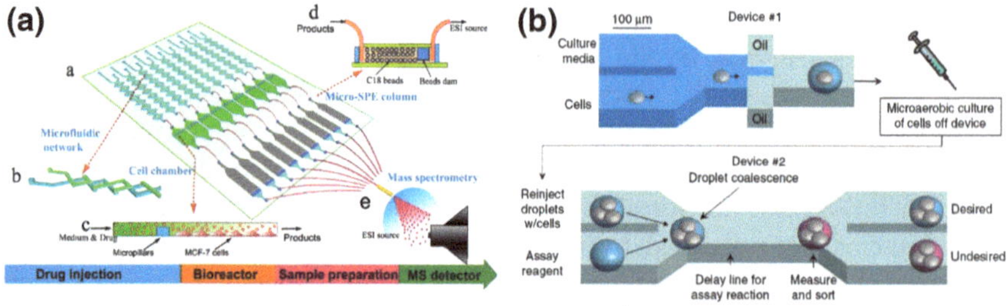

Fig. 2.14 Metabolite analysis on microfluidic chips. **a** Chip-MS platform for drug metabolism analysis; **b** Microfluidic optical detection for cell metabolite analysis. (Reprinted with permission from Ref. [293, 295])

that used microfluidics to compartmentalize individual cells for growth and analysis in monodisperse nanoliter aqueous droplets surrounded by an immiscible fluorinated oil phase (Fig. 2.14b) [295]. Fluorescent assay system was used to measure the concentration of the metabolites (oxidase enzymes), and the assay reaction started when a cell-containing droplet coalesced with an assay droplet. Based on the cellular metabolism behavior, this system was able to identify xylose-overconsuming *Saccharomyces cerevisiae* cells from a population containing one such cell per 10^4 cells and enrich L-lactate-producing *Escherichia coli* clones 5800 × from a population containing one L-lactate producer per 10^4 D-lactate producers.

2.6 Conclusion and Perspective

Microfluidics, combined with advanced molecular, imaging and bioinformatics techniques, constitute a robust 'toolbox' and revolutionize the way for cell biology researches. Multiple processes including cell culture, cell manipulation, cell stimulation and cell analysis can be transferred and integrated in a small single microfluidic device, promoting the development of cell researches. Owing to the unique advantages of microfluidic technology, it has been applied to various biological fields, such as single-cell analysis, cancer research, drug discovery and screening, clinical diagnostics, stem cell research, intra- and inter-cellular signaling, tissue models and microbiology. These applications have been partially introduced in previous sections and will be reviewed in detail in the following chapters in this book.

Despite the rapid development and significant progress of microfluidic technology in recent years, there are still challenges and hurdles that should be addressed for microfluidic cell analysis. One pivotal challenge is the low adoption of novel microfluidic techniques in mainstream biology researches. Most microfluidic-based techniques for cell biology applications recently are only iterative improvements on methods that already exist. And owing to the gaps in expertise between microfluidic engineers and biologists, it is not easy for biologists to handle complex microfluidic systems. Thus the superiority and particularity of microfluidic techniques over traditional methods are not obvious, and biologists will generally prefer to use conventional macroscale methods than to learn and adopt the new microfluidic techniques. To address this challenge, efforts can be made from different directions. First, collaborations between multidisciplinary researchers (physicists, chemists, engineers, molecular and cell biologists, and clinicians) should be further strengthened, which can make the microfluidic-based techniques more biologically compatible, allow direct interaction and feedback between designers and end-users, and accelerate the applications of newly developed techniques. Second, more attentions should be payed to research areas in which microfluidic technology shows unique advantages and irreplaceable roles, such as diagnostic devices for low-resource settings, rapidly assaying biofluids for research and clinical applications, and development of more physiologically relevant in vitro models. Third, simplifying microfluidic devices and operation processes is an effective way to lower the barrier to entry for biologists and expand practical applications. In addition, developing automated microfluidic systems with diverse functions can also improve the adoption of microfluidic technology in mainstream cell researches [7, 15].

In summary, with twenty-year development, microfluidic technology has become one of the most powerful and promising tool for the study of cell biology and cell analysis. It has been extensively utilized in various fields of cell research and is still in rapid progression. We believe that with further advance and maturation, this versatile and robust technology will continue to introduce new paradigm for cell study, and make contributions to the deep cognition and development of both basic biological research and clinic applications.

References

1. Barthes J, Ozcelik H, Hindie M, Ndreu-Halili A, Hasan A, Vrana NE (2014) Cell microenvironment engineering and monitoring for tissue engineering and regenerative medicine: the recent advances. Biomed Res Int 2014:921905. doi:10.1155/2014/921905
2. El-Ali J, Sorger PK, Jensen KF (2006) Cells on chips. Nature 442(7101):403–411. doi:10.1038/nature05063
3. Gattazzo F, Urciuolo A, Bonaldo P (2014) Extracellular matrix: a dynamic microenvironment for stem cell niche. Biochim Biophys Acta 1840(8):2506–2519. doi:10.1016/j.bbagen.2014.01.010
4. Quail DF, Joyce JA (2013) Microenvironmental regulation of tumor progression and metastasis. Nat Med 19(11):1423–1437. doi:10.1038/nm.3394
5. Sun Y, Chen CS, Fu J (2012) Forcing stem cells to behave: a biophysical perspective of the cellular microenvironment. Annu Rev Biophys 41:519–542. doi:10.1146/annurev-biophys-042910-155306
6. Whitesides GM (2006) The origins and the future of microfluidics. Nature 442(7101):368–373. doi:10.1038/nature05058

7. Sackmann EK, Fulton AL, Beebe DJ (2014) The present and future role of microfluidics in biomedical research. Nature 507(7491):181–189. doi:10.1038/nature13118
8. Mark D, Haeberle S, Roth G, von Stetten F, Zengerle R (2010) Microfluidic lab-on-a-chip platforms: requirements, characteristics and applications. Chem Soc Rev 39(3):1153–1182. doi:10.1039/b820557b
9. Nge PN, Rogers CI, Woolley AT (2013) Advances in microfluidic materials, functions, integration, and applications. Chem Rev 113(4):2550–2583. doi:10.1021/cr300337x
10. Livak-Dahl E, Sinn I, Burns M (2011) Microfluidic chemical analysis systems. Annu Rev Chem Biomol Eng 2:325–353. doi:10.1146/annurev-chembioeng-061010-114215
11. Salieb-Beugelaar GB, Simone G, Arora A, Philippi A, Manz A (2010) Latest developments in microfluidic cell biology and analysis systems. Anal Chem 82(12):4848–4864. doi:10.1021/ac1009707
12. Zhuang Q-C, Ning R-Z, Ma Y, Lin J-M (2016) Recent developments in microfluidic chip for in vitro cell-based research. Chin J Anal Chem 44(4):522–532. doi:10.1016/s1872-2040(16)60919-2
13. Andersson H, van den Berg A (2003) Microfluidic devices for cellomics: a review. Sens Actuators B: Chem 92(3):315–325. doi:10.1016/s0925-4005(03)00266-1
14. Xiong B, Ren K, Shu Y, Chen Y, Shen B, Wu H (2014) Recent developments in microfluidics for cell studies. Adv Mater 26(31):5525–5532. doi:10.1002/adma.201305348
15. Duncombe TA, Tentori AM, Herr AE (2015) Microfluidics: reframing biological enquiry. Nat Rev Mol Cell Biol 16(9):554–567. doi:10.1038/nrm4041
16. Priest C (2010) Surface patterning of bonded microfluidic channels. Biomicrofluidics 4 (3):32206. doi:10.1063/1.3493643
17. Li Jeon N, Baskaran H, Dertinger SK, Whitesides GM, Van de Water L, Toner M (2002) Neutrophil chemotaxis in linear and complex gradients of interleukin-8 formed in a microfabricated device. Nat Biotechnol 20(8):826–830. doi:10.1038/nbt712
18. Lucchetta EM, Lee JH, Fu LA, Patel NH, Ismagilov RF (2005) Dynamics of Drosophila embryonic patterning network perturbed in space and time using microfluidics. Nature 434 (7037):1134–1138. doi:10.1038/nature03509
19. Araci IE, Brisk P (2014) Recent developments in microfluidic large scale integration. Curr Opin Biotechnol 25:60–68. doi:10.1016/j.copbio.2013.08.014
20. Melin J, Quake SR (2007) Microfluidic large-scale integration: the evolution of design rules for biological automation. Annu Rev Biophys Biomol Struct 36:213–231. doi:10.1146/annurev.biophys.36.040306.132646
21. Wu J, He Z, Chen Q, Lin J-M (2016) Biochemical analysis on microfluidic chips. TrAC Trends Anal Chem 80:213–231. doi:10.1016/j.trac.2016.03.013
22. Kellogg RA, Gomez-Sjoberg R, Leyrat AA, Tay S (2014) High-throughput microfluidic single-cell analysis pipeline for studies of signaling dynamics. Nat Protoc 9(7):1713–1726. doi:10.1038/nprot.2014.120
23. Shembekar N, Chaipan C, Utharala R, Merten CA (2016) Droplet-based microfluidics in drug discovery, transcriptomics and high-throughput molecular genetics. Lab Chip 16 (8):1314–1331. doi:10.1039/c6lc00249h
24. Joensson HN, Andersson Svahn H (2012) Droplet microfluidics—a tool for single-cell analysis. Angew Chem Int Ed Engl 51(49):12176–12192. doi:10.1002/anie.201200460
25. Rothbauer M, Wartmann D, Charwat V, Ertl P (2015) Recent advances and future applications of microfluidic live-cell microarrays. Biotechnol Adv 33(6 Pt 1):948–961. doi:10.1016/j.biotechadv.2015.06.006
26. Willaert R, Goossens K (2015) Microfluidic bioreactors for cellular microarrays. Fermentation 1(1):38–78. doi:10.3390/fermentation1010038
27. Mehling M, Tay S (2014) Microfluidic cell culture. Curr Opin Biotechnol 25:95–102. doi:10.1016/j.copbio.2013.10.005
28. Halldorsson S, Lucumi E, Gomez-Sjoberg R, Fleming RM (2015) Advantages and challenges of microfluidic cell culture in polydimethylsiloxane devices. Biosens Bioelectron 63:218–231. doi:10.1016/j.bios.2014.07.029

29. Shields CW, Reyes CD, Lopez GP (2015) Microfluidic cell sorting: a review of the advances in the separation of cells from debulking to rare cell isolation. Lab Chip 15(5):1230–1249. doi:10.1039/c4lc01246a
30. Pappas D (2016) Microfluidics and cancer analysis: cell separation, cell/tissue culture, cell mechanics, and integrated analysis systems. Analyst 141(2):525–535. doi:10.1039/c5an01778e
31. Ertl P, Sticker D, Charwat V, Kasper C, Lepperdinger G (2014) Lab-on-a-chip technologies for stem cell analysis. Trends Biotechnol 32(5):245–253. doi:10.1016/j.tibtech.2014.03.004
32. Mach AJ, Adeyiga OB, Di Carlo D (2013) Microfluidic sample preparation for diagnostic cytopathology. Lab Chip 13(6):1011–1026. doi:10.1039/c2lc41104k
33. Eicher D, Merten CA (2011) Microfluidic devices for diagnostic applications. Expert Rev Mol Diagn 11(5):505–519. doi:10.1586/ERM.11.25
34. Dittrich PS, Manz A (2006) Lab-on-a-chip: microfluidics in drug discovery. Nat Rev Drug Discov 5(3):210–218. doi:10.1038/nrd1985
35. Neuzi P, Giselbrecht S, Lange K, Huang TJ, Manz A (2012) Revisiting lab-on-a-chip technology for drug discovery. Nat Rev Drug Discov 11(8):620–632. doi:10.1038/nrd3799
36. Choi NW, Cabodi M, Held B, Gleghorn JP, Bonassar LJ, Stroock AD (2007) Microfluidic scaffolds for tissue engineering. Nat Mater 6(11):908–915. doi:10.1038/nmat2022
37. Young EW, Beebe DJ (2010) Fundamentals of microfluidic cell culture in controlled microenvironments. Chem Soc Rev 39(3):1036–1048. doi:10.1039/b909900j
38. Tehranirokh M, Kouzani AZ, Francis PS, Kanwar JR (2013) Microfluidic devices for cell cultivation and proliferation. Biomicrofluidics 7(5):51502. doi:10.1063/1.4826935
39. Gao D, Liu H, Jiang Y, Lin J-M, Gao D, Liu H, Jiang Y (2012) Recent developments in microfluidic devices for in vitro cell culture for cell-biology research. TrAC Trends Anal Chem 35:150–164. doi:10.1016/j.trac.2012.02.008
40. Gupta N, Liu JR, Patel B, Solomon DE, Vaidya B, Gupta V (2016) Microfluidics-based 3D cell culture models: utility in novel drug discovery and delivery research. Bioeng Transl Med 1(1):63–81. doi:10.1002/btm2.10013
41. Shamir ER, Ewald AJ (2014) Three-dimensional organotypic culture: experimental models of mammalian biology and disease. Nat Rev Mol Cell Biol 15(10):647–664. doi:10.1038/nrm3873
42. Ravi M, Paramesh V, Kaviya SR, Anuradha E, Solomon FD (2015) 3D cell culture systems: advantages and applications. J Cell Physiol 230(1):16–26. doi:10.1002/jcp.24683
43. Edmondson R, Broglie JJ, Adcock AF, Yang L (2014) Three-dimensional cell culture systems and their applications in drug discovery and cell-based biosensors. Assay Drug Dev Technol 12(4):207–218. doi:10.1089/adt.2014.573
44. van Duinen V, Trietsch SJ, Joore J, Vulto P, Hankemeier T (2015) Microfluidic 3D cell culture: from tools to tissue models. Curr Opin Biotechnol 35:118–126. doi:10.1016/j.copbio.2015.05.002
45. Lee DH, Bae CY, Kwon S, Park JK (2015) User-friendly 3D bioassays with cell-containing hydrogel modules: narrowing the gap between microfluidic bioassays and clinical end-users' needs. Lab Chip 15(11):2379–2387. doi:10.1039/c5lc00239g
46. Li XJ, Valadez AV, Zuo P, Nie Z (2012) Microfluidic 3D cell culture: potential application for tissue-based bioassays. Bioanalysis 4(12):1509–1525. doi:10.4155/bio.12.133
47. Sung KE, Su X, Berthier E, Pehlke C, Friedl A, Beebe DJ (2013) Understanding the impact of 2D and 3D fibroblast cultures on in vitro breast cancer models. PLoS ONE 8(10):e76373. doi:10.1371/journal.pone.0076373
48. Chen Q, Utech S, Chen D, Prodanovic R, Lin JM, Weitz DA (2016) Controlled assembly of heterotypic cells in a core-shell scaffold: organ in a droplet. Lab Chip 16(8):1346–1349. doi:10.1039/c6lc00231e
49. Griffin DR, Weaver WM, Scumpia PO, Di Carlo D, Segura T (2015) Accelerated wound healing by injectable microporous gel scaffolds assembled from annealed building blocks. Nat Mater 14(7):737–744. doi:10.1038/nmat4294

50. Frey O, Misun PM, Fluri DA, Hengstler JG, Hierlemann A (2014) Reconfigurable microfluidic hanging drop network for multi-tissue interaction and analysis. Nat Commun 5:4250. doi:10.1038/ncomms5250
51. Misun PM, Rothe J, Schmid YRF, Hierlemann A, Frey O (2016) Multi-analyte biosensor interface for real-time monitoring of 3D microtissue spheroids in hanging-drop networks. Microsyst Nanoeng 2:16022. doi:10.1038/micronano.2016.22
52. Chen YC, Lou X, Zhang Z, Ingram P, Yoon E (2015) High-throughput cancer cell sphere formation for characterizing the efficacy of photo dynamic therapy in 3D cell cultures. Sci Rep 5:12175. doi:10.1038/srep12175
53. Tsutsui H, Yu E, Marquina S, Valamehr B, Wong I, Wu H, Ho CM (2010) Efficient dielectrophoretic patterning of embryonic stem cells in energy landscapes defined by hydrogel geometries. Ann Biomed Eng 38(12):3777–3788. doi:10.1007/s10439-010-0108-1
54. Physiology in perspective: cell-cell interactions: the physiological basis of communication (2014) Physiology (Bethesda) 29(4):220–221. doi:10.1152/physiol.00031.2014
55. Kim SH, Turnbull J, Guimond S (2011) Extracellular matrix and cell signalling: the dynamic cooperation of integrin, proteoglycan and growth factor receptor. J Endocrinol 209(2):139–151. doi:10.1530/JOE-10-0377
56. Zervantonakis IK, Kothapalli CR, Chung S, Sudo R, Kamm RD (2011) Microfluidic devices for studying heterotypic cell-cell interactions and tissue specimen cultures under controlled microenvironments. Biomicrofluidics 5(1):13406. doi:10.1063/1.3553237
57. Guo F, French JB, Li P, Zhao H, Chan CY, Fick JR, Benkovic SJ, Huang TJ (2013) Probing cell-cell communication with microfluidic devices. Lab Chip 13(16):3152–3162. doi:10.1039/c3lc90067c
58. Delamarche E, Tonna N, Lovchik RD, Bianco F, Matteoli M (2013) Pharmacology on microfluidics: multimodal analysis for studying cell-cell interaction. Curr Opin Pharmacol 13(5):821–828. doi:10.1016/j.coph.2013.07.005
59. Nahavandi S, Tang SY, Baratchi S, Soffe R, Nahavandi S, Kalantar-zadeh K, Mitchell A, Khoshmanesh K (2014) Microfluidic platforms for the investigation of intercellular signalling mechanisms. Small 10(23):4810–4826. doi:10.1002/smll.201401444
60. Jeon JS, Bersini S, Gilardi M, Dubini G, Charest JL, Moretti M, Kamm RD (2015) Human 3D vascularized organotypic microfluidic assays to study breast cancer cell extravasation. Proc Natl Acad Sci U S A 112(1):214–219. doi:10.1073/pnas.1417115112
61. Zervantonakis IK, Hughes-Alford SK, Charest JL, Condeelis JS, Gertler FB, Kamm RD (2012) Three-dimensional microfluidic model for tumor cell intravasation and endothelial barrier function. Proc Natl Acad Sci U S A 109(34):13515–13520. doi:10.1073/pnas.1210182109
62. Kimura H, Yamamoto T, Sakai H, Sakai Y, Fujii T (2008) An integrated microfluidic system for long-term perfusion culture and on-line monitoring of intestinal tissue models. Lab Chip 8(5):741–746. doi:10.1039/b717091b
63. Chen Q, Wu J, Zhuang Q, Lin X, Zhang J, Lin JM (2013) Microfluidic isolation of highly pure embryonic stem cells using feeder-separated co-culture system. Sci Rep 3:2433. doi:10.1038/srep02433
64. Liu W, Li L, Wang X, Ren L, Wang X, Wang J, Tu Q, Huang X, Wang J (2010) An integrated microfluidic system for studying cell-microenvironmental interactions versatilely and dynamically. Lab Chip 10(13):1717–1724. doi:10.1039/c001049a
65. Lin X, Chen Q, Liu W, Zhang J, Wang S, Lin Z, Lin JM (2015) Oxygen-induced cell migration and on-line monitoring biomarkers modulation of cervical cancers on a microfluidic system. Sci Rep 5:9643. doi:10.1038/srep09643
66. Tumarkin E, Tzadu L, Csaszar E, Seo M, Zhang H, Lee A, Peerani R, Purpura K, Zandstra PW, Kumacheva E (2011) High-throughput combinatorial cell co-culture using microfluidics. Integr Biol (Camb) 3(6):653–662. doi:10.1039/c1ib00002k
67. Ricoult SG, Goldman JS, Stellwagen D, Juncker D, Kennedy TE (2012) Generation of microisland cultures using microcontact printing to pattern protein substrates. J Neurosci Methods 208(1):10–17. doi:10.1016/j.jneumeth.2012.04.016

68. Khetani SR, Bhatia SN (2008) Microscale culture of human liver cells for drug development. Nat Biotechnol 26(1):120–126. doi:10.1038/nbt1361
69. Cho CH, Park J, Tilles AW, Berthiaume F, Toner M, Yarmush ML (2010) Layered patterning of hepatocytes in co-culture systems using microfabricated stencils. Biotechniques 48(1):47–52. doi:10.2144/000113317
70. Edahiro J, Sumaru K, Ooshima Y, Kanamori T (2009) Selective separation and co-culture of cells by photo-induced enhancement of cell adhesion (PIECA). Biotechnol Bioeng 102 (4):1278–1282. doi:10.1002/bit.22124
71. Gao Y, Broussard J, Haque A, Revzin A, Lin T (2016) Functional imaging of neuron–astrocyte interactions in a compartmentalized microfluidic device. Microsyst Nanoeng 2:15045. doi:10.1038/micronano.2015.45
72. Shin Y, Han S, Jeon JS, Yamamoto K, Zervantonakis IK, Sudo R, Kamm RD, Chung S (2012) Microfluidic assay for simultaneous culture of multiple cell types on surfaces or within hydrogels. Nat Protoc 7(7):1247–1259. doi:10.1038/nprot.2012.051
73. Dura B, Dougan SK, Barisa M, Hoehl MM, Lo CT, Ploegh HL, Voldman J (2015) Profiling lymphocyte interactions at the single-cell level by microfluidic cell pairing. Nat Commun 6:5940. doi:10.1038/ncomms6940
74. Bhatia SN, Ingber DE (2014) Microfluidic organs-on-chips. Nat Biotechnol 32(8):760–772. doi:10.1038/nbt.2989
75. Huh D, Hamilton GA, Ingber DE (2011) From 3D cell culture to organs-on-chips. Trends Cell Biol 21(12):745–754. doi:10.1016/j.tcb.2011.09.005
76. Lee E, Song HG, Chen CS (2016) Biomimetic on-a-chip platforms for studying cancer metastasis. Curr Opin Chem Eng 11:20–27. doi:10.1016/j.coche.2015.12.001
77. Esch EW, Bahinski A, Huh D (2015) Organs-on-chips at the frontiers of drug discovery. Nat Rev Drug Discov 14(4):248–260. doi:10.1038/nrd4539
78. Chrobak KM, Potter DR, Tien J (2006) Formation of perfused, functional microvascular tubes in vitro. Microvasc Res 71(3):185–196. doi:10.1016/j.mvr.2006.02.005
79. Tsai M, Kita A, Leach J, Rounsevell R, Huang JN, Moake J, Ware RE, Fletcher DA, Lam WA (2012) In vitro modeling of the microvascular occlusion and thrombosis that occur in hematologic diseases using microfluidic technology. J Clin Invest 122(1):408–418. doi:10.1172/JCI58753
80. Cho H, Seo JH, Wong KH, Terasaki Y, Park J, Bong K, Arai K, Lo EH, Irimia D (2015) Three-dimensional blood-brain barrier model for in vitro studies of neurovascular pathology. Sci Rep 5:15222. doi:10.1038/srep15222
81. Toh YC, Lim TC, Tai D, Xiao G, van Noort D, Yu H (2009) A microfluidic 3D hepatocyte chip for drug toxicity testing. Lab Chip 9(14):2026–2035. doi:10.1039/b900912d
82. Carraro A, Hsu WM, Kulig KM, Cheung WS, Miller ML, Weinberg EJ, Swart EF, Kaazempur-Mofrad M, Borenstein JT, Vacanti JP, Neville C (2008) In vitro analysis of a hepatic device with intrinsic microvascular-based channels. Biomed Microdevices 10 (6):795–805. doi:10.1007/s10544-008-9194-3
83. Huh D, Fujioka H, Tung YC, Futai N, Paine R 3rd, Grotberg JB, Takayama S (2007) Acoustically detectable cellular-level lung injury induced by fluid mechanical stresses in microfluidic airway systems. Proc Natl Acad Sci U S A 104(48):18886–18891. doi:10.1073/pnas.0610868104
84. Huh D, Leslie DC, Matthews BD, Fraser JP, Jurek S, Hamilton GA, Thorneloe KS, McAlexander MA, Ingber DE (2012) A human disease model of drug toxicity-induced pulmonary edema in a lung-on-a-chip microdevice. Sci Transl Med 4(159):159ra147. doi:10.1126/scitranslmed.3004249
85. Jang KJ, Suh KY (2010) A multi-layer microfluidic device for efficient culture and analysis of renal tubular cells. Lab Chip 10(1):36–42. doi:10.1039/b907515a
86. Wilmer MJ, Ng CP, Lanz HL, Vulto P, Suter-Dick L, Masereeuw R (2016) Kidney-on-a-chip technology for drug-induced nephrotoxicity screening. Trends Biotechnol 34(2):156–170. doi:10.1016/j.tibtech.2015.11.001

87. Agarwal A, Goss JA, Cho A, McCain ML, Parker KK (2013) Microfluidic heart on a chip for higher throughput pharmacological studies. Lab Chip 13(18):3599–3608. doi:10.1039/c3lc50350j
88. Grosberg A, Alford PW, McCain ML, Parker KK (2011) Ensembles of engineered cardiac tissues for physiological and pharmacological study: heart on a chip. Lab Chip 11(24):4165–4173. doi:10.1039/c1lc20557a
89. Park SH, Sim WY, Min BH, Yang SS, Khademhosseini A, Kaplan DL (2012) Chip-based comparison of the osteogenesis of human bone marrow- and adipose tissue-derived mesenchymal stem cells under mechanical stimulation. PLoS ONE 7(9):e46689. doi:10.1371/journal.pone.0046689
90. Grosberg A, Nesmith AP, Goss JA, Brigham MD, McCain ML, Parker KK (2012) Muscle on a chip: in vitro contractility assays for smooth and striated muscle. J Pharmacol Toxicol Methods 65(3):126–135. doi:10.1016/j.vascn.2012.04.001
91. Zheng Y, Chen J, Craven M, Choi NW, Totorica S, Diaz-Santana A, Kermani P, Hempstead B, Fischbach-Teschl C, Lopez JA, Stroock AD (2012) In vitro microvessels for the study of angiogenesis and thrombosis. Proc Natl Acad Sci U S A 109(24):9342–9347. doi:10.1073/pnas.1201240109
92. Huh D, Kim HJ, Fraser JP, Shea DE, Khan M, Bahinski A, Hamilton GA, Ingber DE (2013) Microfabrication of human organs-on-chips. Nat Protoc 8(11):2135–2157. doi:10.1038/nprot.2013.137
93. Huh D, Matthews BD, Mammoto A, Montoya-Zavala M, Hsin HY, Ingber DE (2010) Reconstituting organ-level lung functions on a chip. Science 328(5986):1662–1668. doi:10.1126/science.1188302
94. Benam KH, Villenave R, Lucchesi C, Varone A, Hubeau C, Lee HH, Alves SE, Salmon M, Ferrante TC, Weaver JC, Bahinski A, Hamilton GA, Ingber DE (2016) Small airway-on-a-chip enables analysis of human lung inflammation and drug responses in vitro. Nat Methods 13(2):151–157. doi:10.1038/nmeth.3697
95. Young EW (2013) Cells, tissues, and organs on chips: challenges and opportunities for the cancer tumor microenvironment. Integr Biol (Camb) 5(9):1096–1109. doi:10.1039/c3ib40076j
96. Fan Y, Nguyen DT, Akay Y, Xu F, Akay M (2016) Engineering a brain cancer chip for high-throughput drug screening. Sci Rep 6:25062. doi:10.1038/srep25062
97. Albanese A, Lam AK, Sykes EA, Rocheleau JV, Chan WC (2013) Tumour-on-a-chip provides an optical window into nanoparticle tissue transport. Nat Commun 4:2718. doi:10.1038/ncomms3718
98. Zheng F, Fu F, Cheng Y, Wang C, Zhao Y, Gu Z (2016) Organ-on-a-chip systems: microengineering to biomimic living systems. Small 12(17):2253–2282. doi:10.1002/smll.201503208
99. Yi C, Li C-W, Ji S, Yang M (2006) Microfluidics technology for manipulation and analysis of biological cells. Anal Chim Acta 560(1–2):1–23. doi:10.1016/j.aca.2005.12.037
100. Mu X, Zheng W, Sun J, Zhang W, Jiang X (2013) Microfluidics for manipulating cells. Small 9(1):9–21. doi:10.1002/smll.201200996
101. Yarmush ML, King KR (2009) Living-cell microarrays. Annu Rev Biomed Eng 11:235–257. doi:10.1146/annurev.bioeng.10.061807.160502
102. Jonczyk R, Kurth T, Lavrentieva A, Walter JG, Scheper T, Stahl F (2016) Living cell microarrays: an overview of concepts. Microarrays (Basel) 5(2). doi:10.3390/microarrays5020011
103. Chung J, Kim YJ, Yoon E (2011) Highly-efficient single-cell capture in microfluidic array chips using differential hydrodynamic guiding structures. Appl Phys Lett 98(12):123701. doi:10.1063/1.3565236
104. Lin L, Chu YS, Thiery JP, Lim CT, Rodriguez I (2013) Microfluidic cell trap array for controlled positioning of single cells on adhesive micropatterns. Lab Chip 13(4):714–721. doi:10.1039/c2lc41070b

105. Chung K, Kim Y, Kanodia JS, Gong E, Shvartsman SY, Lu H (2011) A microfluidic array for large-scale ordering and orientation of embryos. Nat Methods 8(2):171–176. doi:10.1038/nmeth.1548
106. Sarioglu AF, Aceto N, Kojic N, Donaldson MC, Zeinali M, Hamza B, Engstrom A, Zhu H, Sundaresan TK, Miyamoto DT, Luo X, Bardia A, Wittner BS, Ramaswamy S, Shioda T, Ting DT, Stott SL, Kapur R, Maheswaran S, Haber DA, Toner M (2015) A microfluidic device for label-free, physical capture of circulating tumor cell clusters. Nat Methods 12 (7):685–691. doi:10.1038/nmeth.3404
107. Lecault V, Vaninsberghe M, Sekulovic S, Knapp DJ, Wohrer S, Bowden W, Viel F, McLaughlin T, Jarandehei A, Miller M, Falconnet D, White AK, Kent DG, Copley MR, Taghipour F, Eaves CJ, Humphries RK, Piret JM, Hansen CL (2011) High-throughput analysis of single hematopoietic stem cell proliferation in microfluidic cell culture arrays. Nat Methods 8(7):581–586. doi:10.1038/nmeth.1614
108. Novo P, Dell'Aica M, Janasek D, Zahedi RP (2016) High spatial and temporal resolution cell manipulation techniques in microchannels. Analyst 141(6):1888–1905. doi:10.1039/c6an00027d
109. Dudani JS, Gossett DR, Tse HT, Di Carlo D (2013) Pinched-flow hydrodynamic stretching of single-cells. Lab Chip 13(18):3728–3734. doi:10.1039/c3lc50649e
110. McGrath J, Jimenez M, Bridle H (2014) Deterministic lateral displacement for particle separation: a review. Lab Chip 14(21):4139–4158. doi:10.1039/c4lc00939h
111. Qian C, Huang H, Chen L, Li X, Ge Z, Chen T, Yang Z, Sun L (2014) Dielectrophoresis for bioparticle manipulation. Int J Mol Sci 15(10):18281–18309. doi:10.3390/ijms151018281
112. Lim B, Reddy V, Hu X, Kim K, Jadhav M, Abedini-Nassab R, Noh YW, Lim YT, Yellen BB, Kim C (2014) Magnetophoretic circuits for digital control of single particles and cells. Nat Commun 5:3846. doi:10.1038/ncomms4846
113. Ahmed D, Ozcelik A, Bojanala N, Nama N, Upadhyay A, Chen Y, Hanna-Rose W, Huang TJ (2016) Rotational manipulation of single cells and organisms using acoustic waves. Nat Commun 7:11085. doi:10.1038/ncomms11085
114. Zhang H, Liu KK (2008) Optical tweezers for single cells. J R Soc Interface 5(24):671–690. doi:10.1098/rsif.2008.0052
115. Warkiani ME, Khoo BL, Wu L, Tay AK, Bhagat AA, Han J, Lim CT (2016) Ultra-fast, label-free isolation of circulating tumor cells from blood using spiral microfluidics. Nat Protoc 11(1):134–148. doi:10.1038/nprot.2016.003
116. Karabacak NM, Spuhler PS, Fachin F, Lim EJ, Pai V, Ozkumur E, Martel JM, Kojic N, Smith K, Chen PI, Yang J, Hwang H, Morgan B, Trautwein J, Barber TA, Stott SL, Maheswaran S, Kapur R, Haber DA, Toner M (2014) Microfluidic, marker-free isolation of circulating tumor cells from blood samples. Nat Protoc 9(3):694–710. doi:10.1038/nprot.2014.044
117. Collins DJ, Morahan B, Garcia-Bustos J, Doerig C, Plebanski M, Neild A (2015) Two-dimensional single-cell patterning with one cell per well driven by surface acoustic waves. Nat Commun 6:8686. doi:10.1038/ncomms9686
118. Voldman J (2006) Electrical forces for microscale cell manipulation. Annu Rev Biomed Eng 8:425–454. doi:10.1146/annurev.bioeng.8.061505.095739
119. Yasukawa T, Nagamine K, Horiguchi Y, Shiku H, Koide M, Itayama T, Shiraishi F, Matsue T (2008) Electrophoretic cell manipulation and electrochemical gene-function analysis based on a yeast two-hybrid system in a microfluidic device. Anal Chem 80 (10):3722–3727. doi:10.1021/ac800143t
120. Park K, Suk HJ, Akin D, Bashir R (2009) Dielectrophoresis-based cell manipulation using electrodes on a reusable printed circuit board. Lab Chip 9(15):2224–2229. doi:10.1039/b904328d
121. Glawdel T, Ren CL (2009) Electro-osmotic flow control for living cell analysis in microfluidic PDMS chips. Mech Res Commun 36(1):75–81. doi:10.1016/j.mechrescom.2008.06.015

122. Geng T, Lu C (2013) Microfluidic electroporation for cellular analysis and delivery. Lab Chip 13(19):3803–3821. doi:10.1039/c3lc50566a
123. Wu W, Qu Y, Hu N, Zeng Y, Yang J, Xu H, Yin ZQ (2015) A cell electrofusion chip for somatic cells reprogramming. PLoS ONE 10(7):e0131966. doi:10.1371/journal.pone.0131966
124. Pethig R (2010) Review article-dielectrophoresis: status of the theory, technology, and applications. Biomicrofluidics 4(2). doi:10.1063/1.3456626
125. Mazutis L, Gilbert J, Ung WL, Weitz DA, Griffiths AD, Heyman JA (2013) Single-cell analysis and sorting using droplet-based microfluidics. Nat Protoc 8(5):870–891. doi:10.1038/nprot.2013.046
126. Yarmush ML, Golberg A, Sersa G, Kotnik T, Miklavcic D (2014) Electroporation-based technologies for medicine: principles, applications, and challenges. Annu Rev Biomed Eng 16:295–320. doi:10.1146/annurev-bioeng-071813-104622
127. Movahed S, Li D (2010) Microfluidics cell electroporation. Microfluid Nanofluid 10(4):703–734. doi:10.1007/s10404-010-0716-y
128. Garcia PA, Ge Z, Moran JL, Buie CR (2016) Microfluidic screening of electric fields for electroporation. Sci Rep 6:21238. doi:10.1038/srep21238
129. Qu B, Eu YJ, Jeong WJ, Kim DP (2012) Droplet electroporation in microfluidics for efficient cell transformation with or without cell wall removal. Lab Chip 12(21):4483–4488. doi:10.1039/c2lc40360a
130. Kang W, Giraldo-Vela JP, Nathamgari SS, McGuire T, McNaughton RL, Kessler JA, Espinosa HD (2014) Microfluidic device for stem cell differentiation and localized electroporation of postmitotic neurons. Lab Chip 14(23):4486–4495. doi:10.1039/c4lc00721b
131. Ogle BM, Cascalho M, Platt JL (2005) Biological implications of cell fusion. Nat Rev Mol Cell Biol 6(7):567–575. doi:10.1038/nrm1678
132. Hu N, Yang J, Joo SW, Banerjee AN, Qian S (2013) Cell electrofusion in microfluidic devices: a review. Sens Actuators B: Chem 178:63–85. doi:10.1016/j.snb.2012.12.034
133. Skelley AM, Kirak O, Suh H, Jaenisch R, Voldman J (2009) Microfluidic control of cell pairing and fusion. Nat Methods 6(2):147–152. doi:10.1038/nmeth.1290
134. Takayama S, Ostuni E, LeDuc P, Naruse K, Ingber DE, Whitesides GM (2001) Subcellular positioning of small molecules. Nature 411(6841):1016. doi:10.1038/35082637
135. Takayama S, Ostuni E, LeDuc P, Naruse K, Ingber DE, Whitesides GM (2003) Selective chemical treatment of cellular microdomains using multiple laminar streams. Chem Biol 10(2):123–130. doi:10.1016/s1074-5521(03)00019-x
136. Lee CY, Romanova EV, Sweedler JV (2013) Laminar stream of detergents for subcellular neurite damage in a microfluidic device: a simple tool for the study of neuroregeneration. J Neural Eng 10(3):036020. doi:10.1088/1741-2560/10/3/036020
137. Au AK, Lai H, Utela BR, Folch A (2011) Microvalves and micropumps for BioMEMS. Micromachines 2(4):179–220. doi:10.3390/mi2020179
138. Ogden S, Klintberg L, Thornell G, Hjort K, Bodén R (2013) Review on miniaturized paraffin phase change actuators, valves, and pumps. Microfluid Nanofluid 17(1):53–71. doi:10.1007/s10404-013-1289-3
139. Iverson BD, Garimella SV (2008) Recent advances in microscale pumping technologies: a review and evaluation. Microfluid Nanofluid 5(2):145–174. doi:10.1007/s10404-008-0266-8
140. Shen J, Cai C, Yu Z, Pang Y, Zhou Y, Qian L, Wei W, Huang Y (2015) A microfluidic live cell assay to study anthrax toxin induced cell lethality assisted by conditioned medium. Sci Rep 5:8651. doi:10.1038/srep08651
141. Taylor RJ, Falconnet D, Niemisto A, Ramsey SA, Prinz S, Shmulevich I, Galitski T, Hansen CL (2009) Dynamic analysis of MAPK signaling using a high-throughput microfluidic single-cell imaging platform. Proc Natl Acad Sci U S A 106(10):3758–3763. doi:10.1073/pnas.0813416106

142. Nguyen EH, Schwartz MP, Murphy WL (2011) Biomimetic approaches to control soluble concentration gradients in biomaterials. Macromol Biosci 11(4):483–492. doi:10.1002/mabi.201000448
143. Dhumpa R, Roper MG (2012) Temporal gradients in microfluidic systems to probe cellular dynamics: a review. Anal Chim Acta 743:9–18. doi:10.1016/j.aca.2012.07.006
144. Chung BG, Choo J (2010) Microfluidic gradient platforms for controlling cellular behavior. Electrophoresis 31(18):3014–3027. doi:10.1002/elps.201000137
145. Toh AGG, Wang ZP, Yang C, Nguyen N-T (2013) Engineering microfluidic concentration gradient generators for biological applications. Microfluid Nanofluid 16(1–2):1–18. doi:10.1007/s10404-013-1236-3
146. Lin F, Butcher EC (2006) T cell chemotaxis in a simple microfluidic device. Lab Chip 6 (11):1462–1469. doi:10.1039/b607071j
147. Englert DL, Manson MD, Jayaraman A (2010) Investigation of bacterial chemotaxis in flow-based microfluidic devices. Nat Protoc 5(5):864–872. doi:10.1038/nprot.2010.18
148. Chung BG, Flanagan LA, Rhee SW, Schwartz PH, Lee AP, Monuki ES, Jeon NL (2005) Human neural stem cell growth and differentiation in a gradient-generating microfluidic device. Lab Chip 5(4):401–406. doi:10.1039/b417651k
149. Dertinger SK, Jiang X, Li Z, Murthy VN, Whitesides GM (2002) Gradients of substrate-bound laminin orient axonal specification of neurons. Proc Natl Acad Sci U S A 99(20):12542–12547. doi:10.1073/pnas.192457199
150. Gao D, Li H, Wang N, Lin JM (2012) Evaluation of the absorption of methotrexate on cells and its cytotoxicity assay by using an integrated microfluidic device coupled to a mass spectrometer. Anal Chem 84(21):9230–9237. doi:10.1021/ac301966c
151. Wu J, Wu X, Lin F (2013) Recent developments in microfluidics-based chemotaxis studies. Lab Chip 13(13):2484–2499. doi:10.1039/c3lc50415h
152. Kim S, Kim HJ, Jeon NL (2010) Biological applications of microfluidic gradient devices. Integr Biol (Camb) 2(11–12):584–603. doi:10.1039/c0ib00055h
153. Haessler U, Pisano M, Wu M, Swartz MA (2011) Dendritic cell chemotaxis in 3D under defined chemokine gradients reveals differential response to ligands CCL21 and CCL19. Proc Natl Acad Sci U S A 108(14):5614–5619. doi:10.1073/pnas.1014920108
154. Nguyen DH, Stapleton SC, Yang MT, Cha SS, Choi CK, Galie PA, Chen CS (2013) Biomimetic model to reconstitute angiogenic sprouting morphogenesis in vitro. Proc Natl Acad Sci U S A 110(17):6712–6717. doi:10.1073/pnas.1221526110
155. Chabaud M, Heuze ML, Bretou M, Vargas P, Maiuri P, Solanes P, Maurin M, Terriac E, Le Berre M, Lankar D, Piolot T, Adelstein RS, Zhang Y, Sixt M, Jacobelli J, Benichou O, Voituriez R, Piel M, Lennon-Dumenil AM (2015) Cell migration and antigen capture are antagonistic processes coupled by myosin II in dendritic cells. Nat Commun 6:7526. doi:10.1038/ncomms8526
156. Boneschansker L, Yan J, Wong E, Briscoe DM, Irimia D (2014) Microfluidic platform for the quantitative analysis of leukocyte migration signatures. Nat Commun 5:4787. doi:10.1038/ncomms5787
157. Vanapalli SA, Duits MH, Mugele F (2009) Microfluidics as a functional tool for cell mechanics. Biomicrofluidics 3(1):12006. doi:10.1063/1.3067820
158. Tee SY, Bausch AR, Janmey PA (2009) The mechanical cell. Curr Biol 19(17):R745–R748. doi:10.1016/j.cub.2009.06.034
159. Polacheck WJ, Li R, Uzel SG, Kamm RD (2013) Microfluidic platforms for mechanobiology. Lab Chip 13(12):2252–2267. doi:10.1039/c3lc41393d
160. Jain A, Graveline A, Waterhouse A, Vernet A, Flaumenhaft R, Ingber DE (2016) A shear gradient-activated microfluidic device for automated monitoring of whole blood haemostasis and platelet function. Nat Commun 7:10176. doi:10.1038/ncomms10176
161. Sundd P, Gutierrez E, Koltsova EK, Kuwano Y, Fukuda S, Pospieszalska MK, Groisman A, Ley K (2012) 'Slings' enable neutrophil rolling at high shear. Nature 488(7411):399–403. doi:10.1038/nature11248

162. Sundd P, Gutierrez E, Pospieszalska MK, Zhang H, Groisman A, Ley K (2010) Quantitative dynamic footprinting microscopy reveals mechanisms of neutrophil rolling. Nat Methods 7 (10):821–824. doi:10.1038/nmeth.1508
163. Miura S, Sato K, Kato-Negishi M, Teshima T, Takeuchi S (2015) Fluid shear triggers microvilli formation via mechanosensitive activation of TRPV6. Nat Commun 6:8871. doi:10.1038/ncomms9871
164. Humphrey JD, Dufresne ER, Schwartz MA (2014) Mechanotransduction and extracellular matrix homeostasis. Nat Rev Mol Cell Biol 15(12):802–812. doi:10.1038/nrm3896
165. Hsieh HY, Camci-Unal G, Huang TW, Liao R, Chen TJ, Paul A, Tseng FG, Khademhosseini A (2014) Gradient static-strain stimulation in a microfluidic chip for 3D cellular alignment. Lab Chip 14(3):482–493. doi:10.1039/c3lc50884f
166. Kollmannsperger A, Sharei A, Raulf A, Heilemann M, Langer R, Jensen KF, Wieneke R, Tampe R (2016) Live-cell protein labelling with nanometre precision by cell squeezing. Nat Commun 7:10372. doi:10.1038/ncomms10372
167. Si F, Li B, Margolin W, Sun SX (2015) Bacterial growth and form under mechanical compression. Sci Rep 5:11367. doi:10.1038/srep11367
168. Wells RG (2008) The role of matrix stiffness in regulating cell behavior. Hepatology 47 (4):1394–1400. doi:10.1002/hep.22193
169. Sundararaghavan HG, Monteiro GA, Firestein BL, Shreiber DI (2009) Neurite growth in 3D collagen gels with gradients of mechanical properties. Biotechnol Bioeng 102(2):632–643. doi:10.1002/bit.22074
170. Garcia S, Sunyer R, Olivares A, Noailly J, Atencia J, Trepat X (2015) Generation of stable orthogonal gradients of chemical concentration and substrate stiffness in a microfluidic device. Lab Chip 15(12):2606–2614. doi:10.1039/c5lc00140d
171. Polacheck WJ, Chen CS (2016) Measuring cell-generated forces: a guide to the available tools. Nat Methods 13(5):415–423. doi:10.1038/nmeth.3834
172. Tan JL, Tien J, Pirone DM, Gray DS, Bhadriraju K, Chen CS (2003) Cells lying on a bed of microneedles: an approach to isolate mechanical force. Proc Natl Acad Sci U S A 100 (4):1484–1489. doi:10.1073/pnas.0235407100
173. du Roure O, Saez A, Buguin A, Austin RH, Chavrier P, Silberzan P, Ladoux B (2005) Force mapping in epithelial cell migration. Proc Natl Acad Sci U S A 102(7):2390–2395. doi:10.1073/pnas.0408482102
174. Fu J, Wang YK, Yang MT, Desai RA, Yu X, Liu Z, Chen CS (2010) Mechanical regulation of cell function with geometrically modulated elastomeric substrates. Nat Methods 7(9):733–736. doi:10.1038/nmeth.1487
175. Ghassemi S, Meacci G, Liu S, Gondarenko AA, Mathur A, Roca-Cusachs P, Sheetz MP, Hone J (2012) Cells test substrate rigidity by local contractions on submicrometer pillars. Proc Natl Acad Sci U S A 109(14):5328–5333. doi:10.1073/pnas.1119886109
176. Trichet L, Le Digabel J, Hawkins RJ, Vedula SR, Gupta M, Ribrault C, Hersen P, Voituriez R, Ladoux B (2012) Evidence of a large-scale mechanosensing mechanism for cellular adaptation to substrate stiffness. Proc Natl Acad Sci U S A 109(18):6933–6938. doi:10.1073/pnas.1117810109
177. Zare RN, Kim S (2010) Microfluidic platforms for single-cell analysis. Annu Rev Biomed Eng 12:187–201. doi:10.1146/annurev-bioeng-070909-105238
178. Chen Y, Li P, Huang PH, Xie Y, Mai JD, Wang L, Nguyen NT, Huang TJ (2014) Rare cell isolation and analysis in microfluidics. Lab Chip 14(4):626–645. doi:10.1039/c3lc90136j
179. Autebert J, Coudert B, Bidard FC, Pierga JY, Descroix S, Malaquin L, Viovy JL (2012) Microfluidic: an innovative tool for efficient cell sorting. Methods 57(3):297–307. doi:10.1016/j.ymeth.2012.07.002
180. Gao Y, Li W, Pappas D (2013) Recent advances in microfluidic cell separations. Analyst 138(17):4714–4721. doi:10.1039/c3an00315a
181. Gossett DR, Weaver WM, Mach AJ, Hur SC, Tse HT, Lee W, Amini H, Di Carlo D (2010) Label-free cell separation and sorting in microfluidic systems. Anal Bioanal Chem 397 (8):3249–3267. doi:10.1007/s00216-010-3721-9

182. Plouffe BD, Murthy SK (2014) Perspective on microfluidic cell separation: a solved problem? Anal Chem 86(23):11481–11488. doi:10.1021/ac5013283
183. Bhagat AA, Bow H, Hou HW, Tan SJ, Han J, Lim CT (2010) Microfluidics for cell separation. Med Biol Eng Comput 48(10):999–1014. doi:10.1007/s11517-010-0611-4
184. Warkiani ME, Wu L, Tay AK, Han J (2015) Large-volume microfluidic cell sorting for biomedical applications. Annu Rev Biomed Eng 17:1–34. doi:10.1146/annurev-bioeng-071114-040818
185. Stott SL, Hsu CH, Tsukrov DI, Yu M, Miyamoto DT, Waltman BA, Rothenberg SM, Shah AM, Smas ME, Korir GK, Floyd FP Jr, Gilman AJ, Lord JB, Winokur D, Springer S, Irimia D, Nagrath S, Sequist LV, Lee RJ, Isselbacher KJ, Maheswaran S, Haber DA, Toner M (2010) Isolation of circulating tumor cells using a microvortex-generating herringbone-chip. Proc Natl Acad Sci U S A 107(43):18392–18397. doi:10.1073/pnas.1012539107
186. Yu M, Bardia A, Wittner BS, Stott SL, Smas ME, Ting DT, Isakoff SJ, Ciciliano JC, Wells MN, Shah AM, Concannon KF, Donaldson MC, Sequist LV, Brachtel E, Sgroi D, Baselga J, Ramaswamy S, Toner M, Haber DA, Maheswaran S (2013) Circulating breast tumor cells exhibit dynamic changes in epithelial and mesenchymal composition. Science 339(6119):580–584. doi:10.1126/science.1228522
187. Chen Q, Wu J, Zhang Y, Lin Z, Lin JM (2012) Targeted isolation and analysis of single tumor cells with aptamer-encoded microwell array on microfluidic device. Lab Chip 12 (24):5180–5185. doi:10.1039/c2lc40858a
188. Herzenberg LA, Parks D, Sahaf B, Perez O, Roederer M, Herzenberg LA (2002) The history and future of the fluorescence activated cell sorter and flow cytometry: a view from Stanford. Clin Chem 48(10):1819–1827
189. Lenshof A, Laurell T (2010) Continuous separation of cells and particles in microfluidic systems. Chem Soc Rev 39(3):1203–1217. doi:10.1039/b915999c
190. Yao B, Luo GA, Feng X, Wang W, Chen LX, Wang YM (2004) A microfluidic device based on gravity and electric force driving for flow cytometry and fluorescence activated cell sorting. Lab Chip 4(6):603–607. doi:10.1039/b408422e
191. Baret JC, Miller OJ, Taly V, Ryckelynck M, El-Harrak A, Frenz L, Rick C, Samuels ML, Hutchison JB, Agresti JJ, Link DR, Weitz DA, Griffiths AD (2009) Fluorescence-activated droplet sorting (FADS): efficient microfluidic cell sorting based on enzymatic activity. Lab Chip 9(13):1850–1858. doi:10.1039/b902504a
192. Sun Y, Lim CS, Liu AQ, Ayi TC, Yap PH (2007) Design, simulation and experiment of electroosmotic microfluidic chip for cell sorting. Sens Actuators A: Phys 133(2):340–348. doi:10.1016/j.sna.2006.06.047
193. Austin Suthanthiraraj PP, Piyasena ME, Woods TA, Naivar MA, Lomicronpez GP, Graves SW (2012) One-dimensional acoustic standing waves in rectangular channels for flow cytometry. Methods 57(3):259–271. doi:10.1016/j.ymeth.2012.02.013
194. Johansson L, Nikolajeff F, Johansson S, Thorslund S (2009) On-chip fluorescence-activated cell sorting by an integrated miniaturized ultrasonic transducer. Anal Chem 81(13):5188–5196. doi:10.1021/ac802681r
195. Wu TH, Chen Y, Park SY, Hong J, Teslaa T, Zhong JF, Di Carlo D, Teitell MA, Chiou PY (2012) Pulsed laser triggered high speed microfluidic fluorescence activated cell sorter. Lab Chip 12(7):1378–1383. doi:10.1039/c2lc21084c
196. Yung CW, Fiering J, Mueller AJ, Ingber DE (2009) Micromagnetic-microfluidic blood cleansing device. Lab Chip 9(9):1171–1177. doi:10.1039/b816986a
197. Hoshino K, Huang YY, Lane N, Huebschman M, Uhr JW, Frenkel EP, Zhang X (2011) Microchip-based immunomagnetic detection of circulating tumor cells. Lab Chip 11 (20):3449–3457. doi:10.1039/c1lc20270g
198. Xia N, Hunt TP, Mayers BT, Alsberg E, Whitesides GM, Westervelt RM, Ingber DE (2006) Combined microfluidic-micromagnetic separation of living cells in continuous flow. Biomed Microdevices 8(4):299–308. doi:10.1007/s10544-006-0033-0

199. Kim S, Han SI, Park MJ, Jeon CW, Joo YD, Choi IH, Han KH (2013) Circulating tumor cell microseparator based on lateral magnetophoresis and immunomagnetic nanobeads. Anal Chem 85(5):2779–2786. doi:10.1021/ac303284u
200. Brown RB, Audet J (2008) Current techniques for single-cell lysis. J R Soc Interface 5(Suppl 2):S131–S138. doi:10.1098/rsif.2008.0009.focus
201. Nan L, Jiang Z, Wei X (2014) Emerging microfluidic devices for cell lysis: a review. Lab Chip 14(6):1060–1073. doi:10.1039/c3lc51133b
202. Hosic S, Murthy SK, Koppes AN (2016) Microfluidic sample preparation for single cell analysis. Anal Chem 88(1):354–380. doi:10.1021/acs.analchem.5b04077
203. Yun SS, Yoon SY, Song MK, Im SH, Kim S, Lee JH, Yang S (2010) Handheld mechanical cell lysis chip with ultra-sharp silicon nano-blade arrays for rapid intracellular protein extraction. Lab Chip 10(11):1442–1446. doi:10.1039/b925244d
204. Kim J, Hee Jang S, Jia G, Zoval JV, Da Silva NA, Madou MJ (2004) Cell lysis on a microfluidic CD (compact disc). Lab Chip 4(5):516–522. doi:10.1039/b401106f
205. Siegrist J, Gorkin R, Bastien M, Stewart G, Peytavi R, Kido H, Bergeron M, Madou M (2010) Validation of a centrifugal microfluidic sample lysis and homogenization platform for nucleic acid extraction with clinical samples. Lab Chip 10(3):363–371. doi:10.1039/b913219h
206. Mellors JS, Jorabchi K, Smith LM, Ramsey JM (2010) Integrated microfluidic device for automated single cell analysis using electrophoretic separation and electrospray ionization mass spectrometry. Anal Chem 82(3):967–973. doi:10.1021/ac902218y
207. Jokilaakso N, Salm E, Chen A, Millet L, Guevara CD, Dorvel B, Reddy B Jr, Karlstrom AE, Chen Y, Ji H, Chen Y, Sooryakumar R, Bashir R (2013) Ultra-localized single cell electroporation using silicon nanowires. Lab Chip 13(3):336–339. doi:10.1039/c2lc40837f
208. Lee C-Y, Lee G-B, Lin J-L, Huang F-C, Liao C-S (2005) Integrated microfluidic systems for cell lysis, mixing/pumping and DNA amplification. J Micromech Microeng 15(6):1215–1223. doi:10.1088/0960-1317/15/6/011
209. Sarkar A, Kolitz S, Lauffenburger DA, Han J (2014) Microfluidic probe for single-cell analysis in adherent tissue culture. Nat Commun 5:3421. doi:10.1038/ncomms4421
210. Yang W, Woolley AT (2010) Integrated multi-process microfluidic systems for automating analysis. JALA Charlottesv Va 15(3):198–209. doi:10.1016/j.jala.2010.01.008
211. Cui F, Rhee M, Singh A, Tripathi A (2015) Microfluidic sample preparation for medical diagnostics. Annu Rev Biomed Eng 17:267–286. doi:10.1146/annurev-bioeng-071114-040538
212. Wu D, Qin J, Lin B (2008) Electrophoretic separations on microfluidic chips. J Chromatogr A 1184(1–2):542–559. doi:10.1016/j.chroma.2007.11.119
213. Cong H, Xu X, Yu B, Yuan H, Peng Q, Tian C (2015) Recent progress in preparation and application of microfluidic chip electrophoresis. J Micromech Microeng 25(5):053001. doi:10.1088/0960-1317/25/5/053001
214. Karlinsey JM (2012) Sample introduction techniques for microchip electrophoresis: a review. Anal Chim Acta 725:1–13. doi:10.1016/j.aca.2012.02.052
215. Liu P, Yeung SH, Crenshaw KA, Crouse CA, Scherer JR, Mathies RA (2008) Real-time forensic DNA analysis at a crime scene using a portable microchip analyzer. Forensic Sci Int Genet 2(4):301–309. doi:10.1016/j.fsigen.2008.03.009
216. Lin X, Chen Q, Liu W, Yi L, Li H, Wang Z, Lin JM (2015) Assay of multiplex proteins from cell metabolism based on tunable aptamer and microchip electrophoresis. Biosens Bioelectron 63:105–111. doi:10.1016/j.bios.2014.07.013
217. Smejkal P, Bottenus D, Breadmore MC, Guijt RM, Ivory CF, Foret F, Macka M (2013) Microfluidic isotachophoresis: a review. Electrophoresis 34(11):1493–1509. doi:10.1002/elps.201300021
218. Schoch RB, Ronaghi M, Santiago JG (2009) Rapid and selective extraction, isolation, preconcentration, and quantitation of small RNAs from cell lysate using on-chip isotachophoresis. Lab Chip 9(15):2145–2152. doi:10.1039/b903542g

219. Tetala KK, Vijayalakshmi MA (2016) A review on recent developments for biomolecule separation at analytical scale using microfluidic devices. Anal Chim Acta 906:7–21. doi:10.1016/j.aca.2015.11.037
220. Wu R, Hu L, Wang F, Ye M, Zou H (2008) Recent development of monolithic stationary phases with emphasis on microscale chromatographic separation. J Chromatogr A 1184(1–2):369–392. doi:10.1016/j.chroma.2007.09.022
221. Lin SL, Lin TY, Fuh MR (2014) Microfluidic chip-based liquid chromatography coupled to mass spectrometry for determination of small molecules in bioanalytical applications: an update. Electrophoresis 35(9):1275–1284. doi:10.1002/elps.201300415
222. Chen ZW, Fuchs K, Sieghart W, Townsend RR, Evers AS (2012) Deep amino acid sequencing of native brain GABAA receptors using high-resolution mass spectrometry. Mol Cell Proteomics 11(1):M111 011445
223. Hwang KY, Kwon SH, Jung SO, Namkoong K, Jung WJ, Kim JH, Suh KY, Huh N (2012) Solid phase DNA extraction with a flexible bead-packed microfluidic device to detect methicillin-resistant Staphylococcus aureus in nasal swabs. Anal Chem 84(18):7912–7918. doi:10.1021/ac3016533
224. Kumar S, Sahore V, Rogers CI, Woolley AT (2016) Development of an integrated microfluidic solid-phase extraction and electrophoresis device. Analyst 141(5):1660–1668. doi:10.1039/c5an02352a
225. Ramsey JD, Collins GE (2005) Integrated microfluidic device for solid-phase extraction coupled to micellar electrokinetic chromatography separation. Anal Chem 77(20):6664–6670. doi:10.1021/ac0507789
226. Mao S, Zhang J, Li H, Lin JM (2013) Strategy for signaling molecule detection by using an integrated microfluidic device coupled with mass spectrometry to study cell-to-cell communication. Anal Chem 85(2):868–876. doi:10.1021/ac303164b
227. Zhang J, Wu J, Li H, Chen Q, Lin JM (2015) An in vitro liver model on microfluidic device for analysis of capecitabine metabolite using mass spectrometer as detector. Biosens Bioelectron 68:322–328. doi:10.1016/j.bios.2015.01.013
228. Gao D, Liu H, Lin JM, Wang Y, Jiang Y (2013) Characterization of drug permeability in Caco-2 monolayers by mass spectrometry on a membrane-based microfluidic device. Lab Chip 13(5):978–985. doi:10.1039/c2lc41215b
229. Hagan KA, Meier WL, Ferrance JP, Landers JP (2009) Chitosan-coated silica as a solid phase for RNA purification in a microfluidic device. Anal Chem 81(13):5249–5256. doi:10.1021/ac900820z
230. Tia S, Herr AE (2009) On-chip technologies for multidimensional separations. Lab Chip 9 (17):2524–2536. doi:10.1039/b900683b
231. Emrich CA, Medintz IL, Chu WK, Mathies RA (2007) Microfabricated two-dimensional electrophoresis device for differential protein expression profiling. Anal Chem 79(19):7360–7366. doi:10.1021/ac0711485
232. Choi JR, Song H, Sung JH, Kim D, Kim K (2016) Microfluidic assay-based optical measurement techniques for cell analysis: a review of recent progress. Biosens Bioelectron 77:227–236. doi:10.1016/j.bios.2015.07.068
233. Chrimes AF, Khoshmanesh K, Stoddart PR, Mitchell A, Kalantar-Zadeh K (2013) Microfluidics and Raman microscopy: current applications and future challenges. Chem Soc Rev 42(13):5880–5906. doi:10.1039/c3cs35515b
234. Perro A, Lebourdon G, Henry S, Lecomte S, Servant L, Marre S (2016) Combining microfluidics and FT-IR spectroscopy: towards spatially resolved information on chemical processes. React Chem Eng. doi:10.1039/c6re00127k
235. Kuswandi B, Nuriman, Huskens J, Verboom W (2007) Optical sensing systems for microfluidic devices: a review. Anal Chim Acta 601(2):141–155. doi:10.1016/j.aca.2007.08.046
236. Rackus DG, Shamsi MH, Wheeler AR (2015) Electrochemistry, biosensors and microfluidics: a convergence of fields. Chem Soc Rev 44(15):5320–5340. doi:10.1039/c4cs00369a

237. Kiilerich-Pedersen K, Rozlosnik N (2012) Cell-Based biosensors: electrical sensing in microfluidic devices. Diagnostics (Basel) 2(4):83–96. doi:10.3390/diagnostics2040083
238. D'hahan NP (2011) Live cell analysis: when electric detection interfaces microfluidics. J Biochips Tissue Chips 01(01). doi:10.4172/2153-0777.s1-001
239. Rossier J, Reymond F, Michel PE (2002) Polymer microfluidic chips for electrochemical and biochemical analyses. Electrophoresis 23(6):858–867. doi:10.1002/1522-2683(200203)23:6<858:AID-ELPS858>3.0.CO;2-3
240. Wang X, Yi L, Mukhitov N, Schrell AM, Dhumpa R, Roper MG (2015) Microfluidics-to-mass spectrometry: a review of coupling methods and applications. J Chromatogr A 1382:98–116. doi:10.1016/j.chroma.2014.10.039
241. Gao D, Liu H, Jiang Y, Lin JM (2013) Recent advances in microfluidics combined with mass spectrometry: technologies and applications. Lab Chip 13(17):3309–3322. doi:10.1039/c3lc50449b
242. Feng X, Liu BF, Li J, Liu X (2015) Advances in coupling microfluidic chips to mass spectrometry. Mass Spectrom Rev 34(5):535–557. doi:10.1002/mas.21417
243. Mao X, Huang TJ (2012) Microfluidic diagnostics for the developing world. Lab Chip 12(8):1412–1416. doi:10.1039/c2lc90022j
244. Chen J, Li J, Sun Y (2012) Microfluidic approaches for cancer cell detection, characterization, and separation. Lab Chip 12(10):1753–1767. doi:10.1039/c2lc21273k
245. Giobbe GG, Michielin F, Luni C, Giulitti S, Martewicz S, Dupont S, Floreani A, Elvassore N (2015) Functional differentiation of human pluripotent stem cells on a chip. Nat Methods 12(7):637–640. doi:10.1038/nmeth.3411
246. Lewis DM, Gerecht S (2016) Microfluidics and biomaterials to study angiogenesis. Curr Opin Chem Eng 11:114–122. doi:10.1016/j.coche.2016.02.005
247. Huang Y, Agrawal B, Sun D, Kuo JS, Williams JC (2011) Microfluidics-based devices: new tools for studying cancer and cancer stem cell migration. Biomicrofluidics 5(1):13412. doi:10.1063/1.3555195
248. Chung S, Sudo R, Vickerman V, Zervantonakis IK, Kamm RD (2010) Microfluidic platforms for studies of angiogenesis, cell migration, and cell-cell interactions. Ann Biomed Eng 38(3):1164–1177. doi:10.1007/s10439-010-9899-3
249. Kim C, Kasuya J, Jeon J, Chung S, Kamm RD (2015) A quantitative microfluidic angiogenesis screen for studying anti-angiogenic therapeutic drugs. Lab Chip 15(1):301–310. doi:10.1039/c4lc00866a
250. Haandbaek N, Burgel SC, Heer F, Hierlemann A (2014) Characterization of subcellular morphology of single yeast cells using high frequency microfluidic impedance cytometer. Lab Chip 14(2):369–377. doi:10.1039/c3lc50866h
251. Kim J, Johnson M, Hill P, Gale BK (2009) Microfluidic sample preparation: cell lysis and nucleic acid purification. Integr Biol (Camb) 1(10):574–586. doi:10.1039/b905844c
252. Chang CM, Chang WH, Wang CH, Wang JH, Mai JD, Lee GB (2013) Nucleic acid amplification using microfluidic systems. Lab Chip 13(7):1225–1242. doi:10.1039/c3lc41097h
253. Wu J, Kodzius R, Cao W, Wen W (2013) Extraction, amplification and detection of DNA in microfluidic chip-based assays. Microchim Acta 181(13–14):1611–1631. doi:10.1007/s00604-013-1140-2
254. Mauk MG, Liu C, Song J, Bau HH (2015) Integrated microfluidic nucleic acid isolation, isothermal amplification, and amplicon quantification. Microarrays (Basel) 4(4):474–489. doi:10.3390/microarrays4040474
255. Zhang R, Li X, Ramaswami G, Smith KS, Turecki G, Montgomery SB, Li JB (2014) Quantifying RNA allelic ratios by microfluidic multiplex PCR and sequencing. Nat Methods 11(1):51–54. doi:10.1038/nmeth.2736
256. Fang X, Chen H, Xu L, Jiang X, Wu W, Kong J (2012) A portable and integrated nucleic acid amplification microfluidic chip for identifying bacteria. Lab Chip 12(8):1495–1499. doi:10.1039/c2lc40055c

257. Liu P, Mathies RA (2009) Integrated microfluidic systems for high-performance genetic analysis. Trends Biotechnol 27(10):572–581. doi:10.1016/j.tibtech.2009.07.002
258. Foudeh AM, Fatanat Didar T, Veres T, Tabrizian M (2012) Microfluidic designs and techniques using lab-on-a-chip devices for pathogen detection for point-of-care diagnostics. Lab Chip 12(18):3249–3266. doi:10.1039/c2lc40630f
259. Horsman KM, Bienvenue JM, Blasier KR, Landers JP (2007) Forensic DNA analysis on microfluidic devices: a review. J Forensic Sci 52(4):784–799. doi:10.1111/j.1556-4029.2007.00468.x
260. Zheng GX, Lau BT, Schnall-Levin M, Jarosz M, Bell JM, Hindson CM, Kyriazopoulou-Panagiotopoulou S, Masquelier DA, Merrill L, Terry JM, Mudivarti PA, Wyatt PW, Bharadwaj R, Makarewicz AJ, Li Y, Belgrader P, Price AD, Lowe AJ, Marks P, Vurens GM, Hardenbol P, Montesclaros L, Luo M, Greenfield L, Wong A, Birch DE, Short SW, Bjornson KP, Patel P, Hopmans ES, Wood C, Kaur S, Lockwood GK, Stafford D, Delaney JP, Wu I, Ordonez HS, Grimes SM, Greer S, Lee JY, Belhocine K, Giorda KM, Heaton WH, McDermott GP, Bent ZW, Meschi F, Kondov NO, Wilson R, Bernate JA, Gauby S, Kindwall A, Bermejo C, Fehr AN, Chan A, Saxonov S, Ness KD, Hindson BJ, Ji HP (2016) Haplotyping germline and cancer genomes with high-throughput linked-read sequencing. Nat Biotechnol 34(3):303–311. doi:10.1038/nbt.3432
261. Ting DT, Wittner BS, Ligorio M, Vincent Jordan N, Shah AM, Miyamoto DT, Aceto N, Bersani F, Brannigan BW, Xega K, Ciciliano JC, Zhu H, MacKenzie OC, Trautwein J, Arora KS, Shahid M, Ellis HL, Qu N, Bardeesy N, Rivera MN, Deshpande V, Ferrone CR, Kapur R, Ramaswamy S, Shioda T, Toner M, Maheswaran S, Haber DA (2014) Single-cell RNA sequencing identifies extracellular matrix gene expression by pancreatic circulating tumor cells. Cell Rep 8(6):1905–1918. doi:10.1016/j.celrep.2014.08.029
262. Kimmerling RJ, Lee Szeto G, Li JW, Genshaft AS, Kazer SW, Payer KR, de Riba Borrajo J, Blainey PC, Irvine DJ, Shalek AK, Manalis SR (2016) A microfluidic platform enabling single-cell RNA-seq of multigenerational lineages. Nat Commun 7:10220. doi:10.1038/ncomms10220
263. Sanchez-Freire V, Ebert AD, Kalisky T, Quake SR, Wu JC (2012) Microfluidic single-cell real-time PCR for comparative analysis of gene expression patterns. Nat Protoc 7(5):829–838. doi:10.1038/nprot.2012.021
264. Streets AM, Zhang X, Cao C, Pang Y, Wu X, Xiong L, Yang L, Fu Y, Zhao L, Tang F, Huang Y (2014) Microfluidic single-cell whole-transcriptome sequencing. Proc Natl Acad Sci U S A 111(19):7048–7053. doi:10.1073/pnas.1402030111
265. Shalek AK, Satija R, Shuga J, Trombetta JJ, Gennert D, Lu D, Chen P, Gertner RS, Gaublomme JT, Yosef N, Schwartz S, Fowler B, Weaver S, Wang J, Wang X, Ding R, Raychowdhury R, Friedman N, Hacohen N, Park H, May AP, Regev A (2014) Single-cell RNA-seq reveals dynamic paracrine control of cellular variation. Nature 510(7505):363–369. doi:10.1038/nature13437
266. Cao Z, Chen C, He B, Tan K, Lu C (2015) A microfluidic device for epigenomic profiling using 100 cells. Nat Methods 12(10):959–962. doi:10.1038/nmeth.3488
267. Bennett MR, Hasty J (2009) Microfluidic devices for measuring gene network dynamics in single cells. Nat Rev Genet 10(9):628–638. doi:10.1038/nrg2625
268. Toriello NM, Douglas ES, Thaitrong N, Hsiao SC, Francis MB, Bertozzi CR, Mathies RA (2008) Integrated microfluidic bioprocessor for single-cell gene expression analysis. Proc Natl Acad Sci U S A 105(51):20173–20178. doi:10.1073/pnas.0806355106
269. Busch W, Moore BT, Martsberger B, Mace DL, Twigg RW, Jung J, Pruteanu-Malinici I, Kennedy SJ, Fricke GK, Clark RL, Ohler U, Benfey PN (2012) A microfluidic device and computational platform for high-throughput live imaging of gene expression. Nat Methods 9(11):1101–1106. doi:10.1038/nmeth.2185
270. Yu J, Zhou J, Sutherland A, Wei W, Shin YS, Xue M, Heath JR (2014) Microfluidics-based single-cell functional proteomics for fundamental and applied biomedical applications. Annu Rev Anal Chem (Palo Alto Calif) 7:275–295. doi:10.1146/annurev-anchem-071213-020323

271. Sun J, Masterman-Smith MD, Graham NA, Jiao J, Mottahedeh J, Laks DR, Ohashi M, DeJesus J, Kamei K, Lee KB, Wang H, Yu ZT, Lu YT, Hou S, Li K, Liu M, Zhang N, Wang S, Angenieux B, Panosyan E, Samuels ER, Park J, Williams D, Konkankit V, Nathanson D, van Dam RM, Phelps ME, Wu H, Liau LM, Mischel PS, Lazareff JA, Kornblum HI, Yong WH, Graeber TG, Tseng HR (2010) A microfluidic platform for systems pathology: multiparameter single-cell signaling measurements of clinical brain tumor specimens. Cancer Res 70(15):6128–6138. doi:10.1158/0008-5472.CAN-10-0076
272. Nguyen CQ, Ogunniyi AO, Karabiyik A, Love JC (2013) Single-cell analysis reveals isotype-specific autoreactive B cell repertoires in Sjogren's syndrome. PLoS ONE 8(3): e58127. doi:10.1371/journal.pone.0058127
273. Bailey RC, Kwong GA, Radu CG, Witte ON, Heath JR (2007) DNA-encoded antibody libraries: a unified platform for multiplexed cell sorting and detection of genes and proteins. J Am Chem Soc 129(7):1959–1967. doi:10.1021/ja065930i
274. Xue M, Wei W, Su Y, Kim J, Shin YS, Mai WX, Nathanson DA, Heath JR (2015) Chemical methods for the simultaneous quantitation of metabolites and proteins from single cells. J Am Chem Soc 137(12):4066–4069. doi:10.1021/jacs.5b00944
275. Ma C, Fan R, Ahmad H, Shi Q, Comin-Anduix B, Chodon T, Koya RC, Liu CC, Kwong GA, Radu CG, Ribas A, Heath JR (2011) A clinical microchip for evaluation of single immune cells reveals high functional heterogeneity in phenotypically similar T cells. Nat Med 17(6):738–743. doi:10.1038/nm.2375
276. Poovathingal SK, Kravchenko-Balasha N, Shin YS, Levine RD, Heath JR (2016) Critical points in Tumorigenesis: a carcinogen-initiated phase transition analyzed via single-cell proteomics. Small 12(11):1425–1431. doi:10.1002/smll.201501178
277. He M, Herr AE (2010) Automated microfluidic protein immunoblotting. Nat Protoc 5 (11):1844–1856. doi:10.1038/nprot.2010.142
278. Hughes AJ, Herr AE (2012) Microfluidic Western blotting. Proc Natl Acad Sci U S A 109 (52):21450–21455. doi:10.1073/pnas.1207754110
279. Hughes AJ, Spelke DP, Xu Z, Kang CC, Schaffer DV, Herr AE (2014) Single-cell western blotting. Nat Methods 11(7):749–755. doi:10.1038/nmeth.2992
280. Kang CC, Yamauchi KA, Vlassakis J, Sinkala E, Duncombe TA, Herr AE (2016) Single cell-resolution western blotting. Nat Protoc 11(8):1508–1530. doi:10.1038/nprot.2016.089
281. Lee JR, Bechstein DJ, Ooi CC, Patel A, Gaster RS, Ng E, Gonzalez LC, Wang SX (2016) Magneto-nanosensor platform for probing low-affinity protein-protein interactions and identification of a low-affinity PD-L1/PD-L2 interaction. Nat Commun 7:12220. doi:10.1038/ncomms12220
282. Lee J, Soper SA, Murray KK (2009) Microfluidic chips for mass spectrometry-based proteomics. J Mass Spectrom 44(5):579–593. doi:10.1002/jms.1585
283. Chao TC, Hansmeier N (2013) Microfluidic devices for high-throughput proteome analyses. Proteomics 13(3–4):467–479. doi:10.1002/pmic.201200411
284. Vollmer M, Hörth P, Rozing G, Couté Y, Grimm R, Hochstrasser D, Sanchez J-C (2006) Multi-dimensional HPLC/MS of the nucleolar proteome using HPLC-chip/MS. J Sep Sci 29 (4):499–509. doi:10.1002/jssc.200500334
285. Lee J, Soper SA, Murray KK (2009) Microfluidics with MALDI analysis for proteomics—a review. Anal Chim Acta 649(2):180–190. doi:10.1016/j.aca.2009.07.037
286. Lee J, Soper SA, Murray KK (2011) A solid-phase bioreactor with continuous sample deposition for matrix-assisted laser desorption/ionization time-of-flight mass spectrometry. Rapid Commun Mass Spectrom 25(6):693–699. doi:10.1002/rcm.4921
287. Rubakhin SS, Romanova EV, Nemes P, Sweedler JV (2011) Profiling metabolites and peptides in single cells. Nat Methods 8(4 Suppl):S20–S29. doi:10.1038/nmeth.1549
288. Kraly JR, Holcomb RE, Guan Q, Henry CS (2009) Review: microfluidic applications in metabolomics and metabolic profiling. Anal Chim Acta 653(1):23–35. doi:10.1016/j.aca.2009.08.037
289. Lin L, Lin JM (2015) Development of cell metabolite analysis on microfluidic platform. J Pharm Anal 5(6):337–347. doi:10.1016/j.jpha.2015.09.003

290. He X, Chen Q, Zhang Y, Lin JM (2014) Recent advances in microchip-mass spectrometry for biological analysis. TrAC Trends Anal Chem 53:84–97. doi:10.1016/j.trac.2013.09.013
291. Liu W, Wang N, Lin X, Ma Y, Lin JM (2014) Interfacing microsampling droplets and mass spectrometry by paper spray ionization for online chemical monitoring of cell culture. Anal Chem 86(14):7128–7134. doi:10.1021/ac501678q
292. Mao S, Gao D, Liu W, Wei H, Lin JM (2012) Imitation of drug metabolism in human liver and cytotoxicity assay using a microfluidic device coupled to mass spectrometric detection. Lab Chip 12(1):219–226. doi:10.1039/c1lc20678h
293. Chen Q, Wu J, Zhang Y, Lin JM (2012) Qualitative and quantitative analysis of tumor cell metabolism via stable isotope labeling assisted microfluidic chip electrospray ionization mass spectrometry. Anal Chem 84(3):1695–1701. doi:10.1021/ac300003k
294. Zhuang Q, Wang S, Zhang J, He Z, Li H, Ma Y, Lin JM (2015) Nephrocyte-neurocyte interaction and cellular metabolic analysis on membrane-integrated microfluidic device. Sci China Chem 59(2):243–250. doi:10.1007/s11426-015-5453-3
295. Wang BL, Ghaderi A, Zhou H, Agresti J, Weitz DA, Fink GR, Stephanopoulos G (2014) Microfluidic high-throughput culturing of single cells for selection based on extracellular metabolite production or consumption. Nat Biotechnol 32(5):473–478. doi:10.1038/nbt.2857

Chapter 3
Microfluidic Cell Isolation and Recognition for Biomedical Applications

Qiushui Chen and Jin-Ming Lin

Abstract There are hundreds of types of cells in human body, and each type of them regulates specific functions in the activities of daily living. Modern studies have revealed that different types of cells exhibit heterogeneity in gene expression, protein secretion and some crucial information between normal and diseased cells. Thus development of efficient technologies for cell isolation and recognition is valuable to achieve disease diagnosis and other fundamental biological researches. Microfluidic technologies have been increasingly emerged as a powerful tool for the studies of cell isolation and recognition in the field of microfluidic cell analysis, which show great promising in the basic and applied biomedical researches. In this chapter, we present an overview of advanced microfluidics for cell isolation and recognition in the past decade. We also discuss current emergence of the biomedical applications in microfluidics including circulating tumor cells (CTCs) detection, cell-based biological assays and stem cell purification. At the end of this Chapter we make a brief summary on the development and challenge of microfluidic technologies in cell isolation and recognition, followed by a future perspective in this study field.

Keywords Microfluidics · Cell isolation · Biological affinity · Biomedical applications

3.1 Introduction

Over the past 10 years, microfluidic system has been rapidly developed to be one of most promising platforms that is able to precisely control cells at the micro-scale level [1–4]. The concepts of 'cells on chips' and 'micro-total-analysis-system' (μTAS) have become very popular in the cell-based basic and applied researches, mainly because microfabricated technologies have put microfluidics within the

Q. Chen · J.-M. Lin (✉)
Department of Chemistry, Tsinghua University, Beijing 100084, People's Republic of China
e-mail: jmlin@mail.tsinghua.edu.cn

J.-M. Lin (ed.), *Cell Analysis on Microfluidics*, Integrated Analytical Systems,
https://doi.org/10.1007/978-981-10-5394-8_3

reach of cell-focused biomedical researches [5–8]. In typical, microfluidic devices can be designed and fabricated by integrating several microchambers and connecting them with desired fluidic networks into a single chip, which is capable to incorporate several steps of an assay task together [9, 10]. Microfluidics allows cell-based experiments to be performed for rapid analysis on very small volumes, and carry out many impossible experiments in the conventional techniques for cell-based studies [11]. For example, conventional handling of cell culture, selection, recognition, and analysis at a controllable precision is difficult, and suffers from the problems of unavoidable losses and poor reproducibility. Moreover, the small integrated microchips have the outstanding advantages of rapid and high throughput analysis, little sample consumption and low cost, robust and portable device [12]. Thus, microfluidic technologies create practical opportunities for developing multi-function integration and point-of-care auto-systems, which are likely to become important in the microfluidic cell analysis [13, 14].

The study of microfluidic technologies for cell isolation and recognition is an important content in the research field of microfluidic cell analysis [15–18]. For instance, in the early 2000s, micro-dam structures were fabricated in the microchannels to physically trap single cells from a large population of cells, and perform cellular analysis at the single-cell level [19]. Further, a microfluidic platform was developed to control cell pairing and fusion, which harvested a surprised high pairing efficiency up to 70% [20]. Microfluidics is also combined with cell-based biological recognition technologies for selective isolation and analysis of circulating tumor cells (CTCs) from whole blood samples for clinical applications [16, 21, 22]. A unique microfluidic technique using antibody-functionalized microchips was reported to efficiently capture CTCs [23]. This work has opened a broad avenue to use the microfluidic chips as portable point-of-care devices in the cancer diagnosis for clinical applications. Meanwhile, aptamer-based cell specific recognition techniques and nano-structured interfaces were developed to improve the efficiency of cell capture based on the principles of cell surface interaction [24]. To date, significant advances have been achieved on the microfluidic platforms for cell isolation and recognition, as well as their biomedical applications [25].

In this chapter, we introduce the basic concepts and recent developments of microfluidic techniques employed for cell isolation and recognition. We focus on the classical technologies in microfluidics that enable to achieve efficient cell isolation by microstructured capture and specific cell recognition. We present an overview of those microfluidic technologies that have been used to enhance cell isolation efficiency, including antibody/aptamer-based immunoaffinity cell capture, and micro/nano-structured surface with cell interactions. We also discuss the prominent biomedical applications in microfluidic cell isolation and recognition reported in the past years, such as circulating tumor cells detection, cell-based biological assays and stem cell studies. In the last part, we give a conclusion on this Chapter, and make a future perspective in this research field.

3.2 Physical Approaches for Cell Isolation in Microfluidics

The technologies for cell isolation and analysis are of significance in the basic and applied biological researches. Using microfluidic techniques, many physical methods are able to perform cell isolation without using cell-specific biomarkers. Typically, size-based separation is a classical method for cell isolation from a complex biological sample [26]. For example, the ability to isolate CTCs from a whole blood sample is based on their larger size than that of the blood cells [27]. The emergence of microfluidic technologies has enabled this technique more prominent for cell isolation through a precisely controlled microfabrication. To date, there are many microfluidic techniques developed for physical traps of target cells, including microfabricated structures, hydrodynamic force, acoustophoresis and dielectrophoresis in the microfluidic devices. In this part, we talk about several most popular techniques for microfluidic cell isolation.

3.2.1 Microfabricated Structures for Cell Trap

The development of microfabrication has allowed realizing microfluidic cell isolation by the fabricated microstructures in the microchannels. In the early 2000s, fabrication of microdocks in microfluidic channels is reported to perform single-cell isolation from a bulk cell suspension for single-cell analysis (Fig. 3.1a) [19]. The microdocks are typically designed and fabricated to match the size of target cells. This study attracts lots of attentions and leads to new insights into precise control of cells in microfluidics for biological analysis. Based on this technique, microfluidic cell pairing and fusion were achieved through passive hydrodynamic cell traps in a microdevice containing a dense array of weirs [20]. Two different types of cells were immobilized and paired into a two-cell-matched microstructure array via a 3-step loading protocol, and then fused and analyzed on-chip with high throughputs and efficiency. Such microfluidic method realizes an excellent control of cell pairing and fusion that cannot be achieved by previous conventional standard fusion techniques.

The size difference among cells has been proposed to achieve selective isolation of cells in microfluidics. For example, a micromachined device was designed to physically fractionate whole blood for rare cell isolation from peripheral circulation blood. Various widths of channel gaps (5, 10, 15, and 20 μm) enable to isolate cells based on size and deformability between CTCs and blood cells in the whole blood sample [28]. In addition, a microfluidic technology that integrates specialized bifurcating traps was developed to capture CTC clusters independently of tumor-specific biomarkers from whole blood samples (Fig. 3.1b) [29]. The CTC clusters were differentiated from single cells in blood through the unique geometries of cellular aggregates in the microfluidic chip. This technique was used to test patients' blood with metastatic breast or prostate cancer or with melanoma.

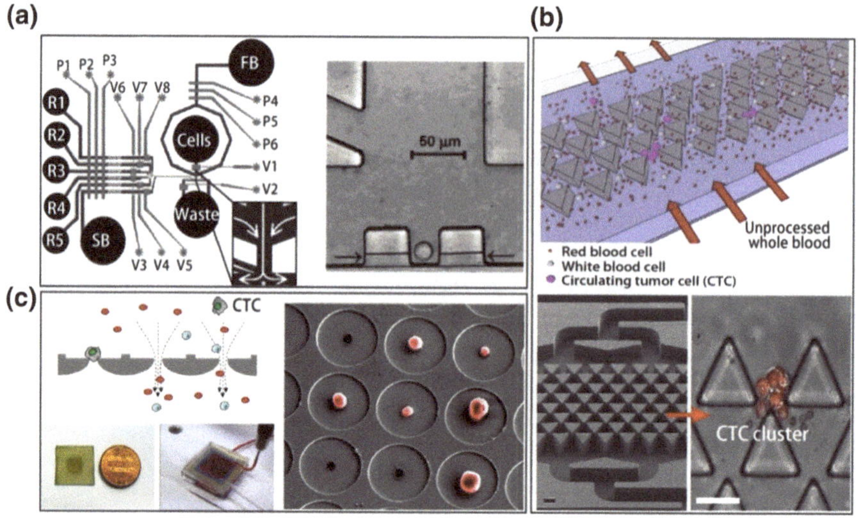

Fig. 3.1 Microfluidic cell capture in the microfabricated structures. **a** Single-cell trap analysis in a microdock in the microchip (Reprinted with permission from Ref. [19]. Copyright American Chemical Society 2003.). **b** Microfluidic capture of CTC clusters through specialized bifurcating traps (Reprinted with permission from Ref. [29]. Copyright Nature Publishing Group 2015.). **c** CTC isolation device equipped with the size-selective microcavity array (Reprinted with permission from Ref. [30]. Copyright American Chemical Society 2010.)

Another development is to fabricate size-selective microcavity arrays in a microfluidic device [30]. Using this microchip, it is convenient to perform highly efficient and rapid isolation of CTCs from the whole blood sample based on the size and deformability difference. Thus, the ability of size-dependent cell separation has been demonstrated to be valuable for microfluidic cell isolation in the complex cell sample.

Apart from the microcavity, designing microfilters becomes an excellent strategy for efficient cell isolation in microfluidics [31, 32]. In a three-dimensional microfilter device, two layers of parylene membranes with pores and gap are precisely integrated into the microchip by photolithography. The top-bottom design of microfilters plays a key role to minimize the cell hurts and increase the cell capture efficiency. A similar method is reported by fabricating silicon microsieves in a microfluidic device to rapidly isolate tumor cells from whole blood samples, following by fluorescence analysis of the captured cells on-chip (Fig. 3.1c) [30]. Furthermore, soft photolithography also allows a more flexible microfabrication to achieve this performance. As such, a PDMS porous membrane integrating with different pore sizes in a microfluidic device was developed to facilitate the cell isolation from a whole blood sample [33, 34]. Then, this technique was further employed for purification and analysis of hematopoietic stem cells from bone marrow [35]. The results have shown that the microfabricated structures are capable to efficiently isolate the cells based on their size difference. Such techniques have obvious advantages in achieving a rapid and simple detection of target cells from a large population of cells from complex samples.

3.2.2 Hydrodynamic Force for Cell Separation

Microfluidic hydrodynamics have emerged as a useful strategy for cell isolation since there are various size and weight among different types of cells. Microfluidic technologies allow operating hydrophoresis in a fluidic network to achieve label-free cell separation by the means of kinetic, equilibrium and elution [36]. For example, a novel strategy for high-throughput and label-free isolation of cells was developed by using hydrodynamic force in a microfluidic system [37]. In this study, different sizes of cells have an irreversible migration into the microscale vortices and thus are trapped in the reservoirs. This physical approach for cell isolation in the microfluidic device is superiors in rapid analysis of cell samples. A development of multi-orifice flow fractionation is also an effective approach for cancer cell separation from blood samples [38]. The design of microstructures and dimensions in the microfluidic system determines the hydrodynamic operations and separation efficiency in the cell isolation performance. Large-sized cancer cells traveled through the side channel at the first stage and directed to the middle channel for cell enrichment in final. Further, a multiple parallel flow fractionation device was developed to continuously isolate the CTCs from the whole blood of breast cancer patients based on the cell size difference and hydrodynamic forces. It is worth to mention that the velocity distribution in the fluidic network can be calculated by a computational simulation, which could facilitate to design an optimized microfluidic system for cell isolation.

Microfluidic centrifugation has been developed as an excellent alternative of hydrodynamic forces for rapid isolation of cells based on the size difference. Two examples of this concept have been demonstrated to isolate CTCs from the whole blood by inherent centrifugal forces in the spiral microchannels (Fig. 3.2) [39, 40]. The proposed microscale centrifugation is tunable and precisely controllable in a predesigned microchip. This technique shows greatly promising in the clinical applications due to its low cost and convenience. In addition, an integrated hydrodynamic microfluidic system was employed for cell isolation by performing deterministic lateral displacement, inertial focusing and magnetophoresis [41–43]. This microfluidic technology enables to isolate the cells by either dependent or independent of tumor membrane epitopes. The microfluidic device can be further combined with automatable cell sorting technology and molecular assays of captured cells for precision clinical cancer diagnosis. Thus, hydrophoresis technology has the advantages of low cost, ease of operation, and marker-free, therefore it has been recognized as a promising strategy for microfluidic cell isolation.

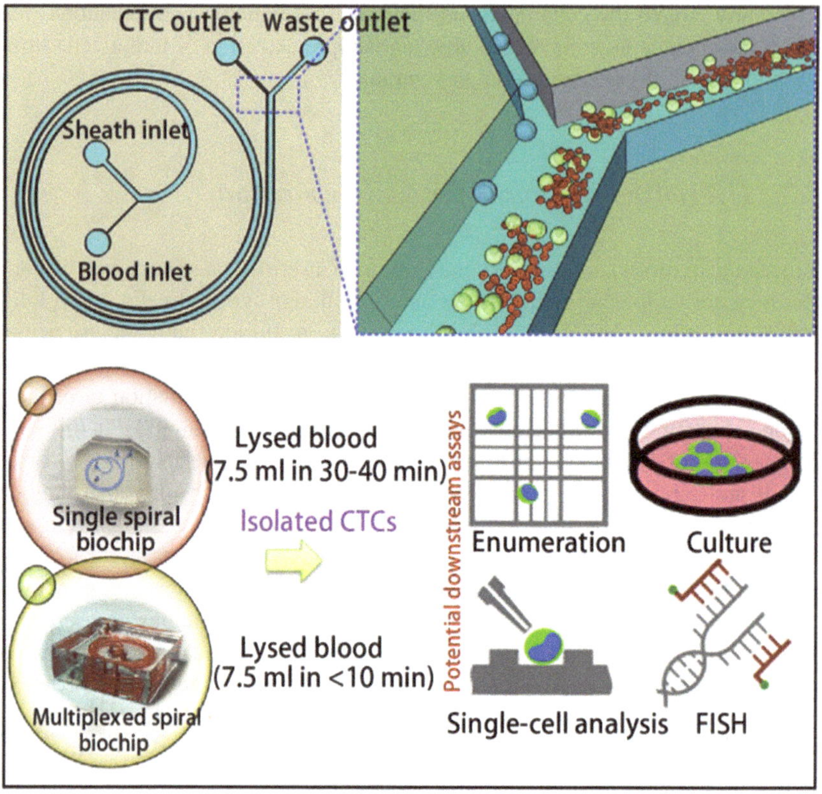

Fig. 3.2 Hydrodynamic isolation of cells in microfluidics. CTC enrichment and analysis from blood samples by a spiral channel (Reprinted with permission from Ref. [39]. Copyright Nature Publishing Group 2015.)

3.2.3 Surface Acoustic Wave-Based Cell Isolation

Surface acoustic wave has been widely used in the microfluidic system to achieve a sensitive and label-free cell isolation [44]. The concept of acoustophoresis has been proposed to operate cells in microfluidics [45]. In a microfluidic acoustophoresis operation, a surface acoustic wave is controlled by an electrical signal onto the integrated piezoelectric materials in the microfluidic device. The surface acoustic wave interacts and changes the mass and viscosity of target cells. Petersson et al. reported a free flow acoustophoresis method to perform continuous separation of mixed cells in a single microchip (Fig. 3.3a) [45]. They used acoustic forces to isolate the target cells based on their difference in size and density. It has led to a convenient, effective and reproducible method for microfluidic cell isolation. The standing surface acoustic wave has also been used for the microfluidic flow cytometer [46, 47]. Using this technique, cells are focused inside a microchannel by

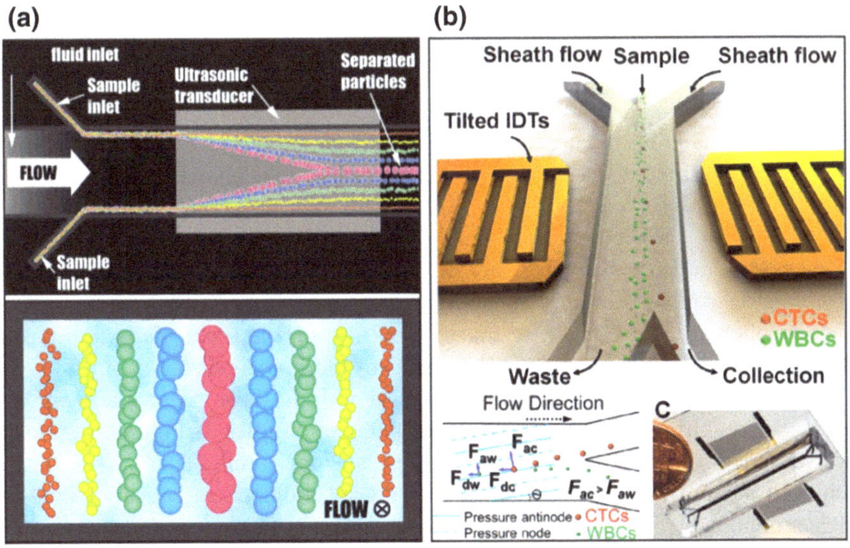

Fig. 3.3 Surface acoustic wave-based microfluidic cell isolation. **a** Free flow acoustophoresis for cell separation (Reprinted with permission from Ref. [45]. Copyright American Chemical Society 2007.). **b** High-throughput tilted-angle standing surface acoustic waves (taSSAW) device for cancer cell separation (Reprinted with permission from Ref. [48]. Copyright National Academy of Sciences, U.S.A. 2015.)

using the standing surface acoustic wave, and then sorted or separated by the signals from a laser-induced fluorescence detection system. As well as we know, this microfluidic cytometer technique has been adopted by Thermo Fisher Scientific Inc. for future commercial applications. Moreover, the acoustic separation of CTCs has been developed as a minimal invasive way for clinical cancer diagnosis (Fig. 3.3b) [48]. This method is demonstrated to be safe, label-free and contactless nature. By combination of other microfluidic technologies and molecular detection techniques, we believe that the acoustic-based approach can be further developed to achieve higher precision and efficiency for microfluidic cell isolation applications in the future [49].

3.2.4 Dielectrophoresis-Based Cell Sorting

In addition to the acoustophoresis, dielectrophoresis is another popular physical technique used for cell isolation, especially by combining with the microfluidic platform [50]. The dielectrophoresis technology relies on the difference of electrical properties of cellular composition, membrane and size. It can be employed to differentiate and isolate the cells from a cell population due to different polarization-induced dielectric forces exerted on cells. A dielectrophoretic device called ApoStream™ was developed to isolate CTCs by exploiting the differences in biophysical characteristics between normal and cancer cells, and healthy blood cells (Fig. 3.4a) [51]. In the microfluidic flow channel, efficient isolation and enrichment of CTCs from blood samples were performed by optimizing the key operating parameters, including frequency, voltage, flow rates, and buffer formulations. The ApoStream technology used to capture viable cancer cells allows achieving further biochemical analysis for precision medicine. This dielectrophoresis method has the advantages of antibody-independent CTCs isolation (Fig. 3.4b) [52, 53]. A contactless dielectrophoresis with low frequency was further developed as a new technique to isolate cells in a microfluidic device [54]. It is the first contactless dielectrophoresis device that is operated below 100 kHz to successfully manipulate the human leukemia cells without the effects of red blood cells.

The dielectrophoresis-based cell isolation device has been successfully applied for high-throughput screening of single cells in microfluidics. Using this technique, droplet-based microfluidics is used to encapsulate single cells. Then cell sorting is achieved in a Y-shared junction where single-cell droplets are actively sorted by dielectrophoresis [55]. This dielectrophoretic device is further developed as a new generation of flow cytometer for conducting gene sequencing [56]. Moreover, the optically-induced-dielectrophoretic force was fully integrated into a microfluidic platform for high-purity and label-free isolation of CTCs [57]. Such device has been demonstrated particularly powerful in exploiting the subsequent cell-based biochemical assays to the captured CTCs. Dielectrophoresis-based cell isolation in microfluidics provides great promising in separating and sorting targeted cells with an electrical integration, easy automation, and high performance, thus it offers lots of opportunities for commercial applications.

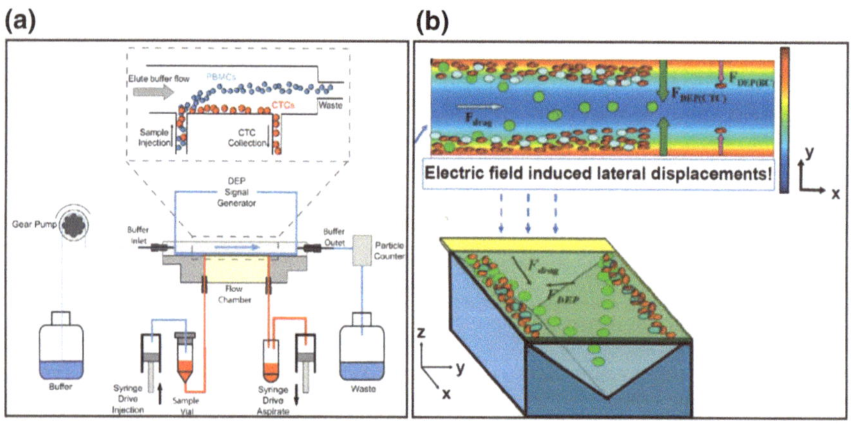

Fig. 3.4 Dielectrophoresis-based cell separation in microfluidic devices. **a** Antibody-independent isolation of viable cancer cells by a dielectrophoretic device (Reprinted with permission from Ref. [51]. Copyright American Institute of Physics 2012.). **b** Antibody-free isolation of CTCs based on 3D lateral dielectrophoresis in a microfluidic chip (Reprinted with permission from Ref. [53]. Copyright The Royal Society of Chemistry 2015.)

3.3 Affinity-Based Cell Isolation in Microfluidics

As physical traps-based cell isolation in microfluidics is often nonspecific, thus combination of biological affinity-based cell separation in the microfluidic devices is very important for their practical biomedical applications. In this part, we will introduce the use of cell-specific antibody, aptamer and bio-nano-interface in microfluidics for efficient cell isolation and recognition.

3.3.1 Antibody-Based Cell Recognition

It is well known that tumor-specific antigens are over-expressed on the surface and inside of cancer cells. Thus it makes researchers possible to specially recognize and isolate target cancer cells from whole blood since there are many types of other cells in such a complex sample [58]. In fact, conventional physical methods for cell capture in the microfluidic device are not able to isolate cancer cells with biological specificity; the combination of affinity-based cell recognition with microfluidic technologies is particularly interesting and useful in the practical cell-based assays. In 2007, Toner et al. firstly reported an antibody-coated microchip for isolation of rare circulating tumour cells (CTCs) from whole blood samples in cancer patients (Fig. 3.5a) [23]. EpCAM antibody were modified on the micropots in the microchannels for specific recognition of metastatic cancer cells (metastatic lung, prostate, pancreatic, breast and colon cancer). The same group further developed a microvortex-generating herringbone-chip to improve the performance efficiency of CTCs isolation (Fig. 3.5b) [59]. It is interesting that a rational design of

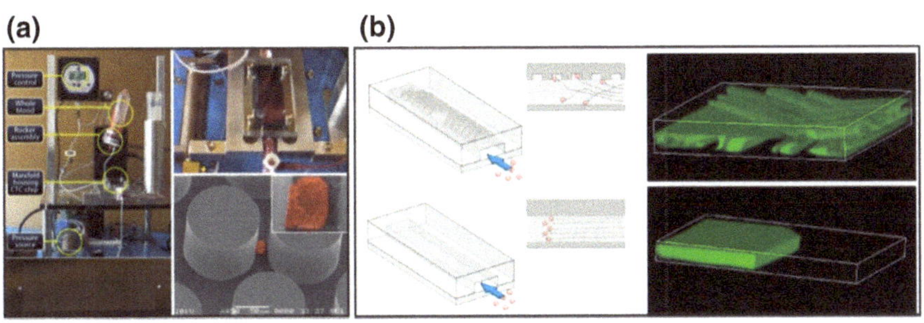

Fig. 3.5 Microfluidic isolation of CTCs from whole blood using an antibody-modified chip. **a** EpCAM antibody-coated microposts for CTCs capture (Reprinted with permission from Ref. [23]. Copyright Nature Publishing Group 2007.). **b** Antibody-combined with herringbone-chip for CTCs isolation in a chip (Reprinted with permission from Ref. [59]. Copyright National Academy of Sciences, U.S.A. 2010)

herringnone-chip enables to significantly increase the interactions between target cancer cells and antibody-modified microchannel surface through a generation of microvortices, thus leading to achieve much higher cell capture capability in microfluidics.

In clinical applications, label-free enumeration of CTCs is useful to assess the CTCs level in the whole blood samples. A conductivity sensor was designed in the microfluidic chip after the CTCs were selectively and specifically isolated through an antibody-functionalized microchip [60]. The detection of CTCs was enumerated when the single tumor cells were traveling through the detection electrodes based on the solution conductivity. The sensitivity and specificity of the CTCs detection can be also improved by introducing graphene oxide nanosheets on a patterned gold surface in the microfluidic device [61]. Functionalized graphene oxide nanosheets on the flower-shaped gold patterns on the substrates can sense the CTCs after they are captured by the coated antibodies in the microfluidic chip. Enumeration, electrokinetic manipulation and molecular analysis of CTCs can also be conducted in an integrated microfluidic device after the CTCs were isolated by the modified antibodies [62]. Such antibody-based multifunctional microchips are very valuable in the clinical disease diagnosis by integrating multi-step analytical processes in a microfluidic system.

3.3.2 Aptamer-Specific Cell Capture

Aptamer is a kind of artificial chemical antibody that has high affinity in molecular recognition [63]. Aptamers are commonly composed of single-stranded oligonucleotides that are sorted from a method called systematic evolution of ligands by exponential enrichment (SELEX) [64, 65]. As aptamers have been demonstrated to be efficient to specifically capture target cancer cells, they have been combined with microfluidics for high-efficiency isolation and recognition of cancer cells [66, 67]. For example, Tan et al. reported an aptamer-modified microfluidic device to capture rare cancer cells (Fig. 3.6a) [68, 69]. High purity and efficiency of capturing cancer cells are achieved since the cell-specific aptamers are immobilized on the microfluidic channels for effective recognition with the cancer cells. Further, this group developed a unique strategy for simultaneous enrichment, sorting and detection of multiple cancer cells from a cell mixture using an aptamer-based microfluidic device [69]. In a typical experiment, three cell-specific aptamers were respectively modified onto the three different regions in the microchannels, which enabled to multiplexed detection of three types of cancer cells from a complex sample in a simple microfluidic chip. In addition to these strategies via aptamer-functionalization on the plane microchannels, a platform approach by synthesizing bioinspired multivalent DNA aptamer 3D-networks in a microfluidic channel was developed to capture and isolate the cells (Fig. 3.6b). In this study, the

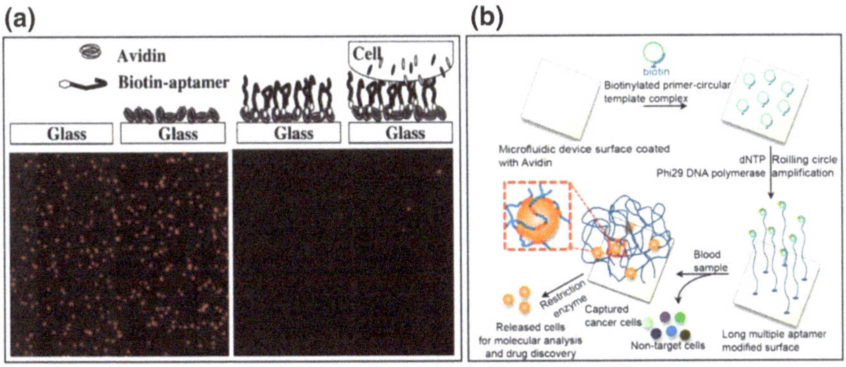

Fig. 3.6 Aptamer affinity-based cell isolation in the microfluidic device. **a** Surface immobilized aptamer selectively captures target cells (Reprinted with permission from Ref. [69]. Copyright American Chemical Society 2009.). **b** Bioinspired multivalent DNA network for capture and release of cells (Reprinted with permission from Ref. [70]. Copyright National Academy of Sciences, U.S.A. 2012.)

cell-specific DNA aptamer 3D-networks were demonstrated to have significantly high cell capture efficiency than the conventional monovalent antibody and aptamer, leading to a high purity of captured cells [70].

To realize a precision cancer diagnosis, it is also very important to further identify the captured cells after cell isolation and recognition in microfluidics. To date, various analytical methods, such as fluorescence, chemiluminescence PCR, immunestaining, have been developed for post-identification. The combination of aptamer-based cell isolation in microfluidic channels and chemiluminescence detection is reported to establish this idea in a convenient way [71]. Although the aptamer-based microfluidics is very promising in cell isolation and recognition for biomedical applications, the limited aptamer library and low sorting efficiency of cell-specific aptamers are still challenged for the future applications.

3.3.3 Cell Capture on Bio-Nano-Interfaces

Although microfluidic hydrodynamics by the microfabricated structure in the chip has been demonstrated to increase the interactions between the cells and antibody-coated surface, the efficiency of cell capture and recognition is still not satisfied. To address this issue, the development of nanotechnology-based bio-interfaces has attracted particular interests [71]. Wang et al. reported the first time to use three-dimensional nano-structured substrates in the microchannels to perform an efficient capture of CTCs [72, 73]. In that study, silicon-nanopillar array was fabricated to enhance local topographic interactions between nanoscale components of the cell surface and the coated antibodies on the three-dimensional nano-interface. This artificial microchip leads to significantly enhance the capability of cell capture in microfluidics. An integration of chaotic micromixers and nano-structured silicon substrates in microfluidic device was further developed to perform the CTCs detection [73]. Such a powerful CTC-capture platform has two functional components: antibody-coated silicon nano-pillar substrates for recognizing and capturing antigen-expressed cancer cells, and a serpentine chaotic mixing channel in the microchip to encourage cell-substrate interactions. Another sample is the use of multivalent DNA nanospheres to enhance cancer cell capture in the microfluidic devices (Fig. 3.7) [74]. This microfluidic technique enables to operate the cells in blood samples at the microscale that is not possible in the conventional technique, because the design of bio-nano-interfaces is very useful for the practical cell isolation and recognition in the experiments [75].

To achieve further analysis of the captured CTCs, a laser microdissection (LMD) technique was combined with polymer nanosubstrates-embedded microchip for single-CTC isolation [76, 77]. Using this method, single-cell genotyping can be performed to verify the CTC's role in the metastasis. The nanoVelcro-LMD technology was also used for mutational analysis and whole exome sequencing based on the cell isolation and recognition in microfluidics. Another advanced development is to graft thermo-responsive polymer brushes onto the silicon nanopillars on the substrates in the microfluidic device [22, 66]. This new generation microfluidic device is able to capture and release CTCs at 37 °C and 4 °C, respectively. Rapid isolation of viable CTCs with high purity can be effectively conducted for further molecular characterization and functional analysis. Thus, the combination of nanotechnology with microfluidics has shown great promising for high-efficiency cell capture and will have commercial biomedical applications for CTCs detection [78].

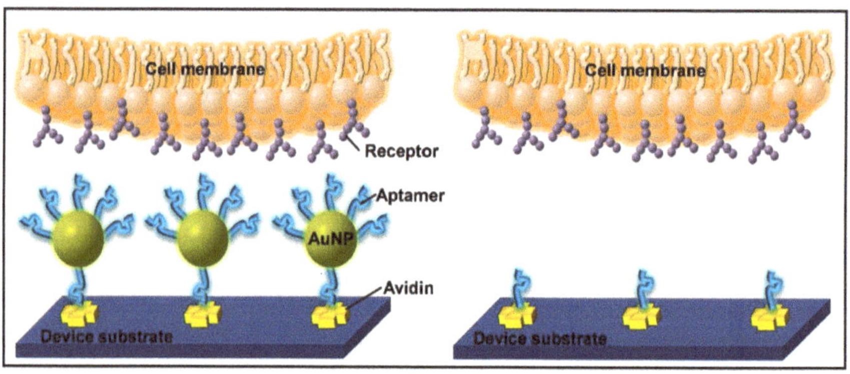

Fig. 3.7 Microfluidic cell capture on bio-nano-interfaces. Enhanced cell capture using AuNP-aptamer modified surface (Reprinted with permission from Ref. [74]. Copyright American Chemical Society 2013.)

3.4 Recent Technologies for Biomedical Applications

The fast development of microfluidic technologies for cell isolation and recognition has promoted the biomedical applications in several research fields. As described in the above sections, microfluidic device enables to effectively isolate target cells from a large population of cells, either by physical traps or biological recognition. These manipulations can be precisely performed in a designed microchip, and several operation steps can be integrated into a single device, including cell introduction, recognition, separation, and post-analysis. This capability has created three important applications in the basic and fundamental researches. In the section, we present three major biomedical applications in this field including CTCs detection, cell-based biological assays and stem cell studies.

3.4.1 CTCs Isolation, Recognition and Detection

Isolation and analysis of rare CTCs is an advanced technique to identify the cancer metastasis in clinical cancer diagnosis. Because there are only a few CTCs in the blood of patients with metastatic cancer, the isolation of several numbers of CTCs from 10^9 hematologic cells is still technically difficult. Therefore, a platform for efficient isolation and recognition of CTCs is highly desired. Recently, microfluidic technologies have been increasingly emerged as a powerful tool to perform isolation and analysis of CTCs from the complex blood samples. The microfluidic platform that enables cell isolation and recognition has been reported to have important biomedical application for CTCs detection.

Affinity-based cell isolation approaches are usually conducted by cell-specific antibodies that can recognize the biomarkers over-expressed on the tumor cell surface. Two commonly used antibodies in the microfluidic CTCs isolation from the peripheral blood are epithelial cell adhesion molecule (EpCAM) and prostate-specific membrane antigen (PSMA) antibody. For example, the first generation of CTC-chip was developed for efficient and selective isolation of viable CTCs by coating the EpCAM antibody onto microposts, which is applicable to diagnose the patients with metastatic prostate, lung, breast and colon cancers [23]. A method for prostate CTCs detection was developed by using a microfluidic device, which used geometrically enhanced immune-capture and an antibody of PSMA to achieve the aims of high efficiency, purity and specificity [79]. Highly efficient isolation and recognition of the CTCs from whole blood samples can be achieved by designing a microvortex-generating herringbone-chip. The nanostructured biological interface coating with cell-specific antibodies has been demonstrated to enhance the interactions between and cell surface biomarkers and microchips, thus significantly improve the CTC capture capability in microfluidics (Fig. 3.8a) [61]. Besides, aptamer-modified microchips are also developed for CTCs isolation and recognition [75, 80, 81]. The programmable DNA aptamers are

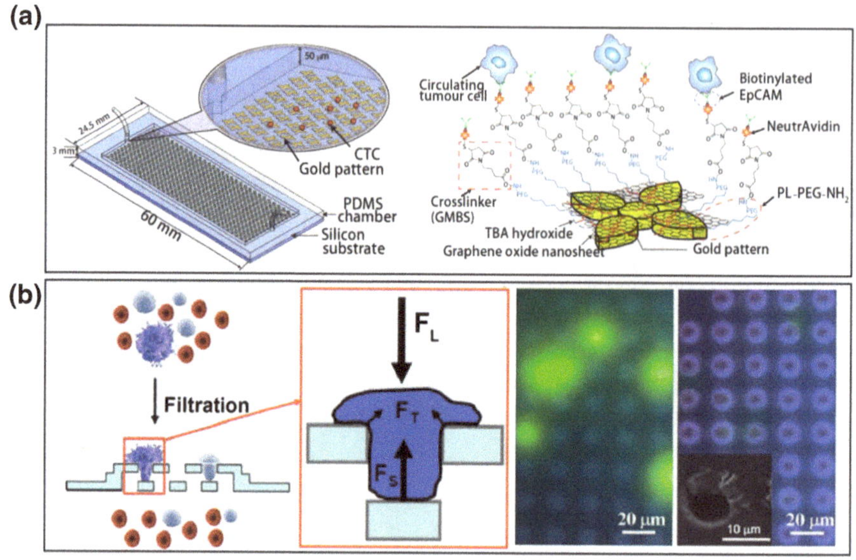

Fig. 3.8 Microfluidic cell isolation platform for CTCs detection. **a** Graphene oxide nanosheets functionalized microchips for sensitive capture of CTCs (Reprinted with permission from Ref. [61]. Copyright Nature Publishing Group 2013.). **b** Microfluidic isolation of visible CTCs using a 3D microfilter device (Reprinted with permission from Ref. [31]. Copyright Springer Science + Business Media, LLC 2011.)

very flexible to be modified in the microchips, which enable for simultaneous isolation and enrichment of multiple CTCs [69]. The combination of aptamers and microfluidic technologies is very promising in clinical applications due to simple preparation procedures and easy functionalization.

In addition to the biological affinity-based cell capture, microfluidic isolation and recognition of CTCs can also be achieved by the physical methods for clinical applications. Such method is convenient, low cost, label free and easy to be applied to rapid separation of CTCs based on the size difference. For example, the CTCs from whole blood samples can be isolated by fabricating microfilters in the microchip [82–84]. The geometry, density and size of the microfilter pores can be precisely controlled to match the CTCs for size-selection and purity during the microfluidic isolation. The captured CTCs can be further analyzed to assist the clinical applications (Fig. 3.8b) [31]. Moreover, the capture of CTCs from blood samples can be also performed by a classical hydrodynamic operation in microfluidic centrifugation device [42, 85, 86]. Using the centrifugal force, the CTCs can be isolated and enriched into a designated microchannel. Compared with the microfilters or micro-centrifugation-based technique, acoustophoresis and dielectrophoresis in microfluidics are two different methods that rely on an automatic mechanical control for operation. Especially, the surface acoustic wave and dielectric force have been integrated for microfluidic flow cytometer that can be employed for fast screening and isolation of the CTCs. These will lead to the development of automatic micro-total-analysis-system for the separation, enumeration and post-analysis of the CTCs in the real sample, thus show great promising for the biomedical applications in cancer diagnosis.

3.4.2 Cell-Based Biological Assays

The developed technologies for microfluidic cell isolation and recognition have enabled cell-based biological assays from a complex sample. In the conventional methods for cell analysis, it is not easy to isolate target cells from a cell mixture, and it is also difficult to precisely operate and control the cells for the following biological assays. Unlike these techniques, microfluidic technologies have created a powerful platform that is possible to simultaneously perform several experimental steps in a single microchip, including cell isolation, pretreatment, intracellular analysis and data acquisition [24]. For example, in the early development, cells can be isolated by hydrodynamic traps in microfabriacted structures in the microdevice [19]. Proof-of-concept applications have been performed for cell-based biological assays including cell viability test, receptor-mediated Ca^{2+} measurements, and ionophore-mediated intracellular Ca^{2+} measurements. High-throughput single-cell analysis was conducted by capturing single cells in microwell arrays [87].

To achieve cell-based biological assays, an aptamer-encoded microwell array was fabricated to realize targeted cell isolation and recognition in the microfluidic chip [88]. Fluorescence kinetics of cellular enzyme in single cells was achieved by using this microfluidic technique. In addition, multiplexed and quantitative analysis of secreted proteins in single cells was achieved based on specific recognition and controlled capture of cells in the microfluidic chip (Fig. 3.9a) [89]. Microfluidic technologies are also very useful in the cell sorting and analysis. Using the microfluidic technique, target tumor cells were selectively captured in the microfluidic device and genetic analysis of cells was conducted based on the microfluidic cell isolation technique [90]. Moreover, the development of automatic microfluidic cell sorting system integrated with dielectrophoresis is also very useful for rapid analysis of single cells (Fig. 3.9b) [55]. High throughput screening applications, including single-cell and single-molecule assays, can be performed by the droplet-based microfluidics [91, 92]. The development of microfluidic technologies has provided a powerful platform for the cell-based biological assays.

3.4.3 Stem Cell Purification and Screening

Stem cell studies have been recognized as a revolutionary opportunity to understand the scientific phenomena for human disease treatment [93]. The basic technique in this research field is to develop an efficient platform for stem cell purification,

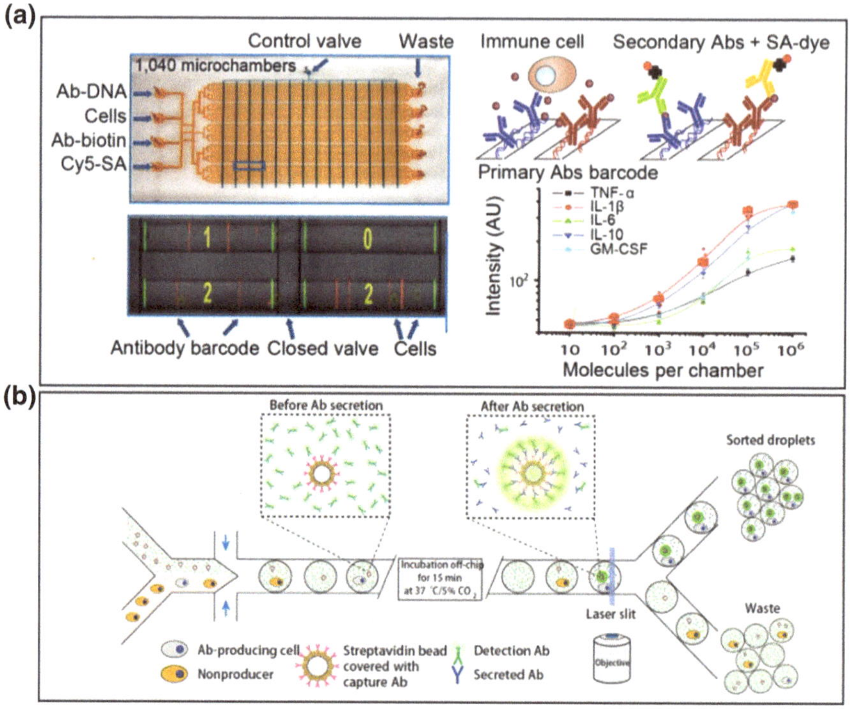

Fig. 3.9 Microfluidic cell isolation and analysis. **a** Single-cell capture and protein secretome analysis on a chip (Reprinted with permission from Ref. [89]. Copyright Nature Publishing Group 2011.). **b** Droplet-based microfluidics for cell sorting and analysis (Reprinted with permission from Ref. [55]. Copyright Nature Publishing Group 2013.)

because the purity of fully reprogrammed induced pluripotent stem cells is not high, and there are still many undifferentiated stem cells in the stem cell culture process. This is a major challenge in the conventional techniques for stem cell studies. Currently, microfluidic cell isolation technologies have been applied for stem cell purification and screening. To study in the heterogeneity in stem cell populations, a microfluidic platform containing large-scale separated nanoliter chambers was developed to culture single primitive mouse hematopoietic cells [94]. This study presents a new avenue to investigate otherwise inaccessible mechanisms in stem cell growth and fate decision. The strategy for label-free isolation of stem cells in a cell differentiation experiment is exploited by using the difference in the adhesion strength between parental and differentiated cells (Fig. 3.10a) [95]. This microfluidic technology has led to an easy strategy for stem cell purification in the routine stem cell culture. In addition, highly pure stem cell isolation was also conducted by designing a porous-membrane integrated microdevice for stem cell cocultures (Fig. 3.10b) [96]. In this design, the stem cells and feeder layer cells were respectively cultured on the top and bottom layers of the 3D microchip, which can serve as an effective and simple method for stem cell purification.

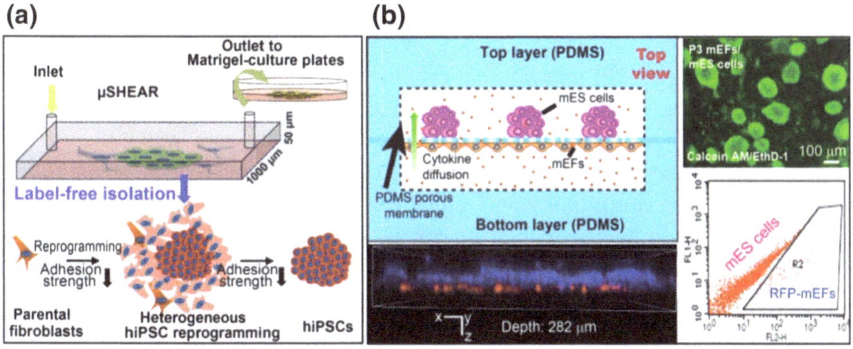

Fig. 3.10 Microfluidic cell purification and analysis for stem cell researches. **a** Label-free isolation of iPSCs based on adhesion strength difference in microfluidic devices **b** PDMS porous membrane-assembled microfluidic co-culture platform for stem cell isolation (Reprinted with permission from Refs. [95, 96]. Copyright Nature Publishing Group 2013.)

Microfluidic cell isolation technology has also been proposed to carry out stem cell analysis. For example, using the size-based cell isolation, hematopoietic stem cells were selectively separated from bone marrow by a PDMS porous membrane-incorporated microfluidic chip [35]. The stem cells were further stained with specific antibodies for further analysis. Using droplets-based microfluidic technology, sequencing analysis of single stem cells was conducted by using an inDrop microfluidic platform. Single stem cells were encapsulated into droplets with RNA sequencing reagents [56], which enables to perform a high-throughput analysis of single cells as the next-generation sequencing. Thus, microfluidic technologies have presented many other advances for stem cell studies and hold great promising in the future.

3.5 Conclusion and Future Perspective

In this Chapter we have introduced the basic concepts and recent development of microfluidic technologies employed for cell isolation and recognition. Significant progress has been made in developing advanced physical approaches and biological affinity strategy for microfluidic cell isolation and recognition in the past years. Recently researchers have made efforts to better understand more efficient cell-microstructure interactions with hydrodynamic traps and biological recognition by the combination of microfabrication, nanotechnology and cell engineering. This will promote the fundamental understanding, device design, and practical application for cell isolation and recognition using microfluidic technologies. Moreover, practical biomedical applications including CTCs detection, cell-based biological assays and stem cell researches, have demonstrated the potential of this technology in the basic and applied biological researches.

Although microfluidic technologies have been well developed, there are still several challenges in achieving highly efficient cell isolation and specific cell recognition. In particular, microfluidic capture of rare circulating tumor cells from whole blood samples is actually very difficult and low efficiency for practical clinical applications. The efforts on improving sensitivity and reliability of rare CTCs detection in microfluidics are much needed. Further development likely requires the use of cell biology and signal amplification technique to identify the cancer biomarkers at the molecular or genetic level. Furthermore, more attentions will also be paid in the growing emphasis on scientific insights into cell-based biological assays. Advanced techniques for cell operation and analysis including optical tweezers and single-cell sequencing could be employed to accurately assay the heterogeneity among cells in a large population [97, 98]. Another challenge is subsequent cell sorting apart from microfluidic stem cell separation. The simple purification of stem cells in cultures is far from for fundamental insights into stem cell biology. One opportunity is the use of droplet-based microfluidics for rapid analysis of stem cell differentiation and molecule expressions.

The future direction is to develop a micro-total-analysis-system framework that is able to create microfluidic devices integrating several steps of an assay into a single chip, including cell selection, cell lysis, and cell analysis. In addition, recent new technologies of 3D printing technology and nanotechnology would be available to mimic complex biochemistries and geometries in microfluidic channels and enhance the capability for cell isolation and recognition [99]. Their current biomedical applications will effectively compete with the conventional laboratory-scale technologies for commercial purposes of cancer diagnosis and cell biology studies [100]. Highly integrated microfluidic devices will offer new commercial opportunities in the applied biomedical and fundamental cell biology studies, and low-cost point-of-care microdevices will be possible to go to the market in the future.

References

1. El-Ali J, Sorger PK, Jensen KF (2006) Cells on chips. Nature 442:403–411. doi:10.1038/nature05063
2. Chen Q, Wu J, Zhang Y, Lin JM (2012) Qualitative and quantitative analysis of tumor cell metabolism via stable isotope labeling assisted microfluidic chip electrospray ionization mass spectrometry. Anal Chem 84:1695–1701. doi:10.1021/ac300003k
3. Wang BL, Ghaderi A, Zhou H, Agresti J, Weitz DA, Fink GR, Stephanopoulos G (2014) Microfluidic high-throughput culturing of single cells for selection based on extracellular metabolite production or consumption. Nat Biotechnol 32:473–478. doi:10.1038/nbt.2857
4. Sackmann EK, Fulton AL, Beebe DJ (2014) The present and future role of microfluidics in biomedical research. Nature 507:181–189. doi:10.1038/nature13118
5. Chen Q, Utech S, Chen D, Prodanovic R, Lin JM, Weitz DA (2016) Controlled assembly of heterotypic cells in a core-shell scaffold: organ in a droplet. Lab Chip 16:1346–1349. doi:10.1039/c6lc00231e

6. Dittrich PS, Manz A (2006) Lab-on-a-chip: microfluidics in drug discovery. Nat Rev Drug Discov 5:210–218. doi:10.1038/nrd1985
7. Chen Q, He Z, Liu W, Lin X, Wu J, Li H, Lin JM (2015) Engineering cell-compatible paper chips for cell culturing, drug screening, and mass spectrometric sensing. Adv Healthc Mater 4:2291–2296. doi:10.1002/adhm.201500383
8. Gobaa S, Hoehnel S, Roccio M, Negro A, Kobel S, Lutolf MP (2011) Artificial niche microarrays for probing single stem cell fate in high throughput. Nat Methods 8:949–955. doi:10.1038/nmeth.1732
9. Schasfoort RB, Schlautmann S, Hendrikse J, van den Berg A (1999) Field-effect flow control for microfabricated fluidic networks. Science 286:942–945. doi:10.1126/science.286.5441.942
10. Tsang VL, Bhatia SN (2004) Three-dimensional tissue fabrication. Adv Healthc Mater 56 (11):1635–1647. doi:10.1016/j.addr.2004.05.001
11. Wu J, He Z, Chen Q, Lin JM (2016) Biochemical analysis on microfluidic chips. Trends Anal Chem 80:213–231. doi:10.1016/j.trac.2016.03.013
12. Thorsen T, Maerkl SJ, Quake SR (2002) Microfluidic large-scale integration. Science 298:580–584. doi:10.1126/science.1076996
13. Gomez-Sjoberg R, Leyrat AA, Pirone DM, Chen CS, Quake SR (2007) Versatile, fully automated, microfluidic cell culture system. Anal Chem 79:8557–8563. doi:10.1021/ac071311w
14. Young EW, Beebe DJ (2010) Fundamentals of microfluidic cell culture in controlled microenvironments. Chem Soc Rev 39:1036–1048. doi:10.1039/b909900j
15. Yu L, Ng SR, Xu Y, Dong H, Wang YJ, Li CM (2013) Advances of lab-on-a-chip in isolation, detection and post-processing of circulating tumour cells. Lab Chip 13:3163–3182. doi:10.1039/c3lc00052d
16. Esmaeilsabzali H, Beischlag TV, Cox ME, Parameswaran AM, Park EJ (2013) Detection and isolation of circulating tumor cells: principles and methods. Biotechnol Adv 31:1063–1084. doi:10.1016/j.biotechadv.2013.08.016
17. Shields CW, Reyes CD, Lopez GP (2015) Microfluidic cell sorting: a review of the advances in the separation of cells from debulking to rare cell isolation. Lab Chip 15:1230–1249. doi:10.1039/c4lc01246a
18. Chen Y, Li P, Huang PH, Xie Y, Mai JD, Wang L, Nguyen NT, Huang TJ (2014) Rare cell isolation and analysis in microfluidics. Lab Chip 14:626–645. doi:10.1039/c3lc90136j
19. Wheeler AR, Throndset WR, Whelan RJ, Leach AM, Zare RN, Liao YH, Farrell K, Manger ID, Daridon A (2003) Microfluidic device for single-cell analysis. Anal Chem 75:3581–3586
20. Skelley AM, Kirak O, Suh H, Jaenisch R, Voldman J (2009) Microfluidic control of cell pairing and fusion. Nat Methods 6:147–152. doi:10.1038/nmeth.1290
21. Gao Y, Li W, Pappas D (2013) Recent advances in microfluidic cell separations. Analyst 138:4714–4721. doi:10.1039/c3an00315a
22. Hou S, Zhao H, Zhao L, Shen Q, Wei KS, Suh DY, Nakao A, Garcia MA, Song M, Lee T, Xiong B, Luo SC, Tseng HR, Yu HH (2013) Capture and stimulated release of circulating tumor cells on polymer-grafted silicon nanostructures. Adv Mater 25:1547–1551. doi:10.1002/adma.201203185
23. Nagrath S, Sequist LV, Maheswaran S, Bell DW, Irimia D, Ulkus L, Smith MR, Kwak EL, Digumarthy S, Muzikansky A, Ryan P, Balis UJ, Tompkins RG, Haber DA, Toner M (2007) Isolation of rare circulating tumour cells in cancer patients by microchip technology. Nature 450:1235–1239. doi:10.1038/nature06385
24. Torres-Garcia W, Ashili S, Kelbauskas L, Johnson RH, Zhang W, Runger GC, Meldrum DR (2012) A statistical framework for multiparameter analysis at the single-cell level. Mol BioSyst 8:804–817. doi:10.1039/c2mb05429a
25. Lin M, Chen JF, Lu YT, Zhang Y, Song J, Hou S, Ke Z, Tseng HR (2014) Nanostructure embedded microchips for detection, isolation, and characterization of circulating tumor cells. Acc Chem Res 47:2941–2950. doi:10.1021/ar5001617

26. Vona G, Sabile A, Louha M, Sitruk V, Romana S, Schutze K, Capron F, Franco D, Pazzagli M, Vekemans M, Lacour B, Brechot C, Paterlini-Brechot P (2000) Isolation by size of epithelial tumor cells: a new method for the immunomorphological and molecular characterization of circulatingtumor cells. Am J Pathol 156:57–63. doi:10.1016/S0002-9440(10)64706-2
27. Yu M, Stott S, Toner M, Maheswaran S, Haber DA (2011) Circulating tumor cells: approaches to isolation and characterization. J Cell Biol 192:373–382. doi:10.1083/jcb.201010021
28. Mohamed H, Murray M, Turner JN, Caggana M (2009) Isolation of tumor cells using size and deformation. J Chromatogr A 1216:8289–8295. doi:10.1016/j.chroma.2009.05.036
29. Sarioglu AF, Aceto N, Kojic N, Donaldson MC, Zeinali M, Hamza B, Engstrom A, Zhu H, Sundaresan TK, Miyamoto DT, Luo X, Bardia A, Wittner BS, Ramaswamy S, Shioda T, Ting DT, Stott SL, Kapur R, Maheswaran S, Haber DA, Toner M (2015) A microfluidic device for label-free, physical capture of circulating tumor cell clusters. Nat Methods 12:685–691. doi:10.1038/nmeth.3404
30. Hosokawa M, Hayata T, Fukuda Y, Arakaki A, Yoshino T, Tanaka T, Matsunaga T (2010) Size-selective microcavity array for rapid and efficient detection of circulating tumor cells. Anal Chem 82:6629–6635. doi:10.1021/ac101222x
31. Zheng S, Lin HK, Lu B, Williams A, Datar R, Cote RJ, Tai YC (2011) 3D microfilter device for viable circulating tumor cell (CTC) enrichment from blood. Biomed Microdevices 13:203–213. doi:10.1007/s10544-010-9485-3
32. Chen W, Huang NT, Oh B, Lam RH, Fan R, Cornell TT, Shanley TP, Kurabayashi K, Fu J (2013) Surface-micromachined microfiltration membranes for efficient isolation and functional immunophenotyping of subpopulations of immune cells. Adv Healthc Mater 2:965–975. doi:10.1002/adhm.201200378
33. Lim LS, Hu M, Huang MC, Cheong WC, Gan AT, Looi XL, Leong SM, Koay ES, Li MH (2012) Microsieve lab-chip device for rapid enumeration and fluorescence in situ hybridization of circulating tumor cells. Lab Chip 12:4388–4396. doi:10.1039/c2lc20750h
34. Wei H, Chueh BH, Wu H, Hall EW, Li CW, Schirhagl R, Lin JM, Zare RN (2011) Particle sorting using a porous membrane in a microfluidic device. Lab Chip 11:238–245. doi:10.1039/c0lc00121
35. Schirhagl R, Fuereder I, Hall EW, Medeiros BC, Zare RN (2011) Microfluidic purification and analysis of hematopoietic stem cells from bone marrow. Lab Chip 11:3130–3135. doi:10.1039/c1lc20353c
36. Gossett DR, Weaver WM, Mach AJ, Hur SC, Tse HT, Lee W, Amini H, Di Carlo D (2010) Label-free cell separation and sorting in microfluidic systems. Anal Bioanal Chem 397:3249–3267. doi:10.1007/s00216-010-3721-9
37. Hur SC, Mach AJ, Di Carlo D (2011) High-throughput size-based rare cell enrichment using microscale vortices. Biomicrofluidics 5:22206. doi:10.1063/1.3576780
38. Moon HS, Kwon K, Hyun KA, Seok Sim T, Chan Park J, Lee JG, Jung HI (2013) Continual collection and re-separation of circulating tumor cells from blood using multi-stage multi-orifice flow fractionation. Biomicrofluidics 7:14105. doi:10.1063/1.4788914
39. Warkiani ME, Khoo BL, Wu L, Tay AK, Bhagat AA, Han J, Lim CT (2016) Ultra-fast, label-free isolation of circulating tumor cells from blood using spiral microfluidics. Nat Protocols 11:134–148. doi:10.1038/nprot.2016.003
40. Hou HW, Warkiani ME, Khoo BL, Li ZR, Soo RA, Tan DS, Lim WT, Han J, Bhagat AA, Lim CT (2013) Isolation and retrieval of circulating tumor cells using centrifugal forces. Sci Rep 3:1259. doi:10.1038/srep01259
41. Ozkumur E, Shah AM, Ciciliano JC, Emmink BL, Miyamoto DT, Brachtel E, Yu M, Chen PI, Morgan B, Trautwein J, Kimura A, Sengupta S, Stott SL, Karabacak NM, Barber TA, Walsh JR, Smith K, Spuhler PS, Sullivan JP, Lee RJ, Ting DT, Luo X, Shaw AT, Bardia A, Sequist LV, Louis DN, Maheswaran S, Kapur R, Haber DA, Toner M (2013) Inertial focusing for tumor antigen-dependent and -independent sorting of rare circulating tumor cells. Sci Transl Med 5(179):179ra47. doi:10.1126/scitranslmed.3005616

42. Karabacak NM, Spuhler PS, Fachin F, Lim EJ, Pai V, Ozkumur E, Martel JM, Kojic N, Smith K, Chen PI, Yang J, Hwang H, Morgan B, Trautwein J, Barber TA, Stott SL, Maheswaran S, Kapur R, Haber DA, Toner M (2014) Microfluidic, marker-free isolation of circulating tumor cells from blood samples. Nat Protocols 9:694–710. doi:10.1038/nprot.2014.044
43. Hyun KA, Kwon K, Han H, Kim SI, Jung HI (2013) Microfluidic flow fractionation device for label-free isolation of circulating tumor cells (CTCs) from breast cancer patients. Biosens Bioelectron 40:206–212. doi:10.1016/j.bios.2012.07.021
44. Ding X, Li P, Lin SC, Stratton ZS, Nama N, Guo F, Slotcavage D, Mao X, Shi J, Costanzo F, Huang TJ (2013) Surface acoustic wave microfluidics. Lab Chip 13:3626–3649. doi:10.1039/c3lc50361e
45. Petersson F, Aberg L, Sward-Nilsson AM, Laurell T (2007) Free flow acoustophoresis: microfluidic-based mode of particle and cell separation. Anal Chem 79:5117–5123. doi:10.1021/ac070444e
46. Chen Y, Nawaz AA, Zhao Y, Huang PH, McCoy JP, Levine SJ, Wang L, Huang TJ (2014) Standing surface acoustic wave (SSAW)-based microfluidic cytometer. Lab Chip 14:916–923. doi:10.1039/c3lc51139a
47. Ding X, Peng Z, Lin SC, Geri M, Li S, Li P, Chen Y, Dao M, Suresh S, Huang TJ (2014) Cell separation using tilted-angle standing surface acoustic waves. PNAS 111:12992–12997. doi:10.1073/pnas.1413325111
48. Li P, Mao Z, Peng Z, Zhou L, Chen Y, Huang PH, Truica CI, Drabick JJ, El-Deiry WS, Dao M, Suresh S, Huang TJ (2015) Acoustic separation of circulating tumor cells. PNAS 112:4970–4975. doi:10.1073/pnas.1504484112
49. Senveli SU, Ao Z, Rawal S, Datar RH, Cote RJ, Tigli O (2016) A surface acoustic wave biosensor for interrogation of single tumour cells in microcavities. Lab Chip 16:163–171. doi:10.1039/c5lc01212k
50. Shafiee H, Caldwell JL, Sano MB, Davalos RV (2009) Contactless dielectrophoresis: a new technique for cell manipulation. Biomed Microdevices 11:997–1006. doi:10.1007/s10544-009-9317-5
51. Gupta V, Jafferji I, Garza M, Melnikova VO, Hasegawa DK, Pethig R, Davis DW (2012) ApoStream(), a new dielectrophoretic device for antibody independent isolation and recovery of viable cancer cells from blood. Biomicrofluidics 6:24133. doi:10.1063/1.4731647
52. Shim S, Stemke-Hale K, Tsimberidou AM, Noshari J, Anderson TE, Gascoyne PR (2013) Antibody-independent isolation of circulating tumor cells by continuous-flow dielectrophoresis. Biomicrofluidics 7:11807. doi:10.1063/1.4774304
53. Cheng IF, Huang WL, Chen TY, Liu CW, Lin YD, Su WC (2015) Antibody-free isolation of rare cancer cells from blood based on 3D lateral dielectrophoresis. Lab Chip 15:2950–2959. doi:10.1039/c5lc00120j
54. Sano MB, Caldwell JL, Davalos RV (2011) Modeling and development of a low frequency contactless dielectrophoresis (cDEP) platform to sort cancer cells from dilute whole blood samples. Biosens Bioelectron 30:13–20. doi:10.1016/j.bios.2011.07.048
55. Mazutis L, Gilbert J, Ung WL, Weitz DA, Griffiths AD, Heyman JA (2013) Single-cell analysis and sorting using droplet-based microfluidics. Nat Protocols 8:870–891. doi:10.1038/nprot.2013.046
56. Klein AM, Mazutis L, Akartuna I, Tallapragada N, Veres A, Li V, Peshkin L, Weitz DA, Kirschner MW (2015) Droplet barcoding for single-cell transcriptomics applied to embryonic stem cells. Cell 161:1187–1201. doi:10.1016/j.cell.2015.04.044
57. Huang SB, Wu MH, Lin YH, Hsieh CH, Yang CL, Lin HC, Tseng CP, Lee GB (2013) High-purity and label-free isolation of circulating tumor cells (CTCs) in a microfluidic platform by using optically-induced-dielectrophoretic (ODEP) force. Lab Chip 13:1371–1383. doi:10.1039/c3lc41256c
58. Galletti G, Sung MS, Vahdat LT, Shah MA, Santana SM, Altavilla G, Kirby BJ, Giannakakou P (2014) Isolation of breast cancer and gastric cancer circulating tumor cells by

use of an anti HER2-based microfluidic device. Lab Chip 14:147–156. doi:10.1039/c3lc51039e
59. Stott SL, Hsu CH, Tsukrov DI, Yu M, Miyamoto DT, Waltman BA, Rothenberg SM, Shah AM, Smas ME, Korir GK, Floyd FP Jr, Gilman AJ, Lord JB, Winokur D, Springer S, Irimia D, Nagrath S, Sequist LV, Lee RJ, Isselbacher KJ, Maheswaran S, Haber DA, Toner M (2010) Isolation of circulating tumor cells using a microvortex-generating herringbone-chip. PNAS 107:18392–18397. doi:10.1073/pnas.1012539107
60. Adams AA, Okagbare PI, Feng J, Hupert ML, Patterson D, Gottert J, McCarley RL, Nikitopoulos D, Murphy MC, Soper SA (2008) Highly efficient circulating tumor cell isolation from whole blood and label-free enumeration using polymer-based microfluidics with an integrated conductivity sensor. JACS 130:8633–8641. doi:10.1021/ja8015022
61. Yoon HJ, Kim TH, Zhang Z, Azizi E, Pham TM, Paoletti C, Lin J, Ramnath N, Wicha MS, Hayes DF, Simeone DM, Nagrath S (2013) Sensitive capture of circulating tumour cells by functionalized graphene oxide nanosheets. Nat Nanotechnol 8:735–741. doi:10.1038/nnano.2013.194
62. Dharmasiri U, Njoroge SK, Witek MA, Adebiyi MG, Kamande JW, Hupert ML, Barany F, Soper SA (2011) High-throughput selection, enumeration, electrokinetic manipulation, and molecular profiling of low-abundance circulating tumor cells using a microfluidic system. Anal Chem 83:2301–2309. doi:10.1021/ac103172y
63. Fang X, Tan W (2010) Aptamers generated from cell-SELEX for molecular medicine: a chemical biology approach. Acc Chem Res 43:48–57. doi:10.1021/ar900101s
64. Tuerk C, Gold L (1990) Systematic evolution of ligands by exponential enrichment: RNA ligands to bacteriophage T4 DNA polymerase. Science 249:505–510
65. Ellington AD, Szostak JW (1990) In vitro selection of RNA molecules that bind specific ligands. Nature 346:818–822. doi:10.1038/346818a0
66. Shen Q, Xu L, Zhao L, Wu D, Fan Y, Zhou Y, Ouyang WH, Xu X, Zhang Z, Song M, Lee T, Garcia MA, Xiong B, Hou S, Tseng HR, Fang X (2013) Specific capture and release of circulating tumor cells using aptamer-modified nanosubstrates. Adv Mater 25:2368–2373. doi:10.1002/adma.201300082
67. Wan Y, Tan J, Asghar W, Kim YT, Liu Y, Iqbal SM (2011) Velocity effect on aptamer-based circulating tumor cell isolation in microfluidic devices. J Phy Chem B 115:13891–13896. doi:10.1021/jp205511m
68. Phillips JA, Xu Y, Xia Z, Fan ZH, Tan W (2009) Enrichment of cancer cells using aptamers immobilized on a microfluidic channel. Anal Chem 81:1033–1039. doi:10.1021/ac802092j
69. Xu Y, Phillips JA, Yan J, Li Q, Fan ZH, Tan W (2009) Aptamer-based microfluidic device for enrichment, sorting, and detection of multiple cancer cells. Anal Chem 81:7436–7442. doi:10.1021/ac9012072
70. Zhao W, Cui CH, Bose S, Guo D, Shen C, Wong WP, Halvorsen K, Farokhzad OC, Teo GS, Phillips JA, Dorfman DM, Karnik R, Karp JM (2012) Bioinspired multivalent DNA network for capture and release of cells. PNAS 109:19626–19631. doi:10.1073/pnas.1211234109
71. Liu W, Wei H, Lin Z, Mao S, Lin JM (2011) Rare cell chemiluminescence detection based on aptamer-specific capture in microfluidic channels. Biosens Bioelectron 28:438–442. doi:10.1016/j.bios.2011.07.067
72. Wang S, Wang H, Jiao J, Chen KJ, Owens GE, Kamei K, Sun J, Sherman DJ, Behrenbruch CP, Wu H, Tseng HR (2009) Three-dimensional nanostructured substrates toward efficient capture of circulating tumor cells. Angew Chem Int Ed 48:8970–8973. doi:10.1002/anie.200901668
73. Wang S, Liu K, Liu J, Yu ZT, Xu X, Zhao L, Lee T, Lee EK, Reiss J, Lee YK, Chung LW, Huang J, Rettig M, Seligson D, Duraiswamy KN, Shen CK, Tseng HR (2011) Highly efficient capture of circulating tumor cells by using nanostructured silicon substrates with integrated chaotic micromixers. Angew Chem Int Ed 50:3084–3088. doi:10.1002/anie.201005853

74. Sheng W, Chen T, Tan W, Fan ZH (2013) Multivalent DNA nanospheres for enhanced capture of cancer cells in microfluidic devices. ACS Nano 7:7067–7076. doi:10.1021/nn4023747
75. Zhu J, Nguyen T, Pei R, Stojanovic M, Lin Q (2012) Specific capture and temperature-mediated release of cells in an aptamer-based microfluidic device. Lab Chip 12:3504–3513. doi:10.1039/c2lc40411g
76. Hou S, Zhao L, Shen Q, Yu J, Ng C, Kong X, Wu D, Song M, Shi X, Xu X, OuYang WH, He R, Zhao XZ, Lee T, Brunicardi FC, Garcia MA, Ribas A, Lo RS, Tseng HR (2013) Polymer nanofiber-embedded microchips for detection, isolation, and molecular analysis of single circulating melanoma cells. Angew Chem Int Ed 52:3379–3383. doi:10.1002/anie.201208452
77. Zhao L, Lu YT, Li F, Wu K, Hou S, Yu J, Shen Q, Wu D, Song M, OuYang WH, Luo Z, Lee T, Fang X, Shao C, Xu X, Garcia MA, Chung LW, Rettig M, Tseng HR, Posadas EM (2013) High-purity prostate circulating tumor cell isolation by a polymer nanofiber-embedded microchip for whole exome sequencing. Adv Mater 25:2897–2902. doi:10.1002/adma.201205237
78. Qian W, Zhang Y, Chen W (2015) Capturing cancer: emerging microfluidic technologies for the capture and characterization of circulating tumor cells. Small 11:3850–3872. doi:10.1002/smll.201403658
79. Gleghorn JP, Pratt ED, Denning D, Liu H, Bander NH, Tagawa ST, Nanus DM, Giannakakou PA, Kirby BJ (2010) Capture of circulating tumor cells from whole blood of prostate cancer patients using geometrically enhanced differential immunocapture (GEDI) and a prostate-specific antibody. Lab Chip 10:27–29. doi:10.1039/b917959c
80. Sheng W, Chen T, Kamath R, Xiong X, Tan W, Fan ZH (2012) Aptamer-enabled efficient isolation of cancer cells from whole blood using a microfluidic device. Anal Chem 84:4199–4206. doi:10.1021/ac3005633
81. Oh SS, Qian J, Lou X, Zhang Y, Xiao Y, Soh HT (2009) Generation of highly specific aptamers via micromagnetic selection. Anal Chem 81:5490–5495. doi:10.1021/ac900759k
82. McFaul SM, Lin BK, Ma H (2012) Cell separation based on size and deformability using microfluidic funnel ratchets. Lab Chip 12:2369–2376. doi:10.1039/c2lc21045b
83. Adams DL, Zhu P, Makarova OV, Martin SS, Charpentier M, Chumsri S, Li S, Amstutz P, Tang CM (2014) The systematic study of circulating tumor cell isolation using lithographic microfilters. RSC Adv 9:4334–4342. doi:10.1039/C3RA46839A
84. Tang Y, Shi J, Li S, Wang L, Cayre YE, Chen Y (2014) Microfluidic device with integrated microfilter of conical-shaped holes for high efficiency and high purity capture of circulating tumor cells. Sci Rep 4:6052. doi:10.1038/srep06052
85. Gorkin R, Park J, Siegrist J, Amasia M, Lee BS, Park JM, Kim J, Kim H, Madou M, Cho YK (2010) Centrifugal microfluidics for biomedical applications. Lab Chip 10(14):1758–1773. doi:10.1039/b924109d
86. Burger R, Kirby D, Glynn M, Nwankire C, O'Sullivan M, Siegrist J, Kinahan D, Aguirre G, Kijanka G, Gorkin RA 3rd, Ducree J (2012) Centrifugal microfluidics for cell analysis. Curr Opin Chem Biol 16:409–414. doi:10.1016/j.cbpa.2012.06.002
87. Liu C, Liu J, Gao D, Ding M, Lin JM (2010) Fabrication of microwell arrays based on two-dimensional ordered polystyrene microspheres for high-throughput single-cell analysis. Anal Chem 82:9418–9424. doi:10.1021/ac102094r
88. Chen Q, Wu J, Zhang Y, Lin Z, Lin JM (2012) Targeted isolation and analysis of single tumor cells with aptamer-encoded microwell array on microfluidic device. Lab Chip 12:5180–5185. doi:10.1039/c2lc40858a
89. Ma C, Fan R, Ahmad H, Shi Q, Comin-Anduix B, Chodon T, Koya RC, Liu CC, Kwong GA, Radu CG, Ribas A, Heath JR (2011) A clinical microchip for evaluation of single immune cells reveals high functional heterogeneity in phenotypically similar T cells. Nat Med 17:738–743. doi:10.1038/nm.2375
90. Autebert J, Coudert B, Champ J, Saias L, Guneri ET, Lebofsky R, Bidard FC, Pierga JY, Farace F, Descroix S, Malaquin L, Viovy JL (2015) High purity microfluidic sorting and

analysis of circulating tumor cells: towards routine mutation detection. Lab Chip 15:2090–2101. doi:10.1039/c5lc00104h
91. Guo MT, Rotem A, Heyman JA, Weitz DA (2012) Droplet microfluidics for high-throughput biological assays. Lab Chip 12:2146–2155. doi:10.1039/c2lc21147e
92. Spencer SJ, Tamminen MV, Preheim SP, Guo MT, Briggs AW, Brito IL, David AW, Pitkanen LK, Vigneault F, Juhani Virta MP, Alm EJ (2016) Massively parallel sequencing of single cells by epicPCR links functional genes with phylogenetic markers. ISME J 10:427–436. doi:10.1038/ismej.2015.124
93. Hattori F, Chen H, Yamashita H, Tohyama S, Satoh YS, Yuasa S, Li W, Yamakawa H, Tanaka T, Onitsuka T, Shimoji K, Ohno Y, Egashira T, Kaneda R, Murata M, Hidaka K, Morisaki T, Sasaki E, Suzuki T, Sano M, Makino S, Oikawa S, Fukuda K (2010) Nongenetic method for purifying stem cell-derived cardiomyocytes. Nat Methods 7:61–66. doi:10.1038/nmeth.1403
94. Lecault V, Vaninsberghe M, Sekulovic S, Knapp DJ, Wohrer S, Bowden W, Viel F, McLaughlin T, Jarandehei A, Miller M, Falconnet D, White AK, Kent DG, Copley MR, Taghipour F, Eaves CJ, Humphries RK, Piret JM, Hansen CL (2011) High-throughput analysis of single hematopoietic stem cell proliferation in microfluidic cell culture arrays. Nat Methods 8:581–586. doi:10.1038/nmeth.1614
95. Singh A, Suri S, Lee T, Chilton JM, Cooke MT, Chen W, Fu J, Stice SL, Lu H, McDevitt TC, Garcia AJ (2013) Adhesion strength-based, label-free isolation of human pluripotent stem cells. Nat Methods 10:438–444. doi:10.1038/nmeth.2437
96. Chen Q, Wu J, Zhuang Q, Lin X, Zhang J, Lin JM (2013) Microfluidic isolation of highly pure embryonic stem cells using feeder-separated co-culture system. Sci Rep 3:2433. doi:10.1038/srep02433
97. Wang X, Chen S, Kong M, Wang Z, Costa KD, Li RA, Sun D (2011) Enhanced cell sorting and manipulation with combined optical tweezer and microfluidic chip technologies. Lab Chip 11:3656–3662. doi:10.1039/c1lc20653b
98. Wang MM, Tu E, Raymond DE, Yang JM, Zhang H, Hagen N, Dees B, Mercer EM, Forster AH, Kariv I, Marchand PJ, Butler WF (2005) Microfluidic sorting of mammalian cells by optical force switching. Nat Biotechnol 23:83–87. doi:10.1038/nbt1050
99. Au AK, Huynh W, Horowitz LF, Folch A (2016) 3D-Printed Microfluidics. Angew Chem Int Ed 55:3862–3881. doi:10.1002/anie.201504382
100. Wu J, Chen Q, Lin JM (2017) Microfluidic technologies in cell isolation and analysis for biomedical applications. Analyst 142:421–441. doi:10.1039/C6AN01939K

Chapter 4
Cell Culture and Observation on Microfluidics

Linglu Yi and Jin-Ming Lin

Abstract Recent advances in microfluidic systems have led to the development of on-chip culture and observation. The on-chip culture offers several advantages over conventional in vitro culture methods: Reduced cell and reagent consumption, more accurate mimicking of in vivo cellular microenvironment, integrating function modules and controllable mechanical and chemical factors. These advantages support the microfluidic applications in biological science, especially for cellular dynamic study. Based on transparent characteristics of chip substrate, the microfluidic devices can be easily coupled to cell imaging methods and realize bio-application in cellomics. In this chapter, we introduce the development of microfluidic culture and how the microfluidic devices work in cell observations, especially for cell biology.

Keywords Microfluidic cell culture · Non-destructive observations
Cell biology applications

4.1 Introduction

Conventional cell culture and observation is limited by the number of experiments for the lack of high-throughput handling. In addition, these formats are not sufficient to support understanding of the cell response dynamically in a physiologically relevant way. Despite that small animal models are commonly applied, they typically can only provide an end-point evaluation. Microfluidics has been used as in vitro culture system for facilitating nutrient supply and waste removal, and it has the potential to address these challenges for its superiority in miniaturization, automation and integration. The ability to maintain, manipulate, and assay cells

L. Yi
School of Science, Beijing University of Chemical Technology, Beijing 100029, China

L. Yi · J.-M. Lin (✉)
Department of Chemistry, Tsinghua University, Beijing 100084, People's Republic of China
e-mail: jmlin@mail.tsinghua.edu.cn

J.-M. Lin (ed.), *Cell Analysis on Microfluidics*, Integrated Analytical Systems,
https://doi.org/10.1007/978-981-10-5394-8_4

proves a very useful application for microfluidic device in establishment of in vitro model. This model has advantages including flexible control and handing of cells, long-term culture, compartmentalized culture, co-culture, perfusion culture and real-time analysis. Microfluidic device has been successfully applied in tissue engineering. Several in vitro models have been established based on microfluidic device, are able to recapitulate physical properties of the in vivo microenvironment, allowing further investigation of tumor cell behavior in details. In this chapter, a review involved cell culture, manipulation and observation methods on microfluidic device are introduced. The overall aim is to present research in cell culture and observation on microfluidic chip for readers.

4.2 Types of Cell Culture

Cell culture refers to the controllable growth and maintenance of cells in the environment of lab [1]. Such culture models are the mainstream for experiment in cell biology (Fig. 4.1). Microfluidic cell culture is developed based on techniques and devices to culture, maintain and analyze cells in micro-scale volumes [2]. Compared with macroscopic culture, microfluidic cell culture has several advantages as follows: (1) Flexibility of device design, (2) experimental flexibility and control, (3) a low number of cells is sufficient, (4) single cell handling, (5) real-time/on-chip analysis, (6) automation, (7) direct coupling to downstream analysis systems, (8) ability to perform perfusion culture, (9) controlled co-culture, and (10) reduced reagent consumption [1].

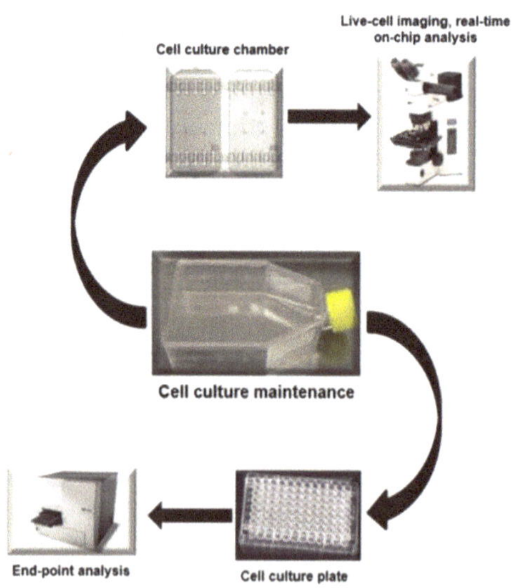

Fig. 4.1 Overview of both macroscopic and microfluidic cell culture (reprinted with permission from Ref. [1])

Even though microfluidic cell culture is limited by present technologies to show some challenges: (1) Non-standard culture protocols, (2) novel culture surface (e.g. PDMS), (3) small volumes, (4) challenging subsequent analytical chemistry and (5) complex operational control and chip design [1]. Still microfluidic cell culture is promising. The design of microfluidic devices is of great flexibility, and can be tailored to in terms of different types of cell culture models [2].

4.2.1 Two-Dimensional (2D) Culture

Two-dimensional (2D) cell culture includes cell culture in suspension and in monolayer. There are two kinds of cells need culturing environment in fluid suspension: They are mammalian blood cells and yeast cells. Besides, most cell types need to be cultured in monolayer to remain important phenotypic characteristics.

The effect of micro channel dimension has been studied on the proliferation of suspended insect cells, and also the effects of exogenous growth factors, cell density, and media change frequency of normal murine mammary gland cells on their growth rate [3]. One of the first studies in microfluidic channels of adherent cell culture was performed for primary rat hepatocytes co-cultured with fibroblasts. These cells were cultured in channels with either a polycarbonate top or a gas exchange membrane top [4]. Other studies that indicate the environmental differences between conventional and culture microfluidic systems come from the researches for embryo [5].

An important factor that affects cell attachment in microfluidic channels is surface treatment and coating, because most substrates for microfluidic devices, such as polydimethylsiloxane (PDMS), are hydrophobic material and must be rendered hydrophilic before they are available for cell adhesion. There are variety of methods that have been developed for reducing hydrophobicity, among which oxidizing with oxygen plasma or UV treatment is widely used. However, due to uncross-linked PDMS polymeric side-chains (low molecular weight) diffusing from the bulk to the surface, PDMS shows hydrophobic recovery, which means that PDMS surface again becomes hydrophobic within one week after these treatments. This limits the storage time of treated devices or the viability of surface oxidization for long-term culture experiments.

To reduce hydrophobicity, the other way is modified surface with extra cellular matrix proteins (e.g. laminin, fibronectin and collagen). These polar substrates easily interact with hydrophobic PDMS through polar–polar interactions or hydrogen bonding. Proteins will also adsorb to an untreated surface, interacting with PDMS by their methyl or alkyl groups via van der Waals forces.

The use of charged molecules like poly-D-lysine for PDMS coating is the third possible way. Due to the positive charged amino group on the end, each lysine is hydrophilic and may easily interact with negatively charged ions on cell membrane [6]. Unwanted cell attachment may occur due to non-specific protein adsorption.

To overcome this, surfactants are used to modify the surface to prevent protein adsorption. Poly (ethylene oxide)-terminated triblock polymers are the most commonly used surfactants, which can develop a stable adsorption layer on the PDMS surface that is hydrophobic. Polysaccharides were also used to coat native PDMS by a photo catalyzed surface modification method. The carboxymethyl cellulose coating repels both negatively and positively charged proteins, while still allows cell attachment, migration and proliferation [7].

4.2.2 Three-Dimensional (3D) Culture

2D culture can be applied to manipulate and observe mammalian cells, also laid the basis for molecular and cell biology. Nevertheless, 2D culture does not completely simulate extracellular matrix (ECM) within organs and tissues and recapitulate the 3D organization of cells. The 3D models perform better than 2D models to reproduce the in vivo-like environment. The microfluidic system provides spatially control, integrate perfusion, flow generated and controlled over gradients; therefore, it is promising in 3D culture.

A diverse number of 3D culture substrate materials are capable of incorporating into microfluidic chips and developing into hydrogel-based pattern and gel-free pattern [8, 9]. The hydrogel-based pattern is mainly used in 3D culture. By guiding structures such as ridges, posts or pillars [10–12], hydrogels can be spatially controlled in microfluidic chip, a cell suspension was then added to adhere to the hydrogel (Fig. 4.2a).The hydrogels can be individually addressed by phaseguides (Fig. 4.2b). In case of 3D cell culture, vasculature models benefit most from these microfluidics tools for their advantages including spatial co-culture, temporal and spatial control of gradients and perfusion. Therefore, this in vitro model is specifically fit for tumor angiogenesis, which studies how the factors in the microenvironment affect vascularization around the tissue. These factors include remodeled ECM and immobilized molecular cues from tumor cells supporting EC recruitment and morphogenesis [13]. Vascular models are typically established based on hydrogel, with endothelial cells seeded in or against it [14, 15].

Hydrogels are nature hydrophilic materials that are commonly used in creating 3D in vitro models. Hydrogels can be used to tune the chemical structures and mechanical strength of cellular environment. Studies have shown that varying crosslinking densities create different stiffness of gels, which affect the survival, proliferation, and migration of the cells in the gels and induce specific differentiation of stem cells [16, 17]. Chemically, hydrogels can be modified with enhanced cell-attaching sites, such as Arg-Gly-Asp, RGD, and MMP-degradable sites. These sites are very important for endothelial migration, tumor progression and tumor angiogenesis [13, 18, 19]. In the recent, complex blood vessels are developed by combination of hydrogels with technologies like lithography, showing advantages in more clinically relevant and complex tumor angiogenesis models (Fig. 4.2c) [14, 20, 21]. The gel-based pattern is complex to embed cells in hydrogels, thus may

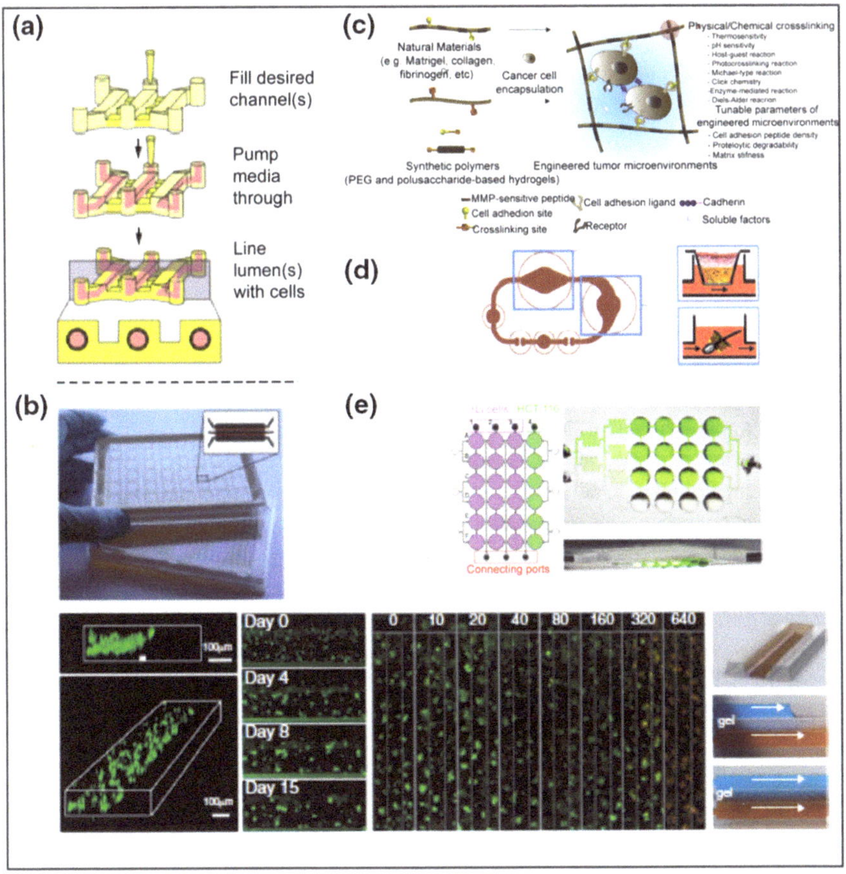

Fig. 4.2 Microfluidic techniques for 3D cell culture (reprinted with permission from Ref. [25]). **a** Cell patterning inside a hydrogel, exploiting the microfluidic properties and differences in viscosity and pressure. **b** 96 microfluidic culture chambers integrated underneath a microtiter plate. Extracellular matrix gels are selectively patterned in the chamber by phaseguides to obtain a layered profile. **c** A schematic representation of engineered tumor microenvironment using natural/synthetic hydrogel materials for in vitro tumor models (reprinted with permission from Ref. [26]). **d** A whole tissue perfusion system. Tissues are isolated and cultured on membrane inserts. Microfluidic channels interconnect the tissue chambers, allowing continuous perfusion and paracrine cell signaling. **e** A hanging droplet system. Since the columns and rows are addressable, flexible co-culture setups are possible. This picture demonstrated the hanging droplet array in combination with a gradient generator

hinder mass transfer. Besides, it is not a proper pattern for establishing constructs with cell-dense and ECM-poor characteristics. 3D culture can also be realized without hydrogels by culturing whole tissues in microfluidic chip (Fig. 4.2d) or generate a hanging drop system (Fig. 4.2e). In general, microchambers or droplets are used for free-gel 3D culture, in which suspended cells reside and aggregate to form spheroids [22, 23]. Micro-fabricated pillar arrays combined with transient inter-cellular polymeric linker was also used for gel-free 3D culture to immobilize and maintain 3D multi-cellular clusters in a microchannel [8, 9]. A recently developed 3D human central nervous system model in gel-free pattern has attracted great attention by collecting cells from a rat's cortical layers and seeding them in microwell arrays, to form neurospheroids. This gel-free 3D model better mimics the in vivo environment [24].

4.3 Cell Manipulation

Cell manipulation in microfluidics means high spatiotemporal control strategies. These strategies are divided into two kinds. One is about spatial-control of cells in micro channels, and the other one is related to temporal control of stimulation and quenching on cell reaction. Previously, there have emerged many different microfluidic technologies for cell manipulation. These technologies can be categorized into active, passive and integrated strategies (Fig. 4.3). Structure design, one method that belongs passive strategies, can be used in rationally ways to deflect cells' trajectories, and ultimately, control the cells' position within microfluidic channels [28, 29].

The manipulation can also be realized by force resulted from actuator [30, 31], as a result of cells' electrical [32, 33], mechanical and magnetic properties. These are the examples for active strategies [34, 35]. The combination of different strategies

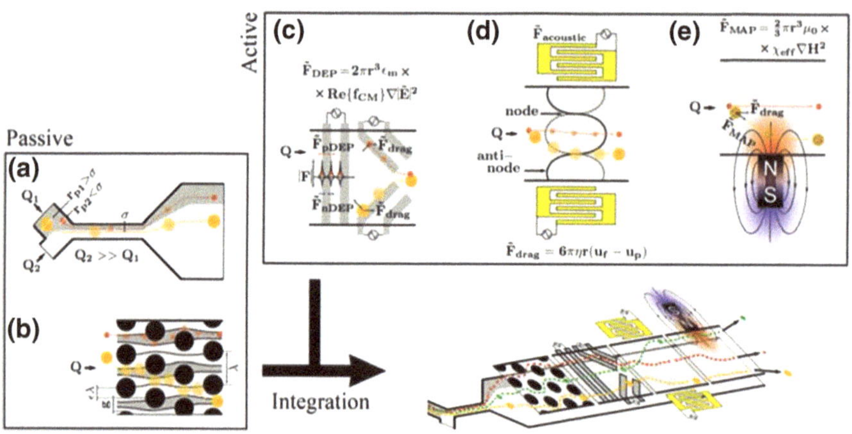

Fig. 4.3 Passive, active and integrated microfluidic strategies for cell manipulation (reprinted with permission from Ref. [27]). **a** STE controlled cell solution exchange. **b** DLD separation of cells with different diameters. **c** "Electrode cell highways" by DEP. **d** Separation of cells with different geometrical and mechanical properties by acoustophoresis. **e** Isolation of cells labeled with magnetic particles using magnetophoresis. *Q* indicates the inlet flow. In the drag force equation, $\vec{F}$ drag, η stands for the fluid viscosity, and uf and up for fluid and particle velocity, respectively

offers advantages to improve performance and functionalities within integrated systems [36–38].

However, combination is challenging especially when consider the simplicity of handling, such as the necessity for alignment of structures in micrometer size or fluidic and electronic connections [39, 40]. Cell stimulation techniques make use of cell adhering at specific position of the microfluidic channels. It is important to note that more skills and time are required for microfluidic culture methods than that for traditional culture methods to successfully cultivate, adhere and maintain cells. Besides, microfluidics permits automation in a high-throughput and well-controlled way [41, 42].

If fast solution switching is required, two important limitations have to be concerned. One is that cells that are sensitive for shear stress cannot permit high-flow-rate condition. Fast switching without contamination is a time limiting factor. Recent works have been reported about cell adhesion models, namely methods relying on perforated membranes [43], electro wetting-digital microfluidics [44, 45] and microchamber-based microfluidics for sequential and continuous cell stimulation [46, 47]. There are two main obstacles to transfer microfluidic cell manipulation into real applications. One is to ensure reproducibility of operation in real application rather than in the lab experiment, with no need for specialists. The other obstacle is to make a working prototype industrially available. Therefore, although microfluidic device is quite promise in fundamental research, there is still a lot of work before its commonly use in real applications and further interdisciplinary development.

In the last decade, optical manipulation has been widely used as a versatile tool in the field of life sciences. Optical manipulation and trapping is a side product from research for studying the effect of angular and linear momentum of laser light on small neutral particles. Multiple optical traps show possibilities in single-cell analysis for good statistics provided by parallel measurements. Multifunctional optical tweezers can be used for study of cell heterogeneity in the whole, and force measurements can be used to demonstrate the mechanical features of individual cells. Optical manipulation is promising in research of tissue engineering and cell signaling. When combined with microfluidic systems, the chemical environment of cells can be controlled precisely via integration of optical manipulation microfluidic systems. Optical manipulation can be easily implemented in microfluidic chip due to the transparent characteristics of the chip material. Obviously, the integration of optical tweezers into microfluidic systems develops a powerful tool to investigate cell-responses to environmental stimuli at single-cell level. Cells can either be trapped by optical tweezers even when they are exposed to changed condition, such as continuous flow of buffers with varying chemical properties, or be transferred between different parts of a microfluidic device that contain different substances [48]. In a laminar flow, an optically immobilized cell can also be moved reversibly between two different media [49]. Another application of optical tweezers can be found in cell sorting in a micro-cultivation assay for monitoring the movements and growth of isolated cells, which were induced with different nutrients [50]. In addition, optical scalpels and electroporation have also been integrated into microfluidic devices [51]. The chemical environment of cells can be controlled due to these integrations, allowing real-time investigation of the influence of salt concentration, pH, temperature and drugs.

4.4 Cellular Microenvironment Control

The microenvironment is crucial for studies in cell biology, such as the study of cancer, development, and stem cells. There are four distinct factors that govern cellular phenotype and direct cell fate, differentiation, and function, that is, physical factors (e.g., 3D organization, temperature and shear stress), chemical factors (e.g., growth factors glucose, drugs and toxins), cell-cell interactions (e.g., homotypic or heterotypic interactions, cell-cell paracrine communication and contact), and cell-extracellular matrix (ECM) interaction (e.g., RDG peptide and collagen type) (Fig. 4.4). Microfluidic cell cultures win over conventional culture methods in controlling these factors temporally and spatially and processing analysis for single-cell studies.

4.4.1 Gradient of Physical Factors

Microfluidic systems can facilitate elucidating the physical factors (shear stress, confinement and temperature) that impact cells' behavior. The impact of oxygen had been investigated on migration of cancer cells on an integrated microfluidic device [57]. The Caski cells showed slower migration rate under 15% of oxygen than that under 5% oxygen. The impact of shear stress is mainly discussed on vascularized tissues. Various microfliuidic devices have previously been described for governing the shear stress imposed on cells. The two-independent-channel types, which separated by a gel or a scaffold network, have been usually utilized to control [58] or study [59] the effect of shear stress on cell behaviors. The microfluidics shows advantages for precisely controlling the flow in the channel and across the gel compartment as well [60]. The influence of shear stress can be limited by separating the cells from the flow with a barrier, such as hydrogels [61, 62], nanoporous membranes [63], and microchannels [64], that allows passive diffusion of biomolecules.

The in vivo microenvironment is a confined space for cells to live. The confinement has impact on cell structure and behavior. Microfluidic devices enable readily achievements of rebuilt of this confinement, especially for physical confinement encountered by cells during migration. There are several successful reports about cell motility, for example, study the regulation of dendritic cell migration [65]. Researchers have found that mechanical confinement itself has impact on cell motility [66]. Temperature is another important factor that affects cellular behavior. To study the impact of temperature on microfluidic systems, the liquid that has

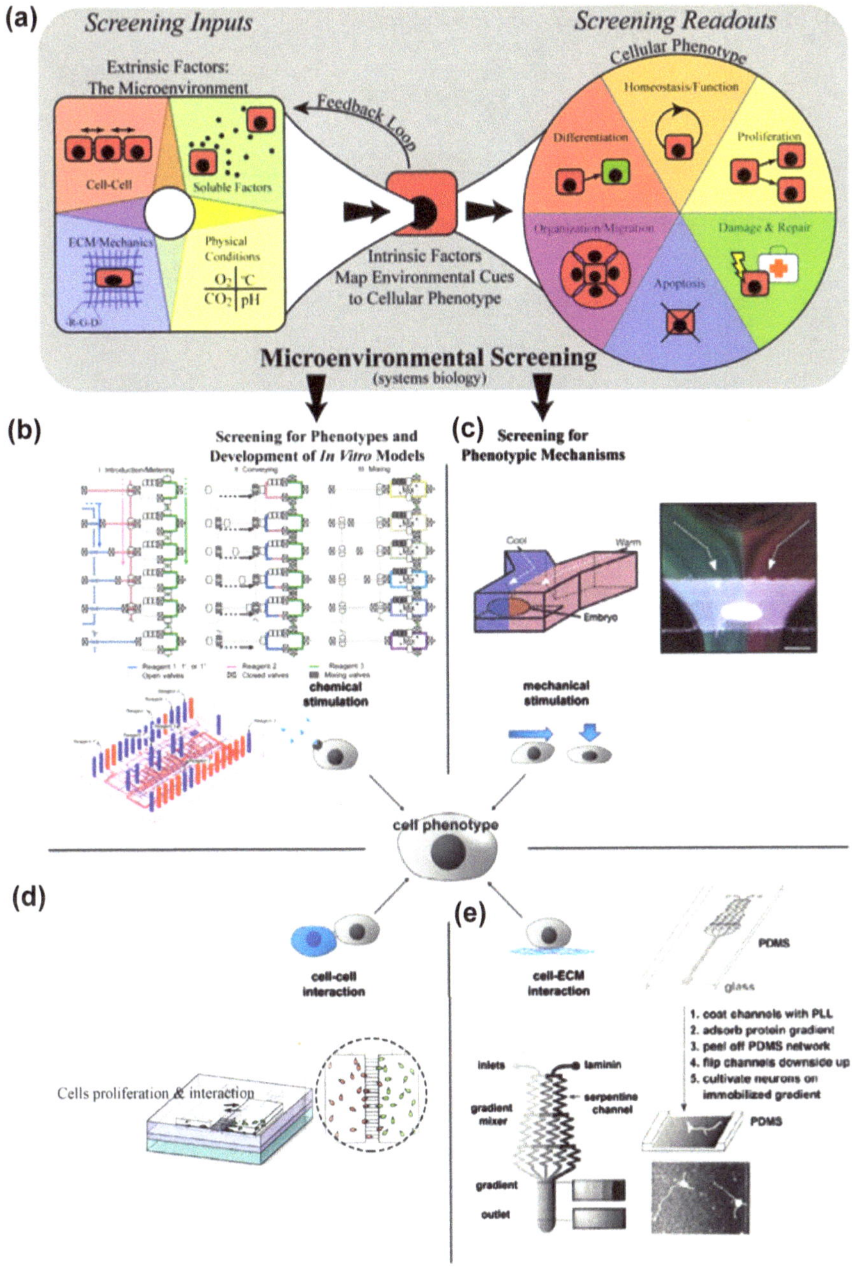

Fig. 4.4 Microfluidic system for cell environment studies. **a** Microenvironmental screening and some of its applications (reprinted with permission from Ref. [52]). **b** Device for the study of soluble factor on cells. Step-by-step process (*top*) and schematic representation (*bottom*) of a microfluidic chip for wide range of concentration gradient of soluble factor. *Blue line* represent the fluidic channels, *red lines* represent the control channels (reprinted with permission from Ref. [53]). **c** Laminar flow-based device for mechanical stimulation of cells. Schematic representation (*left*) and photograph (*right*) of a *Drosophilia* embryo submitted to different temperature (reprinted with permission from Ref. [54]). **d** Device for cell-cell interaction studies. (reprinted with permission from Ref. [55]). **e** Device for the formation of gradient of laminin and the study of neurons laminin-based orientation (reprinted with permission from Ref. [56])

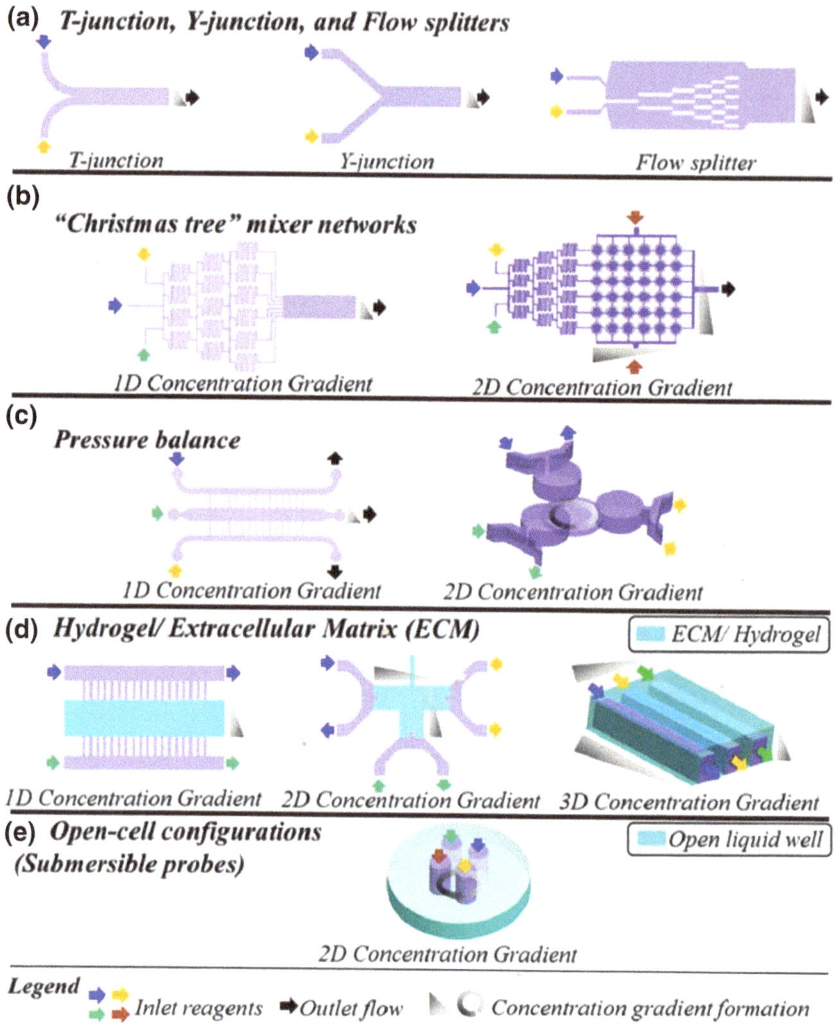

Fig. 4.5 Microfluidics-based gradient generator designs: **a** (*left*) Y junction, (*centre*) T-junction [69], and (*right*) flow splitter designs [70]; **b** (*left*) "Christmas tree" design for 1D [71] and (*right*) 2D gradient generation [72]; **c** (*left*) 1D gradient generation [73] and (*right*) 2D gradient generation through pressure balance designs [74]; **d** Moulded gels for (*left*) 1D [75], (*centre*) 2D [76], and (*right*) 3D gradient generation [77], and **e** Submersible microfluidic probes [78]. Reprinted with permission from Ref. [70])

controlled temperature were introduced into the systems. The control of temperature is realized by two different laminar flow systems [67, 68]. This approach has been applied on revealing the dependence of temperature for development rate of embryo [67] (Fig. 4.4).

4.4.2 Gradient of Chemical Factors

Chemical concentration gradients are regulated to control a lot of fundamental cell functions and biological processes such as cancer metastasis [79, 80], cellular chemotaxis [63, 81] and migration [82, 83], differentiation [84], development [85, 86], immune response [76, 87], wound healing [88, 89] and embryogenesis [90].

Early in vitro platforms to study the influence of chemical gradient, like Boyden chamber [91], Dunn slide chamber [92], Zigmond chamber [93] and agarose/petri dish [94], are limited to recapitulate the length scales of biological cells. Microfluidic device overcomes these shortages, offers high gradient resolutions and provides well-controlled conditions for mass transportation and hydrodynamics. The flow-based microfluidic devices win over widely available methods for temporal and spatial control for gradients of immobilized factors on surfaces or soluble factors. The effect of flow on neutrophils has been conducted alone. Additionally, mechanical activation by shear generated by laminar flow in micro channels was also demonstrated [95].

Gradients can typically be classified into flow-based or non-flow-based, dynamic or static, and 3-D or 2-D. And the flow field, fluid flow and fluid flow velocities are three factors that determine mass transport phenomena in microfluidic gradient generators. In general, the transport of solute concentrations can be detected equivalently to hydraulic circuits in pressure-driven gradient generators [96]. The transport of a solute species through mass convection and/or diffusion is controlled by the convection–diffusion (C–D) equation, while the concentration gradient relies on the flow field in convection-based gradient generators. Besides, the time scale of gradient generation is also crucial to determine how the biological experiment can be implemented. Usually, the time scale of a microfluidic gradient generator is decided by the featured length and the transport parameter of the device. Alicia et al. [97] reviewed the efficient methods for creating concentration gradients in biological applications, which include the studies that chemical gradients affect cell migration and proliferation, the observation of cell dynamics, and cell in a three-dimensional microenvironment. T-sensor devices are the simplest design for generating gradients. Based on T-sensor devices, more complex pre-mixer gradient generators have been developed (Fig. 4.5).

For example, a laminar-flow-based microfluidic system was developed to optimize cell differentiation and proliferation [84]. Human neural stem cells (hNSCs) were stimulated constantly by a continuous gradient of a growth factor (GF) mixture. The hNSCs responded to varied concentrations of the GFs with different proportions of proliferated and differentiated astrocytes. Concentration gradients are conventionally produced via serial diluting of solution by pipetting, which is possible to shield influence on the precision of the diluting steps. Accurate logarithmic-scale concentration gradients were developed by Hong's group (Fig. 4.5b). The valves were placed along microchannels at certain distances, allowing precise filling of liquid within the channels.

4.4.3 Cell-Cell and Cell-Extracellular Matrix (ECM) Interactions

Cell-cell interaction is realized by cell-cell junctions or paracrine signaling mechanisms, and determines the response to stimuli and cellular phenotypes. Microfluidics offers opportunity to isolate specific signals at single cell level and understand clearly of phenotypic and genotypic variation among similar cell types by high-throughput screening. By directing interactions between cell surfaces, specific heterotypic or homotypic cell-cell interactions can be studied [99, 100]. Paracrine communication is usually studied on a two-channel types that separated by an intervening gel or a gel channel. In this model, interstitial flow application induced 3D tissue-like structures, which proved the enhancement of cell-cell cohesion [55].

Unlike 2D culture, 3D culture model typically needs to premix cells with matrix. The 3D structure provides the foundation of complex human tissue network and is suitable for serving in microfluidic studies. Other designs of chips for cell-cell interactions include microvalves, microchannels and membrane as shown in reference [101–104] and to control administration of metabolites from one kind of cells to the receptor cells, our group had once presented a so called surface tension plug on a microfluidic device. Based on this device, the signal pathway had been investigated between 293 and L-02 cells [105].

Co-culture is a way to study cell-cell interaction. To study of cell-cell interactions, the first step is seeding the cells separately and then observing the cell interactions dynamically. In microfluidic device, this goal can be realized in several ways. To regulate the co-culture interaction, cell patterning is applied for varying the spacing between different cell types. Micro-scale of cell patterning can be completed via surface alterations [106], cell settling, laminar flow [107], or even via mechanical means [108]. Other methods include mixing culture (2D/3D) [99], microvalves [109] and microgaps [110] application, membrane insertion [111], droplets technology [112]. There are numerous co-culture types in microfluidic chip (Table 4.1).

The physical features and composition of ECM directly affect a variety of cellular processes including cell shape, life, death, motility, differentiation and polarization [113, 114]. Therefore, the ECM surrounding cells is critical to determine the cell fate. Various ECM materials have been used including naturally-derived polymers (collagen, hyaluronan and fibrin) and artificial polymers (polyethylene glycol (PEG) , poly (lactic-co-glycolic) acid (PLGA) and alginate). Naturally-derived ECM materials are biocompatible but limited in cost and repeatability, while artificial polymers are readily controllable in architecture, stiffness, porosity and shape. Fischbach's group had reported the use of arginylglycylaspartic acid (RGD)-peptide that incorporated into a 3D alginate ECM, which increased angiogenic activity of cancer cells [115].

Table 4.1 Common co-culture modes in microfluidics and their features (reprinted with permission from Ref. [98])

Co-culture modes	Typical research area	Typical microfluidics	Cell types
Vascular system	Angiogenesis	3D structure	Epithelial cells Fibroblasts, Pericytes
	Hemodynamics	Membrane insert	Muscle cells
	Cancer effect	Microgaps	
	Wound healing		
Cancer research	EMT	Microvalves application	Cancer cell lines
	Inflammatory response	Microgaps (channels)	Fibroblasts
			Epithelial cells
			Immune cells
Neurosciences	Neuroprotection	Microchannels	Neuron
	Neurological disorders		Astrocyte
Eukaryocyte and bacteria	Intestinal microenvironment	Membrane insert	Bacterial
	Bacterial cancer targeting	Microgaps	Epithelial cells
Tissue engineering	Organs on chip (liver, lung, kidney et al.)	3D structure	Tissue related cells
		Membrane insert	

Laminar-flow microfluidics is an attractive approach to produce a linear gradient by use of small amounts of molecules, and thus is suitable for studies on cell-ECM interactions specifically [56, 116]. For example, Dertinger et al. [56] showed microfluidic application on haptotaxis study based on a T-sensor network, and analysis showed that axon specification was oriented by the surface density of laminin. In 3D microfluidic vessel model, luminal combined with interstitial flow are major factors to regulate cell-cell and cell-ECM signaling. The interstitial flow refers to extracellular fluid existing in the interstitial spaces between tissue spaces, and it permits directing of cancer cell migration (Fig. 4.6) [117]. Generally, to control the chemical and mechanical composition of ECM is a consideration in the field of biochemistry. Creating surface density gradients of 2-D ECMs is a choice to combine with screening ECM composition [118–120]. Besides, cells embedded in micro-hydrogel "beads" or spots can be a way to incorporate ECM and cells into microfluidic systems [121, 122]. Another option for this purpose is to use microfluidics upon ECMs [123]. In the latter case, channels were formed using PDMS molds in gelatin and used for micro-scale culture.

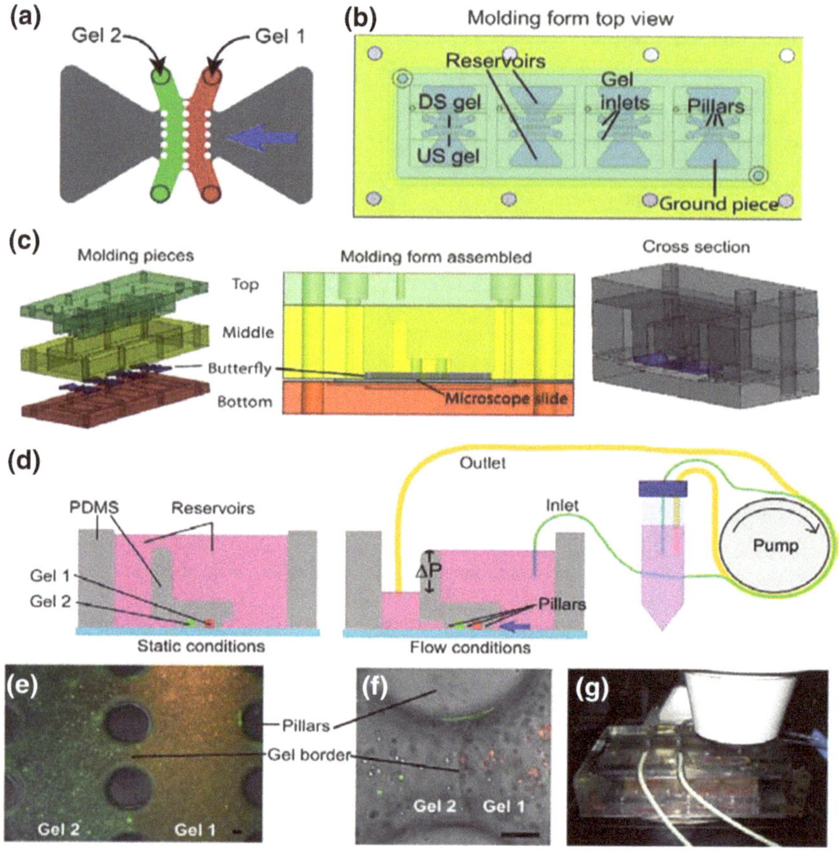

Fig. 4.6 Injection molding allows easy and robust set up of 4 parallel experiments on a conventional scope. **a** A single unit containing two gel compartments [upstream (US) and downstream (DS)] and medium reservoirs with different pressure heads that drive flow. **b** Top view of device shows four units on one slide, allowing parallel experimentation with a motorized microscope stage. **c** Flow chamber modules consisting of top, middle and bottom part defining the reservoirs and the butterfly-shaped ground piece creating gel channels in different perspectives as labelled. **d** Schematic of flow setup. For static conditions (*left*), both reservoirs are filled to the top; for flow conditions, a pressure head of 7 mm H_2O is maintained throughout the experiment with a peristaltic pump setup as shown. **e** Fluorescence microscopy image overlaid with phase of the interface between two fluorescently-labeled cell-loaded gels confined by PDMS pillars. **f** Close-up of (**e**) showing the gel border. **g** Chamber setup with tubing casket on an inverted microscope. Scale bars: 100 μm. Reprinted with permission from Ref. [117]

4.5 Non-destructive Observation Methods on Microfluidic Devices

The microfluidic device serves as a miniaturization platform to save reaction reagent, however, it challenges the methods for micro analysis. Unlike DNA/RNA-based molecular techniques, bioanalytical detections of samples in small volume lack a significant signal amplification step. The main microfluidic detection methods are categorized into three main kinds, they are optical methods, mass spectrometric (MS) methods and electrochemical methods. Mass spectrometry (MS) provides structural information for unknown analytes and allows label-free identification; therefore, it presents a detection method with the richest possible information. Perhaps MALDI MS and nanoelectrospray ionization (nanoESI) MS are the most suitable methods for microanalysis. But MS is omitted in this part to focus on in situ observation for intact cell structure. Optical imaging methods show high sensitivity and leave out pretreatment for purification and enrichment, thus are especially suitable for real-time on-chip study. These methods are presented as main body in this part. Besides, other methods like those based on bio-impedance is also briefly introduced (Fig. 4.7).

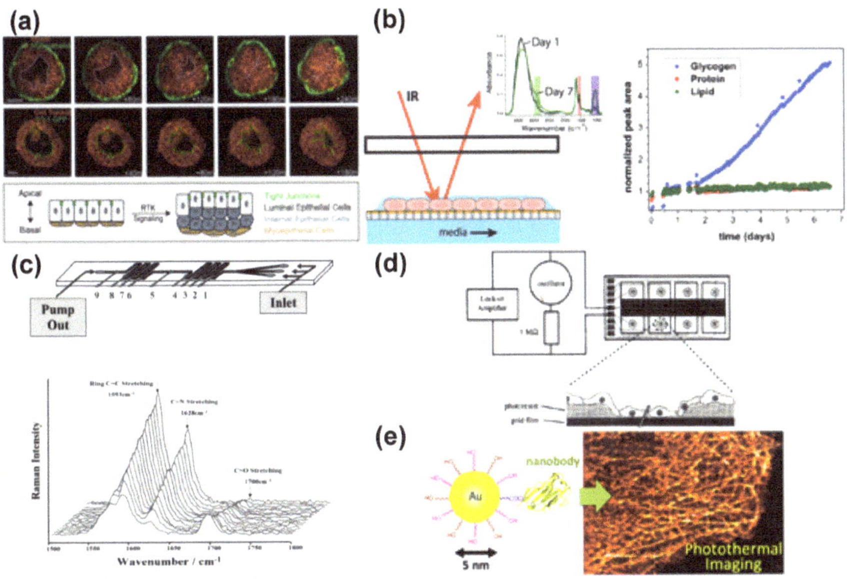

Fig. 4.7 Real-time and in-situ observation methods on microfluidic devices. **a** Mammary stratification generates an internal population of luminal epithelial cells lacking tight junctions. Still images of an organoid undergoing stratification with ZO-1-GFP marking tight junctions and Cell Tracker Red staining the cytosol, and a cartoon depiction of mammary epithelial stratification showing the generation of an internal luminal epithelial cell population lacking tight junctions (reprinted with permission from Ref. [124]). Scale bars: 20 μm. **b** Open-Channel microfluidic membrane device for long-term FT-IR spectromicroscopy of live adherent cells (reprinted with permission from Ref. [125]). **c** Time-dependent Raman spectra of imine formation in a glass microfluidic channel (reprinted with permission from Ref. [126]). **d** Schematic of the experimental setup to perform electric cell–substrate impedance sensing (reprinted with permission from Ref. [127]). **e** Photothermal imaging by a highly specific gold nanoprobe (reprinted with permission from Ref. [128])

4.5.1 Optical Methods

Microfluidic devices have benefits optical aspect over smaller multi-well plates, for instance, 96-well plates. To visualize cells by either fluorescence or phase in a 96-well plate is hardly to be realized, because the thickness of the plastic bottom is not fit for high magnification fluorescence images and interferes with phase contrast (meniscus of fluid in the well). Optical methods are the most widely used imaging methods combined with microfluidic chip, and are especially powerful in dynamic tracking cells and molecules. Yi et al. [129] realized dynamic imaging of specific mRNAs for G1 cell cycle arrest in live cells.

There are various kinds of optical methods that have been introduced in microfluidic chip, such fluorescence, absorbance, infrared (IR) and surface plasma resonance (SPR), optical emission spectroscopy, Raman detection, chemiluminescence and electrochemiluminescence. Among these optical methods, fluorescence-based method, because of its high compatibility and sensitivity at micro-space level, has found widely utilities in microfluidic systems to label cells [130].

Microfluidic devices for culture allow easy visualization via phase contrast along the length of the channel. In additional, most microfluidic devices are capable of being placed on any kinds of substrate, e.g., glass to be used for devices when need for fluorescent observation and analysis. Different microscope systems are able to be applied for different purpose, performing complementary roles to each other. Differential interference contrast is suitable for label-free visualization and imaging of collagen I fibers and optical sectioning in thick gels made for ECM. The confocal multicolor 3D time-lapse imaging make it possible to reach a 70–100 μm imaging depth in the tissue [131], and two-photon microscopy enables deeper visualization of ECM by the use of second harmonic signals. However, two-photon microscopy is limited for expanded applications due to high-energy deposition focal plane as well as high cost. Imaging by fluorescence labeled cells in 3D cultures facilitates a continuous description for cells and molecules in tissue-level development that can last for days to weeks [132]. An inverted microscope is typically used for this imaging, which also requires robust control of humidity, temperature, evaporation and CO_2. A major obstacle of this method is the scattering of light in thick gels used in 3D culture models. For the purpose of scattering minimization, one way is to simplify the culture, such as by using the core tissue organoids instead of whole organs [131, 133]. A second way is to adjust the index of refraction of the immersion medium in accordance with that of the culture medium. Third, the working distance of the lens should be fully regulated to match the image size and structure of the entire culture area.

Apart from fluorescence, IR spectroscopy and SPR have also been applied in on-chip cell study: For example, to monitor the spatial and chemical changes in chemical aspect in living PC12 cells by spatially resolved IR spectroscopy [125], and to monitor cell proliferation process of HeLa cells in a two-compartment microfluidic device [134].

In addition to these optical methods, samples come from 3D culture can be prepared for scanning and transmission electron microscopy to gain more information at subcellular level [126, 135, 136].

Thermal lens spectrometry (TLS), one of the photothermal techniques, relies on non-radiative relaxation processes and conversion from absorbed energy to heat. In theory, TLS can detect any substance absorbing the light, except that fluorescence quantum yield equals unity. TLS provides high sensitivity that is equal to spectrofluorimetric methods, enabling detection limit down to 10^{-7}AU (absorbance units). TLS is a method of molecular absorption spectroscopy, which covers a larger spectrum of substances than fluorescence technologies. Beyond that it takes all the advantages of conventional spectrophotometry. Besides, TLS is fit for integrating into an optical microscope to develop the thermal lens microscope (TLM) due to its fast and small-volume detection features. TLM has been developed into a temporal and spatial detector for microfluidic analysis. Because of lower interference and an increased efficiency of reaction in micro-scale space, TLM has better measure sensitivity and selectivity for the same path length of sample compared with conventional apparatus in macro space. The detection limit in absorbance can be increased several orders of magnitude in context of the use of thermal lens effect, especially when using nonpolar solvents [137].

To track single molecules in micro-environment, specific nanoparticles are used to facilitate photothermal imaging. These nanoparticles have extremely week luminescence yield, therefore, can convert all the absorbed energy into heat. The variation of the refraction is leaded by the increase of temperature. For long-time observation of small labels at nano-scale, researchers used green fluorescence protein (GFP) (<10 nm) functionally modified gold nanoparticles (5 nm) in environments such as cytoskeletal structures and adhesion sites [128]. Fast visualization of moving objects is unavailable to be clearly captured in the process of a raster scan. A triangulation-based tracking scheme was designed to resolve this problem. Due to the high stability of photothermal signal, this tracking method permits fully recording for the history of proteins, including intermediate states, inside cells [138].

4.5.2 Electrochemical Methods

Electrochemical methods are very sensitive and are the choice in the case that cells are captured on functional modified electrodes surface. These methods can be carried out in field-portable devices. These label-free quantitative methods, including single-cell dielectric spectroscopy, impedance spectroscopy and electric cell-substrate impedance sensing (ECIS) [139–141], were able to monitor cellular processes that implicate cell size or shape changes like cell growth, division, apoptosis and adhesion. Integrated with microfluidic devices, these methods are capable to measure cell–cell variations and on individual cells. Among these electrochemical methods, impedance is powerful, responding to modifications arisen from the target cells and using periodic small AC amplitudes. Bio-impedance monitoring has previously been applied to study cellular kinetics [142], cancer drug screening [143], or parasitized cells [144, 145]. Coupled to microfluidic chip, the sensing performance of impedance spectroscopy can be greatly improved by the use of microfluidics to take both advantages of greater flexibility and effective liquid manipulation. Zhou et al. [146] processed single cells studies by electrical impedance measurements for mouse embryonic stem cell differentiation. They analyzed the varying in electrical properties of cells quantitatively by evaluating the magnitude of the cell impedance in terms of their size. Manczak et al. [147] reported improved detecting sensitivity on target cells by a impedimetric microfluidic device. The CD14 antibody was modified on the surface of micro electrode to trap monocytes, and the sensitivity was assessed based on the trapping capacity. By detecting impedance, it was also successfully realized screen of ovarian cancer cells from whole blood [148].

4.6 Applications in Cellular Biology and Metabolomics

Bio-microfluidics have successfully been applied in cellular biology (Fig. 4.8) for study various cellular behaviors including cell growth [149], differentiation [152], signal transduction [153], protein secretion and transport [154], gene expression [155], cell and ECM behaviors [151], and cytoskeletal dynamics [156].

4.6.1 Signal Transduction

On a microfuidic chip system that suitable for migration model, the endothelium-derived paracrine signaling that affects vascular events important for adaptive remodeling was studied [59]. The migration and moving of smooth muscle cells (SMCs) indicated that biomechanical forces control cell-cell interactions and cellular phenotype. For on-chip observation of cellular reactions, a microfluidic device was designed with cell control and diluting functions, while the latter function is used to produce a concentration gradient [157]. Based on the ATP-dependent calcium uptake reaction model in HL-60 cells, the on-chip measurement was conducted for the threshold of ATP concentration, which induces significant signal of intracellular calcium. The results prove the high feasibility of microfluidic system application for real-time observation for cellular processes when treated with a concentration gradient of changed solution. Livestock embryo manipulation needs the handling of individual embryos, and is becoming conventional for the development of in vitro fertilization, gene manipulation and cloning techniques. A microfluidic system was designed to transport individual, pre-implantation mouse embryos to selected locations through a channel network [158]. Another microsystem for cell experiments including a cell culture microchip and a scanning

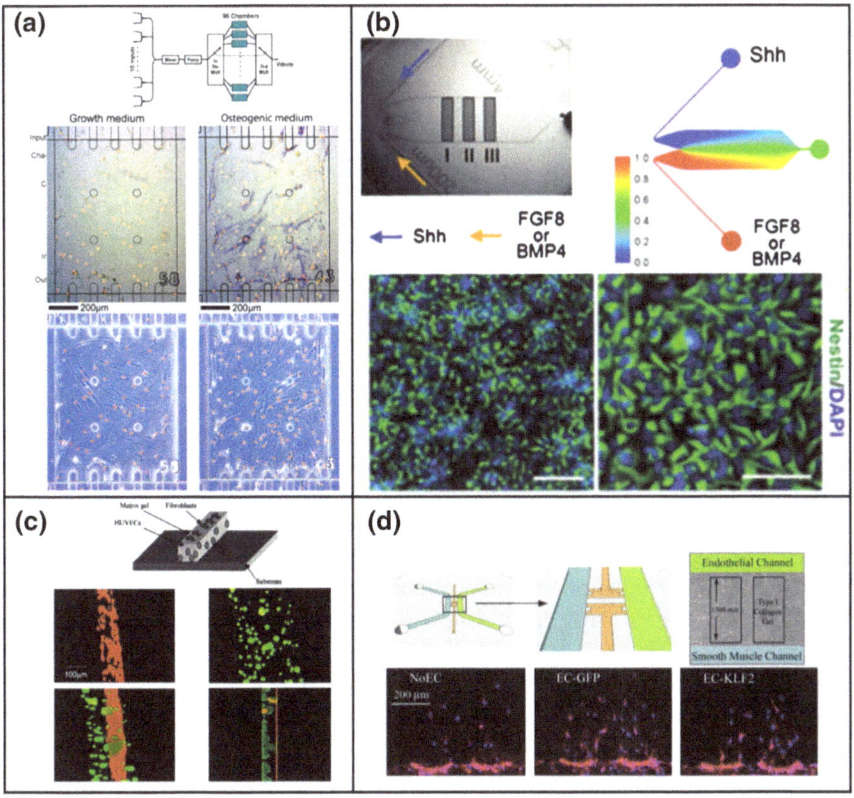

Fig. 4.8 Microfluidic applications in cellular biology. **a** Effect of osteogenic stimulation on differentiation and proliferation during transient stimulation (reprinted with permission from Ref. [149]). **b** Differentiation of human ESC-derived neural progenitor cells in the microfluidic gradient chip (reprinted with permission from Ref. [150]). **c** Microfluidic Patterning of Cells in Extracellular Matrix Biopolymers (reprinted with permission from Ref. [151]). **d** EC expressing KLF2 inhibit EC-directed SMC migration (reprinted with permission from Ref. [59])

thermal lens microscope detection system has been described by Tamaki [159]. For the target of high throughput screening, living cells expressed engineered fluorescent protein biosensors were selectively localized into microarrays [160]. Hediger and his group developed a microchip for epithelial cell culture and electrical characterization, realizing a functionally integrated device for cell culture and characterization of potentiometric and impedimetric simultaneously [161]. Cai's group integrated microelectrode sensors with SU8 gel chambers in a microsystem, performing electrochemical measurements of lactate dynamically in the process of cell permeabilization of single heart of cells [162]. Heuschkel's group reported a microfluidic device having buried microchannels for study on neural cells. This device realized culture, stimulation and detection on a single device [163]. The microchannels permitted fast and local drug delivering to the cells. Chen et al. [164] developed gated pinched-flow (GPF), a new microfluidic approach, to couple with optical imaging for cell-to-cell signaling dynamics investigation. It enables precise on-chip chemical perfusion or stimulation upon target cells among cells populations temporally and spatially, showing advantages of both conventional hydrodynamic focusing and gating. By GPF, stimulation can be localized on single cells, avoiding interference with adjacent cells. Guided by GPF, seeded cells in the channel were imaged with high repeatability for their ATP-activated propagation of intercellular calcium waves. Inhibition investigations demonstrated that the calcium signals between cells were transferred via gap junctions in dependence of direct cytosolic transfer of molecules.

4.6.2 Gene Expression

Typically, the on–chip detection for gene expression is realized by amplifying DNA or RNA [155]. Microfluidic devices have been utilized in single cells to measure gene network dynamics. The ways have been reviewed, in which microfluidic devices can be used from the standpoint of synthetic biology to examine the dynamics of gene regulation and intracellular signaling pathways in yeast, bacteria and higher eukaryotes [165]. This review presents various designs of the ways to lead to non-trivial environmental perturbations (e.g. spatial gradients and time-dependent fluctuations of media concentrations) in microfluidic devices.

In embryonic whole-lung explants, real-time tracking has been used to illustrate the localization specificity of polarity protein Scribble. This protein is required for lumen morphogenesis in mam-malian lung rather than in bud bifurcation [166]. The results revealed that the misalignment of distal epithelial cells and mislocalization of junction proteins result from the depletion of Scribble in lungs. In combination with genetic manipulation (siRNA knockdown), live-cell fluorescence imaging method provides information about the signals required for salivary gland clefting [167, 168]. Specific genes had been identified to play key role in lung morphogenesis and salivary gland clefting. This real-time tracking approach is promising for identification of the genes involved in additional cell processes in other organs, for example, dynamic tracking exclusion of engineered epithelial cells. These cells lack RET from the tips of elongating ureteric buds. This process demonstrates that the receptor tyrosine kinase (RTK) RET is necessary in the process of branching morphogenesis in mouse kidney [169].

4.6.3 Metabolic Function Analysis

The study of metabolomics focuses on high-throughput small-molecule screening in research of pharmaceutical and environmental toxicology [170]. Metabolic function is another factor that reflects chemical toxicity and the development of disease, which is directly correlated with mitochondrial function. Metabolic function analysis is a comprehensive evaluation of factors (such as oxygen uptake rate, glucose uptake, lactate production, and ATP production), which reflect the state of oxidative phosphorylation and anaerobic glycolysis. Coupled to microfluidic chip, dynamic and prognostic information for metabolic function can be a achieved through controllable microfluidic switchboard. A report by Do et al. [171] described an analyzer system, which is capable of detecting multiple metabolic parameters in human blood simultaneously, combines three parts including an interface socket, a smart biochip cartridge, and an ASIC electronic/control circuit.

Buehler et al. [172] detected respiration and acidification by Clark-type electrodes and ion-sensitive field-effect transistors (ISFETs), cell adhesion by interdigitated electrode structures (IDES), which are integrated in the mircrosensor chip. Similar function analysis had also been realized in heart-on-chip model [173].

Bavli et al. [174] quantified the dynamics of cellular adaptation to mitochondrial damage on a liver-on-a-chip device. They provided real-time analysis for these metabolic parameters by non-destructive methods. For example, they monitored oxygen uptake rate in real time using two-frequency phase. The change of glucose and lactate were measured contiguous by electrochemical method via a computer-controlled switch system.

4.7 Perspectives

The microfluidic systems that coupled to imaging or other methods (e.g. lasers, mass spectrometry, scanning probe microscopes and nuclear magnetic resonance spectroscopy) will be a powerful tool in real-time dynamic stimulation and detection. For this goal, microfluidics should be easily and widely used. In other word, it must have the ability to serve for users like cell biologists, clinicians, public health officials or police officers who are not experts in fluid physics. For these applications, commercial systems must be commonly used and should be inexpensively available.

References

1. Halldorsson S, Lucumi E, Gómez-Sjöberg R et al (2015) Advantages and challenges of microfluidic cell culture in polydimethylsiloxane devices. Biosens Bioelectron 63:218–231
2. Meyvantsson I, Beebe DJ (2008) Cell culture models in microfluidic systems. Annu Rev Anal Chem (Palo Alto Calif) 1:423–449
3. Yu H, Alexander CM, Beebe DJ (2007) Understanding microchannel culture: parameters involved in soluble factor signaling. Lab Chip 7:726–730
4. Tilles AW, Baskaran H, Roy P et al (2001) Effects of oxygenation and flow on the viability and function of rat hepatocytes cocultured in a microchannel flat-plate bioreactor. Biotechnol Bioeng 73:379–389

5. Raty S, Walters EM, Davis J et al (2004) Embryonic development in the mouse is enhanced via microchannel culture. Lab Chip 4:186–190
6. Wang L, Sun B, Ziemer KS et al (2010) Chemical and physical modifications to poly (dimethylsiloxane) surfaces affect adhesion of Caco-2 cells. J Biomed Mater Res A 93A:1260–1271
7. Yang L, Li L, Tu Q et al (2010) Photocatalyzed surface modification of poly(dimethylsiloxane) with polysaccharides and assay of their protein adsorption and cytocompatibility. Anal Chem 82:6430–6439
8. Ong SM, Zhang C, Toh YC et al (2008) A gel-free 3D microfluidic cell culture system. Biomaterials 29:3237–3244
9. Choi J, Kim S, Jung J et al (2011) Wnt5a-mediating neurogenesis of human adipose tissue-derived stem cells in a 3D microfluidic cell culture system. Biomaterials 32:7013–7022
10. Chung S, Sudo R, Mack PJ et al (2009) Cell migration into scaffolds under co-culture conditions in a microfluidic platform. Lab Chip 9:269–275
11. Bersini S, Jeon JS, Dubini G et al (2014) A microfluidic 3D in vitro model for specificity of breast cancer metastasis to bone. Biomaterials 35:2454–2461
12. Kim S, Lee H, Chung M et al (2013) Engineering of functional, perfusable 3D microvascular networks on a chip. Lab Chip 13:1489–1500
13. Weis SM, Cheresh DA (2011) Tumor angiogenesis: molecular pathways and therapeutic targets. Nat Med 17:1359–1370
14. Zheng Y, Chen J, Craven M et al (2012) In vitro microvessels for the study of angiogenesis and thrombosis. Proc Natl Acad Sci U S A 109:9342–9347
15. Lee H, Kim S, Chung M et al (2014) A bioengineered array of 3D microvessels for vascular permeability assay. Microvasc Res 91:90–98
16. Baker BM, Trappmann B, Stapleton SC et al (2013) Microfluidics embedded within extracellular matrix to define vascular architectures and pattern diffusive gradients. Lab Chip 13:3246–3252
17. Park YK, Tu TY, Lim SH et al (2014) In vitro microvessel growth and remodeling within a three-dimensional microfluidic environment. Cell Mol Bioeng 7:15–25
18. Tourovskaia A, Fauver M, Kramer G et al (2014) Tissue-engineered microenvironment systems for modeling human vasculature. Exp Biol Med (Maywood) 239:1264–1271
19. Mu X, Zheng W, Xiao L et al (2013) Engineering a 3D vascular network in hydrogel for mimicking a nephron. Lab Chip 13:1612–1618
20. Drury JL, Mooney DJ (2003) Hydrogels for tissue engineering: scaffold design variables and applications. Biomaterials 24:4337–4351
21. Hoffman AS (2012) Hydrogels for biomedical applications. Adv Drug Deliv Rev 64:18–23
22. Kwapiszewska K, Michalczuk A, Rybka M et al (2014) A microfluidic-based platform for tumour spheroid culture, monitoring and drug screening. Lab Chip 14:2096–2104
23. Ziolkowska K, Stelmachowska A, Kwapiszewski R et al (2013) Long-term three-dimensional cell culture and anticancer drug activity evaluation in a microfluidic chip. Biosens Bioelectron 40:68–74
24. Choi YJ, Park J, Lee SH (2013) Size-controllable networked neurospheres as a 3D neuronal tissue model for Alzheimer's disease studies. Biomaterials 34:2938–2946
25. van Duinen V, Trietsch SJ, Joore J et al (2015) Microfluidic 3D cell culture: from tools to tissue models. Curr Opin Biotechnol 35:118–126
26. Song HH, Park KM, Gerecht S (2014) Hydrogels to model 3D in vitro microenvironment of tumor vascularization. Adv Drug Deliv Rev 79–80:19–29
27. Novo P, Dell'Aica M, Janasek D et al (2016) High spatial and temporal resolution cell manipulation techniques in microchannels. Analyst 141:1888–1905
28. Chiang YY, Haeri S, Gizewski C et al (2013) Whole cell quenched flow analysis. Anal Chem 85:11560–11567
29. Holmes D, Whyte G, Bailey J et al (2014) Separation of blood cells with differing deformability using deterministic lateral displacement(dagger). Interface Focus 4:20140011

30. Laurell T, Petersson F, Nilsson A (2007) Chip integrated strategies for acoustic separation and manipulation of cells and particles. Chem Soc Rev 36:492–506
31. Bruus H (2012) Acoustofluidics 10: scaling laws in acoustophoresis. Lab Chip 12:1578–1586
32. Pohl HA, Crane JS (1971) Dielectrophoresis of cells. Biophys J 11:711
33. Salmanzadeh A, Romero L, Shafiee H et al (2012) Isolation of prostate tumor initiating cells (TICs) through their dielectrophoretic signature. Lab Chip 12:182–189
34. Karabacak NM, Spuhler PS, Fachin F et al (2014) Microfluidic, marker-free isolation of circulating tumor cells from blood samples. Nat Protoc 9:694–710
35. Del Giudice F, Madadi H, Villone MM et al (2015) Magnetophoresis 'meets' viscoelasticity: deterministic separation of magnetic particles in a modular microfluidic device. Lab Chip 15:1912–1922
36. Collins DJ, Alan T, Neild A (2014) Particle separation using virtual deterministic lateral displacement (vDLD). Lab Chip 14:1595–1603
37. Beech JP, Jonsson P, Tegenfeldt JO (2009) Tipping the balance of deterministic lateral displacement devices using dielectrophoresis. Lab Chip 9:2698–2706
38. Adams JD, Thevoz P, Bruus H et al (2009) Integrated acoustic and magnetic separation in microfluidic channels. Appl Phys Lett 95:254103
39. Novo P, Volpetti F, Chu V et al (2013) Control of sequential fluid delivery in a fully autonomous capillary microfluidic device. Lab Chip 13:641–645
40. Novo P, Chu V, Conde JP (2014) Integrated optical detection of autonomous capillary microfluidic immunoassays: a hand-held point-of-care prototype. Biosens Bioelectron 57:284–291
41. Hughes AJ, Spelke DP, Xu Z et al (2014) Single-cell western blotting. Nat Methods 11:749–755
42. Denervaud N, Becker J, Delgado-Gonzalo R et al (2013) A chemostat array enables the spatio-temporal analysis of the yeast proteome. Proc Natl Acad Sci U S A 110:15842–15847
43. Chingozha L, Zhan M, Zhu C et al (2014) A generalizable, tunable microfluidic platform for delivering fast temporally varying chemical signals to probe single-cell response dynamics. Anal Chem 86:10138–10147
44. Ng AH, Dean Chamberlain M, Situ H et al (2015) Digital microfluidic immunocytochemistry in single cells. Nat Commun 6:7513
45. Fair RB (2007) Digital microfluidics: is a true lab-on-a-chip possible? Microfluid Nanofluid 3:245–281
46. Solari FA, Dell'Aica M, Sickmann A et al (2015) Why phosphoproteomics is still a challenge. Mol BioSyst 11:1487–1493
47. Nguyen N-T, Wu Z (2004) Micromixers—a review. J Micromech Microeng 15:R1
48. Enger J, Goksor M, Ramser K et al (2004) Optical tweezers applied to a microfluidic system. Lab Chip 4:196–200
49. Eriksson E, Enger J, Nordlander B et al (2007) A microfluidic system in combination with optical tweezers for analyzing rapid and reversible cytological alterations in single cells upon environmental changes. Lab Chip 7:71–76
50. Umehara S, Wakamoto Y, Inoue I et al (2003) On-chip single-cell microcultivation assay for monitoring environmental effects on isolated cells. Biochem Biophy Res Co 305:534–540
51. Andersson H, van den Berg A (2003) Microfluidic devices for cellomics: a review. Sensor Actuat B-Chem 92:315–325
52. Warrick JW, Murphy WL, Beebe DJ (2008) Screening the cellular microenvironment: a role for microfluidics. IEEE Rev Biomed Eng 1:75–93
53. Yun JY, Jambovane S, Kim S-K et al (2011) Log-scale dose response of inhibitors on a chip. Anal Chem 83:6148–6153
54. Lucchetta EM, Lee JH, Fu LA et al (2005) Dynamics of Drosophila embryonic patterning network perturbed in space and time using microfluidics. Nature 434:1134–1138
55. Li R, Lv X, Zhang X et al (2016) Microfluidics for cell-cell interactions: a review. Front Chem Sci Eng 10:90–98

56. Dertinger SKW, Jiang X, Li Z et al (2002) Gradients of substrate-bound laminin orient axonal specification of neurons. Proc Natl Acad Sci U S A 99:12542–12547
57. Lin X, Chen Q, Liu W et al (2015) Oxygen-induced cell migration and on-line monitoring biomarkers modulation of cervical cancers on a microfluidic system. Sci Rep 5:9643
58. Vickerman V, Blundo J, Chung S et al (2008) Design, fabrication and implementation of a novel multi-parameter control microfluidic platform for three-dimensional cell culture and real-time imaging. Lab Chip 8:1468–1477
59. Mack PJ, Zhang Y, Chung S et al (2009) Biomechanical regulation of endothelium-dependent events critical for adaptive remodeling. J Biol Chem 284:8412–8420
60. Song JW, Munn LL (2011) Fluid forces control endothelial sprouting. Proc Natl Acad Sci U S A 108:15342–15347
61. Mosadegh B, Huang C, Park JW et al (2007) Generation of stable complex gradients across two-dimensional surfaces and three-dimensional gels. Langmuir 23:10910–10912
62. Wu H, Huang B, Zare RN (2006) Generation of complex, static solution gradients in microfluidic channels. J Am Chem Soc 128:4194–4195
63. Abhyankar VV, Lokuta MA, Huttenlocher A et al (2006) Characterization of a membrane-based gradient generator for use in cell-signaling studies. Lab Chip 6:389–393
64. Saadi W, Rhee SW, Lin F et al (2007) Generation of stable concentration gradients in 2D and 3D environments using a microfluidic ladder chamber. Biomed Microdevices 9:627–635
65. Faure-Andre G, Vargas P, Yuseff M-I et al (2008) Regulation of dendritic cell migration by CD74, the MHC class II-associated invariant chain. Science 322:1705–1710
66. Irimia D, Toner M (2009) Spontaneous migration of cancer cells under conditions of mechanical confinement. Integr Biol-UK 1:506–512
67. Lucchetta EM, Lee JH, Fu LA et al (2005) Dynamics of *Drosophila* embryonic patterning network perturbed in space and time using microfluidics. Nature 434:1134–1138
68. Pearce TM, Wilson JA, Oakes SG et al (2005) Integrated microelectrode array and microfluidics for temperature clamp of sensory neurons in culture. Lab Chip 5:97–101
69. Long T, Ford RM (2009) Enhanced transverse migration of bacteria by chemotaxis in a porous T-sensor. Environ Sci Technol 43:1546–1552
70. Toh AGG, Wang ZP, Yang C et al (2014) Engineering microfluidic concentration gradient generators for biological applications. Microfluid Nanofluid 16:1–18
71. Jeon NL, Dertinger SK, Chiu DT et al (2000) Generation of solution and surface gradients using microfluidic systems. Langmuir 16:8311–8316
72. Hung PJ, Lee PJ, Sabounchi P et al (2005) Continuous perfusion microfluidic cell culture array for high-throughput cell-based assays. Biotech Bioeng 89:1–8
73. Irimia D, Charras G, Agrawal N et al (2007) Polar stimulation and constrained cell migration in microfluidic channels. Lab Chip 7:1783–1790
74. Atencia J, Cooksey GA, Locascio LE (2012) A robust diffusion-based gradient generator for dynamic cell assays. Lab Chip 12:309–316
75. Ahmed T, Shimizu TS, Stocker R (2010) Bacterial chemotaxis in linear and nonlinear steady microfluidic gradients. Nano Lett 10:3379–3385
76. Kothapalli CR, Van Veen E, De Valence S et al (2011) A high-throughput microfluidic assay to study neurite response to growth factor gradients. Lab Chip 11:497–507
77. Haessler U, Pisano M, Wu M et al (2011) Dendritic cell chemotaxis in 3D under defined chemokine gradients reveals differential response to ligands CCL21 and CCL19. Proc Natl Acad Sci U S A 108:5614–5619
78. Qasaimeh MA, Gervais T, Juncker D (2011) Microfluidic quadrupole and floating concentration gradient. Nat Commun 2:464
79. Wang S-J, Saadi W, Lin F et al (2004) Differential effects of EGF gradient profiles on MDA-MB-231 breast cancer cell chemotaxis. Exp Cell Res 300:180–189
80. Nandagopal S, Wu D, Lin F (2011) Combinatorial guidance by CCR7 ligands for T lymphocytes migration in co-existing chemokine fields. PLoS ONE 6:e18183

81. Jeon NL, Baskaran H, Dertinger SK et al (2002) Neutrophil chemotaxis in linear and complex gradients of interleukin-8 formed in a microfabricated device. Nat Biotechnol 20:826–830
82. Barkefors I, Le Jan S, Jakobsson L et al (2008) Endothelial cell migration in stable gradients of vascular endothelial growth factor a and fibroblast growth factor 2 effects on chemotaxis and chemokinesis. J Biol Chem 283:13905–13912
83. Irimia D, Liu S-Y, Tharp WG et al (2006) Microfluidic system for measuring neutrophil migratory responses to fast switches of chemical gradients. Lab Chip 6:191–198
84. Chung BG, Flanagan LA, Rhee SW et al (2005) Human neural stem cell growth and differentiation in a gradient-generating microfluidic device. Lab Chip 5:401–406
85. Park JY, Hwang CM, Lee SH et al (2007) Gradient generation by an osmotic pump and the behavior of human mesenchymal stem cells under the fetal bovine serum concentration gradient. Lab Chip 7:1673–1680
86. Gupta K, Kim D-H, Ellison D et al (2010) Lab-on-a-chip devices as an emerging platform for stem cell biology. Lab Chip 10:2019–2031
87. Li GN, Liu J, Hoffman-Kim D (2008) Multi-molecular gradients of permissive and inhibitory cues direct neurite outgrowth. Ann Biomed Eng 36:889–904
88. Lin F (2009) A microfluidics-based method for analyzing leukocyte migration to chemoattractant gradients. Method Enzymol 461:333–347
89. van der Meer AD, Vermeul K, Poot AA et al (2010) A microfluidic wound-healing assay for quantifying endothelial cell migration. Am J Physiol-Heart C 298:H719–H725
90. Heo Y, Cabrera L, Bormann C et al (2010) Dynamic microfunnel culture enhances mouse embryo development and pregnancy rates. Hum Reprod 25:613–622
91. Boyden S (1962) The chemotactic effect of mixtures of antibody and antigen on polymorphonuclear leucocytes. J Exp Med 115:453–466
92. Zicha D, Dunn GA, Brown AF (1991) A new direct-viewing chemotaxis chamber. J Cell Sci 99:769–775
93. Zigmond SH, Hirsch JG (1973) Leukocyte locomotion and chemotaxis new methods for evaluation, and demonstration of a cell-derived chemotactic factor. J Exp Med 137:387–410
94. Kim M, Kim T (2010) Diffusion-based and long-range concentration gradients of multiple chemicals for bacterial chemotaxis Assays. Anal Chem 82:9401–9409
95. Yap B, Kamm RD (1985) Mechanical deformation of neutrophils into narrow channels induces pseudopod projection and changes in biomechanical properties. J Appl Physiol 98:1930–1939
96. Oh KW, Lee K, Ahn B, Furlani EP (2012) Design of pressure-driven microfluidic networks using electric circuit analogy. Lab Chip 12:515–545
97. Toh AG, Wang Z, Yang C et al (2014) Engineering microfluidic concentration gradient generators for biological applications. Microfluid Nanofluid 16:1–18
98. Li R, Lv X, Zhang X, Saeed O, Deng Y (2016) Microfluidics for cell-cell interactions: a review. Front Chem Sci Eng 10:90–98
99. Huang CP, Lu J, Seon H et al (2009) Engineering microscale cellular niches for three-dimensional multicellular co-cultures. Lab Chip 9:1740–1748
100. van der Meer AD, Orlova VV, ten Dijke P et al (2013) Three-dimensional co-cultures of human endothelial cells and embryonic stem cell-derived pericytes inside a microfluidic device. Lab Chip 13:3562–3568
101. Jie M, Li H-F, Lin L et al (2016) Integrated microfluidic system for cell co-culture and simulation of drug metabolism. Rsc Adv 6:54564–54572
102. Chen Q, Wu J, Zhuang Q et al (2013) Microfluidic isolation of highly pure embryonic stem cells using feeder-separated co-culture system. Sci Rep 3:2433
103. Gao D, Liu H, Lin JM et al (2012) Characterization of drug permeability in Caco-2 monolayers by mass spectrometry on a membrane-based microfluidic device. Lab Chip 13:978–985

104. Wu J, Jie M, Dong X et al (2016) Multi-channel cell co-culture for drug development based on glass microfluidic chip-mass spectrometry coupled platform. Rapid Commun Mass Spectrom 30:80–86
105. Mao S, Zhang J, Li H et al (2013) Strategy for signaling molecule detection by using an integrated microfluidic device coupled with mass spectrometry to study cell-to-cell communication. Anal Chem 85:868–876
106. Kane RS, Takayama S, Ostuni E et al (1999) Patterning proteins and cells using soft lithography. Biomaterials 20:2363–2376
107. Abhyankar VV, Beebe DJ (2007) Spatiotemporal micropatterning of cells on arbitrary substrates. Anal Chem 79:4066–4073
108. Chen CS, Jiang X, Whitesides GM (2005) Microengineering the environment of mammalian cells in culture. MRS Bull 30:194–201
109. Liu W, Li L, Wang X et al (2010) An integrated microfluidic system for studying cell-microenvironmental interactions versatilely and dynamically. Lab Chip 10:1717–1724
110. Businaro L, De Ninno A, Schiavoni G et al (2013) Cross talk between cancer and immune cells: exploring complex dynamics in a microfluidic environment. Lab Chip 13:229–239
111. Ostrovidov S, Sakai Y, Fujii T (2011) Integration of a pump and an electrical sensor into a membrane-based PDMS microbioreactor for cell culture and drug testing. Biomed Microdevices 13:847–864
112. Chin L, Luo K, Park W et al (2012) Double-layer hepatocyte tumor co-culture using hydrogel for drug effectivity and specificity analysis. In: Micro electro mechanical systems (MEMS), 2012 IEEE 25th international conference on, IEEE, pp. 808–811
113. Discher DE, Janmey P, Y-l Wang (2005) Tissue cells feel and respond to the stiffness of their substrate. Science 310:1139–1143
114. Trappmann B, Gautrot JE, Connelly JT et al (2012) Extracellular-matrix tethering regulates stem-cell fate. Nat Mater 11:642–649
115. Fischbach C, Kong HJ, Hsiong SX, Evangelista MB, Yuen W, Mooney DJ (2009) Cancer cell angiogenic capability is regulated by 3D culture and integrin engagement. Proc Natl Acad Sci U S A 106:399–404
116. Gunawan RC, Silvestre J, Gaskins HR et al (2006) Cell migration and polarity on microfabricated gradients of extracellular matrix proteins. Langmuir 22:4250–4258
117. Haessler U, Teo JCM, Foretay D et al (2012) Migration dynamics of breast cancer cells in a tunable 3D interstitial flow chamber. Integr Biol 4:401–409
118. Terrill RH, Balss KM, Zhang Y et al (2000) Dynamic monolayer gradients: Active spatiotemporal control of alkanethiol coatings on thin gold films. J Am Chem Soc 122:988–989
119. Hypolite CL, McLernon TL, Adams DN et al (1997) Formation of microscale gradients of protein using heterobifunctional photolinkers. Bioconjugate Chem 8:658–663
120. Herbert CB, McLernon TL, Hypolite CL et al (1997) Micropatterning gradients and controlling surface densities of photoactivatable biomolecules on self-assembled monolayers of oligo(ethylene glycol) alkanethiolates. Chem Biol 4:731–737
121. Tan WH, Takeuchi S (2007) Monodisperse alginate hydrogel microbeads for cell encapsulation. Adv Mater 19:2696–2701
122. Koh WG, Pishko MV (2006) Fabrication of cell-containing hydrogel microstructures inside microfluidic devices that can be used as cell-based biosensors. Anal Bioanal Chem 385:1389–1397
123. Wu H, Huang B, Zare RN (2006) Generation of complex, static solution gradients in microfluidic channels. J Am Chem Soc 128:4194–4195
124. Huebner RJ, Lechler T, Ewald AJ (2014) Developmental stratification of the mammary epithelium occurs through symmetry-breaking vertical divisions of apically positioned luminal cells. Development 141:1085–1094
125. Loutherback K, Chen L, Holman HY (2015) Open-channel microfluidic membrane device for long-term FT-IR spectromicroscopy of live adherent cells. Anal Chem 87:4601–4606

126. Chen L, Choo J (2008) Recent advances in surface-enhanced Raman scattering detection technology for microfluidic chips. Electrophoresis 29:1815–1828
127. Wegener J, Keese CR, Giaever I (2000) Electric cell–substrate impedance sensing (ECIS) as a noninvasive means to monitor the kinetics of cell spreading to artificial surfaces. Exp Cell Res 259:158–166
128. Leduc C, Si S, Gautier J et al (2013) A highly specific gold nanoprobe for live-cell single-molecule imaging. Nano Lett 13:1489–1494
129. Yi L, Lin X, Li H et al (2017) Dynamic imaging of MYC and CDKN1A mRNAs as an indicator of cell G1-phase arrest. Chem Commun 53:1900–1903
130. Okagbare PI, Soper SA (2009) High throughput single molecule detection for monitoring biochemical reactions. Analyst 134:97–106
131. Huebner RJ, Lechler T, Ewald AJ (2014) Developmental stratification of the mammary epithelium occurs through symmetry-breaking vertical divisions of apically positioned luminal cells. Development 141:1085–1094
132. Ewald AJ (2013) Practical considerations for long-term time-lapse imaging of epithelial morphogenesis in three-dimensional organotypic cultures. Cold Spring Harb Protoc 2013:100–117
133. Ewald AJ, Brenot A, Duong M et al (2008) Collective epithelial migration and cell rearrangements drive mammary branching morphogenesis. Dev Cell 14:570–581
134. Deng S, Yu X, Liu R et al (2016) A two-compartment microfluidic device for long-term live cell detection based on surface plasmon resonance. Biomicrofluidics 10:044109
135. Underwood JM, Imbalzano KM, Weaver VM et al (2006) The ultrastructure of MCF-10A acini. J Cell Physiol 208:141–148
136. Ewald AJ, Huebner RJ, Palsdottir H et al (2012) Mammary collective cell migration involves transient loss of epithelial features and individual cell migration within the epithelium. J Cell Sci 125:2638–2654
137. Kitamori T, Tokeshi M, Hibara A et al (2004) Peer reviewed: thermal lens microscopy and microchip chemistry. Anal Chem 76:52 A–60 A
138. Lasne D, Blab GA, Berciaud S et al (2006) Single nanoparticle photothermal tracking (SNaPT) of 5-nm gold beads in live cells. Biophy J 91:4598–4604
139. Bagnaninchi PO, Drummond N (2011) Real-time label-free monitoring of adipose-derived stem cell differentiation with electric cell-substrate impedance sensing. Proc Natl Acad Sci U S A 108:6462–6467
140. Gawad S, Cheung K, Seger U et al (2004) Dielectric spectroscopy in a micromachined flow cytometer: theoretical and practical considerations. Lab Chip 4:241–251
141. Chen N-C, Chen C-H, Chen M-K et al (2014) Single-cell trapping and impedance measurement utilizing dielectrophoresis in a parallel-plate microfluidic device. Sens Actuat B-Chem 190:570–577
142. Wegener J, Keese CR, Giaever I (2000) Electric cell-substrate impedance sensing (ECIS) as a noninvasive means to monitor the kinetics of cell spreading to artificial surfaces. Exp Cell Res 259:158–166
143. Caviglia C, Zor K, Canepa S et al (2015) Interdependence of initial cell density, drug concentration and exposure time revealed by real-time impedance spectroscopic cytotoxicity assay. Analyst 140:3623–3629
144. Liu Q, Wu C, Cai H et al (2014) Cell-based biosensors and their application in biomedicine. Chem Rev 114:6423–6461
145. Reybier K, Ribaut C, Coste A et al (2010) Characterization of oxidative stress in Leishmaniasis-infected or LPS-stimulated macrophages using electrochemical impedance spectroscopy. Biosens Bioelectron 25:2566–2572
146. Zhou Y, Basu S, Laue E et al (2016) Single cell studies of mouse embryonic stem cell (mESC) differentiation by electrical impedance measurements in a microfluidic device. Biosens Bioelectron 81:249–258

147. Manczak R, Fouet M, Courson R et al (2016) Improved on-chip impedimetric immuno-detection of subpopulations of cells toward single-cell resolution. Sens Actua B-Chem 230:825–831
148. Nwankire CE, Venkatanarayanan A, Glennon T et al (2015) Label-free impedance detection of cancer cells from whole blood on an integrated centrifugal microfluidic platform. Biosens Bioelectron 68:382–389
149. Gómez-Sjöberg R, Leyrat AA, Pirone DM et al (2007) Versatile, fully automated, microfluidic cell culture system. Anal Chem 79:8557–8563
150. Park JY, Kim SK, Woo DH et al (2009) Differentiation of neural progenitor cells in a microfluidic chip-generated cytokine gradient. Stem Cells 27:2646–2654
151. Tan W, Desai TA (2003) Microfluidic patterning of cells in extracellular matrix biopolymers: effects of channel size, cell type, and matrix composition on pattern integrity. Tissue Eng 9:255–267
152. Tourovskaia A, Figueroa-Masot X, Folch A (2005) Differentiation-on-a-chip: a microfluidic platform for long-term cell culture studies. Lab Chip 5:14–19
153. Liedert A, Kaspar D, Blakytny R et al (2006) Signal transduction pathways involved in mechanotransduction in bone cells. Biochem Biophy Res Commun 349:1–5
154. Huang W-H, Cheng W, Zhang Z et al (2004) Transport, location, and quantal release monitoring of single cells on a microfluidic device. Anal Chem 76:483–488
155. Toriello NM, Douglas ES, Thaitrong N et al (2008) Integrated microfluidic bioprocessor for single-cell gene expression analysis. Proc Natl Acad Sci U S A 105:20173–20178
156. Fletcher DA, Mullins RD (2010) Cell mechanics and the cytoskeleton. Nature 463:485–492
157. Yang M, Li CW, Yang J (2002) Cell docking and on-chip monitoring of cellular reactions with a controlled concentration gradient on a microfluidic device. Anal Chem 74:3991–4001
158. Glasgow IK, Zeringue HC, Beebe DJ et al (1998) Individual embryo transport and retention on a chip. In: Micro total analysis systems' 98, Springer, pp. 199–202
159. Tamaki E, Sato K, Tokeshi M et al (2002) Single-cell analysis by a scanning thermal lens microscope with a microchip: direct monitoring of cytochrome c distribution during apoptosis process. Anal Chem 74:1560–1564
160. Kapur R, Giuliano KA, Campana M et al (1999) Streamlining the drug discovery process by integrating miniaturization, high throughput screening, high content screening, and automation on the cellchip™ system. Biomed Microdevices 2:99–109
161. Hediger S, Sayah A, Horisberger J et al (2001) Modular microsystem for epithelial cell culture and electrical characterisation. Biosens Bioelectron 16:689–694
162. Cai X, Klauke N, Glidle A et al (2002) Ultra-low-volume, real-time measurements of lactate from the single heart cell using microsystems technology. Anal Chem 74:908–914
163. Heuschkel MO, Guérin L, Buisson B et al (1998) Buried microchannels in photopolymer for delivering of solutions to neurons in a network. Sens Actuat B-Chem 48:356–361
164. Chen P, Feng X, Chen D et al (2016) Investigating intercellular calcium waves by microfluidic gated pinched-flow. Sens Actuat B-Chem 234:583–592
165. Bennett MR, Hasty J (2009) Microfluidic devices for measuring gene network dynamics in single cells. Nat Rev Genet 10:628–638
166. Yates LL, Schnatwinkel C, Hazelwood L et al (2013) Scribble is required for normal epithelial cell–cell contacts and lumen morphogenesis in the mammalian lung. Dev Biol 373:267–280
167. Onodera T, Sakai T, Hsu JC-f et al (2010) Btbd7 regulates epithelial cell dynamics and branching morphogenesis. Science 329:562–565
168. Sakai T, Larsen M, Yamada KM (2003) Fibronectin requirement in branching morphogenesis. Nature 423:876–881
169. Chi X, Michos O, Shakya R et al (2009) Ret-dependent cell rearrangements in the Wolffian duct epithelium initiate ureteric bud morphogenesis. Dev Cell 17:199–209
170. Shintu L, Rg Baudoin, Navratil V et al (2012) Metabolomics-on-a-chip and predictive systems toxicology in microfluidic bioartificial organs. Anal Chem 84:1840–1848

171. Do J, Lee S, Han J et al (2008) Development of functional lab-on-a-chip on polymer for point-of-care testing of metabolic parameters. Lab Chip 8:2113–2120
172. Buehler S, Stubbe M, Gimsa U et al (2011) A decrease of intracellular ATP is compensated by increased respiration and acidification at sub-lethal parathion concentrations in murine embryonic neuronal cells: Measurements in metabolic cell-culture chips. Toxicol Lett 207:182–190
173. Wang G, McCain ML, Yang L et al (2014) Modeling the mitochondrial cardiomyopathy of Barth syndrome with iPSC and heart-on-chip technologies. Nat Med 20:616
174. Bavli D, Prill S, Ezra E et al (2016) Real-time monitoring of metabolic function in liver-on-chip microdevices tracks the dynamics of mitochondrial dysfunction. Proc Natl Acad Sci U S A 113:E2231–E2240

Chapter 5
Cell Migration with Microfluidic Chips

Jinxin Dou and Jin-Ming Lin

Abstract Cell migration plays important roles for a variety of physiological and pathological processes including but not limited to host defense, cancer metastasis and embryogenesis, wound healing and inflammation, most of which can be guided by diverse biochemical and biophysical factors. A detailed understanding of individual factor or integrated factors in controlled condition that induces directed cell migration is essential for studying the dynamic and complex behaviors in vivo and capable of improved tissue engineering and therapeutic strategies. Based on this, a number of in vitro assays have been developed to study the complex guiding mechanisms for cell migration. However, those assays are typically end-point detection which is limited by the inability to sustain the stimuli for a long time. Another limitation with traditional assays is high through-put detection. Recently, the design of the standard cell migration assay has incorporated microfluidic platforms, which have opened up new possibilities for exploring the migration in a better control of cellular microenvironment. Cell-attractive and cell-repellant factors involved in cell migration, single or integrated, can be precisely studied in microfluidic devices, some of which have already been developed for medical and commercial applications. In this part, we present recent developments in microfluidic-based cell migration studies and potential trends in this field.

Keywords Microfluidics · Cell migration · Chemotaxis · Electrotaxis

5.1 Introduction

The milieu where cells reside is very complex in which cells are regulated by biochemical, biophysical and physicochemical properties such as soluble factors, cell-matrix interactions and cell-cell contacts, pH, oxygen tension, temperature and

J. Dou · J.-M. Lin (✉)
Department of Chemistry, Tsinghua University, Beijing 100084,
People's Republic of China
e-mail: jmlin@mail.tsinghua.edu.cn

J.-M. Lin (ed.), *Cell Analysis on Microfluidics*, Integrated Analytical Systems,
https://doi.org/10.1007/978-981-10-5394-8_5

osmotic pressure, etc. (Fig. 5.1). These biochemical, biophysical and physicochemical factors contribute to the cell microenvironment. Cell microenvironment regulates the structure, function, and behavior of the cell, and then influence the growth, development and repair of the surrounding tissue [1]. The microenvironment of different cells is distinct. The microenvironment of stem cells, defined as stem cell niche, plays an important role in the survival, self-renewal, and differentiation of stem cells [2]. For tumor cells, the microenvironment holds the key to induce and give rise for metastasis under certain conditions [3].

Field of biochemical and biophysical microenvironment attract the most attention in experimental biology. Specific cell process regulated by single or multiple effects is determined and the mechanisms are proposed gradually. The biochemical microenvironment, constituted by cytokines, growth factors, hormones and other biomolecules, existed in vivo mainly through autocrine and paracrine. Both of the two kinds of processes are diffusion dominated. Autocrine refers to the diffusion of molecular to the same type of cells in the surrounding. Paracrine is defined as the different type of cells that accept the diffused molecular. The soluble factors signaling occurs not only from adjacent cells but also from distant locations from where the molecule is secreted and subsequently transported along the circulatory system to the target cells, which we called endocrine. The presence of controlled gradients existed in the surrounding cells depends on the concentration, half time and binding affinities between the soluble factors and the receptor of target cells. The biochemical microenvironment involves in the regulation of cell function and behavior.

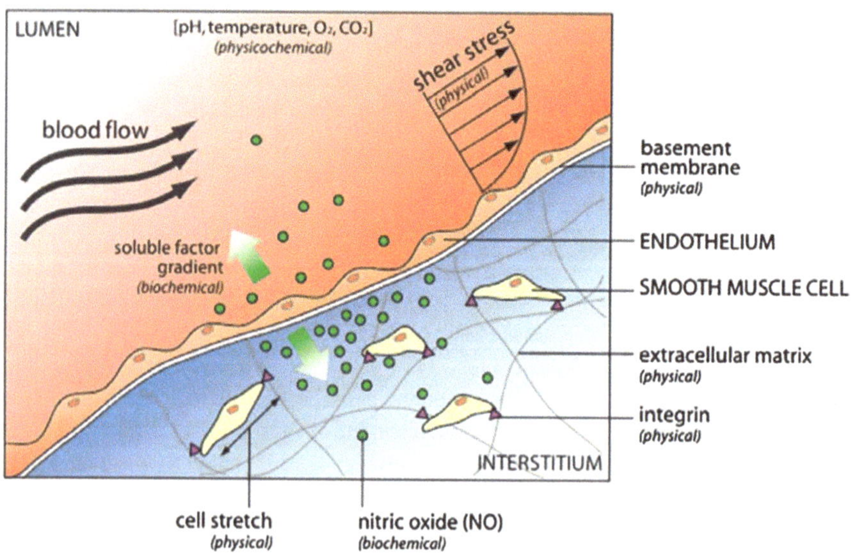

Fig. 5.1 Biophysical, biochemical and physicochemical factors consisted in the cell microenvironment. Reprinted with permission from Ref. [1]

While in the biophysical microenvironment, the gradients in matrix stiffness or ECM composition are essentially due to the heterogeneity of the tissue structure. Most of the cells must be anchored to the extracellular matrix (ECM) by integrin not only lead the physical connection, but also involved in sensing and transducing biophysical signals from focal adhesion sites to cells, which activate a series of mechanisms and pathways related to cytoskeleton and subsequently drive various cellular possesses such as migration, proliferation, differentiation and apoptosis. Whether the microenvironment is biophysical or biochemical leading, they should be carefully considered and precisely regulated when modeling cells and tissues within a simulated microenvironment in vitro [1].

The ability of a cell to migrate directly through a 3-dimension ECM is a highly ordered process orchestrated mainly by a variety of interactions [4, 5]. Such dynamic interactions are involved in a broad range of physiological and pathological processes including angiogenesis, cancer metastasis, inflammatory immune and wound repair [6–9]. Several mechanisms of cell migration have been studied thoroughly including single unattached cells or multicellular groups. There are two subdivisions in directed cell migration of single cells: mesenchymal migration and amoeboid migration. The main difference between the two type is proteolytic degradation [10, 11] or amoeboid-like deformation through holes of the ECM [12, 13]. Directed multicellular migration can be subdivided into collative migration and cell steaming. Cell involved in collective migration is in tight contact with each other while in the latter type cells are not always in tight direct physical contact. The division of the migration is dependent on cell types, microenvironmental factors and the property of the ECM (Table 5.1) [14]. Both of these mechanisms are induced by the biochemical and biophysical properties of the ECM such as the composition, structure and a variety of cytokine binding sites [15–18]. For instance, in tumor cells, four types of directed cell migration have been observed including chemotaxis, electrotaxis, haptotaxis and durotaxis [19]. Among these different cell migration types, chemotaxis is defined the directed migration towards a soluble chemotactic agent. Migration in response to changes in electric fields is defined as electrotaxis. Haptotaxis refer to the migration towards a substrate-bound agent. Migration that is induced by an ECM rigidity is called durotaxis. Because of the high biological and physiological relevance of biochemical and biophysical signal involved in cell migration, there is an increasing need for understanding the mechanisms intensively by in vitro models, which has been an important field of cell biology.

Among various traditional assays, 2D and 3D models have been designed respectively to study cell migration [20]. The wound assay [21, 22] and Teflon fence assay [23] are typical assays to study cell migration in 2D. These assays pioneer new research fields into the cellular and molecular mechanisms of cell migration while are limited to better mimic the microenvironment in vivo. The Boyden chamber/transwell chamber, by contrast, has been the key instrument for cell migration research [24, 25]. By allowing the relative simplicity of operation and systematic study of isolated stimuli on cells, the Boyden chamber has been commonly used to study 3D migration and provided valuable prospect into the metastatic process. However, such assays rely on typical endpoint readout-the number of

Table 5.1 Different modes of cell migration. Reprinted with permission from Ref. [14]

	Single cell migration		Multicellular migration	
	Amoeboid	Mesenchymal	Collective or chain migration	Streaming
Movement type	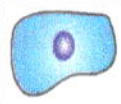	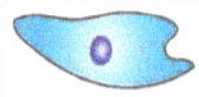	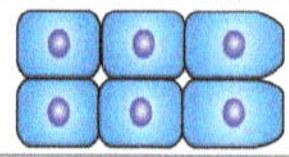	
Cell types	*Dictyostelium discoideum*	Fibroblasts	Border cells in *Drosophila melanogaster*	*D. discoideum*
	Leukocytes	Neural crest endothelium	Epithelial-like cells in zebrafish and *D. discoideum*	Tumour cells
	Tumour cells	Tumour cells	Mammalian cells involved in gastrulation, wound healing and cancer	Developing neural crest cells
Microenvironmental factors influencing migration	Chemokine and growth factor gradients	Chemokine and growth factor gradients, proteolysis	Chemokine and growth factor gradients, reorganization of extracellular matrix	Chemokine and growth factor gradients, reorganization of extracellular matrix, paracrine signaling
Extracellular matrix degradation required?	Sometimes	Yes	Yes	Probably
Can cells switch between types of migration?	Amoeboid to mesenchymal	Mesenchymal to amoeboid, mesenchymal to collective	Collective to single cell migration (amoeboid to mesenchymal)	Not determined
Cell shape during migration	Pseudopodia dominate the shape	Elongated to spindle-shaped	The leader cell is elongated, the cells in the attached sheet are cuboidal	Mixed: cells transition between pseudopodia and elongated
Cell-cell junction maintained?	No	No	Yes	No
Processes where types of migration is observed	Immune response	Development	Development and morphogenesis	Development
	Cancer invasion	Immune response	Vascular sprouting	Cancer invasion and intravation
		Cancer invasion	Tissue repair	
			Cancer invasion	
Types of cancer	Breast and prostate cancers and melanoma	Breast and prostate cancers, melanoma, lung and squamous cell carcinomas	Breast, prostate, lung and colorectal cancers, melanoma and squamous cell carcinoma	Breast cancer

cells that cross a membrane after a certain time-depended on multiple factors including the percentage, velocity, resistance and the direction of the cells. Besides, information of heterogeneous cells cannot be captured. In addition to that, a number of other methods such as the Zigmond chamber [26], the Dunn chamber [27], the micropipette-based assay [28, 29], the under agarose gel assay and scratch assay [30] have been developed, however, these assays are typically complicated to operate with more reagent consumption and the sensitivity needs to be improved. The lack of the ability to maintain and control the stimuli more precisely also restrict its development. Furthermore, some of these assays cannot achieve real-time quantification of cell migration at the single-cell level. It should be noted that important information useful for drug screening and new treatment strategies exploitation may be lost in traditional endpoint assays which give an average result. For example, there is increasing understanding of the significance of the minority cell populations in cancer and immunity which show the difference and play a major role in important response such as intravasation and extravasation. It is the very small percentage of tumor cells that maintain resistance to cancer drug and later evolve to form more aggressive tumors [31]. Along with the increase of people's medical demand for tumor inhibition and wound healing, traditional cell migration assays are no longer able to meet this point.

Microfabrication and microfluidic technology have the potential to overcome these challenges and revolutionize the way we approach to cell biology research by allowing for better and simultaneous control small volume of fluid in micron-sized channels, capillaries, and other geometries and manipulate fluid behaviors such as chemical transport, reaction and their mechanical and electrical properties [32]. Recently, microfluidic devices are become more and more integrated and high-throughput, and achieve the potential goal of lab-on-a-chip system gradually. Although microfluidic device was originally designed for studies in chemistry by analyses of physics at the microscale with the often-cited advantages of faster response times, lower reagent volumes and potential for integration [32], the integration of microfluidic chip and cell biology has become the mainstream of research within the scientific community. Microfluidic chips for cell research has the following unique advantages [33]. (1) The channel size of microfluidic chip is comparable to the physical scale of biological cells, thus can be suitably manipulated, tested and probed in microfluidic environments using tools developed with microscale technology [1]. (2) Factors such as the mass concentration, temperature and pH can be accurately controlled multi-dimension network channel structure. (3) High-throughput analysis can be realized. (4) The vast majority of cell research modules including cell migration and capture, cell sorting, cell-cell interaction can be integrated on a chip by flexible combination of multiple operating units. In order to make a better combination of microfluidic technology and cell biology, therefore, it is urgently important for studying dynamic and complex behaviors of cells induced by environmental signals, how it differs across physical scales in vitro, and how best to control it, using benefits of the microscale [34]. For example, in chemotaxis research, the chemical gradients need to be accurately constructed and maintained in a stable manner. Compared with the traditional method [24, 26, 35,

36], microfluidic chip technology can provide more accurate chemical gradient conditions for chemotaxis studies [37]. Another example is the study of bio mechanical signals. Microfabricated patterns are designed to elaborate that higher stiffness of the surface is more conducive to cell migration [38, 39]. In addition to that, by constructing biological chemical gradient or biomechanical signals in the bottom or the side surface of the microchannel(2-dimensional) and within gel scaffolds(3-dimensional) other researches also induce and monitor the directed cell migration [38, 40–45].

Although the microfluidic technology has a series of unique advantages, however, from the current point of view, the vast majority of cell migration research on microfluidic chips still concentrated on single factor at a time, proved that there is a great gap from stimulation of the real cell migration microenvironment in vivo. The system that can stimulate the realistic microenvironment with precisely controlled gradient and flow to thoroughly research the complicated mechanisms which regulate cell migration is still in pressing demand. With the rapid development of microfluidic-based cell migration research over the past few years, numbers of review articles have been published to summary related works of cell migration, however, current reviews merely focus on the research work of single form of cell migration, such as chemotaxis, but not the systematic integration of all forms of cell migration. In this chapter, different types of directed cell migration will be discussed.

5.2 Microfluidic-Based Chemotaxis Research

Chemotaxis is a well-orchestrated event by which the movement of cells is directed in response to an extracellular chemokine gradient. For example, bacteria sense chemical signals from food molecules or repellants in the environment to change their swim direction [46, 47]. Over 100 years of research have demonstrated that chemotaxis represents an important class of cell-cell communication and plays a crucial role in a number of physiological and pathological processes such as cancer spreading [48], wound healing and morphogenesis [49, 50] in the eukaryotic cells; and animal infection [51], plant infection [52], and carbon cycling in the ocean [53] in the prokaryotic cells [54]. As another typical example, white blood cells such as neutrophils and lymphocytes sense and mobilize chemical signals, interleukin 8 for instance to perform their immune functions [9, 55, 56]. Numerous diseases and disorders such as chronic obstructive pulmonary disease (COPD), inflammatory bowel disease (IBD) and multiple sclerosis (MS) are closely related to the abnormal migration of white blood cells [57].

5.2.1 *Microfluidic Gradient Generator*

Microfluidic gradient generator devices can be categorized into two major classes; one is the convective flow-based device which relies on laminar flow mixing in microfluidic channels to fabricate highly controlled chemical gradient profiles. The other is diffusive based device, gradients are produced by free diffusion in a static fluidic environment. Both types of devices are capable of generating well-defined gradients.

5.2.1.1 Convective Flow-Based Microfluidic Gradient Generator

As the majority of convective flow-based microfluidic gradient generator, one-dimensional (1-D) devices provide the powerful feature for gradient control, while they simplify the in vivo chemical gradient environment. To address the problem, various two-dimensional (2-D) devices have been developed. We divide this chapter into three parts: 1-D devices, 2-D devices, and application in cell migration study. The respective advantages and disadvantages of each kind of model and the development in recent years will also be discussed.

The first 1-D chemical gradient generator is introduced by Whiteside's lab [58]. In this work, diffusive mixing of species in solutions is controlled inside a network of microchannel at low Reynold number. Using this method, gradients in the compositions of solutions and in the topography of the surfaces have been generated. More than that, gradients have been manipulated to manage different size, resolution, and shape. Due to the simplicity and high adaptability of this method, it has been wildly used to create molecular gradient flowing in the channel for chemotaxis research or anchoring on the surface of a substrate for haptotaxis research (Fig. 5.2a).

However, these devices require an additional chemical perfusion system, reducing the portability of the device. In order to solve this problem, Gao et al. demonstrated a microfluidic device using a passive pumping method based on gravity (Fig. 5.2b). Aiming at the balance of pressure between the two inlets, the researchers designed a fluidic circuit module before the downstream network channel. A linear concentration gradient can be quickly established and maintained at the downstream, which facilitates optical imaging at high resolution [59]. Du et al. established a rapid and controllable long range concentration gradients generator in a microfluidic device based on forward and backward flow (Fig. 5.2c). The forward flow was induced by passive pump while the backward flow was induced by evaporation. In the drug test, a gradient of cardiac toxin concentration was applied to detect the pharmacodynamic effect of HL-1 cardiac cells. The difference of morphology and viabilities which responded to the concentration gradient of drug correlated with the toxicity data obtained from multi-plate wells [60]. Recently, Xu et al. [61] optimized and developed a flow-based microfluidic gradient generator by numerically simulating the flow behavior. Based on computational analyses, possible factors that cause non-negligible error in the traditional microfluidic gradient generator devices have been determined. Then the geometrical design of the conventional 2-inlet microfluidic gradient generator devices was optimized and the gradients were more accurate.

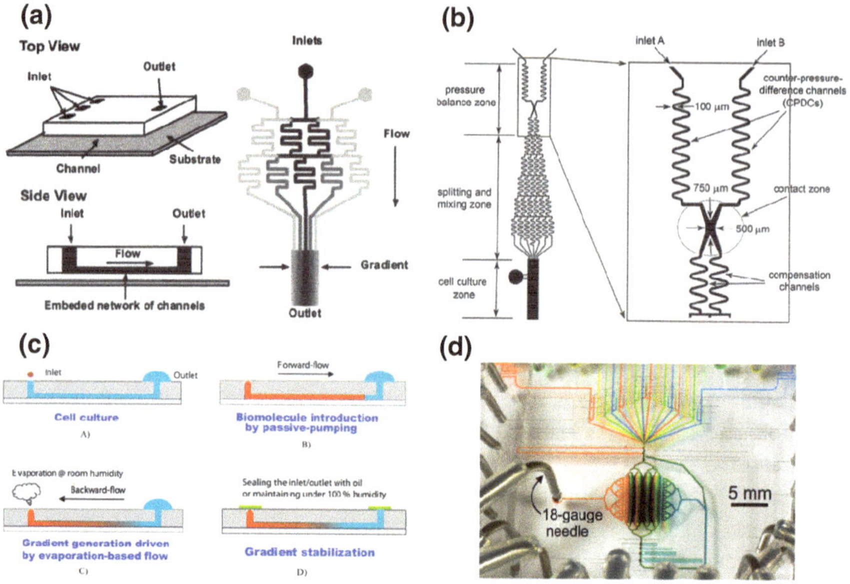

Fig. 5.2 One-dimensional convective flow-based microfluidic gradient generator. **a** Scheme of the microfluidic gradient generator (*upper left* and *lower left*) and the network fabrication design (*right*). Reprinted with permission from Ref. [58]. **b** The design of the microfluidic gradient generator and the amplified representation of pressure balance zone. Different concentration fluids flow into the contact zone to balance the pressure. Reprinted with permission from Ref. [59]. **c** The microfluidic gradient generator combines passive-pump-induced forward flow and evaporation-induced backward flow. Stabilization of the gradient can be realized by sealing or maintaining 100% humidity. Reprinted with permission from Ref. [60]. **d** Photograph of the perfusion system. Different colored fluids from sixteen inlets flowed into the central chamber. 18-gauge needles with 90-degree bend were used as the joint between the device and attachments (inlets and pneumatic lines). Reprinted with permission from Ref. [64]

Compared with the simple gradient generator devices mentioned above, devices capable of generating complex gradient process greater advantages and application prospects. Dertinger et al. [62] developed a microfluidic network capable of gradient generation with complex shapes in solution. A range of gradient shapes including linear, parabolic and periodic was established and maintained for periods of hours spatiotemporally based on appropriately designed networks. Campbell et al. designed architecture of microfluidic networks that generate complex concentration profiles with shape of any planar monotonic function. The source solutions from the two inlets flowed through a series of $k = \log_2(N - 1)$ stages to get split and mixed adequately, then delivered to a set of N outlets. The solutions from separate outlets were combined to fabricate the desired shape of the concentration profile such as linear and parabola. Study of chemotaxis and dose response of live cells validated the employability of the device [63]. Cooksey et al. reported a perfusion system for modulating the gradients over a large population of cells and test the response of cell in well-defined conditions such as medium compositions and flow shear (Fig. 5.2d). The device contained sixteen inlets and a large cell culture chamber. Users could choose to combine several microfluidic tools with the perfusion system to create complex gradient patterns and sequential perfusion for high-throughput and multi-parameter cell culture experiments including toxicology and drug developments. In contrast to previous work, the μFR design had been improved in two aspects. (1) different fluid driving pressures need not recalibration. (2) flow rates were easily extended to wider ranges [64]. Lee et al. [65] developed a network-based device capable of generating on-demand combinatorial dilutions in the range of a 3D simple-centroid. The initial concentration control module was used for generation diluted concentration of each sample in response to variable inlets. Then a full set of seven combinations of the diluted samples could be generated in the integrated combinatorial dilution module downstream in the range of the 3D simple-centroid. The applicability of the device was demonstrated by combinatorial cytotoxicity screening experiment. Using the passive pumping based microfluidic with a balance zone, Wu et al. [66] evaluating neutrophil chemotaxis induced by sputum samples from Chronic Obstructive Pulmonary Disease (COPD) patients.

Limited to the simulation of the chemical gradient in vivo using 1D microfluidic device, more researchers have been attracted to fabricate the desired 2D and 3D chemical concentration gradients using microfluidic devices over recent years. Cheng et al. presented a transparent cell-culture microchamber that combined a variably controlled concentration gradient generator and flow field rectifier (Fig. 5.3). Laser-induced backside wet etching was used for the ablating of a piece of indium tin oxide (ITO) coated glass, then the transparent heater strips could be integrated with the glass substrate. The temperature for cell culturing could be regulated by uniting of the ITO heater and the PMMA microfluidic chamber. The chamber with flow field rectifiers and reagent effusion hole was fabricated by a CO_2 laser. A various chemical gradient was generated by combining lateral flow and effusion flow and the stability was maintained in a long time for cell growth study [67].

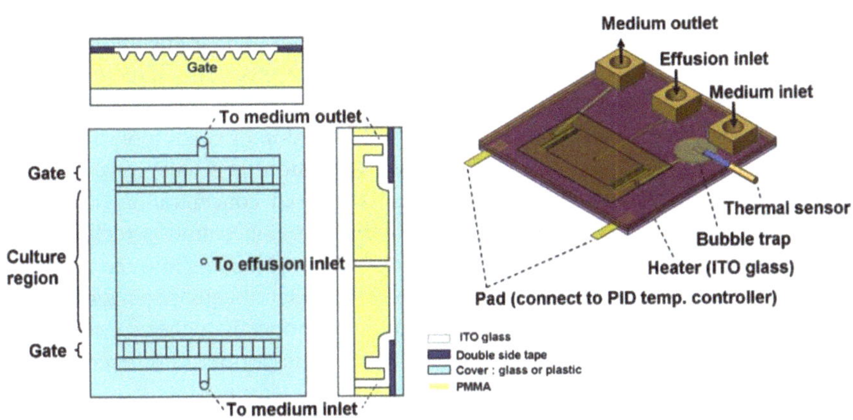

Fig. 5.3 Two-dimensional convective flow-based microfluidic gradient generator. Schematic of the transparent cell-culture microchamber. Reprinted with permission from Ref. [67]

The gap still exists between 2D culture and microenvironment in vivo. Therefore, it is vitally important to develop 3D method for cell migration study. A number of 3D microfluidic gradient generators were integrated by 1D devices and ECM, such as natural hydrogels and synthetic hydrogels, however, migration of cells in 3D microenvironment always involves the squeezing through ECM. This process is dominated by diffusion rather than convective flow and it will be discussed in the next chapter.

5.2.1.2 Diffusion-Based Microfluidic Gradient Generator

Different from convective-based microfluidic gradient generator, device in this part establishes the gradient based on molecular diffusion which does not contact with the cell directly. Nowadays, a number of diffusion-based microfluidic gradient generators have been developed and we review them in four aspects: 1-D devices, 2-D devices, 3-D devices and application in cell migration study. The respective advantages and disadvantages of each kind of model and the development in recent years will also be discussed.

A typical diffusion based gradient generator was presented by Abhyankar et al. [68] to study cell signaling (Fig. 5.4a). A membrane-covered source region and a large volume sink region were integrated with a microfluidic channel. The effect of fluid flow was eliminated by the membrane while the diffusive transport of a chemical species into the channel was not be hindered. By adding sample over the membrane in the source region, the chemical stimulus was established easily. The stability of the gradient was maintained spatiotemporally with the help of large volume sink region at the end of the microfluidic channel. The establishing time of the gradient was adjusted by modifying the system parameters. As a critical piece of the system, the property of membrane was optimized including the hydrophilicity, membrane thickness, protein binding capacity and pore size.

Kim et al. [69] fabricated a hydrogel based porous membrane into microfluidic chip to generate diffusion-driven and long-range concentration gradient for bacterial chemotaxis. The core design of device is the H-shaped microfluidic channel structure, where the agarose gel was injected into the microchannel near the reservoirs to prevent convection flows but allow diffusion of small molecules from the inlet to the center chamber. The critical concentrations and their corresponding chemotactic responses of cells, including the threshold concentration (minimum concentration that induces chemotaxis), the maximum concentration (cells do not show chemotaxis) and the most preferred concentration (cells show the strongest chemotaxis) were characterized. The formation of bacterial cells population bands was observed and the migration of population bands towards higher concentration till the equilibrium point was characterized. Besides, the priority of different carbon sources that induced bacterial cells chemotaxis was determined.

The fabrication of physical barriers by introducing membrane or hydrogel to block convective flow achieved a good result, while it left two drawbacks to solve, the complexity of the fabrication process and the time needed to establish gradient

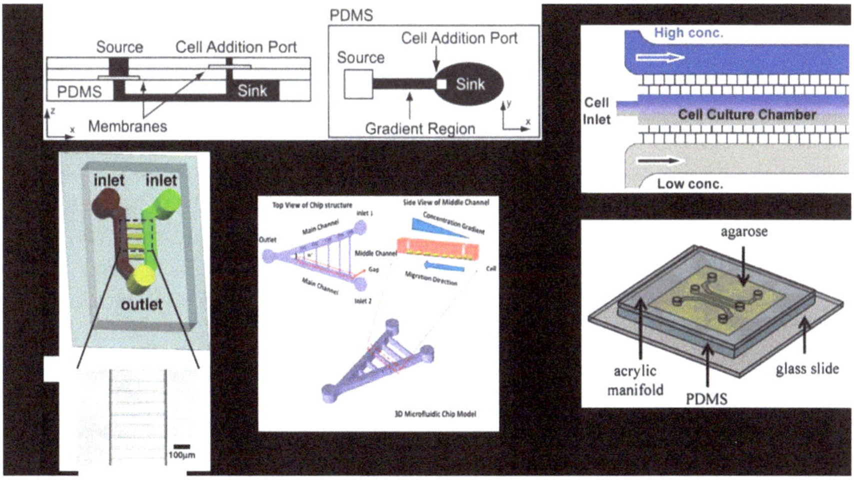

Fig. 5.4 One-dimensional diffusion-based microfluidic gradient generator. **a** Schematic representation of the diffusion based microfluidic gradient generator. A membrane is integrated in the inlet and outlet of the channel to prevent the convective flow from the cell chamber. Chemical flows with different concentration are introduced into the inlet of the device. Cells are seeded through the cell addition port. Reprinted with permission from Ref. [68]. **b** Schematic of the Ladder Chamber for chemical concentration gradient generation from parallel source and sink channels connected by a series of horizontal microgrooves of 3–10 μm. Reprinted with permission from Ref. [45]. **c** The V-shaped microfluidic chip consisted of two main channels and five parallel connecting channels with two gaps at the interface allowed for the entrapment of suspension cells. Cells and medium were introduced from inlet 1 and chemokine was perfused from inlet 2. Reprinted with permission from Ref. [73]. **d** Presentation of the microfluidic device that allowed the generation of morphogenetic gradients across the central chamber designed for cell culture. By introducing the high and low concentration flows on both sides and an array of micropillars between each channel and central channel, the diffusion-based gradient could be established successfully. Reprinted with permission from Ref. [74]. **e** Presentation of the agarose-based three channels microfluidic chemoattractant gradient generator. The device consisted of four layers: glass slide, PDMS substrate, the agarose-based three channel layer and the acrylic manifold layer connected to inlets and outlets. Reprinted with permission from Ref. [76]

increases. To solve the mentioned problem above, various membrane-free or gel-free devices have been developed [70, 71]. Berthier et al. [70] described a microscale assay based gradient generator for patient diagnosis. Surface tension induced passive fluidic methods were adopted to generate a reliable microfluidic gradient. Combined with arrayed imaging and advanced cell tracking algorithms, the throughput increased 50-fold over other methods. However, the gradient is difficult to maintain for a long time in these devices.

In order to improve the performance of current flow-free gradient generators, a number of ladder chamber based devices were developed and have formed a distinct class among flow-free gradient generators. The first ladder-shaped device was presented by Saadi et al. [45] for generating stable gradients in 2D and 3D environment (Fig. 5.4b). The gradients were set up across the horizontal microgrooves between the parallel source and sink channels using a two-compartment diffusion system. Both channels were modified with fibronectin to mimic the presentation ligands on the endothelium. A steady gradient was established within 20 min and could be maintained for 3 h. Migration of cells towards a gradient of chemoattractant (IL-8) was studied using this device successfully. However, the gradient was not well-proportioned across the microgrooves because only the cells localized close to the border of the microgroove became chemotactic response.

Irimia et al. [72] reported a channel network based microfluidic platform capable of polar stimulation and constrained cell migration. A common channel that merged the inlets was designed to balance the pressured of the two streams. This system allowed localized drug treatment to the anterior and posterior of moving neutrophils. The cytoskeletal dynamics could be observed by tracking the localization of actin and microtubules. The device enabled precise and passive balancing of flow for the characterization of morphology of neutrophils which crawled through narrow channels, while the cells were cultured in the hard polymer that was different with the cell-matrix interaction in vivo.

As the size of microgrooves is comparable to cells, the actual gradient profile has a great difference with the theoretical simulation. To diminish the interference of cells, improved ladder chamber and three parallel channels based microfluidic gradient devices have been developed separately. A V-shaped microfluidic device capable of generating multiple stable gradients has been developed by Zou et al. [73] for studying chemotaxis behavior of single adhesive cell (Fig. 5.4c). The microfluidic chip consisted of five parallel connecting channels with different lengths. The gaps between the main channel and connecting channels were designed for the entrapment of suspension cells. The microfluidic platform was applied for studying the chemotaxis of lung cancer stem cell (LCSC) and differentiated LCSC (dLCSC).

Channels on both sides are used as inlet and outlet while cells are cultured in the central channel. Lee et al. [74] presented a single-cell assay platform capable of studying chemotaxis of single cell (Fig. 5.4d). The flow-induced perturbation on cell orientation was eliminated by integrating an array of micropillars near the cell culture chamber. With the help of computer, the gradient profile could be regulated and established at any given time which allowed quantitative monitoring of cell morphology. The expression and localization of specific marker proteins during the polarization process were also determined. Using this device, researchers compared the polarized signaling response of two types cells: wild-type and orientation-deficient *farl-H7* mutant cells, which had been proved to express an abnormal protein. The results demonstrated that Farl protein involved in the establishment of cell polarity.

Keenan et al. [75] described the use of open-chamber microfluidic gradient generator for studying gradient-induced neutrophil desensitization. The device consisted of sink, source channel and cell culture reservoir separated by microjet arrays which processed the fluid resistance. Individual neutrophil chemotaxis responded to homologous desensitization environments followed by heterologous desensitization environments was tested. The device has the potential to be applied to a wide variety of gradient-responsive cell types.

By introducing hydrogel into microfluidic chip, various microgroove-free gradient generators have been reported. Soluble molecules are capable of diffusing directly through the hydrogel barrier to establish the gradient, which simplifies the fabrication of microfluidic chips. An agarose-based three channel microfluidic device presented by Kim et al. [76] was designed to generate a gradient of a chemoattractant, L-aspartate, for testing the behavior of bacteria-propelled beads (Fig. 5.4e). A PDMS positive mold was used to create a reservoir for hot agarose and then removed after gelation of agarose. Different bacteria-attached bead sizes with different behavior were determined to draw the conclusion that the beads exhibit a clear indication of directionality and steering control dominated by the bacteria on the bead when presented in the gradient of chemoattractant.

Cheng et al. [77] developed a hydrogel-based microfluidic chip for the studies of directed cell migration. Chemical solution flows in the source channel while buffer flows in the sink channel. A steady and long term linear concentration gradient was generated in the microfluidic channel with the diffusion of chemical through the agarose gel membrane. The device has been proved to be capable to study the chemotactic responses of wildtype *Escherichia coli* and differentiated HL-60 cells. Choi et al. [78] reported an agarose gel-based microfluidic device for quantitative analysis of the chemotaxis of a green alga, *Chlamydomonas reinhardtii*, towards bicarbonate. Two chambers which served as cell reservoir and chem-effector reservoir on the both ends of the device were connected with 5 agarose gel channels and concentration gradient generation channel. A design of separated observation region allowed for quantitative analysis and a long cell migration channel allowed for monitoring the profile of population band which was chemical concentration and exposure time dependent. The device was validated to be suitable for long-term chemotaxis assays. Another strategy for producing 1D gradients such as microinjection devices [79].

To generate more complex gradient profiles, numerous 2D chemical concentration microfluidic gradient generators have been developed for cell chemotaxis study. Atencia et al. [54] presented a device named "microfluidic palette" to generate multiple chemical 2D gradients inside a microfluidic chamber with spatiotemporal control. The match of flow rates at inlet and outlet diminished convective flow across the main palette while allowing diffusion for gradient generation. The microfluidic palette consisted of three parts: A microfluidic glass chamber, a PMMA rigid layer, and a PDMS gasket. The multiple layer design separated the gradient chamber and the chemical infusion channel and allowed the generation of multiple gradients that can be superimposed in the flow-free chamber. The gradient has been demonstrated to establish and stabilize in about 15 min for small molecular and can be maintained over long periods. This device allowed generation and maintenance of a various chemical gradient profiles (Fig. 5.5).

Qasaimeh et al. [80] developed a two-dimensional microfluidic quadrupole based concentration gradient generator. The gradient was generated and stabilized within seconds. Besides, it experienced no flow and no shear stresses was continuously fed by injecting and aspirating fluids from two pairs of opposing apertures. With the help of theoretical analysis, there was a stagnation point formed at the

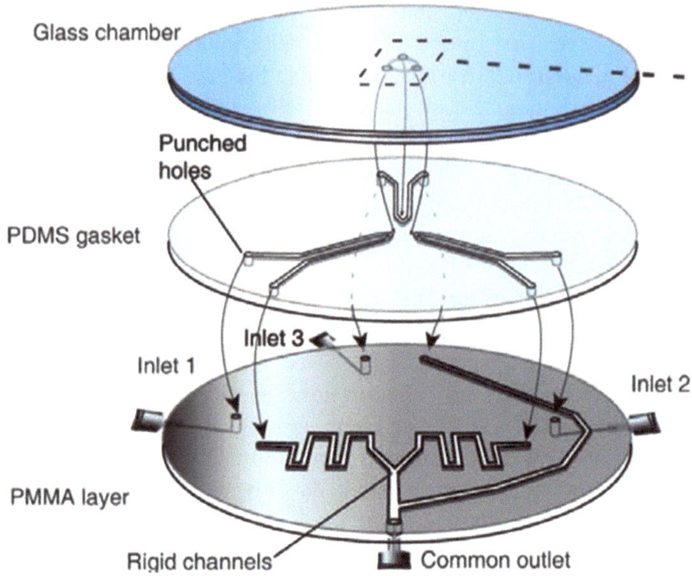

Fig. 5.5 Two-dimensional diffusion-based microfluidic gradient generator. Schematic of the microfluidic palette which consisted of three layers. The *top* layer was a microfluidic chamber sandwiched between two glass wafers. The *middle* layer was a PDMS gasket and the *bottom* layer was made by rigid PMMA that connected three different inlets and a common outlet. Reprinted with permission from Ref. [54]

center of the quadrupole whose position could be adjusted rapidly. A floating concentration gradient was fabricated at the stagnation point after the injection of a solute through one of the poles. Wright et al. [81] designed an on-chip open microfluidic device that produced passive chemoattractant gradients without an external micropipette system for chemotaxis studies. An open environment allowed cells loading directly into the device using a micropipette. Several ports were integrated into this device for high throughput chemotaxis assays. The device was demonstrated that cells can be recruited into the ports along the passive attractant gradient. In addition, precise control of the amount and the location of drugs or reagents could be realized using this device. Moussavi-Harami et al. [82] developed a 3D hydrogel-based microfluidic device to generate a dual chemotactic gradient. The microfluidic chip contained two regions: the gel region and cell region. The gel was injected into the gel region first. After the gelatin of the gel, chemoattractant was loaded into the gel region to establish the dual gradient. The device was able to study the chemotaxis of the neutrophil-like cell line PLB-985 to gradients of fMLP.

A decisive advantage of diffusion-based microfluidic chemical concentration gradient generator compared to convective-based device is that 3D gradient profiles can be generated only by diffusion. A 3D microfluidic gradient device refers to a device where the chemical concentration gradient is applied to a 3D matrix with the encapsulated cells. By integrating 3D cell-embedded matrix hydrogels into the channels, almost all 1D or 2D diffusion-based gradient generators are able to extend

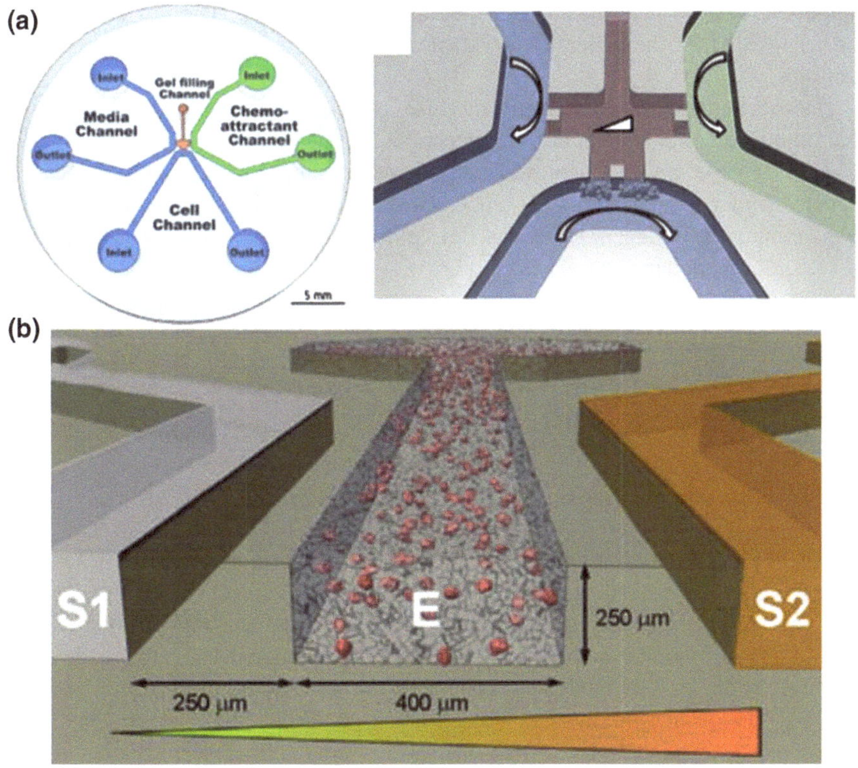

Fig. 5.6 Three-dimensional diffusion-based microfluidic gradient generator. **a** Design of the three-channel microfluidic device. The gel filling channel was used for the injection of Collagen-I while the three V-shaped channels were used for introducing chemo-attractant, basal media and cell. The gradient profile across the 3D scaffolds and the media flow in each channel were illustrated in the amplified central region. Reprinted with permission from Ref. [84]. **b** A three channel unit of the 3D microfluidic chemotaxis device. Cells were encapsulated into collagen in the center channel. The flows in the source and sink channels generated chemical gradient across the center channel. Reprinted with permission from Ref. [87]

to 3D devices easily. As for the biomatrix chose for 3D devices, the compatibility with cells need to be taken into consideration thoroughly. Nowadays, a number of hydrogels have been introduced into 3D microfluidic chemotaxis device including gelatin [83], agarose [77], collagen [84], etc. Kothapalli et al. fabricated the 3D scaffolds of the microfluidic assay by injecting the Collagen-I into the gel-filling channel and allowing to polymerize (Fig. 5.6a) [84]. Three V-shaped channels connected to the 3D scaffolds were designed for chemo-factor of interest, basal media, and neurons. The microfluidic assay was capable of establishing a steep gradient within 30 min and stabilizing for up to 2 d. Using this device, they studied the migration of neurons to chemical gradient.

The structure of the microfluidic chip was then optimized by the same group [85, 86]. Mixtures of cells and Collagen-I were injected into the gel filling port which connected with chemoattractant channel and media channel. The device has been further verified for the generation of stable and controlled gradient through the 3D scaffolds.

Haessler et al. [87] extended the use of the three-channel agarose-based microfluidic chemotaxis device to generate chemical gradient in a 3D matrix (Fig. 5.6b). The murine dendritic cells were injected together into the center channel with collagen and were embedded in a 3D environment with the polymerization of the collagen. Using this device, the chemotactic response to a gradient of CCL19 was determined.

5.2.2 *Convective Flow- and Diffusion-Based Cell Chemotaxis Study*

5.2.2.1 Convective Flow-Based Cell Chemotaxis Study

Convective flow-based microfluidic chemical concentration gradient generators have been developed and improved for creating controlled and complex molecular gradients in the microchannel. Chemotaxis and chemical sensing of different cell types, such as immune and tumor cells, have been studied using these devices. Hermark et al. [88] created the gradient of the chemoattractant, f-Met-Leu-Phe (fMLP) in microfluidic devices and studied the chemotaxis of neutrophil-differentiated HL60 (dHL60) cells . Using a fluorescent marker mixed with fMLP, the actual shape of gradients was determined and C_s for each individual cell trajectory was analyzed. A conclusion could be drew that chemotactic prowess depends on both the variation in attractant concentration in space (represented by ΔC) and the ambient concentration of attractant (represented by C_s). By theoretical analysis of occupied receptors and difference in the receptor occupancy across on cell diameter in an exponential gradient, the researchers demonstrated that neutrophil-like dHL60 cells were capable of sensing the differences of chemical attractant concentrations across their diameters in relatively shallow gradients and deciding whether or not to migrate after the spatial gradient signal was transduced. A model that chemotactic prowess is an increasing function of the absolute difference in receptor occupancy between the front and the back of the cell was verified by experimental data. In addition to that, the major findings of this work was validated microfluidic device independent. A device microfabricated by soft lithography was presented by Jeon et al. to study the neutrophil chemotaxis in linear and complex gradients of interleukin-8 (IL-8) [40]. A network of microfluidic channels for the immobilization of human neutrophils was designed to generate controlled gradients of chemotactic factors spatiotemporally and the migration towards simple and complex IL-8 gradients was detected. The cells showed strong directional migration tendency responded to increasing concentration of IL-8 in simple linear gradients module. While the gradient became complex, cliff gradient (the chemoattractant concentration drop into zero suddenly) for example, the migration of neutrophil was interrupted abruptly. When encountered with a hill gradient (a gradient increase and decrease in chemoattractant), neutrophil migrated further over the maximum concentration before reversing direction. Based on the results that neutrophils were capable of simple or complex migration, researchers conjecture that chemoattractant were likely to exhibit complex gradients at sites of inflammation, such as periodic shapes. Mosadegh et al. employed a microchamber capable of CXCL12 gradient generation. Based on this device, the migration of MDA-MB-231 breast cancer cell affected by epidermal growth factor was characterized, without the interface of autocrine/paracrine signaling [89]. Chemokinesis was evaluated separately to eliminate the interference of chemotactic by testing the migration of MDA-MB-231 in a uniform stimulus of CXCL12 in combination with EGF or CM. It was demonstrated that the migration distances and number of motile cells were increased significantly when the uniform concentration of CXCL12 was supplemented with EGF or CM, compared to the control group. Similar trend was observed when cells were presented with a CXCL12 gradient. In the meanwhile, the directional orientation of cell exhibited dependence towards the concentration of EGF. Based on the experimental results, a cooperation model involved EGF and CXCL12 was proposed that EGF stimulated the motility "machinery" of the cells and CXCL12 gradient served as the "compass" of the cells to determine their directional orientation. Ricart et al. [90] presented murine bone-marrow dendritic cells (DCs) to single and competing chemokine gradients in a controlled, time-invariant microenvironment using a microfluidic device, which was verified to be well suited to studying moderately adhesive cell types because of the low shear stresses and stable gradient spatiotemporally. Chemotaxis of DCs response to single gradients of CCL19 and CCL21 (ligands of CCR7) and CXCL12 (ligands of CXCR4) was studied firstly. Compared to CXCL12, the DCs responded more strongly to CCR7 ligands which was then proved to be receptor expression dependent using a CCR7 knockout mouse. It was demonstrated in the competition assay that DCs were attracted towards CCR19 gradient rather than CCL21 and CXCL12 gradient, both of which were in equal potency. With the use of competing gradient, the central homing of a cell population was studied. Similar central homing behavior was observed when two different chemokines engage single or multiple receptors. The speed of crawling was demonstrated to be pertussis toxin independent while the directed motion was proved to be independent to actin and myosin inhibitors. Wang et al. presented a microfluidic chemotaxis chambers to study differential effects of EGF gradient profiles on metastatic breast cancer cells (MDA-MB-231) chemotaxis [91]. The movement patterns of single cells in uniform concentration and polynomial profiles (linear and nonlinear) was investigated and the extent to which the movement was directed in each gradient was quantified. In uniform EGF concentration profile chemotaxis assay, the migration of cells was demonstrated to be dependent on the concentration of EGF while it was restrained when cells were presented to high concentration of EGF. In EGF concentration gradient profile chemotaxis assay, nonlinear concentration gradients of different ranges, especially gradient of 0–50 ng/ml, were more effective in inducing the movement of MDA-MB-231 cells compared to linear gradients.

Compared with the conventional migration platforms which treat populations in aggregate, devices with single-cell resolution attracts more and more attention. Nowadays, Chen et al. [92] developed a single-cell migration chip to study the individual cells' migration behavior (Fig. 5.7). Using this device, heterogeneous population of cells could be sorted upon the heterogeneity chemotactic phenotype. By optimizing the hydrodynamic resistances, an 85% capture rate in 5 different mammalian cell lines could be achieved. The platform has been demonstrated to

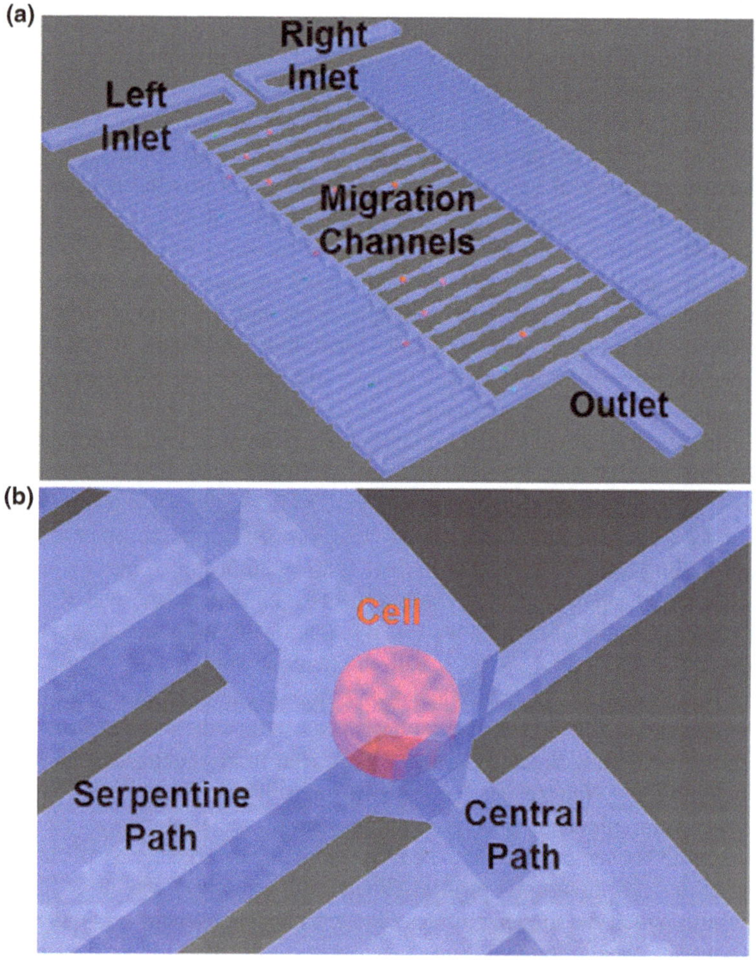

Fig. 5.7 3D schematic of the microfluidic chip for single-cell migration with enlarged cell capture site. The process of cell migration consisted of three steps. A cell was captured in the central path by the hydrodynamic force in the first step. After the attachment to the substrate, the cell migrated into the channel. Reprinted with permission from Ref. [92]

retrieve the highly chemotactic and non-chemotactic cells separately for further propagation and analysis.

In short, the main advantage of the convective flow-based chemotaxis device is that the gradient can be established as rapid as a few milliseconds. While unwanted fluid flows in the cell chamber limited the further development of such devices. The adverse effects induced by fluid flow are reflected in three aspects. First of all, the shear press accompanied with fluid flows exerting on the cells has been proved to be capable of regulating the behavior of the cell [93]. Secondarily, the auto/paracrine signals secreted by the cells would be washed away, which play an important role in cell migration and survival [94, 95]. The third aspect lies in the remodeling of the matrix in the case of a 3D cell culture system [93, 96]. In addition to that, the problem of Taylor dispersion should be taken into consideration within a microfluidic channel. It has been proved that the actual gradient across the cell can be significantly different from theoretical simulation in a convective flow-based device [97].

5.2.2.2 Diffusion-Based Cell Chemotaxis Study

The innovation of diffusion-based gradient generator paves the way for numerous cell chemotaxis studies, especially 3D studies. For example, using the three channel agarose-based microfluidic chemotaxis device, Haessler et al. [98] established a stable gradient of two ligands of CCR7, CCL19 and CCL21 in 3D matrices rapidly and showed that dendritic cells could respond to both gradients as small as 0.4%. In addition to that, the dendritic cells preferentially migrated towards CCL21 when presented to an equal but opposite gradients of CCL21 and CCL19. The results demonstrated that in a 3D environment in vitro, CCL21 was a more effective chemoattractant that inducing dendritic cell migration.

Suh et al. [99] reported a diffusion-based microfluidic platform for separation of similarly-sized particles (Fig. 5.8). Two kinds of particles with different surface properties were introduced into the work area and one group of particles were selectively assembled onto motile *Escherichia coli* bacteria through either specific or non-specific adhesion. After the chemoattractant gradient was established using gravity driven flow, the *Escherichia coli* bacteria chemotaxis-induced separation could be realized swiftly. By optimizing the concentration gradient, a high sorting efficiency could be achieved.

To simplify current microfluidic chemotaxis systems and shorten the time to develop a chemokine gradient, Walsh et al. presented a paperfluidic device capable of establishing the gradient and examining cell chemotaxis over a short time [100]. The paperfluidic device possessed two main advantages: the gradient could be generated on the order of seconds while the shear effects were removed. The device was validated of studying the directed migration of human pan-T cells to the chemokine gradient.

In summary, diffusion-based chemical gradient generator owns various advantages. The effects of unwanted fluid flows could be diminished. The diffusion-based

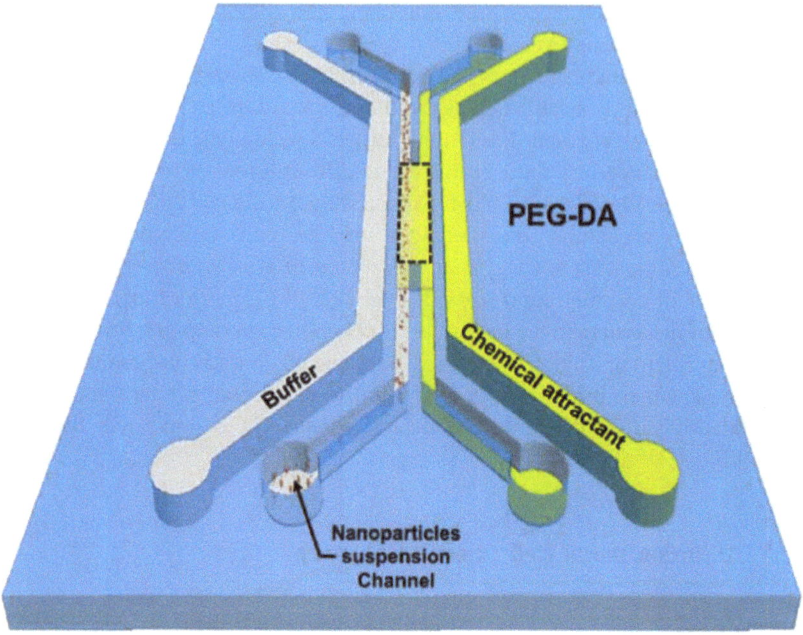

Fig. 5.8 Schematic of the PEGDA-based microfluidic sorting platform with the work area marked by *black dotted line*. The chemoattractant gradient was established across the work area between the buffer channel and chemical attractant channel. The separation of freely-diffusing particles and particle-bacteria assemblies mixtures was carried by chemotactic bacteria. Reprinted with permission from Ref. [99]

devices are suitable for 3D chemotaxis studied extended from 1D or 2D platforms by introducing 3D matrixes. However, what needs to be explained is that most diffusion-based microfluidic chips are inseparable from specialized engineering equipment especially the convective flow system. Although these devices have been proved to be a powerful tool for establishing physiologically gradients, further development is still required for a better control of chemical gradient.

5.3 Microfluidic-Based Electrotaxis Research

Electrotaxis, also known as galvanotaxis, is the applied direct current electric field (dcEF)-induced directional migration of adherent motile cells. Endogenous dcEFs were observed firstly by ED Bois–Reymond at small epidermal wounds created in human fingers over 150 years ago. It has been demonstrated that dcEFs play crucial physiological roles in a variety of cellular activities including embryonic development, neurogenesis, and wound healing [101, 102]. Among these dcEF-induced effects, dcEF-directed cell migration has attracted enough attention and has been studied thoroughly. For example, it has been observed that cells are capable of directing their movement toward the wounds to repair the damage by sensing such dcEFs [103]. In vitro studies have proved that a variety of cell types and organisms show electrotactic response to dcEFs [104]. Most of the cell types migrate toward the cathode including neural crest cells [105], fibroblasts [106–108], rat prostate cancer cells [109], and corneal epithelial cells [110, 111], while only a few towards anodes such as many endothelial cell types [112, 113]. Nowadays, the electrotactic behavior of various types of cancer cell has been reported. Endogenous dcEF with the strengths of few mV/mm to a few of hundreds of Mv/mm originate from the difference in transepithelial potential (TEP) in vivo, which relates to the differential distribution of ion channels in polarized epithelial cells [114]. For example, the change of TEP during cell transformation generates a voltage gradient between the cancerous tissue and normal tissue [109].

The mechanisms of cell electrotaxis have been studied by many groups over the past few decades and a series of models have been presented such as asymmetrical distribution of cell surface receptors and re-establishment of cell polarity [101] driven by dcEF-induced in situ electrophoresis [115] or electro-osmosis [116]. In addition, several signal transduction pathways including dcEF-induced calcium [117], PI(3)Kγ and PTEN signaling pathways [118] have been reported to be involved in electrotaxis. More efforts need to be paid to understand and integrate the mechanisms of electrotaxis, determine the physiological roles and biomedical applications.

Conventionally, experimental investigation of cell electrotaxis is performed in a centimeter-sized electrotactic flow chamber built by integrating cover glasses in Petri dishes where cells are cultured [119]. However, this method has drawbacks including the unavoidable evaporation of culture medium and exposure of possible electrode-induced toxic products to cells. Some improvement schemes have been proposed, for example, of using a pair of agar salt bridge to seal the chamber and prevent the direct contact of electrodes with the cell culture [109, 120, 121]. The visualization of cell migration is able to be realized by using a microscope. Limitation of these dish-based experimental setups including low throughput, poorly controlled microenvironment, and reagent/sample consumption. Other traditional devices were developed to study cell electrotaxis such as the transwell-based electrotaxis assay [122]. By placing two platinum electrode to the top and the bottom well of the Transwell, An EF was applied across the transwell. The number of electrotactic cells migrated from the top well to the bottom well through the membrane was determined. The transwell-based electrotaxis assay allowed for high throughput analysis and post-assay analysis while the control of electric field remained unimproved.

Recently, microfluidic devices have been developed for electrotaxis study with many advantages including better control of the uniformity and stability of the electric field, reducing the Joule heating effect and high throughput analysis. In this part, we review recent researches of electrotaxis from two aspects according to the different type of electric field that applied to the microfluidic device.

5.3.1 Microfluidic-Based Electrotaxis Research in Single Electric Field

Many studies have shown that immune cells such as lymphocyte were able to migrate toward the cathode when presented into an applied physiologically relevant dcEFs. To understand the mechanism of immune cell electrotaxis, a PDMS microfluidic device and a glass microfluidic device were developed by Li et al. to study the migration of activated T lymphocyte toward the cathode of dcEFs at the single cell level [123] (Fig. 5.9a). In the PDMS microfluidic device, two platinum electrodes were coupled into two medium reservoirs and the conducting wires were connected to a DC power. The applied electric field in the microfluidic channel was regulated by computer simulations. In the glass microfluidic device, an array of platinum electrodes was integrated across the channel directly. Using the two microfluidic devices, the comparison of electrotaxis and chemotaxis of the same cell type was realized in the same experimental platform.

Rezai et al. [124] fabricated a PDMS microfluidic environment to study the electrotaxis of *Caenorhabditis elegans* (*C. elegans*) (Fig. 5.9b). The microfluidic device consisted of three parts: the straight macrochannel, length scale measurement setup and electrodes at reservoirs. The *C. elegans* were loaded into the microchannels and positioned in the central section. The wide of microchannel was optimized to guide movement along the channel axis while eliminating the perpendicular motion of the *C. elegans*. The length scale alongside the microchannel allowed for the measure of speed under different dcEF strength. Using this device, several stages of animals responded to different dcEF strength were determined.

The microfluidic devices have been showed to control the electrical field in a better way, however, the integration degree of current experimental setup is still

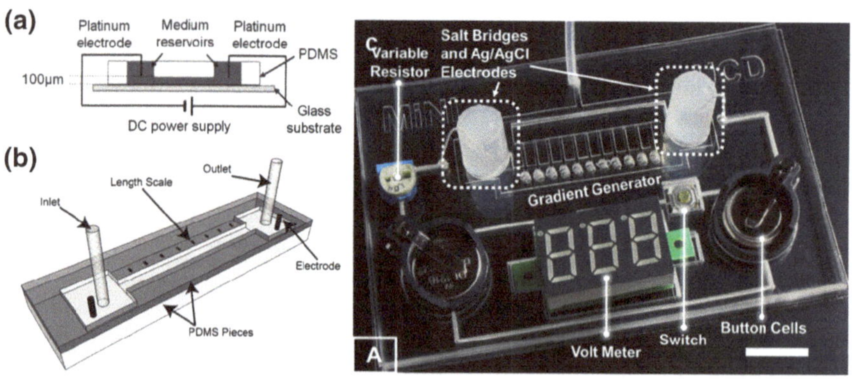

Fig. 5.9 Single electric field-based electrotaxis research. **a** Schematic representation of the PDMS microfluidic electrotaxis device with two platinum electrodes placed into the medium reservoirs. Reprinted with permission from Ref. [123]. **b** Illustration of the microchannel device with two electrodes embedded into the reservoir area. The length scale alongside the microchannel was used for the calculation of speed. Reprinted with permission from Ref. [124]. **c** Schematic of the integrated three PDMS layers-based ECT. Reprinted with permission from Ref. [125]

very low because a series of modules such as external power supply and EF controlling systems are still in need. In addition to that, the high throughput analysis is difficult to achieve using current microfluidic devices. Zhao et al. [125] demonstrated an independently operable microfluidic platform referred to as ElectroTaxis-on-a-Chip (ETC) for EF-directed cell migration (Fig. 5.9c). The ETC device consisted of three layers. An EF gradient and a vacuum network were integrated into the bottom layer while the batteries, a variable resistor, a voltmeter and a switch were fabricated into the top and middle layer. A series of circular holes were designed to monitor the EF. The ETC device was then evaluated under different voltages and showed similar cell electrotaxis behaviors compared with the traditional electrotais chamber. The high-throughput EF-directed screening of human corneal epithelial cells was realized using this device.

5.3.2 *Microfluidic-Based Electrotaxis Research in Multiple Electric Field*

The conventional microfluidic device for electrotaxis is limited to apply single EF strength in a single experiment, which is time-consuming. Huang et al. [126] presented a tightly sealed cell culture chamber for long-term electrotaxis study of lung cancer cell (Fig. 5.10a). Three channels with equal length but different widths were sequentially aligned in the microchannel to generate multiple electric field. CL1-5 and CL1-0 cells were cultured in the microchannel and the electrotaxis behaviors were tested. The results showed that CL1-5 cells, the highly metastatic cells, behaved stronger electrotactic response than CL1-0 cells, the weakly metastatic cells. Using this microfluidic device, the EF-directed cell migration toward three different EF strength was studied in one experiment.

An obvious drawback of the method with three segments of different widths aligned in the microchannel is the heterogeneity of flow field which may cause interference to the experimental results. To solve this problem, Tsai et al. [127] reported the multiple electric field with uniform flow chip (MFUF chip) to study the electrotaxis of oral squamous cell carcinoma cells (Fig. 5.10b). The multiple electric field strength was created and applied to the chip while the flow field was uniform. The main channel of the ladder-shaped microfluidic device was divided into four main segments by a series of vertical channels. The culture medium was injected across the main channel via inlet 1 to the outlet while the electric field was introduced via the ports that combined inlet 2 and outlet, respectively. The separation of flow field and electric field allowed for the distribution of electric field across the main channel with uniform flow field. The MFUF chip was validated by lung adenocarcinoma cells and used to study the electrotaxis of HSC-3 oral squamous cell carcinoma cells. It is worth mentioning that the control condition was able to be fabricated in the main channel.

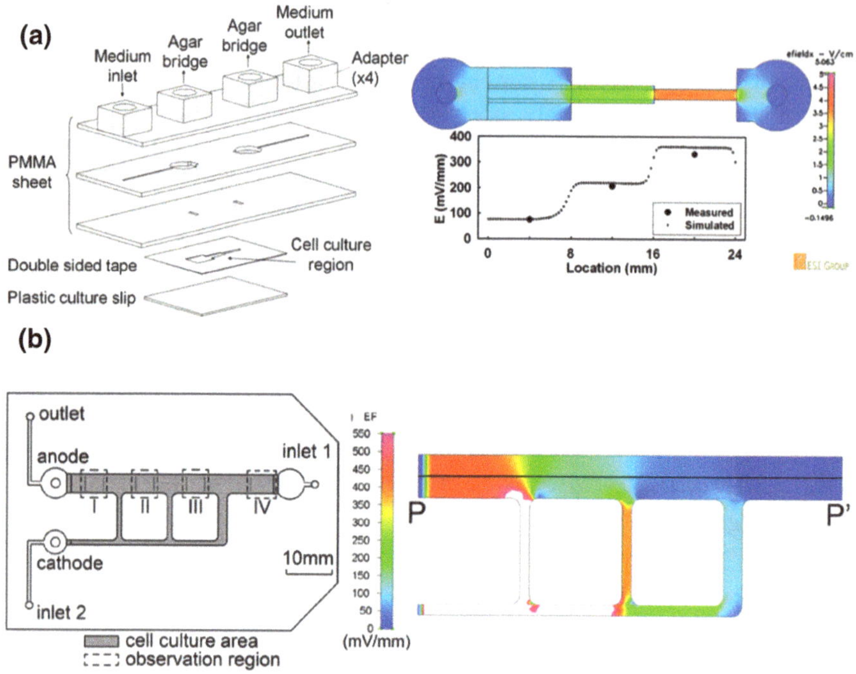

Fig. 5.10 Multiple electric field-based electrotaxis research. **a** A 3D illustration showing the multiple electrotactic chip. Simulated and measured E along the cell culture region. Reprinted with permission from Ref. [126]. **b** Schematic of the multiple electric field with uniform flow chip. The simulation of electric field across the main channel. Reprinted with permission from Ref. [127]

In summary, microfluidic-based electrotaxis device has made great progress in recent years. Compared with conventional assays, microfluidic-based device increases the throughput while reduces the consumption of reagent/sample. In addition to that, the electric field can be controlled in a better way. Extended from chemotaxis device, it is easy to carry with electrotaxis and chemotaxis study in a chip which can achieve the more realistic simulation of the microenvironment in vivo.

5.4 Conclusions

In this review, we have summarized the recent developments of cell migration in two parts: chemotaxis and electrotaxis. The microfluidic-based gradient generator devices are talked about firstly. The convective flow-based devices were easy to be fabricated which mainly rely on laminar flow, while the 3D culture cannot be realized on these devices. As for the diffusion-based microfluidic gradient generators, cells were encapsulated into natural or synthetic materials and the gradient mainly relies on diffusion. We suggest the convective flow-based devices for some kind of specific cells which sense the flow in vivo, such as endothelial cells and epithelial cells. Then the application of gradient generator on chemotaxis and electrotaxis was discussed. A variety of approaches were explored for simple, stable and multi-functional gradient, on the one hand, these designed explorations suggest the advantage of microfluidic; on the another hand, the design microfluidic chip control should be unified as soon as possible in order to obtain reproducible results in the application.

References

1. Young EW, Beebe DJ (2010) Fundamentals of microfluidic cell culture in controlled microenvironments. Chem Soc Rev 39(3):1036–1048. doi:10.1039/b909900j
2. Discher DE, Mooney DJ, Zandstra PW (2009) Growth factors, matrices, and forces combine and control stem cells. Science 324(5935):1673–1677. doi:10.1126/science.1171643
3. Fidler IJ (2002) The organ microenvironment and cancer metastasis. Differentiation 70 (70):498–505. doi:10.1046/j.1432-0436.2002.700904.x
4. Vicente-Manzanares M, Webb DJ, Horwitz AR (2005) Cell migration at a glance. J Cell Sci 118(Pt 21):4917–4919. doi:10.1242/jcs.02662
5. Berzat A, Hall A (2010) Cellular responses to extracellular guidance cues. EMBO J 29 (16):2734–2745. doi:10.1038/emboj.2010.170
6. Jin T, Xu X, Hereld D (2008) Chemotaxis, chemokine receptors and human disease. Cytokine 44(1):1–8. doi:10.1016/j.cyto.2008.06.017
7. Zhao M (2009) Electrical fields in wound healing-An overriding signal that directs cell migration. Semin Cell Dev Biol 20(6):674–682. doi:10.1016/j.semcdb.2008.12.009
8. Campbell JJ, Butcher EC (2000) Chemokines in tissue-specific and microenvironment-specific lymphocyte homing. Curr Opin Immunol 12(3):336–341. doi:10.1016/S0952-7915(00)00096-0
9. Sallusto F, Baggiolini M (1998) Chemokines and leukocyte traffic. Nature 392(6676): 565–568. doi:10.1038/33340
10. Wolf K, Wu YI, Liu Y, Geiger J, Tam E, Overall C, Stack MS, Friedl P (2007) Multi-step pericellular proteolysis controls the transition from individual to collective cancer cell invasion. Nat Cell Biol 9(8):893–904. doi:10.1038/ncb1616
11. Kim HD, Guo TW, Wu AP, Wells A, Gertler FB, Lauffenburger DA (2008) Epidermal growth factor-induced enhancement of glioblastoma cell migration in 3D arises from an intrinsic increase in speed but an extrinsic matrix- and proteolysis-dependent increase in persistence. Mol Biol Cell 19(10):4249–4259. doi:10.1091/mbc.E08-05-0501
12. Lammermann T, Bader BL, Monkley SJ, Worbs T, Wedlich-Soldner R, Hirsch K, Keller M, Forster R, Critchley DR, Fassler R, Sixt M (2008) Rapid leukocyte migration by integrin-independent flowing and squeezing. Nature 453(7191):51–55. doi:10.1038/nature06887
13. Ilina O, Bakker G-J, Vasaturo A, Hoffman RM, Friedl P (2011) Two-photon laser-generated microtracks in 3D collagen lattices: principles of MMP-dependent and -independent collective cancer cell invasion. Phys Biol 8(2):029501. doi:10.1088/1478-3975/8/2/029501
14. Roussos ET, Condeelis JS, Patsialou A (2011) Chemotaxis in cancer. Nat Rev Cancer 11 (8):573–587. doi:10.1038/nrc3078
15. Cukierman E, Bassi DE (2010) Physico-mechanical aspects of extracellular matrix influences on tumorigenic behaviors. Semin Cancer Biol 20(3):139–145. doi:10.1016/j.semcancer.2010.04.004

16. Fraley SI, Feng Y, Krishnamurthy R, Kim DH, Celedon A, Longmore GD, Wirtz D (2010) A distinctive role for focal adhesion proteins in three-dimensional cell motility. Nat Cell Biol 12(6):598–604. doi:10.1038/ncb2062
17. Griffith LG, Swartz MA (2006) Capturing complex 3D tissue physiology in vitro. Nat Rev Mol Cell Biol 7(3):211–224. doi:10.1038/nrm1858
18. Patel DD, Koopmann W, Imai T, Whichard LP, Yoshie O, Krangel MS (2001) Chemokines have diverse abilities to form solid phase gradients. Clin Immunol 99(1):43–52. doi:10.1006/clim.2000.4997
19. Petrie RJ, Doyle AD, Yamada KM (2009) Random versus directionally persistent cell migration. Nat Rev Mol Cell Biol 10(8):538–549. doi:10.1038/nrm2729
20. †PAD, Quinn JA, Albelda SM, Lauffenburger DA (1992) Measurement of individual cell migration parameters for human tissue cells. Aiche J 38(7):1092–1104. doi:10.1002/aic.690380712
21. He X, Shen Y (1999) Vascular endothelial growth factor induced endothelial cell migration and proliferation depend on a nitric oxide mediated decrease in protein kinase C activity. Circ Res 85(3):247–256. doi:10.1083/jcb.140.4.947
22. Rojas JD, Sennoune SR, Maiti D, Bakunts K, Reuveni M, Sanka SC, Martinez GM, Seftor EA, Meininger CJ, Wu G, Wesson DE, Hendrix MJ, Martinez-Zaguilan R (2006) Vacuolar-type H + -ATPases at the plasma membrane regulate pH and cell migration in microvascular endothelial cells. Am J Physiol Heart Circ Physiol 291(3):H1147–1157. doi:10.1152/ajpheart.00166.2006
23. Sagnella SM, Kligman F, Anderson EH, King JE, Murugesan G, Marchant RE, Kottke-Marchant K (2004) Human microvascular endothelial cell growth and migration on biomimetic surfactant polymers. Biomaterials 25(7–8):1249–1259. doi:10.1016/s0142-9612(03)00634-3
24. Boyden S (1962) The chemotactic effect of mixtures of antibody and antigen on polymorphonuclear leucocytes. J Exp Med 115(115):453–466. doi:10.1084/jem.115.3.453
25. Albini A, Iwamoto Y, Kleinman HK, Martin GR, Aaronson SA, Kozlowski JM, Mcewan RN (1987) A rapid in vitro assay for quantitating the invasive potential of tumor cells. Cancer Res 47(47):3239–3245
26. Zigmond SH (1977) Ability of polymorphonuclear leukocytes to orient in gradients of chemotactic factors. J Cell Biol 75(75):606–616
27. Zicha D, Dunn GA, Brown AF (1991) A new direct-viewing chemotaxis chamber. J Cell Sci 99(Pt 4) (4):769–775
28. Servant G, Weiner OD, Herzmark P, Balla T, Sedat JW, Bourne HR (2000) Polarization of chemoattractant receptor signaling during neutrophil chemotaxis. Science 287(5455):1037–1040. doi:10.1126/science.287.5455.1037
29. Wong K, Pertz O, Hahn K, Bourne H (2006) Neutrophil polarization: spatiotemporal dynamics of RhoA activity support a self-organizing mechanism. Proc Natl Acad Sci U S A 103(10):3639–3644. doi:10.1073/pnas.0600092103
30. Liang CC, Park AY, Guan JL (2007) In vitro scratch assay: a convenient and inexpensive method for analysis of cell migration in vitro. Nat Protoc 2(2):329–333. doi:10.1038/nprot.2007.30
31. Hanahan D, Weinberg RA (2011) Hallmarks of cancer: the next generation. Cell 144 (5):646–674. doi:10.1016/j.cell.2011.02.013
32. Beebe DJ, Mensing GA, Walker GM (2002) Physics and applications of microfluidics in biology. Annu Rev Biomed Eng 4:261–286. doi:10.1146/annurev.bioeng.4.112601.125916
33. Mehling M, Tay S (2014) Microfluidic cell culture. Curr Opin Biotechnol 25:95–102. doi:10.1016/j.copbio.2013.10.005
34. Chung BG, Choo J (2010) Microfluidic gradient platforms for controlling cellular behavior. Electrophoresis 31(18):3014–3027. doi:10.1002/elps.201000137
35. Lohof AM, Quillan M, Dan Y, Poo MM (1992) Asymmetric modulation of cytosolic cAMP activity induces growth cone turning. J Neurosci 12(4):1253–1261

36. Nelson RD, Quie PG, Simmons RL (1975) Chemotaxis under agarose: a new and simple method for measuring chemotaxis and spontaneous migration of human polymorphonuclear leukocytes and monocytes. J Immunol 115(6):1650–1656
37. Kim S, Kim HJ, Jeon NL (2010) Biological applications of microfluidic gradient devices. Integr Biol (Camb) 2(11–12):584–603. doi:10.1039/c0ib00055h
38. Chicurel M (2002) Cell migration research is on the move. Science 295(5555):606–609. doi:10.1126/science.295.5555.606
39. Selmeczi D, Mosler S, Hagedorn PH, Larsen NB, Flyvbjerg H (2005) Cell motility as persistent random motion: theories from experiments. Biophys J 89(2):912–931. doi:10.1529/biophysj.105.061150
40. Li Jeon N, Baskaran H, Dertinger SK, Whitesides GM, Van de Water L, Toner M (2002) Neutrophil chemotaxis in linear and complex gradients of interleukin-8 formed in a microfabricated device. Nat Biotechnol 20(8):826–830. doi:10.1038/nbt712
41. Chung BG, Flanagan LA, Rhee SW, Schwartz PH, Lee AP, Monuki ES, Jeon NL (2005) Human neural stem cell growth and differentiation in a gradient-generating microfluidic device. Lab Chip 5(4):401–406. doi:10.1039/b417651k
42. Tourovskaia A, Figueroa-Masot X, Folch A (2005) Differentiation-on-a-chip: a microfluidic platform for long-term cell culture studies. Lab Chip 5(1):14–19. doi:10.1039/b405719h
43. Rafael Gómezsjöberg, † AAL, ‡ DMP, And CSC, ‡, † SRQ (2007) Versatile, fully automated, microfluidic cell culture system. Anal Chem 79(22):8557–8563. doi:10.1021/ac071311w
44. Frisk T, Rydholm S, Liebmann T, Svahn HA, Stemme G, Brismar H (2007) A microfluidic device for parallel 3-D cell cultures in asymmetric environments. Electrophoresis 28 (24):4705–4712. doi:10.1002/elps.200700342
45. Saadi W, Rhee SW, Lin F, Vahidi B, Chung BG, Jeon NL (2007) Generation of stable concentration gradients in 2D and 3D environments using a microfluidic ladder chamber. Biomed Microdevices 9(5):627–635. doi:10.1007/s10544-007-9051-9
46. Chang YC, Funamoto S, Firtel RA (2001) Signaling pathways controlling cell polarity and chemotaxis. Trends Biochem Sci 26(9):557–566. doi:10.1016/S0968-0004(01)01934-X
47. Horwitz AR, Parsons JT (1999) Cell migration–movin' on. Science 286(5442):1102–1103. doi:10.1126/science.286.5442.1102
48. Doerr ME, Jones JI (1996) The roles of integrins and extracellular matrix proteins in the insulin-like growth factor I-stimulated chemotaxis of human breast cancer cells. J Biol Chem 271(5):2443–2447. doi:10.1074/jbc.271.5.2443
49. Keenan TM, Folch A (2008) Biomolecular gradients in cell culture systems. Lab Chip 8 (1):34–57. doi:10.1039/b711887b
50. Meyvantsson I, Beebe DJ (2008) Cell culture models in microfluidic systems. Annu Rev Anal Chem (Palo Alto Calif) 1:423–449. doi:10.1146/annurev.anchem.1.031207.113042
51. Terry K, Williams SM, Connolly L, Ottemann KM (2005) Chemotaxis plays multiple roles during Helicobacter pylori animal infection. Infect Immun 73(2):803–811. doi:10.1128/IAI.73.2.803-811.2005
52. Hawes MC, Smith LY (1989) Requirement for chemotaxis in pathogenicity of Agrobacterium tumefaciens on roots of soil-grown pea plants. J Bacteriol 171(10):5668–5671. doi:10.1128/jb.171.10.5668-5671.1989
53. Stocker R, Seymour JR, Samadani A, Hunt DE, Polz MF (2008) Rapid chemotactic response enables marine bacteria to exploit ephemeral microscale nutrient patches. Proc Natl Acad Sci U S A 105(11):4209–4214. doi:10.1073/pnas.0709765105
54. Atencia J, Morrow J, Locascio LE (2009) The microfluidic palette: a diffusive gradient generator with spatio-temporal control. Lab Chip 9(18):2707–2714. doi:10.1039/b902113b
55. Luster AD, Alon R, von Andrian UH (2005) Immune cell migration in inflammation: present and future therapeutic targets. Nat Immunol 6(12):1182–1190. doi:10.1038/ni1275
56. Baggiolini M, Dewald B, Moser B (1997) Human chemokines: an update. Annu Rev Immunol 15(1):675–705. doi:10.1146/annurev.immunol.15.1.675

57. Yonekawa K, Harlan JM (2005) Targeting leukocyte integrins in human diseases. J Leukoc Biol 77(2):129–140. doi:10.1189/jlb.0804460
58. Jeon NL, Dertinger SKW, Chiu DT, Choi IS, And ADS, Whitesides GM (2000) Generation of solution and surface gradients using microfluidic systems. Langmuir 16(22):8311–8316. doi:10.1021/la000600b
59. Gao Y, Sun J, Lin WH, Webb D, Li D (2012) A compact microfluidic gradient generator using passive pumping. Microfluid Nanofluid 12(6):887–895. doi:10.1007/s10404-011-0908-0
60. Du Y, Shim J, Vidula M, Hancock MJ, Lo E, Chung BG, Borenstein JT, Khabiry M, Cropek DM, Khademhosseini A (2009) Rapid generation of spatially and temporally controllable long-range concentration gradients in a microfluidic device. Lab Chip 9(6):761–767. doi:10.1039/b815990d
61. Xu Z, Huang X, Wang P, Wang H, Weitz DA (2016) Optimization and development of a universal flow-based microfluidic gradient generator. Microfluid Nanofluid 20(6). doi:10.1007/s10404-016-1749-7
62. Dertinger SKW, Chiu DT, Jeon NL, Whitesides GM (2001) Generation of gradients having complex shapes using microfluidic networks. Anal Chem 73(6):1240–1246. doi:10.1021/ac001132d
63. Campbell K, Groisman A (2007) Generation of complex concentration profiles in microchannels in a logarithmically small number of steps. Lab Chip 7(2):264–272. doi:10.1039/b610011b
64. Cooksey GA, Sip CG, Folch A (2009) A multi-purpose microfluidic perfusion system with combinatorial choice of inputs, mixtures, gradient patterns, and flow rates. Lab Chip 9 (3):417–426. doi:10.1039/b806803h
65. Lee K, Kim C, Kim Y, Ahn B, Bang J, Kim J, Panchapakesan R, Yoon Y-K, Kang JY, Oh KW (2011) Microfluidic concentration-on-demand combinatorial dilutions. Microfluid Nanofluid 11(1):75–86. doi:10.1007/s10404-011-0775-8
66. Wu J, Hillier C, Komenda P, Lobato de Faria R, Levin D, Zhang M, Lin F (2015) A microfluidic platform for evaluating neutrophil chemotaxis induced by sputum from COPD patients. PLoS ONE 10(5):e0126523. doi:10.1371/journal.pone.0126523
67. Cheng JY, Yen MH, Kuo CT, Young TH (2008) A transparent cell-culture microchamber with a variably controlled concentration gradient generator and flow field rectifier. Biomicrofluidics 2(2):24105. doi:10.1063/1.2952290
68. Abhyankar VV, Lokuta MA, Huttenlocher A, Beebe DJ (2006) Characterization of a membrane-based gradient generator for use in cell-signaling studies. Lab Chip 6(3):389–393. doi:10.1039/b514133h
69. Kim M, Kim T (2010) Diffusion-based and long-range concentration gradients of multiple chemicals for bacterial chemotaxis assays. Anal Chem 82(22):9401–9409. doi:10.1021/ac102022q
70. Berthier E, Surfus J, Verbsky J, Huttenlocher A, Beebe D (2010) An arrayed high-content chemotaxis assay for patient diagnosis. Integr Biol (Camb) 2(11–12):630–638. doi:10.1039/c0ib00030b
71. Kong Q, Majeska RJ, Vazquez M (2011) Migration of connective tissue-derived cells is mediated by ultra-low concentration gradient fields of EGF. Exp Cell Res 317(11):1491–1502. doi:10.1016/j.yexcr.2011.04.003
72. Irimia D, Charras G, Agrawal N, Mitchison T, Toner M (2007) Polar stimulation and constrained cell migration in microfluidic channels. Lab Chip 7(12):1783–1790. doi:10.1039/b710524j
73. Zou H, Yue W, Yu WK, Liu D, Fong CC, Zhao J, Yang M (2015) Microfluidic platform for studying chemotaxis of adhesive cells revealed a gradient-dependent migration and acceleration of cancer stem cells. Anal Chem 87(14):7098–7108. doi:10.1021/acs.analchem.5b00873
74. Lee SS, Horvath P, Pelet S, Hegemann B, Lee LP, Peter M (2012) Quantitative and dynamic assay of single cell chemotaxis. Integr Biol (Camb) 4(4):381–390. doi:10.1039/c2ib00144f

75. Keenan TM, Frevert CW, Wu A, Wong V, Folch A (2010) A new method for studying gradient-induced neutrophil desensitization based on an open microfluidic chamber. Lab Chip 10(1):116–122. doi:10.1039/b913494h
76. Kim D, Liu A, Diller E, Sitti M (2012) Chemotactic steering of bacteria propelled microbeads. Biomed Microdevices 14(6):1009–1017. doi:10.1007/s10544-012-9701-4
77. Cheng SY, Heilman S, Wasserman M, Archer S, Shuler ML, Wu M (2007) A hydrogel-based microfluidic device for the studies of directed cell migration. Lab Chip 7 (6):763–769. doi:10.1039/b618463d
78. Choi HI, Kim JY, Kwak HS, Sung YJ, Sim SJ (2016) Quantitative analysis of the chemotaxis of a green alga, Chlamydomonas reinhardtii, to bicarbonate using diffusion-based microfluidic device. Biomicrofluidics 10(1):014121. doi:10.1063/1.4942756
79. Chung BG, Lin F, Jeon NL (2006) A microfluidic multi-injector for gradient generation. Lab Chip 6(6):764–768. doi:10.1039/b512667c
80. Qasaimeh MA, Gervais T, Juncker D (2011) Microfluidic quadrupole and floating concentration gradient. Nat Commun 2:464. doi:10.1038/ncomms1471
81. Wright GA, Costa L, Terekhov A, Jowhar D, Hofmeister W, Janetopoulos C (2012) On-chip open microfluidic devices for chemotaxis studies. Microsc Microanal 18(4):816–828. doi:10.1017/S1431927612000475
82. Moussavi-Harami SF, Pezzi HM, Huttenlocher A, Beebe DJ (2015) Simple microfluidic device for studying chemotaxis in response to dual gradients. Biomed Microdevices 17 (3):9955. doi:10.1007/s10544-015-9955-8
83. Paguirigan A, Beebe DJ (2006) Gelatin based microfluidic devices for cell culture. Lab Chip 6(3):407–413. doi:10.1039/b517524k
84. Kothapalli CR, van Veen E, de Valence S, Chung S, Zervantonakis IK, Gertler FB, Kamm RD (2011) A high-throughput microfluidic assay to study neurite response to growth factor gradients. Lab Chip 11(3):497–507. doi:10.1039/c0lc00240b
85. Srinivasan P, Zervantonakis IK, Kothapalli CR (2014) Synergistic effects of 3D ECM and chemogradients on neurite outgrowth and guidance: a simple modeling and microfluidic framework. PLoS ONE 9(6):e99640. doi:10.1371/journal.pone.0099640
86. Kothapalli CR, Honarmandi P (2014) Theoretical and experimental quantification of the role of diffusive chemogradients on neuritogenesis within three-dimensional collagen scaffolds. Acta Biomater 10(8):3664–3674. doi:10.1016/j.actbio.2014.05.009
87. Haessler U, Kalinin Y, Swartz MA, Wu M (2009) An agarose-based microfluidic platform with a gradient buffer for 3D chemotaxis studies. Biomed Microdevices 11(4):827–835. doi:10.1007/s10544-009-9299-3
88. Herzmark P, Campbell K, Wang F, Wong K, El-Samad H, Groisman A, Bourne HR (2007) Bound attractant at the leading vs. the trailing edge determines chemotactic prowess. Proc Natl Acad Sci U S A 104(33):13349–13354. doi:10.1073/pnas.0705889104
89. Mosadegh B, Saadi W, Wang SJ, Jeon NL (2008) Epidermal growth factor promotes breast cancer cell chemotaxis in CXCL12 gradients. Biotechnol Bioeng 100(6):1205–1213. doi:10.1002/bit.21851
90. Ricart BG, John B, Lee D, Hunter CA, Hammer DA (2011) Dendritic cells distinguish individual chemokine signals through CCR7 and CXCR4. J Immunol 186(1):53–61. doi:10.4049/jimmunol.1002358
91. Wang SJ, Saadi W, Lin F, Minh-Canh Nguyen C, Li Jeon N (2004) Differential effects of EGF gradient profiles on MDA-MB-231 breast cancer cell chemotaxis. Exp Cell Res 300 (1):180–189. doi:10.1016/j.yexcr.2004.06.030
92. Chen YC, Allen SG, Ingram PN, Buckanovich R, Merajver SD, Yoon E (2015) Single-cell migration chip for chemotaxis-based microfluidic selection of heterogeneous cell populations. Sci Rep 5:9980. doi:10.1038/srep09980
93. Ng CP, Hinz B, Swartz MA (2005) Interstitial fluid flow induces myofibroblast differentiation and collagen alignment in vitro. J Cell Sci 118(Pt 20):4731–4739. doi:10.1242/jcs.02605

94. Shamloo A, Ma N, Poo MM, Sohn LL, Heilshorn SC (2008) Endothelial cell polarization and chemotaxis in a microfluidic device. Lab Chip 8(8):1292–1299. doi:10.1039/b719788h
95. Walker GM, Zeringue HC, Beebe DJ (2004) Microenvironment design considerations for cellular scale studies. Lab Chip 4(2):91–97. doi:10.1039/b311214d
96. Lee P, Lin R, Moon J, Lee LP (2006) Microfluidic alignment of collagen fibers for in vitro cell culture. Biomed Microdevices 8(1):35–41. doi:10.1007/s10544-006-6380-z
97. Beta C, Frohlich T, Bodeker HU, Bodenschatz E (2008) Chemotaxis in microfluidic devices–a study of flow effects. Lab Chip 8(7):1087–1096. doi:10.1039/b801331d
98. Haessler U, Pisano M, Wu M, Swartz MA (2011) Dendritic cell chemotaxis in 3D under defined chemokine gradients reveals differential response to ligands CCL21 and CCL19. Proc Natl Acad Sci U S A 108(14):5614–5619. doi:10.1073/pnas.1014920108
99. Suh S, Traore MA, Behkam B (2016) Bacterial chemotaxis-enabled autonomous sorting of nanoparticles of comparable sizes. Lab Chip 16(7):1254–1260. doi:10.1039/c6lc00059b
100. Walsh DI 3rd, Lalli ML, Kassas JM, Asthagiri AR, Murthy SK (2015) Cell chemotaxis on paper for diagnostics. Anal Chem 87(11):5505–5510. doi:10.1021/acs.analchem.5b00726
101. McCaig CD, Rajnicek AM, Song B, Zhao M (2005) Controlling cell behavior electrically: current views and future potential. Physiol Rev 85(3):943–978. doi:10.1152/physrev.00020.2004
102. Yao L, Pandit A, Yao S, McCaig CD (2011) Electric field-guided neuron migration: a novel approach in neurogenesis. Tissue Eng Part B Rev 17(3):143–153. doi:10.1089/ten.TEB.2010.0561
103. Jin T, Hereld D (eds) (2009) Chemotaxis: methods and protocols. Humana Press, New York
104. Robinson KR, Messerli MA (2003) Left/right, up/down: the role of endogenous electrical fields as directional signals in development, repair and invasion. BioEssays 25(8):759–766. doi:10.1002/bies.10307
105. Cooper MS, Keller RE (1984) Perpendicular orientation and directional migration of amphibian neural crest cells in dc electrical fields. Proc Natl Acad Sci 81(1):160–164
106. Guo A, Song B, Reid B, Gu Y, Forrester JV, Jahoda CA, Zhao M (2010) Effects of physiological electric fields on migration of human dermal fibroblasts. J Invest Dermatol 130 (9):2320–2327. doi:10.1038/jid.2010.96
107. Finkelstein E, Chang W, Chao P-HG, Gruber D, Minden A, Hung CT, Bulinski JC (2004) Roles of microtubules, cell polarity and adhesion in electric-field-mediated motility of 3T3 fibroblasts. J Cell Sci 117(8):1533–1545. doi:10.1242/jcs.00986
108. Jennings J, Chen D, Feldman D (2008) Transcriptional response of dermal fibroblasts in direct current electric fields. Bioelectromagnetics 29(5):394–405. doi:10.1002/bem.20408
109. Mba D, Mycielska M, Madeja Z, Fraser SP, Korohoda W (2001) Directional movement of rat prostate cancer cells in direct-current electric field: involvement of voltagegated Na + channel activity. J Cell Sci 114(Pt 14):2697–2705
110. Zhao M, Agius-Fernandez A, Forrester JV, McCaig CD (1996) Directed migration of corneal epithelial sheets in physiological electric fields. Invest Ophth Vis Sci 37(13):2548–2558
111. Zhao M, Agius-Fernandez A, Forrester JV, Mccaig CD (1996) Orientation and directed migration of cultured corneal epithelial cells in small electric fields are serum dependent. J Cell Sci 109(Pt 6) (6):1405–1414
112. Rapp B, De B-CA, Gruler H (1988) Galvanotaxis of human granulocytes. Eur Biophys J Biophy 16(5):313–319. doi:10.1007/BF00254068
113. Zhao M, Bai H, Wang E, Forrester JV, Mccaig CD (2004) Electrical stimulation directly induces pre-angiogenic responses in vascular endothelial cells by signaling through VEGF receptors. J Cell Sci 117(3):397–405. doi:10.1242/jcs.00868
114. Biber TUL, Sanders ML (1973) Influence of transepithelial potential difference on the sodium uptake at the outer surface of the isolated frog skin. J Gen Physiol 61(5):529–551. doi:10.1085/jgp.61.5.529
115. Jaffe LF (1977) Electrophoresis along cell membranes. Nature 265(5595):600–602. doi:10.1038/265600a0

116. McLaughlin S, Poo MM (1981) The role of electro-osmosis in the electric-field-induced movement of charged macromolecules on the surfaces of cells. Biophys J 34(1):85–93. doi:10.1016/S0006-3495(81)84838-2
117. Mycielska ME, Djamgoz MB (2004) Cellular mechanisms of direct-current electric field effects: galvanotaxis and metastatic disease. J Cell Sci 117(Pt 9):1631–1639. doi:10.1242/jcs.01125
118. Zhao M, Song B, Pu J et al (2006) Electrical signals control wound healing through phosphatidylinositol-3-OH kinase-gamma and PTEN. Nature 442(7101):457–460. doi:10.1038/nature04925
119. Song B, Gu Y, Pu J, Reid B, Zhao Z, Zhao M (2007) Application of direct current electric fields to cells and tissues in vitro and modulation of wound electric field in vivo. Nat Protoc 2(6):1479–1489. doi:10.1038/nprot.2007.205
120. Li X, Kolega J (2002) Effects of direct current electric fields on cell migration and actin filament distribution in bovine vascular endothelial cells. J Vasc Res 39(5):391–404. doi:10.1159/000064517
121. Sato MJ, Ueda M, Takagi H, Watanabe TM, Yanagida T, Ueda M (2007) Input–output relationship in galvanotactic response of Dictyostelium cells. Biosystems 88(3):261–272. doi:10.1016/j.biosystems.2006.06.008
122. Lin F, Baldessari F, Gyenge CC, Sato T, Chambers RD, Santiago JG, Butcher EC (2008) Lymphocyte electrotaxis in vitro and in vivo. J Immunol 181(4):2465–2471. doi:10.4049/jimmunol.181.4.2465
123. Li J, Nandagopal S, Wu D, Romanuik SF, Paul K, Thomson DJ, Lin F (2011) Activated T lymphocytes migrate toward the cathode of DC electric fields in microfluidic devices. Lab Chip 11(7):1298–1304. doi:10.1039/C0LC00371A
124. Rezai P, Siddiqui A, Selvaganapathy PR, Gupta BP (2010) Electrotaxis of Caenorhabditis elegans in a microfluidic environment. Lab Chip 10(2):220–226. doi:10.1039/B917486A
125. Zhao S, Zhu K, Zhang Y, Zhu Z, Xu Z, Zhao M, Pan T (2014) ElectroTaxis-on-a-Chip (ETC): an integrated quantitative high-throughput screening platform for electrical field-directed cell migration. Lab Chip 14(22):4398–4405. doi:10.1039/C4LC00745J
126. Huang CW, Cheng JY, Yen MH, Young TH (2009) Electrotaxis of lung cancer cells in a multiple-electric-field chip. Biosens Bioelectron 24(12):3510–3516. doi:10.1016/j.bios.2009.05.001
127. Tsai H-F, Peng S-W, Wu C-Y, Chang H-F, Cheng J-Y (2012) Electrotaxis of oral squamous cell carcinoma cells in a multiple-electric-field chip with uniform flow field. Biomicrofluidics 6(3):034116. doi:10.1063/1.4749826

Chapter 6
Biomaterial-Based Microfluidics for Cell Culture and Analysis

Ruizhi Ning, Qichen Zhuang and Jin-Ming Lin

Abstract Microfluidic devices can integrate drug injection, bioreactors, sample separation and detection on a individual platform with properties of fluidic incubation, programmable three-dimensional (3D) channels in micro scale, low reagent consumption, and on-line detection, which makes them popular in in vitro cell culture and analysis. Since the appearance of the initial generation of microfluidics in 1970s, materials for microfluidics experienced the age of inorganics such as silicon and glasses, and now have entered the period of organics, in which main materials used in microfluidics are polymers. Also, polymers constructed the chips gradually undertake much more functions rather than a basic supporting plane through applying functional and smart biomaterials. In this chapter, biomaterials utilized in microfluidics will be elaborated from the perspectives of the fabrication of microfluidics, cell culture and analysis on chips.

Keywords Biomaterials · Microfluidic chip · Cell culture · Scaffolds
Cell analysis · Cell observation

6.1 Introduction

Passed through 20-year development, microfluidic chips now occupy a significant position in cell biology and cell analysis. Comparing with traditional plate culture methods, the on-chip cell culture has incomparable superiorities. As an important example, microchips can realize the simulation of in vivo microenvironment, which makes on-chip cell experiments reliable and accurate. However, the application of microfluidic chips in cell research still has some limitations. Various methods have been developed to increase the feasibility of microfluidic devices, such as employing functional biomaterials or optimizing the design of chip structures to

R. Ning · Q. Zhuang · J.-M. Lin (✉)
Department of Chemistry, Tsinghua University, Beijing 100084,
People's Republic of China
e-mail: jmlin@mail.tsinghua.edu.cn

J.-M. Lin (ed.), *Cell Analysis on Microfluidics*, Integrated Analytical Systems,
https://doi.org/10.1007/978-981-10-5394-8_6

realize better manipulation of cells and fluids. Making use of appropriate biomaterials is a very effective and flexible way of improving the biocompatibility of microfluidic devices and offering functions for on plane and three-dimensional space (3D) cell culture and analysis, which has great significance for drug evaluation, diagnosis and therapy, biological research and fields based on cells. In this chapter, we will introduce materials used in the microfluidic chips construction and on-chip cell culture and analysis.

6.2 Materials of Making Microfluidic Chips

Microfluidic devices have programmable channels and chambers with the lowest resolution in micrometer, which makes them helpful in chemical analysis, clinical evaluation, environmental monitoring, and biology. In particular, due to the dimensional consistence of cells and micro-scale channels, microfluidic devices are extraordinarily suitable for cell research to realize cell culture, manipulation, and analysis at a continuous flowing state [1]. Also, microanalysis on microfluidic platforms only requires a small reagent and analyte volume, which can determine cellular and extracellular secretions at low amount, even for a single cell.

To achieve biofunctions for cell culture and analysis on microfluidics, choosing appropriate materials is the key point, because materials have direct contact with cells and will certainly influence cell state and behaviors such as cell viability, cell proliferation and cell metabolism. Moreover, functional groups on biomaterial surfaces can also be utilized to immobilize proteins, aptamers, or DNA in microfluidics, which further assist in cell adhesion, capture, and analysis. Here, materials used in microfluidics establishment will be elaborated in three parts, the inorganic materials that are mainly employed in the first generation microfluidics, and organic materials containing plastics, elastomers and hydrogels and papers.

6.2.1 Inorganic Materials

With the maturity of micro-electromechanical system (MEMS) preparation technology, some inorganic materials, such as silicon and glass, were employed in constructing the first generation microfluidic chips. The initial microfluidic device was a gas analysis system mainly fabricated in silicon by photolithography and chemical etching techniques [2]. This portable gas chromatography successfully separated gaseous hydrocarbon mixtures within 10 s, which brought a revolution for micro-scaled manipulation and analysis of chemical micromolecule and biological macromolecule from then on. In the fabrication of MEMS devices, four main techniques are involved, which are thin-film deposition, lithograph, etching, and packaging in turn.

The thin-film deposition is the procedure to create a layer on inorganic substrates for further lithography and etching [3, 4]. Chemical vapor deposition (CVD), physical vapor deposition (PVD), electrodeposition and substrate oxidation are some commonly used methods. In the lithography step, channel patterns are designed on a photomask by computer-aided design and then transferred on the deposited layer of inorganic substrates to act as a mask in the next etching step [5]. The etching process can ordinarily be divided into wet etching [6] and dry etching [7]. Wet etching is based on chemical reactions between the corrosion solution and substrates. The etching process has the same reaction rate in all directions in a homogeneous medium, thus, the wet etching is always isotropic. Due to the isotropy, channels with vertical sidewalls can be hardly obtained on amorphous substrates, such as glass. To solve this problem, single-crystal silicon wafers are commonly adopted in anisotropic etching because silicon etchants have anisotropic corrosion along the crystallization direction. As shown in Fig. 6.1, the wet etching was performed on single-crystal silicon surfaces and grooves with different cross-sections shapes were obtained when applying diverse etchants. Compared with wet etching, the dry way has fewer limitations and better anisotropic corrosion, which makes the fabrication of micro-scaled patterns on multiple substrates possible. High-pressure plasma etching, reactive ion etching, and ion milling are some frequently-used techniques in the dry etching. Besides, microchannels with good vertical sidewalls can be obtained at a high aspect ratio through a dry etching process [8]. The final step, packaging process, which is also called the bonding process, ordinarily employs an adhesive layer or thermal fusion treatment to seal the microchannels with substrates [9].

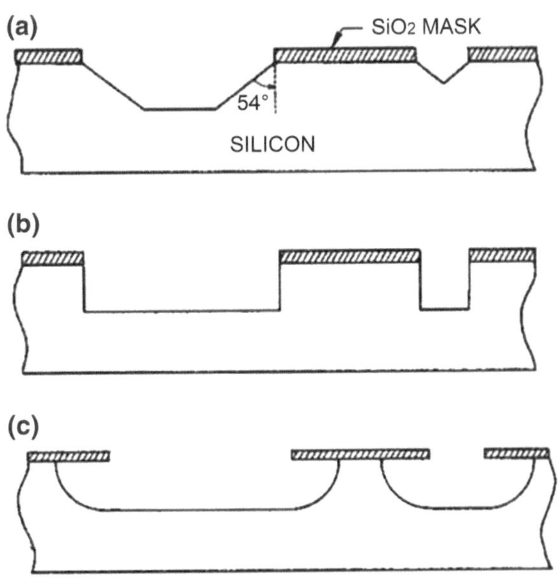

Fig. 6.1 Cross sections of grooves etched in silicon. **a** KOH etchant in (100) silicon, **b** KOH in (110) silicon, **c** HF-HNO_3 etchant. (Reprinted with permission from Ref. [2])

In recent years, silicon and glass are gradually replaced by polymers in constructing microchannel structures. In spite of this, glass is still the main substrate material used in microfluidic devices, because it is widely used in biological monitoring which requires high transparency and good surface modification activity. Except for the glass, other inorganic materials also have good performance as substrates. For instance, quartz-based microfluidic devices were proved to be efficient in cell identification and protein separation. In Dochow's work of Raman-activated cell sorting (RACS) [10], when at near-infrared laser excitation, quartz substrates exhibited fairly low Raman background in the fingerprint region, which solved the problems in RACS when adopting glass as substrates in cell identification. Moreover, quartz also improved protein separation in the way of acting as a substrate with good heat dissipation that leads to larger electric fields in the isoelectric focusing (IEF) of proteins [11].

6.2.2 Polymeric Materials

Inorganic materials are efficient in chemical analysis but have some formidable defects when applied in biology. For instance, necessary gases in cell incubation, such as oxygen and carbon dioxide, are difficult to penetrate through the inorganic layer and produce negative effect on cells. Also, opaque materials like silicon wafers are not suitable for cell observation using microscopes. Meanwhile, polymeric materials have developed rapidly in both diversity and multi-function that can meet the demands of microfluidic devices. Because of the lowe Young's modulus, polymers became the main material used for microfluidic chip construction in biological applications. Besides, polymers can easily be processed in various ways with low cost, and the diversity of polymeric materials makes the surface modification flexible and facile. According to the classification of polymers, polymeric materials constructing microfluidic chips can be divided into two categories, that is, plastics and elastomers.

6.2.2.1 Plastics

Plastic is defined as the polymeric materials with low elastic deformation ability, which can further be divided into two categories, thermoplastics and thermosetting plastics, based on whether the materials can melt and be repeatedly molded when heated or dissolved in solvents.

Thermoplastic materials always have linear molecular chains and low glass transition temperature (T_g), which results in good thermal processability, such as hot embossing, thermal molding and bonding. Several kinds of thermoplastics are commonly used as commercial products in biology. For instance, polystyrene (PS) is the most widely adopted material in culture dishes; polymethyl methacrylate (PMMA), and polyethylene terephthalate (PET) are ideal candidates for cell observation on microfluidic platforms due to their good biocompatibility and transparency, as evidenced by their extensive applications in agriculture, food industry, and medicine [12–14]. Battle et al. fabricated a PMMA microfluidic solid-phase extraction (μSPE) device through hot embossing. In this device, abundant micropillars integrated in the main channel had high efficiency in enriching and purifying membrane proteins from whole blood lysate [15]. However, the linear polymers could easily release low molecular weight compounds into the solution and can be dissolved by organic solvents. Also, the low gas permeability of plastics limits their employment in constructing the principal parts of microfluidic chips. In Midwoud's work, as shown in Fig. 6.2, biocompatibility of PMMA, PS, polycarbonate (PC), and cyclic olefin copolymer (COC) were tested to determine their applicability as the principal material of microfluidic devices [12]. As a result, compared with polydimethylsiloxane (PDMS) which is the chief materials in constructing microfluidic devices at present, no apparent improvement was observed on these thermoplastics.

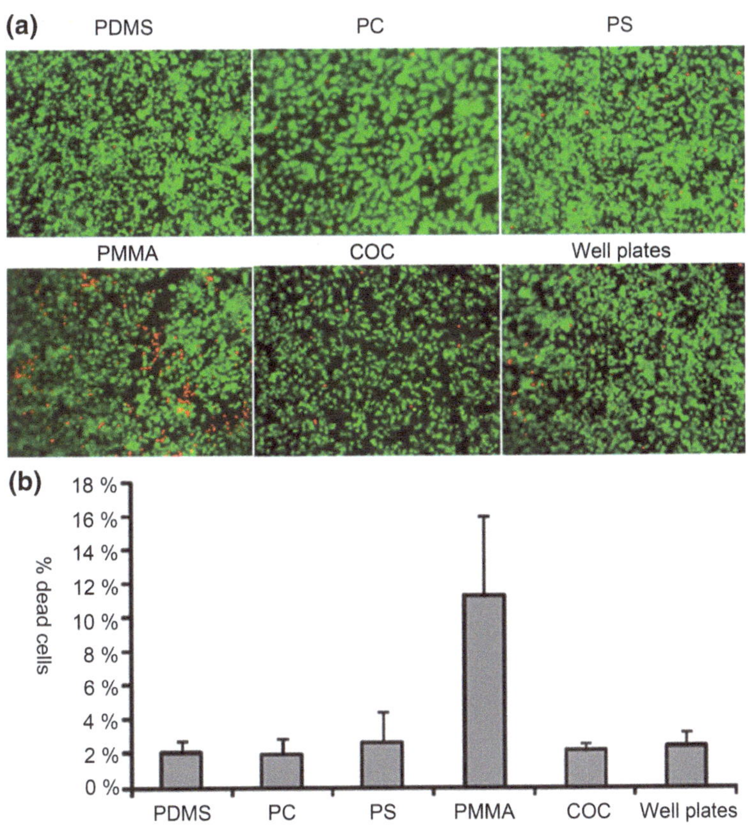

Fig. 6.2 HepG2 cells cultured on different thermoplastics and PDMS after UV-ozone treatment and collagen coating. **a** Fluorescent photographs stained by Line/Dead kit; **b** Cell mortality. (Reprinted with permission from Ref. [12])

In case of thermosetting plastics, molecular chains cross-linked into network structures give the material with resistance to heating and solvents. Once formed the network, stiffness of the thermosetting plastics will increase and it becomes difficult to remold the material. Therefore, few thermosetting plastics are considered to be employed as the principle material for microfluidic chips. However, in microfluidic chips fabrication, the thermosetting plastic SU-8 negative photoresist is frequently used and can be utilized to establish a free-standing microstructure [16].

6.2.2.2 Elastomers

Elastomers are materials that can stretch more than half of the original length along the force direction when sustaining strong external load, and all deformations can recover after the force is removed. The high elastic deformation ability of elastomers makes them flexible and soft. Elastomers are ideal candidates for cell culture and analysis, because soft surfaces lead to better cell viability than rigid surfaces [17]. At present, PDMS is the most commonly used polymer as the principle materials in fabricating microfluidic chips, because it is a flexible material with high transparency, good gas permeability and biocompatibility, which is beneficial to cell activities. Among fabrication methods of PDMS microfluidic devices, the soft lithography technique has developed into a routine way. In particular, glass/PDMS hybrid microfluidic devices, in which PDMS constructs the main part of microfluidic chips and glass is employed as the substrate, are popular in cell culture and analysis. The soft lithography contains replica molding and rapid prototyping. Once a mask is prepared, PDMS casting can be repeatedly performed using the mask.

The process of PDMS microchannel preparation is illustrated in Fig. 6.3. In the first step, a microchannel mask is prepared through UV illumination on photoresist covered by a printed photomask with designed channel patterns. Then, a mixture of PDMS prepolymer and curing agent is poured on the microchannel mask with a convex plain form. Before baking, removing air bubbles is necessary. The principle part of microfluidic chips, microchannels, can be achieved after baking, peeling, and punching holes in turn. The obtained PDMS microchannels can be bonded to other silicon-based surfaces (e.g., silicon, glass, quartz, PDMS) though covalent bonding.

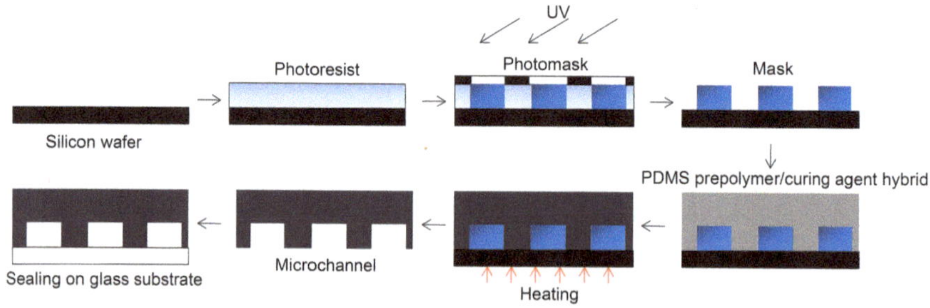

Fig. 6.3 Schematic diagram of PDMS-based microchip fabrication through soft lithography

Although PDMS-based microfluidic devices are widely applied in biological research, their hydrophobic surfaces will lead to non-specific adhesion of proteins or analytes. To address the issue of non-specific adhesion and integrate functional biomolecules, a hydrophilic treatment is always needed. Generally, the hydrophilization is achieved through convenient surface modifications [18], for instance, plasma cleaning [19], silanization treatment [20], and surface coating [21]. After surface modification, PDMS-based microfluidic devices have better performances in biomolecular separation, protein/cell enrichment and release, and cell culture.

In addition to PDMS, other synthesized elastomers are also used in the fabrication of microfluidic devices. For example, Roy et al. replaced PDMS with styrenic thermoplastic elastomers (TPEs) in building multilayer microfluidic devices [22], because PDMS is inadequate for industrial-scale fabrication and application. Human cells were successfully cultured on surfaces of the isothermal TPE microdevice after sealing with DNA and protein solutions. Perfluoropolyether (PFPE) can also be used to build microfluidic chips. Rolland et al. [23] reported a preparation method of microfluidic devices with resistance to organic solvents by using the photo-curing "liquid Teflon" material, which extended the application of microfluidic devices to novel fields.

6.2.3 Hydrogels and Papers

Although hydrogel and paper are essentially polymer materials, by the reason of their particularity and great development potential, here we purposely pick them out as an individual portion to introduce.

6.2.3.1 Hydrogels

Hydrogels have network structures consisted by coadjacent polymeric chains formed under certain conditions, in which the liquid padding the interspace acts as dispersion media. This kind of material containing hydrophilic groups and cross-linked macromolecular networks leads to both high water content and the ability of shape preserving. Different from absorbent materials such as cotton and sponge that the water absorbing capacity comes from the capillary phenomenon, water molecules in hydrogels bond with the hydrophilic groups on networks through chemical bonds, which gives them strong water retention capacity with few water loss even under pressure [24–26]. Therefore, hydrogel is a stable high water-content material with plenty holes between the polymeric chains. The character resulted in hydrogels possessing properties such as easy mass transportation, good cell viability and proliferation and making them an ideal candidate in native extracellular matrix (ECM) imitation.

Both natural and synthesized hydrogels are employed in preparing microfluidic chips. Natural hydrogels such as collagen, calcium alginate, cellulose, gelation, and chitosan are commonly used biomaterials in microfluidics [27, 28], however, due to their low mechanical durability and microstructure reproducibility, few of them were employed in constructing the main part of microfluidic devices. They are frequently used as scaffolds in microchips for 3D cell culture and organ simulation, which will be described in the section "3D culture" (see Sect. 6.3.2) and "Organ-on-chip system" (see Sect. 6.4). Although artificial hydrogels are rarely used in fabricating the main part of microfluidic chips neither, the polyethylene glycol (PEG)-based hydrogel is an attractive material in bioanalytical devices because PEG has strong non-biofouling characters and versatile processability with tunable mechanical strength to support the structure stability. The idea of adopting PEG hydrogels in the fabrication of microfluidic devices comes from a frequently used surface modification method in which PEG is modified on silicon-based surfaces to solve the problem of non-specific adsorption of biomolecules. Though the modification can be realized through physical adsorption and chemical immobilization, inadequacies such as inherent defects on the modified layer and loss of resistance to biofouling in long-term fluid flush gradually invalidate the modification and bring back the hydrophobicity. The fabrication of PEG microfluidic chips mainly use soft lithography, which is similar to that of PDMS devices. Kim et al. [29] employed PEG, PEG-diacrylate (DA) and PEG-dimethacrylate (DMA) in making microchannels through UV curing. In the result, low molecular weight of PEG-DA ($M = 258$) and PEG-DMA ($M = 330$) had good resistance to water swelling, thus were used for the microchannel fabrication. Finally, a PEG microfluidic device entirely made up by cross-linked PEG chains were fabricated within merely 10 min and can be sealed to another PEG layer irreversibly by UV irradiation. In addition of using pure natural or synthesized hydrogels, recently, composite materials of natural and artificial hydrogels have been widely applied in biology. Cellulose hydrogel composites can be used as the principle material to fabricate microfluidic devices after blending with synthesized polymers or changing the solution concentration, structure, and properties of the cellulose hydrogel. For instance, as shown in Fig. 6.4, Pei et al. [30] succeeded in fabricating cross-linked cellulose (RCC) hydrogel and cellulose–collagen (RCC/C) hybrid hydrogel-based integrated microfluidic chips, which has well-controlled pore size, high transparency, good mechanical durability, and good biocompatibility in both 2D and 3D cell culture.

6.2.3.2 Papers

Paper is a sheet consisted of wood fibers deposited layer by layer. This close packing of layered structure of fibers gives paper integrated multi-functions, such as foldability, filterability, hydroscopicity and surface modifiability. Also, the capillary action of papers makes paper-based microfluidic devices have wide applications in rapid diagnostic tests [31]. In paper-based microfluidic chip fabrication, choosing

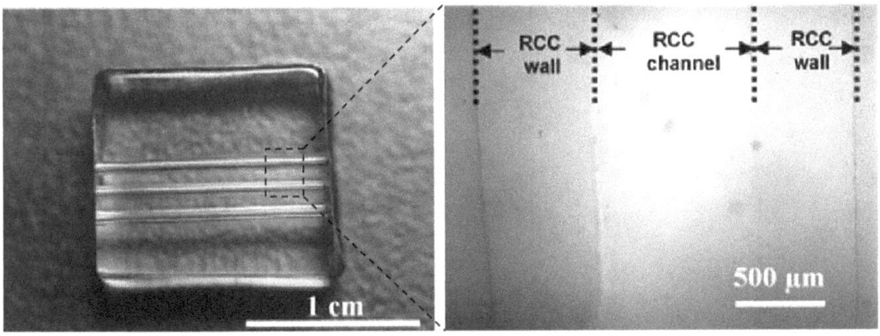

Fig. 6.4 Microchannels fabricated by cellulose hydrogels. Digital images of RCC hydrogel with three channels (*Left*) and optical images of channels in bright field (*Right*). (Reprinted with permission from Ref. [30])

an appropriate paper that can accomplish the requirements of assays is of great significance, because different kinds of papers have obvious differences in chemical and physical properties, such as capillary flow rate, pore size distribution and porosity, and some other factors that have great influence on paper performances. Cellulosic material is the most frequently employed paper in microfluidic devices [32–34], which includes filter paper, chromatography paper, and nitrocellulose paper. The generation of fluidic channels on papers is another key point in preparing a paper-based microfluidic chip. Through making hydrophobic barriers in horizontal or vertical directions, the reagents can be leaded into separated channels for further detections. The channel fabrication can be realized on papers easily and flexibly through combining hydrophobic substance coating (e.g., wax pattering, alkyl ketene dimer printing, PDMS plotting) and paper shaping (e.g., embossing, fold, clipping). Thuo et al. [35] built up a microfluidic device using omniphobic papers through embossing and "cut-and-stack" assembling methods, as illustrated in Fig. 6.5. Fluids in this device behaved similarly as they did in open-channel devices. Laminar flow and droplet generation can be realized using this paper chip, which can be utilized in environmental monitoring and droplet-based synthesis and analysis.

6.3 Cell Culture

Before human trials, preliminary assays should be performed on animal models. However, animal experiments are expensive, cumbersome, restricted, and the obtained results are suspectable for human data due to genetic diversities existed between different species. The emergence of cell clone technology deals with the problems in animal experiments and cell incubation plays a very important role in pharmacology, medicine, tissue engineering, and so on. Commercial culture dishes

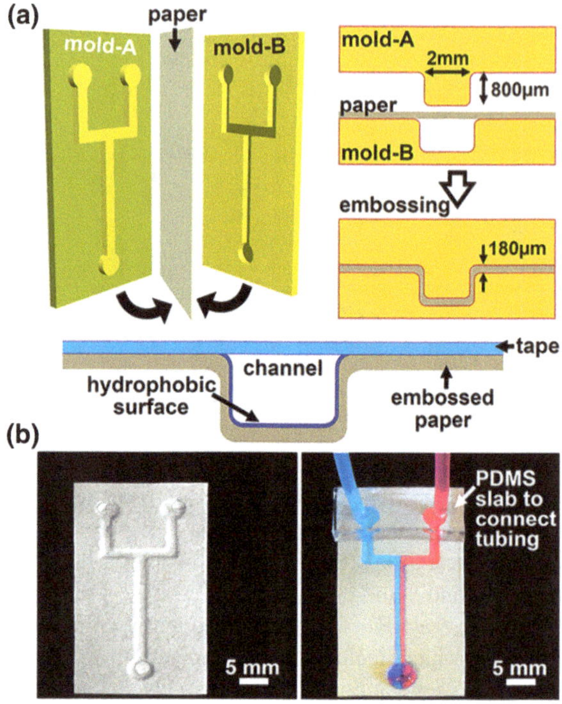

Fig. 6.5 An open-channel paper microfluidic device. **a** Schematic diagram of the construction; **b** Photographs of laminar flow in the device. (Reprinted with permission from Ref. [35])

and bottles merely support static culture in general. However, providing cells an environment similar to their in vivo microenvironment which contains fluid stimulus, substance exchange and cellular interactions is of great importance and has powerful impacts on cellular behaviors and functions. Therefore, microfluidic chips are the most appropriate candidate for microenvironment simulation, because controllable culture medium supply, 3D cell culture, drug stimulation, metabolite extraction and analysis can be integrated on this programmable fluid platform. With microfluidic chips, some tough missions, such as constructing the important nidus structures, now find possible solutions.

6.3.1 *Plane Cell Culture*

Cell culture performed in microfluidic devices are primarily on plane microchannel surfaces for experiments focused on investigating fluid impacts on cells. Most of the materials for on-chip cell culture are polymeric materials, such as PS, PC, TPE, and PDMS. Although these materials can be applied for cell incubation, manipulation and analysis, their inherent hydrophobic surfaces will lead to irreversible adhesion of abundant albumin, which in turn prevents cell adhesion and causes adverse effects on cells. Accordingly, the hydrophilic modification is essential. Numerous methods have been developed to convert the hydrophobicity [36–38], which includes physical protein adsorption, chemical modifications, and modifying biomaterials.

The physical protein adsorption can produce a protein matrix layer on polymer substrates. In general, hydrophobic surfaces of polymer substrates can non-specifically adsorb serum proteins easily from the culture medium, which decreases the affinity of substrates for cell adhesion. However, extracellular matrix (ECM) proteins have good affinity with cells, thus importing ECM protein solutions in microfluidics is one of the most convenient way to achieve protein modification. Fibronectin (FN), laminin, collagen are some frequently employed proteins in physical modification to improve cell adhesion and proliferation. The forces between proteins and substrate surfaces are mainly some weak interactions, therefore, the adsorbed protein can easily detach from the interacting surface. As a result, the physical adsorption of proteins is only valid for a limited duration to maintain cell adhesion.

The chemical modification is capable of creating powerful covalent bond linkages between proteins and the substrates, which is more effective and stable than the physical adsorption. Convenient plasma cleaning can used to introduce hydrophilic groups on substrate surfaces, but the generated hydrophilicity is not very stable and will soon recover to hydrophobicity after a certain duration time or heating [39]. Chemical modifications formed by covalent bonding between proteins and substrate surfaces are very stable. Through this way, ECM proteins are efficiently immobilized on material surfaces. For instance, Kuddannaya et al. [40] succeeded in modifying FN and collagen type I (C1) on PDMS by the chemical reaction with (3-aminopropyl) trirthoxy silane (APTES) and cross-linker glutaraldehyde (GA), as exhibited in Fig. 6.6. After surface modification, mesenchymal stem cells (MSCs) were incubated to evaluate the validity of the modification. Results of cell adhesion and cell viability proved that the surface modified with APTES, GA, and proteins

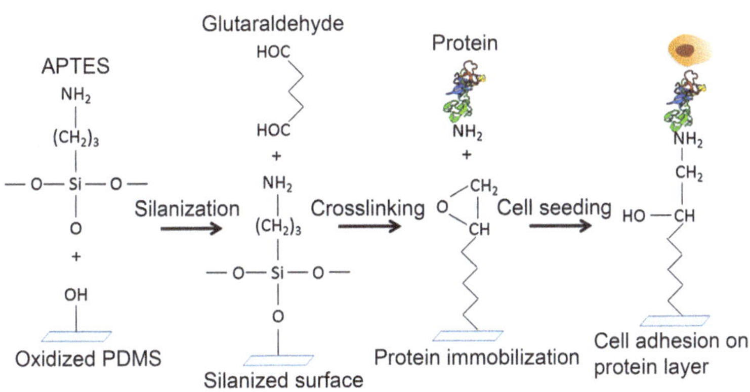

Fig. 6.6 Schematic diagram of PDMS surface modification. (Reprinted with permission from Ref. [40])

effectively reduced the hydrophobicity of PDMS, which is beneficial to cell adhesion and proliferation.

In addition to natural materials (e.g., ECM proteins, peptides, carbohydrates), synthesized biomaterials can also be utilized in improving the hydrophilicity of substrates. Among those, smart biomaterials draw much more attentions because they have response regulation of the surface wettability to environmental changes. The wettability can be controllably adjusted through such as pH value, temperature, light, and solvents. These factors have their advantages under different use requirements, but for on-chip cell culture, temperature is thought to be the most appropriate environmental factor to adjust the hydrophilicity. In many thermal-responsive systems developed for cell culture, poly (*N*-isopropylacrylamide) (PNIPAAm) is the most frequently employed biomaterial [41, 42], because the PNIPAAm chain undergoes a reversible conformational transition near its lower critical solution temperature (LCST) [43] that is acceptable by cells. As illustrated by Fig. 6.7a, when environment temperature is below LCST, a large number of complex H bonds are exposed on surfaces of molecular chains and arrange along

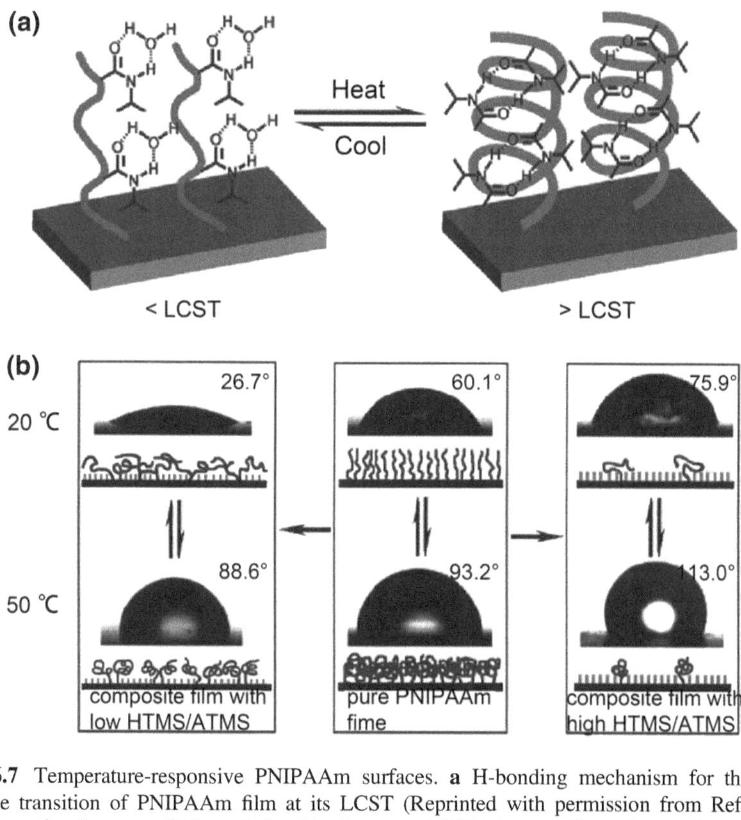

Fig. 6.7 Temperature-responsive PNIPAAm surfaces. **a** H-bonding mechanism for the coil–globule transition of PNIPAAm film at its LCST (Reprinted with permission from Ref. [43]). **b** Schematic diagram of the relationship between PNIPAAm configuration and responsive wettability. (Reprinted with permission from Ref. [44])

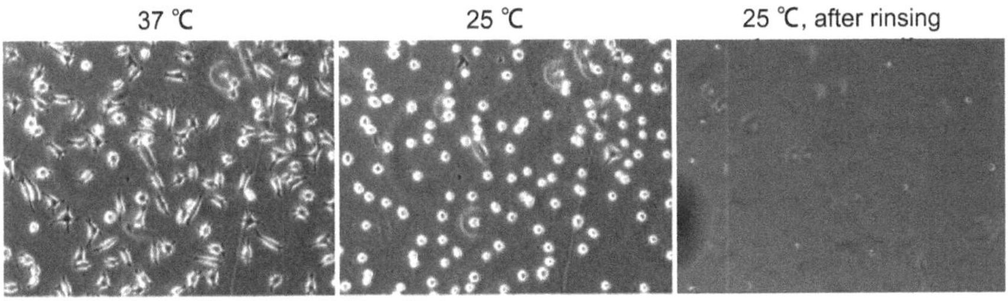

Fig. 6.8 Temperature-responsive cell culture surfaces. Phase contrast microscopy images of L929 mouse fibroblast cells after incubation at 37 °C on PNIPAM microgel films and 20 min after cooling the surface to 25 °C. At 25 °C gentle rinsing nearly removed more than 90% of the cells from the surface. (Reprinted with permission from Ref. [45])

the chains. However, when above LCST, molecular chains distort and cover a part of the H bonds, making wettability of the surface changed (Fig. 6.7b) [44]. As shown in Fig. 6.8, Schmidt et al. achieved cell adhesion conversion on PNIPAA microgel films [45]. Moreover, physicochemical parameters and the cell adhesion performance of the PNIPAA films before and after temperature variation were investigated. The results proved the efficiency of PNIPAA microgel film in regulating cell adhesion and release through temperature control, and the temperature slightly above LCST was found to be the optimum temperature for cell adhesion and proliferation.

6.3.2 3D Cell Culture

Simulation of the in vivo microenvironment on cell culture is expected by drug evaluation, oncotherapy, and biological research. In spite of the plane cell culture can easily be proceed on various substrates, extensive studies proved that the spatial arrangement of cells has significant influence on cellular behaviors and functions [46–48]. In this background, the proposal of developing 3D cell culture system on chip was put forward. Generally, there are two mainly ways achieving spatial cell culture on microfluidic chips. One is through multi-layer chip design, and the other one is making use of scaffold materials, especially hydrogels. In order to achieve better results, the multi-layer structure design and scaffold employment are often used in combination.

Using the multi-soft lithography technique, microchannels with different width, depth, and shape can be integrated on a single microfluidic device. 3D cell culture in these microdevices can be realized through placing cells on target channel surfaces at different layers under control. The spatial cell culture in microfluidic chips can be achieved in various ways through delicate design, such as preparing microwells, height-gradient chambers, and making use of intercellular porous films. The microwell and height-gradient structures are often fabricated by multiple ultraviolet exposures with a group of photomasks. Figure 6.9 exhibits Kang's research of developing a multilayer microfluidic device with concave microwells and plane chambers to culture embryonic stem (ES) cells and regulate the formation of uniform-sized embryoid bodies [49]. Lin's group also did many researches of on-chip 3D cell culture and analysis. For example, Wu constructed a 3D microfluidic chip with height-gradient chambers to imitate the diffusion process between blood vessels and tissues [50], and cytotoxicity of quantum dots (QDs) was investigated on this established drug evaluation platform using HepG2 cells. As another way of building 3D cell culture models, sandwich-structured systems with a porous thin film used as the intercellular film are commonly adopted in 3D cell co-culture to investigate cellular interactions between two or more types of cells [51]. Figure 6.10 exhibits the typical structure of the "sandwich" coculture microfluidic platform. Normal mouse embryonic fibroblasts (mEFs) (without inactivation) and mouse embryonic stem (mES) cells were cultured on two sides of a PDMS porous layer for feeder-separated coculture, which was proved in the tomography by laser confocal microscopy (Fig. 6.10c). The mEFs resulted in the formation of mES cell colonies with excellent undifferentiated state confirmed by the expression of Nanog, octamer binding protein 4 (Oct-4) and alkaline phosphatase (ALP), which demonstrates the significant advantages in efficiency and simplicity of the microfluidic device.

Above methods achieved spatial cell culture to some extent, however, they are not the very real 3D cell culture, because cells cultured on these platforms still attach to microchannel surfaces or the porous membranes that present plane culture state. To fill microchannels with cells, introducing scaffold materials in channel space is necessary to assist in cell proliferation and migration. Meanwhile, porous structures of scaffold materials can guarantee the mass transportation in chip. Generally, biomaterial hydrogels can be used as ideal scaffolds [52, 53], which has been mentioned in the previous portion (see Sect. 6.2.3.1). Some ECM proteins, such as collagen, fibrin, matrigel, FN, and mixture of these proteins were primarily considered as the principal part of hydrogel scaffolds to encapsulate

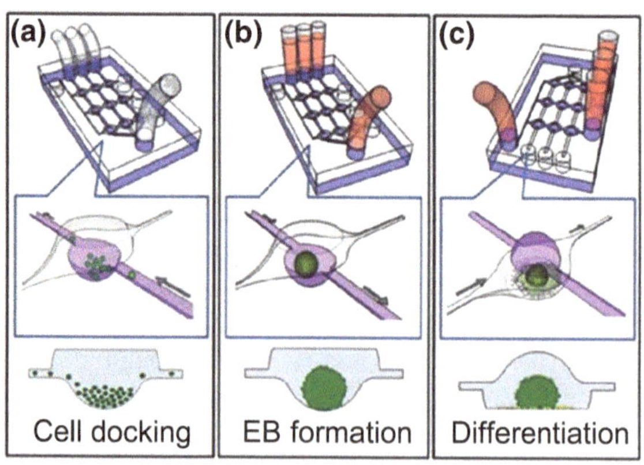

Fig. 6.9 A multi-layer microfluidic device containing concave microwells and flat cell-culture chambers. **a** Schematic of the cell docking in concave microwells; **b** Schematic of the EB formation; **c** Retrieval of EBs from the concave microwells after converting the microfluidic device. (Reprinted with permission from Ref. [49])

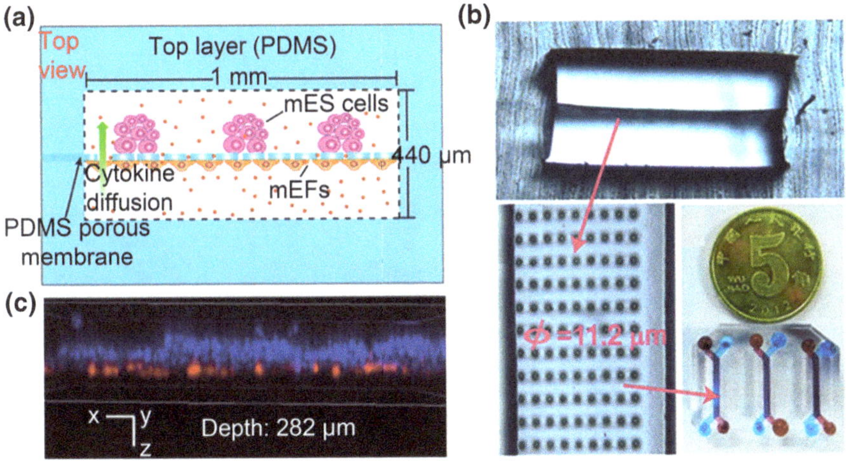

Fig. 6.10 The typical "sandwich" coculture microfluidic platform. **a** Cross section of the device; **b** Photographs of the assembled microchannel, PDMS porous membrane and fabricated microdevice; **c** Confocal image showing the mES cells/mEFs coculture system. (Reprinted with permission from Ref. [51])

dispersed cells. For example, Kamm et al. accomplished 3D cell culture with collagen [28, 54–56]. They fabricated a 3D microfluidic assay filled with collagen gels to determine the influence of endothelial KLF2 expression on smooth muscle cell (SMC) migration [54]. SMC cells co-cultured with endothelial KLF2-expressing endothelial cells (ECs) showed an apparent decrease in migration. They also adopted collagen as 3D scaffolds in investigating the neurite transforming under a growth factor gradient generated by microfluidics [28]. More remarkably, they introduced physiologically relevant 3D capillary morphogenesis on hydrogel microfluidic platforms stimulated by growth factors [55] and cells [56]. Calcium alginate and gelatin also have good biocompatibility as scaffolds in cell culture. For instance, Choong Kim succeeded in embedding fibroblast cells in alginate beads [57]. The cell beads can influence the EC monolayer cultured in collagen scaffolds and leaded to the formation of circular lumen-like structures. In addition to ECM proteins, cellulose, the most abundant biomaterial in natural, shows great potential in fabricating 3D microenvironment for cells on chip. Cellulose is one of the crystalline structural polysaccharides, a linear polymer connected by β-1, 4-D-glucoside bond, which can be produced by wood, tunicate, and bacterial. Among celluloses from different sources, the bacterial cellulose has wide applications in personal care, pharmaceutical, bioscience [58], due to its very special structures and properties (nanometer networks, high water retention, good biosynthesis controllability and biocompatibility). The nanostructure and morphology of bacterial cellulose make it very suitable for being used as the scaffolds for cell immobilization and migration [59, 60] (Fig. 6.11), and forming ECM. Also, as shown in Fig. 6.12, hydrogels of bacterial cellulose is an ideal material for tissue

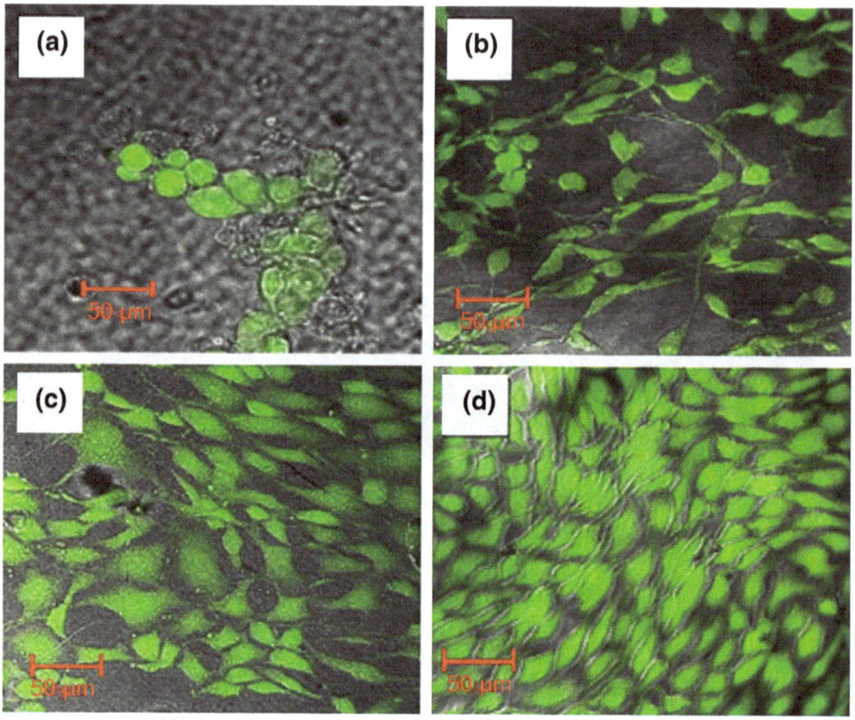

Fig. 6.11 Bacteria cellulose in cell culture. Confocal laser microscopic images of 6 days old fibroblast NIH/3T3 cells grown on **a** AGM_Ca, **b** BC gel, **c** BC_M and **d** GTA_GM. (Reprinted with permission from Ref. [60])

engineering[61], which has been applied in skin repair, wound dressing, tissue regeneration and periodontal treatment. As a famous work, artificial blood vessels made of bacterial cellulose have been employed in clinical experiments [58]. This kind of artificial blood vessel has stable inner cavity, good stitching stability and good permeability of water, ions, and small molecules. Besides, with the development of nanoscience, several techniques are capable of processing cellulose into cellulose nanofibril (CNF), such as TEMPO-mediated oxidation followed by mechanical disintegration treatment [62], and the mechanism and obtained nanofibers are exhibited in Fig. 6.13. Nanocelluloses also exhibit great potentials in biomedicine [63]. For instance, NFC has shear-thinning property, and based on this feature, Bhattacharya et al. prepared a type of injectable hydrogel by making use of plant-sourced CNF [64]. Cytotoxic tests of the CNF hydrogel using HepaRG and HepG2 cells proved the biocompatibility of this material, and mitochondria activity of the cells cultured in CNF hydrogels had no apparent distinction compared with standard 2D culture cell incubation on CNF. Also, a gyroid scaffold consisted of

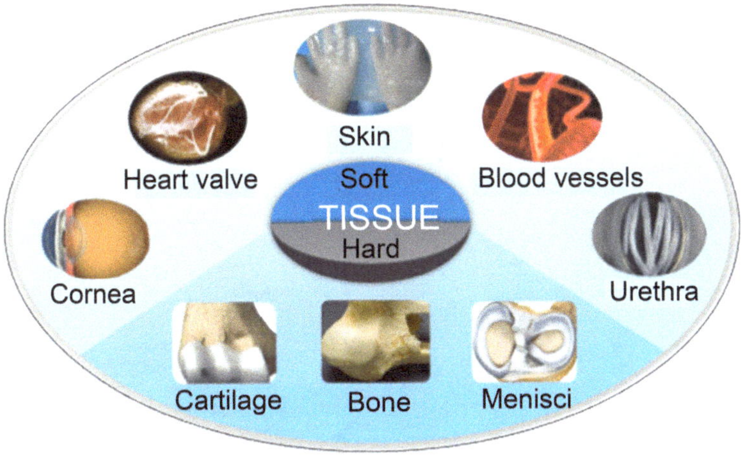

Fig. 6.12 Biomedical applications of BC-based materials in tissue engineering. (Reprinted with permission from Ref. [61])

nanocellulose and nanochitin hydrogel was fabricated using a template-assisted method [65]. The hydrogel possessed shape recovery behavior even after supercritical drying. In investigations of the effectiveness of the produced hydrogel as bone tissue scaffolds, human muscle satellite cells (HMSCs) showed good adhesion and spread on CNF, which may due to the electrostatic interactions between CNF surfaces and collagen 1 at pH 7.4 in phosphate buffer solutions (PBS). Consequently, the cellulose nanomaterial hydrogel is a promising material that can be used in microfluidic devices as scaffolds for 3D cell culture.

Hydrogels, especially natural hydrogels, have some drawbacks in 3D cell culture. For instance, materials of different batches exhibit inconsistent properties. Also, the transportation of nutrients and generated oxygen concentrations have gradient distributions along the thickness of hydrogels. Furthermore, some experiments require dense cell concentrations, such as a multicellular tumor ball, which can hardly be constructed in hydrogel scaffolds. Therefore, gel-free 3D culture was put forward to settle the problems. One way is employing intercellular polymeric linkers to generate cell aggregations[66]. As shown in Fig. 6.14, Mo et al. used a positively charged dendrimeric intercellular linker in assembly of C3A cells. As a consequence, the fabricated cell aggregation had high viability, proliferation, and 3D cell morphology. In other ways, microwells [67] and microbubbles [68] fabricated in chips can assist in forming 3D cell culture structures. For example, Liu et al. prepared quasi-spherical microwells on PDMS substrates using an ice lithography-based benchtop method [69]. The concave microwells can generate dense and homogeneous multicellular tumor spheroids that can be removed from the microwells during cell incubation. Moreover, the long-term in-chip culture can be achieved.

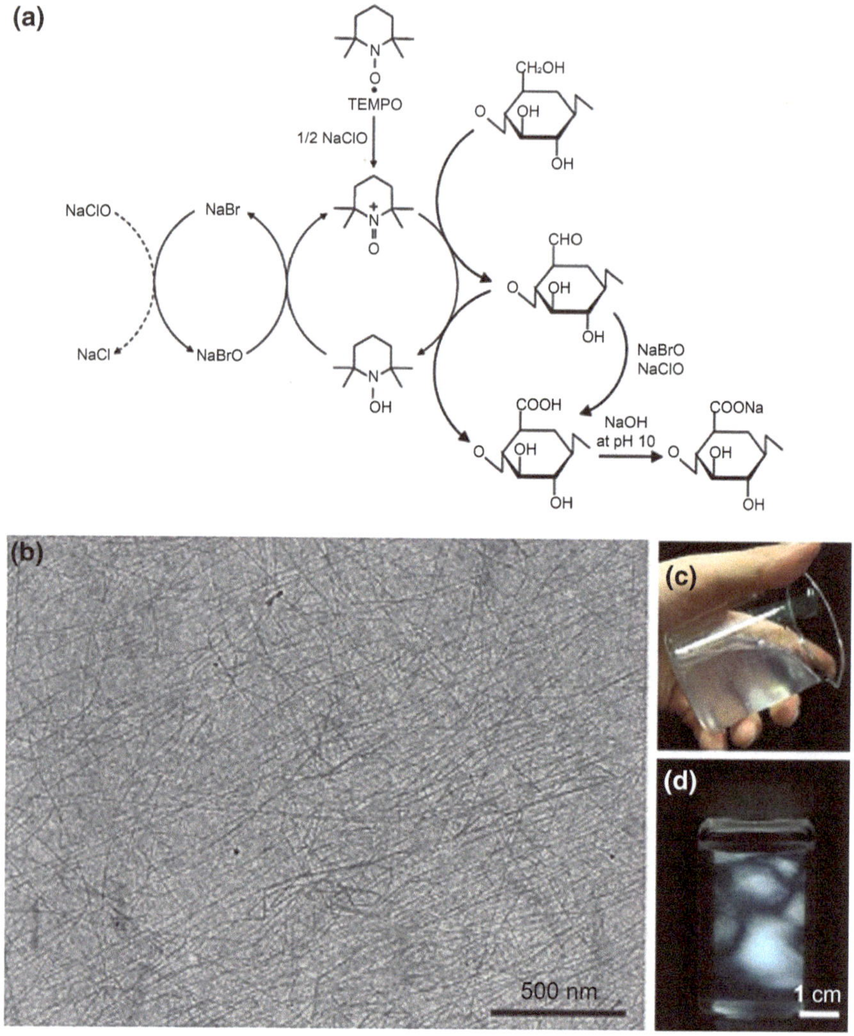

Fig. 6.13 Cellulose nanofibers prepared from wood pulp by TEMPO-mediated oxidation. **a** The mechanism of TEMPO/NaBr/NaClO oxidation in water at pH 10–11; **b** TEM image of a dried dispersion of TEMPO-oxidized hardwood cellulose; **c** Photographs of a typical dispersion and **d** when observed between cross polarizes. (Reprinted with permission from Ref. [62])

6.4 Organ-on-Chip System

Realizing artificial organs for clinical therapy and medical research services is always a hot topic. Cells are the fundamental unit to maintain functions of life activity and metabolism. Basically, organs are the aggregates of various cells that orderly arrange and distribute in a certain space. Therefore, the organ fabrication is

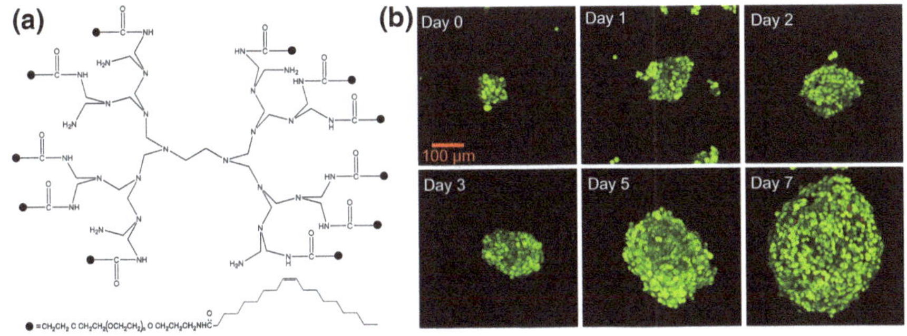

Fig. 6.14 Linker—engineered multi-cellular structures exhibit good viability and proliferation. **a** Schematic diagram of synthesis of oleyl-PEG conjugated DAB dendrimeric linker; **b** Confocal images of C3A cells stained by Live/Dead kit in multi-cellular structures during 7-day culture. (Reprinted with permission from Ref. [66])

essentially a simulation of the microenvironment and the composition and spatial structure of cell aggregations in organs. However, constructing an organ system is of great challenge, because the manipulation of different cells and maintaining their 3D structures with sustaining supplements is a harsh task. The emergence of microfluidic chips settled the issue of creating a blood supply system by designable microchannels. Also, the development of on-chip 3D cell culture makes microfluidic chips promising in fabricating organs, which has great significance in pharmaceutical research, tissue engineering, medicine and biology.

Organ-on-chip technique aims at mimicking in vivo physiological activities, functions, and understanding the operation mechanism of organs on a 3D cell culture microfluidic platform. 3D cell culture of one specific cell type can hardly inspire its functions. The intercellular communication is the key factor of maintaining normal cell activities [70, 71]. Therefore, multicellular culture systems with controllable cell distribution are significant in the construction of organ-on-chip devices. Dongeun Huh developed a very typical lung-on-a-chip microsystem [72], as illustrated by Fig. 6.15. They succeeded in detecting organ-level responses to bacteria and inflammatory cytokine carried on alveolar-capillary interfaces using this organ-on-chip device. As another example, a high throughput "heart on a chip" system was successfully prepared using submillimeter-sized thin-film cantilevers made of soft elastomers. Subsequently, muscular thin films (MTFs) were obtained through building anisotropic cardiac microtissues on the cantilevers [73], as shown in Fig. 6.16.

According to the preparation characteristic of microfluidic chips in constructing organ structures, organ-on-chip systems can mainly be divided into two categories: bottom-up stacking and 3D hydrodynamic flow focusing.

The bottom-up stacking can simulate tissue interfaces in the similar way as "sandwich" structures, that is, from the bottom to the top, accumulating layer by layer. In bottom-up microfluidic devices, two or more types of cells are incubated in different chambers that are separated by porous membranes. This kind of

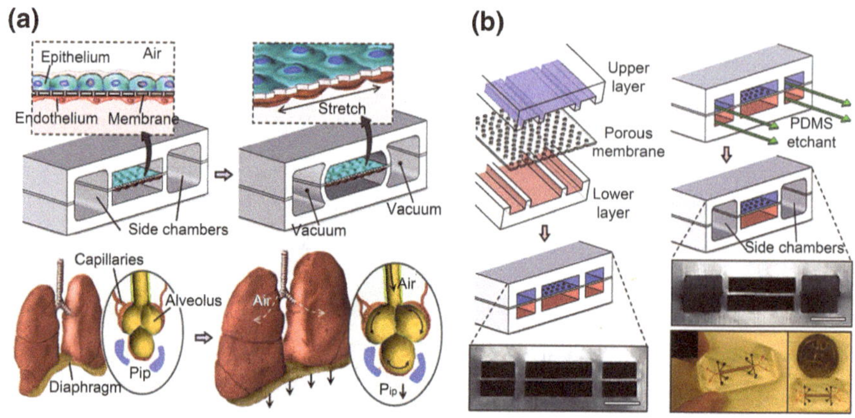

Fig. 6.15 Biomimic design of a human breathing lung on microchip. **a** Simulation of an alveolar-capillary barrier on microdevice; **b** Construction and photograph of the microchip. (Reprinted with permission from Ref. [72])

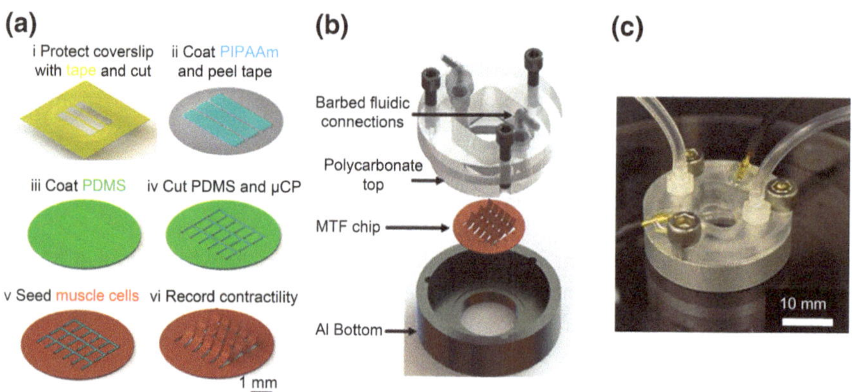

Fig. 6.16 Heart on a chip microdevice. **a** Schematic of fabrication process of MTF chip; **b** Assembly of the artificial heart fluidic device; **c** Photograph of the actual device in action. (Reprinted with permission from Ref. [73])

microfluidic chip is capable of providing fluidic shear stress, through which the lung alveolar-capillary interface and some unique structures in vivo, such as blood-brain carrier (BBB) [74], can be better simulated. The BBB is a unique feature of the central nervous system. Tight junctions of endothelial cells in BBB can prevent exogenetic substances from entering brain tissues. Figure 6.17 exhibits a microfluidic-built BBB structure. Two exogenetic flow channels were designed in one microchip, and EC cells and astrocytes were cultured on two sides of the membrane respectively. EC cells showed good viability of long time incubation, with high expression of tight junctions. C8-D1A (astrocyte) cells on the porous PC

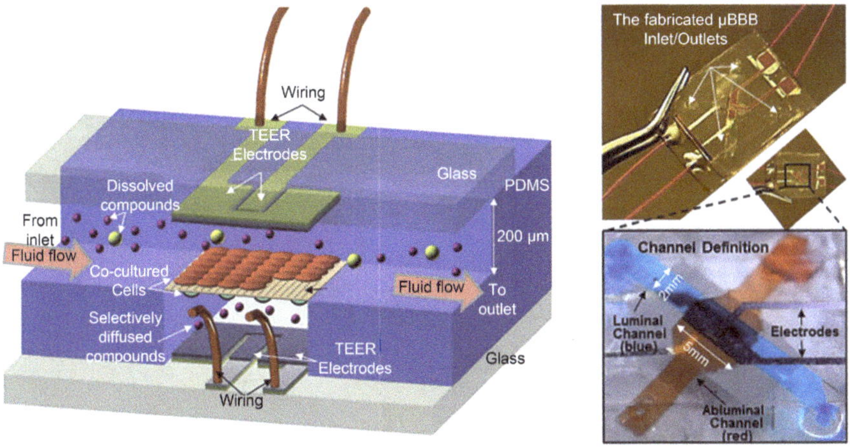

Fig. 6.17 Structure and design of the developed μBBB platform. (Reprinted with permission from Ref. [74])

membrane exhibited typical astrocytic morphology. Detection of the trans-endothelial electrical resistance (TEER) of the artificial microchip and transwells showed TEER value of the dynamic system (exceeded 250 Ω cm^2) was significantly higher than that of the static system (merely 25 Ω cm^2), which could due to the fluidic shear stress generated mecho-transductive effect on endothelial molecular pathways.

In 3D hydrodynamic flow focusing, specific cell arrangements such as spherical [75] and fibrous cell aggregations [76] can be achieved. Spherical and fibrous structures have particular importance in organ construction, because these configurations can be commonly observed in tissues and organs. Spherical structures find significant applications in mimicking tumor tissues and investigating intercellular communications between tumor cells and their neighboring cells. As shown in Fig. 6.18a, Alessanbri et al. [75] fabricated a coextrusion microfluidic device that can produce cell-based assays. They used 3D hydrodynamic flow focusing technique in preparing alginate microcapsules, and CT26 cell spheroids were successfully generated by gel-free and gel-encapsulating ways. To form spheroids, a side flow channel perpendicular to the fluidic direction is always needed to provide shear force in cutting the flow into droplets. Fibrous structures are fabricated in the same way as spheroids in terms of the fabricating mechanism. Both of them applied the coextrusion technique to form a coaxial geometry, but comparing with spheroid fabrication, synthesis of continues fibrous structures do not need the shear force. Cell-embedded hydrogel fibers embedded with cells have potential applications in mimicking fibrous structures in vivo, such as vessels, nerve fibers, and muscle fibers. As shown in Fig. 6.18b, Takeuchi's research group fabricated several meter-long core–shell structure biological fibers with using the hydrodynamic

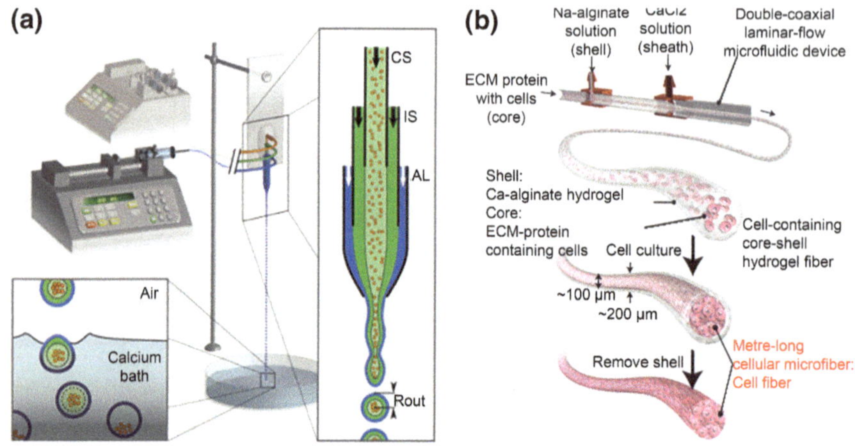

Fig. 6.18 Hydrodynamic flow focusing in forming 3D cell structures. **a** Spherical preparation system (Reprinted with permission from Ref. [75]) and **b** fibrous cell aggregations fabrication with coextrusion microdevices. (Reprinted with permission from Ref. [76])

focusing technique [76]. The coaxial fibers were continuously extruded out from a double-coaxial microfluidic chip. Cells in ECM proteins served as the core and the Ca-alginate hydrogel acted as the shell of the fibers. Using this platform, Cardiomyocyte-Fib, HUVEC-ACol, and Cortical cell-PCol fibers were successfully fabricated. Furthermore, these functional biological microfibers can be collected by waving or reeling, which can be used to prepare fiber-shaped tissues and organs.

On-chip tissues or organs with multi-cellular structures also can be realized through optimizing the chip design. For example, Lee et al. developed a 3D liver-on-chip system [77] in studying paracrine effect of hepatic stellate cells (HECs) on hepatocytes. Hepatocyte spheroids were generated in concave microwells on chip. Although no direct contact existed between the two types of cells, intercellular communications happened as a result of culture medium flow. Through comparing the spheroid structures and the level of albumin and cytochrome P450 reductase of mono-cultured hepatocyte spheroids or co-cultured spheroids, paracrine effect of HSCs was found to have positive effects on maintaining the functions and structures of hepatocyte spheroids.

6.5 Simulation and Manipulation of Cell Microenvironment on Chips

In vivo microenvironment contains many complex chemical and mechanical factors, which leads to various specific physical and chemical properties (e.g., temperature, osmotic pressure, oxygen content, and pH), soluble factors, and intercellular or cell-matrix interactions [78]. Besides, the in vivo microenvironment is a dynamic system, in which every factor is dynamically balanced to maintain the state stable for lives. One unique advantage of the on-chip cell culture is that microchips can mimic the cell environment with manipulation in continuous flow.

In drug screening research, the microfluidic technology can meet demands such as achieving parallel tests with concentration gradients through structure design. A typical microchip design for producing concentration gradients is called the "Xmas-tree" structure. In this structure, laminar flowing effect is utilized to make better mixing of two different solutions and produce concentration gradients along the direction perpendicular to parallel channels [79, 80]. Gao et al. [81] integrated the "Xmas-tree" structure in their microfluidic device to generate drug solutions with gradually changed concentrations. They analyzed cytotoxicity, drug absorption and metabolites with high throughput. Chen et al. [82] fabricated a network structure on chip to generate concentration gradients for cell culture and drug screening. Chen used an isotope-labeled internal standard method in detection, because the mass spectrometry can only obtain semi-quantitative results. This method improved the linear exponential and expanded the linearity range, which achieved real-time qualitative and quantitative detection of metabolites. Scherber et al. [83] found endothelial cells can "perceive" their way along the concentration gradient of endothelial growth factors (EGF) by using a microfluidic chip.

Hypoxia environment proved to be able to facilitate the tumor growth, invasion and metastasis. Accordingly, the simulation of in vivo hypoxia microenvironment and the investigation of hypoxia microenvironment's impact on tumor development are hot topics in tumor research [84]. Wang et al. [85] investigated the cell response to drug stimulation in hypoxia microenvironment fabricated on an oxygen concentration gradient producing microfluidic device. Lin et al. [86] fabricated a double-layer microfluidic chip to mimic the hypoxia environment in tumor. In their work, aptamers were fixed on both the two layers, through this, real-time detection of vascular endothelial growth factor (VEGF) 165 was completed (Fig. 6.19). Besides, the distance influence on endothelial-tumor intercellular interactions and oxygen concentration gradient influence on cell migration were investigated. This investigation of signaling pathways and cell migration behavior in tumor angiogenesis has significant reference values for investigations of the formation and development of cervical cancer.

Cell density can affect cell behavior through the way of leading to intercellular matrix differences and local concentrations of different autocrine or paracrine cytokines. This series of changes caused by cell density lead to diverse in vivo cell reactions, such as intercellular interactions and tissues functions [87]. Wu et al. [88] designed a microfluidic device with a microvoid-embedded channel to generate cell density with high accuracy and good repeatability. Through regulating the microvoid number in channel, cell density can be controlled. Moreover, the cell density had linear proportional relationship with the number of designed microvoids. The obtained cell density showed good consistency with the theoretical calculation. Subsequently, an investigation on the cytotoxicity of carboxyl group coated CdTe quantum dots (QDs) under cell density gradients was performed. As a result,

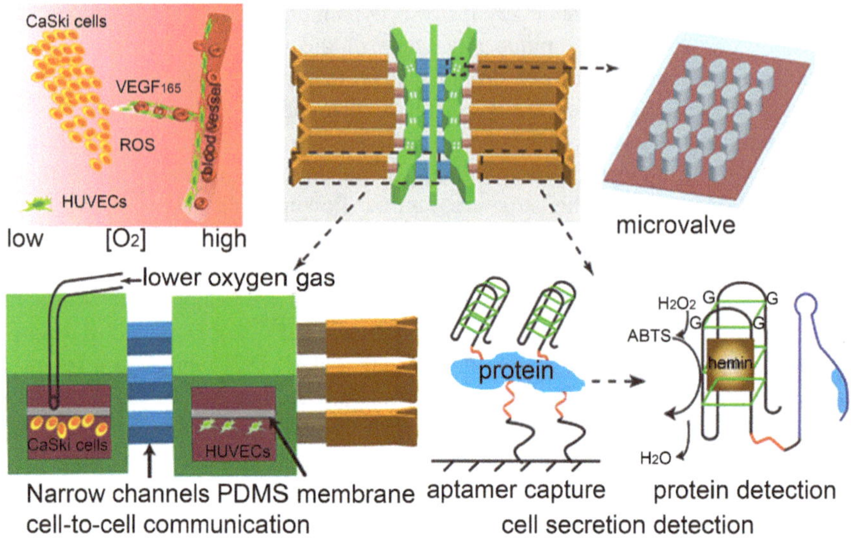

Fig. 6.19 Investigation of hypoxia microenvironment's impact on tumor development. An oxygen concentration gradient producing microfluidic cell co-culture device was constructed for on-line determination of protein $VEGF_{165}$. The $VEGF_{165}$ was qualitatively and semi-quantitatively analyzed by the functional nucleic acid, hemin, ABTS and peroxide system. (Reprinted with permission from Ref. [86])

quantum dots produced lower cytotoxicity under high cell density than the condition of low cell density.

Microfluidic chips made great contributions to investigations of some important cell behaviors (e.g. chemotaxis [89] and electrotaxis [90] of fibroblasts) and understanding of the roles of some proteins involved in cellular activities, for example, the function of selectin in neutrophil adhesion during inflammatory response [91]. Microfluidic platforms are also capable of regulating the cell microenvironment, which can be utilized in investigating paracrine and autocrine signals. For instance, mES-secreted factor regulation has been successfully performed on microfluidic platforms, which has significant importance for the development of dynamic investigation of signaling molecules [92, 93].

6.6 On-Chip Cell Observation

Main applications of microfluidic devices are in cell research, because microchannels have size accordance with cells and possess superiorities of low regent consumption and high detection sensitivity meanwhile. Different form static cell culture on culture dishes or bottles, the dynamic on-chip cell experiments can carry on real-time monitoring under specific flowing environment. The flow culture creates more realistic simulations for cell experiments, but gives several challenges for cell observation at the same time, for example, cells have difficulties in adhering on chip surfaces after cell suspensions were injected into microchannels. Therefore, cell immobilization is a key step in on-chip cell observation and analysis. After successful cell immobilization and incubation in microchannels, the cell morphology and behavior can be observed through various cell imaging methods, in which the fluorescent imaging is the most commonly used method.

6.6.1 Cell Immobilization

In general, the inefficiency and poor regional location capability of cell immobilization lead to great challenges to the on-chip cell observation. At present, cells isolation methods have three main categories: cell adhesion enhancement, porous membranes-assisted isolation, and cell encapsulation using hydrogels. It is important to choose an appropriate way of cell immobilization based on specific features and applicability of different methods.

Cell adhesion enhancement

Microchannels with good cell affinity can be achieved by altering the chemical and physical properties of channel surfaces. Surface modifications can efficiently improve the cell adhesion through adjusting the surface charge property, wettability, and functional groups of microfluidic channels. Most cells have negatively charged cell membranes. Therefore, cell adhesion enhancement can be achieved in the way of coating cationic species on channel surfaces, such as ECM proteins, poly-L-lysine, and APTES. As we have mentioned in the previous parts, the hydrophobicity of PDMS-based microfluidic chips leads to poor affinity between cells and the matrix surfaces. Proper wettability of the matrix surfaces can be prepared by physical protein adsorption and covalent surface chemical modification. In addition, changes of surface morphology of microchannels also work effectively in cell isolation. For example, Wu subtly applied microwell arrays as cell density generators to controllably store cell suspensions by altering the number of microwells [88] (Fig. 6.20).

Porous membranes-assisted isolation

Porous membranes-assisted isolation is developed based on size distinctions between different cells. Cell filtration with a porous membrane can easily achieve cell isolation in a short time. For example, circulating tumor cells (CTCs) enrichment can be realized by using porous membranes. Lin et al. [94] used a parylene membrane-based microdevice in isolating CTCs from human peripheral blood, as shown in Fig. 6.21. More than 90% recovery was achieved when only five tumor cells were seeded in 7.5 mL of blood. The porous membranes-assisted isolation is also often applied in 2D/3D hybrid systems, that is, the before mentioned

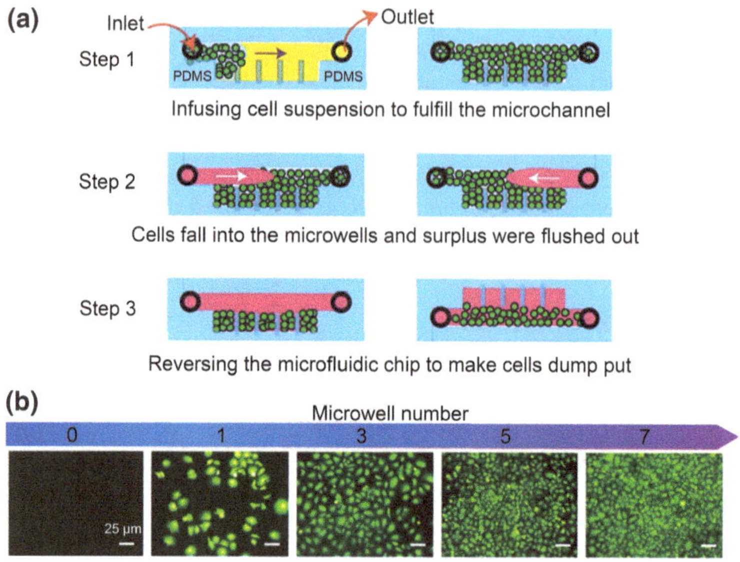

Fig. 6.20 Microwell-based cell density gradient generation microdevice. **a** Schematic of cell density gradient generation on the microfluidic chip; **b** Live/Dead kit staining fluorescent images of cells on microfluidic chips with different microwell numbers. (Reprinted with permission from Ref. [88])

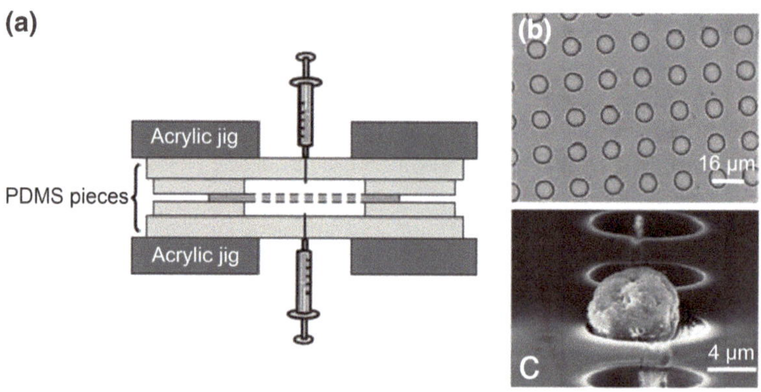

Fig. 6.21 Porous membrane-based cell isolation microdevice. **a** Device assembly; **b** Bright field image of the optically transparent parylene filter; **c** SEM image of the captured tumor cell on the membrane. (Reprinted with permission from Ref. [94])

"sandwich" structures. In the hybrid system, porous membranes are used as filter membranes for target cell isolation in 2D systems, but the whole microchip is a 3D system with two or more functional zones distributed on the two opposite sides of the membrane.

Cell encapsulation in hydrogels

Hydrogels, which have porous structures and are widely employed in 3D cell culture, are generally used in the way of cell encapsulation in cell isolation. Hydrogel-based cell-trapping technique can immobilize target cells in flow and provide a 3D microenvironment for cell proliferation and analysis. Gao et al. [95] obtained cell encapsulation of HepG2 cells and A549 cells simultaneously in PEG hydrogels using a photolithography approach (Fig. 6.22). The immobilized cell arrays can further be utilized to evaluate the anticancer drugs influence on cell viability and intracellular redox parameters.

6.6.2 Cell Imaging

Most cells are transparent and colorless under lamps, which makes direct observations of cytomembrane behaviors and intracellular components very difficult. For this reason, proper cell imaging techniques are necessary in real-time monitoring or detection

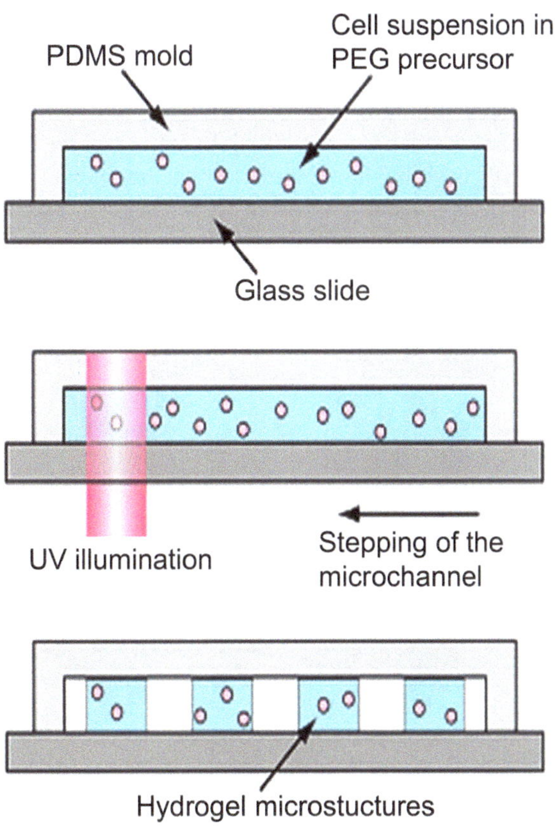

Fig. 6.22 Schematic diagram of the cell immobilization through cell encapsulation in hydrogel by photopolymerization. (Reprinted with permission from Ref. [95])

[96]. Cell imaging techniques are generally noninvasive and highly sensitive, such as photoluminescence (PL), magnetic resonance imaging (MRI), and surface-enhanced Raman scattering (SERS). Here, the fluorescent imaging is selected to elaborate how biomaterials are applied in cell imaging because the fluorescent imaging which contains fluorescent probe dyeing and subsequently appropriate fluorescence detections have been extensively studied and widely used in cell imaging.

Although some kinds of cells have green fluorescent protein (GFP), which makes them visible through emitting fluorescence under the excitation of certain wavelength, biomaterials are still necessary to assist cells in generating fluorescence in some cases, because only small parts of cells have GFP. Metals, polymers, and composite biomaterials, nano-biomaterials in particular, attracted great interests in cell imaging to label cell membrane, cytoplasm, or nucleus.

Inorganic nanomaterials, especially QDs, have been widely studied and applied because of their excellent fluorescence emission performance and tunable emission wavelength through diameter adjustment. Metal and transition metal QDs, graphenes and carbon dots, are some frequently used fluorescent-emitting materials, which has been reviewed in detail by Li et al. [97]. However, these inorganic nanomaterials have drawbacks of quenching and poor hydrophilicity. These problems can be solved in ways of polymer coating [98] or aptamer modification [99]. Recently, a kind of inorganic nanomaterial with excellent energy-converting properties, rare earth (RE) based upconversion nanoparticle (UCNP), was synthesized and applied in biological imaging (Fig. 6.23) with low cytotoxicity [100]. The hydrophilic hollow NaREF4 (RE = Y, Yb, and Lu) nanoparticle (NP) was synthesized through a facile liquid–liquid two-phase method. In vitro tests proved the UCNPs have bright-red emission without apparent noise background, which confirmed the suitability of using UCNPs in cell imaging.

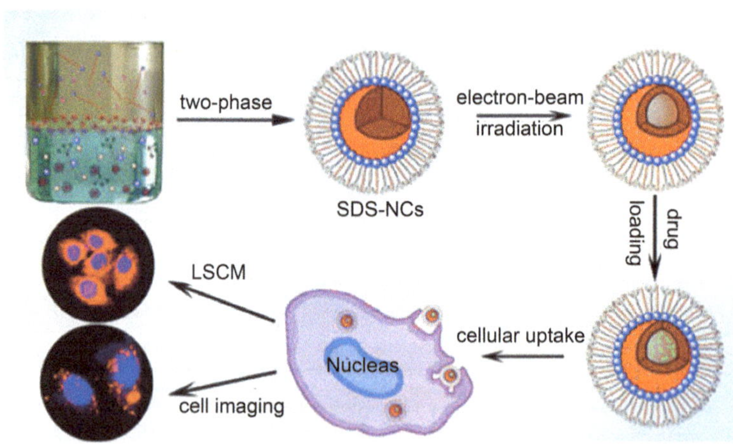

Fig. 6.23 An inorganic cell imaging biomaterial nanoparticle. The synthesis of small NaREF4 (RE = Nd, Sm–Lu, Y) and hydrophilic hollow NaREF4 (RE = Y, Yb, and Lu) NPs for application in cell imaging. (Reprinted with permission from Ref. [100])

Polymers and their nanocomposites are splendid imaging regents for optical detection, because polymeric materials have good transparency, biocompatibility, and easy bio-functionalization property. Fluorescent-emitting NPs, fluorescent-conjugated NPs, and degradable polymers encapsulating fluorophores are commonly used for cell imaging. For example, Zheng et al. [101] employed a single-step assembly and nanoprecipitation approach in preparing folate receptor-targeted (FA) indocyanine green (ICG) dye-doped poly(D,L-lactide-co-glycolide) (PLGA) liquid NPs for cell imaging with good biocompatibility and excellent stability of photobleaching resistance. Some commonly existing defects of routine fluorescent materials, such as the poor aqueous stability, target specificity and rapid elimination of ICG were overcome with using the functional FA-ICG-PLGA-liquid NPs. Comparing with in folate receptor-negative A549 cells, the endocytosis of biomaterial shells in folate receptor-overexpressed MCF-7 cells was found to be more efficient.

6.7 On-Chip Cell Analysis

Microfluidic chip is a multifunctional platform that can integrate cell isolation, cell culture, metabolite enrichment and analysis in a very small area. In biological research, the aim of lab-on-chip experiments is cell analysis, which provides theoretical information to research on cytotoxicity, metabolic pathways, drug evaluation, and intercellular interactions.

6.7.1 Population Cell Analysis

Microfluidic chips have advantages in cell analysis due to their low demand of reagents and cell quantity. However, this property also brings difficulties for analysis due to the limited chamber capacity and the continuous flow lead to very low metabolite concentration. Also, the metabolites always have very complex compose and structures for further analysis. Therefore, accumulating and purifying target metabolites with high efficiency is very important for on-chip cell analysis. The solid-phase extraction (SPE) is frequently adopted in liquid chromatogram (LC). Bacause microfluidic chips are potable LC devices in micro-scale with high throughput, the SPE is also feasible for on-chip analysis. Gao et al. [81] built an online electrospray ionization quadrupole time-of-flight mass spectrometer (ESI-Q-TOF MS) to determine drug absorption and cytotoxicity. Two functional parts, cell culture chambers combined with a drug solution flow and an on-chip SPE column, were connected by PE tubes. In the cell culture part, surface modification of 0.1% poly-L-lysine (PLL) was carried out to improve the cell adhesion. After providing drug stimulation to the cells, the SPE purification was used to determine the drug absorption. Also, SPE columns participated in the investigation of drug permeability [102]. As shown in Fig. 6.24, a "sandwich" microchip was fabricated with using a semi-permeable PC membrane. The model drug releasing was performed at one side of the membrane to observe the drug permeation to the other side of the channel. ESI-Q-TOF MS detection was used to determine the permeated curcumin concentration. The results showed that only 30 min is necessary to accomplish the analysis on this platform, and only 6 μL of the drug solution is required for one permeation test, which proved the integrated microfluidic device had high sensitivity and efficiency for drug discovery and development. Moreover,

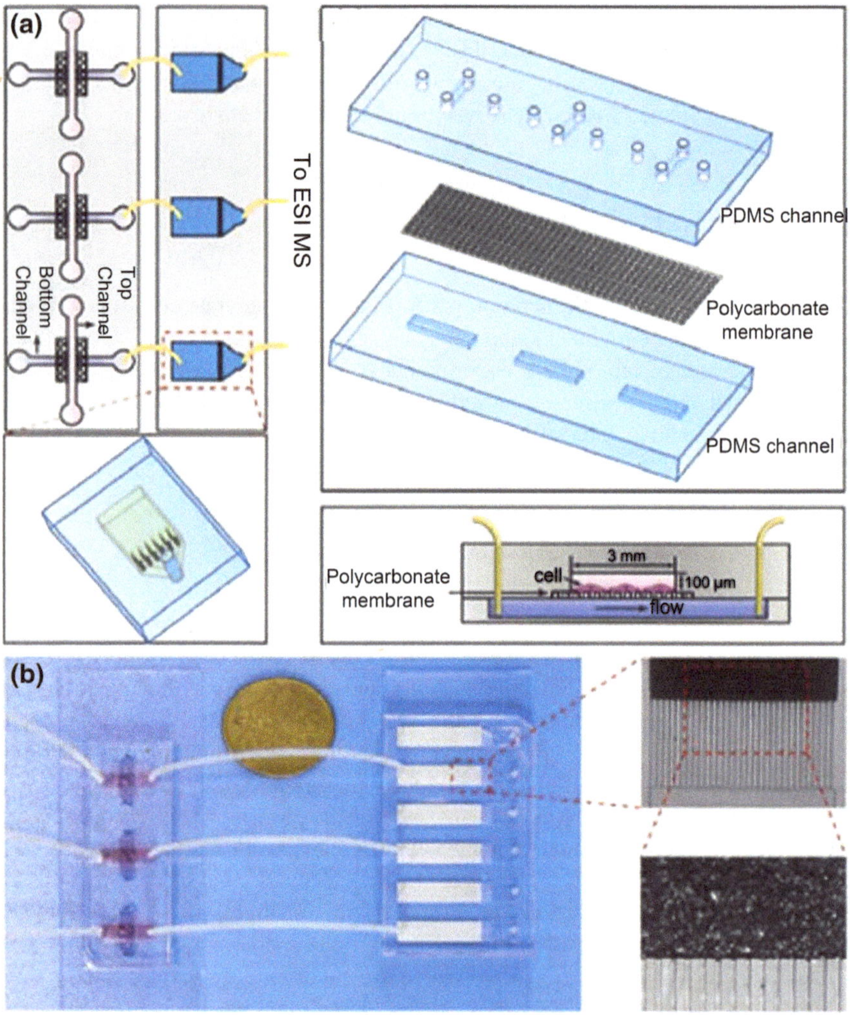

Fig. 6.24 A real-time online drug permeability assay with direct coupling to MS. **a** The integrated microdevice contained a "sandwich" cell culture section, and a sample pretreatment section consisted by μSPE columns prior to MS detection; **b** Photograph of the microfluidic device. (Reprinted with permission from Ref. [102])

Lin et al. adopted porous polymeric monolithic columns in SPE and chemiluminescence (CL) detection [103]. A porous polymeric column made of ethylenediamine modified poly(glycidyl methacrylate-co-ethylene dimethacrylate) was used in catechin enrichment. Then, the CL was generated through the reaction of catechin and potassium permanganate. This method has advantages of elution skipping, low limit of detection (LOD, 1.0×10^{-9} M), and high recovery ranging from 90 to 110%. Lin also synthesized poly(ethyleneglycol)diacrylate (PEG-DA) micro-column arrays in microchip channels as the probe for proteins and glucose detection [104].

6.7.2 Single Cell Analysis

In general cell analysis, analytic results are collected from a large amount of cells exposed to specific culture conditions, thus the results represent overall cell states in testing system. However, each cell is unique and not identical to the overall cell groups. Besides, recent research proved that there are obvious differences existing between individual cells. Therefore, developing single cell manipulation, screening, analysis, and cell heterogeneity investigation methods are meaningful for biological medicine. Microfluidic devices have superiority in low volume control with high accuracy, which is suitable for single-cell manipulation, such as cell sorting, cytomixis, transfection, gene sequencing and labeling [105].

At present, cell separation and capture techniques are commonly used to isolate intact single cell from the cell population and further intracellular substance analysis is performed after the cell lysis treatment. As one of the most direct physical method, some special microstructure design of microchips can realize cell separation and capture [106–109]. Microwell arrays-assisted cell trapping is one efficient way of single cell isolation and analysis. Chen et al. [106] achieved specific capture of single cell on chip through combining microvoid arrays with aptamer immobilization (Fig. 6.25). In this method, 3D microvoid structures assisted in enhancing the interaction between the aptamer-immobilized microvoid surfaces and cells. As a result, high single-cell occupancy of 88.2% was obtained. Using this highly specific cell capture microchip, target cells were isolated from the complex cell suspensions and formed cell arrays. Further single-cell level enzyme reaction dynamics analysis revealed that metabolic differences existing between different individual cells. Liu et al. [107] fabricated PDMS microvoid array on glass surface by employing the self-assembly of polystyrene microspheres (Fig. 6.26) for single-cell trapping. Subsequently, the real-time single-cell enzyme activity analysis confirmed the dissimilarity of viability and conditions among the trapped cells by analyzing the fluorescence intensity change of calcein AM-stained HeLa cells.

Aside from the microwell trapping ways, encapsulation, current separation, and some other methods are also effective in single-cell capture and analysis. For instance, Liu developed a modified microscope projection photolithography system that can activate the photopolymerization of PEGDA under control and finally

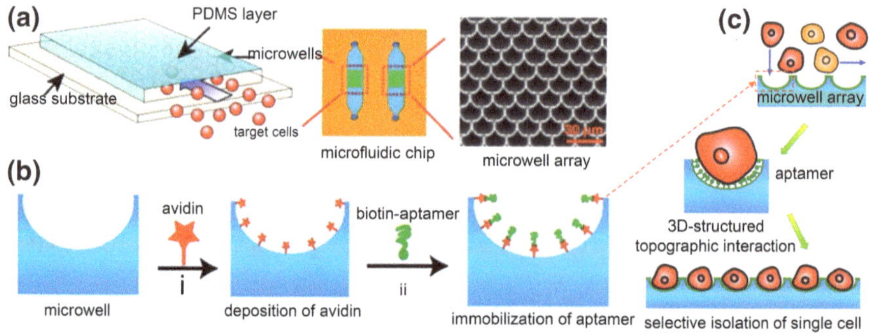

Fig. 6.25 A single-cell specific trapping microdevice. **a** Schematic illustration of the microwell arrays-assisted single target cells isolation microfluidic platform; **b** Functionalization of glass surface was conducted by the deposition of avidin and the immobilization of biotin–aptamer in turn; **c** Single-cell specific isolation was accomplished through the combination of microwell trapping and biotin–aptamer immobilization on glass substrates. (Reprinted with permission from Ref. [106])

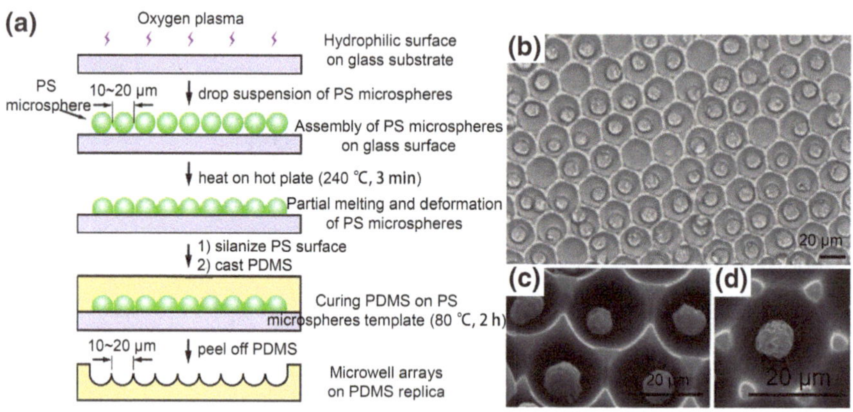

Fig. 6.26 Cell arrays generated on microwells for single cell enzyme activity analysis. **a** Schematic illustration for the fabrication of PDMS microwell arrays; **b** Dense trapping of Ramos cells; (**c**, **d**) SEM images of Ramos cells isolated in microwells. (Reprinted with permission from Ref. [107])

achieved single-cell encapsulation with an efficiency of 80%. Mach et al. [110] developed a microfluidic centrifuge for cell isolation based on eddy current separation. Wang completed single cell distribution in aqueous droplets surrounded by oil phase [111]. They further carried out cell incubation on a microfluidics platform with high throughput and chemiluminescence was employed for cell analysis. By this mean, only saccharomyces cerevisiae cells which excessively consumed xylose was successfully isolated from 10^4 cells, and the genomic variation resulting in excessive consumption of xylose was also analyzed.

With the development of microfluidic technology, single-cell analysis will develop quickly, and cell heterogeneity which covered by massive cells will be disclosed in the future.

6.7.3 Chip-MS Technology

Microfluidics and capillary electrophoresis combining with chemiluminescence, laser-induced fluorescence and electrochemical detection are some mature and commonly used cell analysis methods, but their testing targets in separation or detection still limited in molecules with simple or known structures so far [112, 113], which only contains less species. Lin et al. [114] realized rapid determination of mini-short tandem repeat (miniSTR) loci by means of combining polymerase chain reaction (PCR) with microchip capillary electrophoresis (MCE). Under the optimized conditions, merely 0.001 ng of DNA templates was necessary for miniSTRs generation, and the produced 15 miniSTRs had good narrow distribution with the relative standard deviations in the scope of 0.49–4.41%. Also, the miniSTR detection method was workable in practical samples, such as human hair, which has great potentials in criminal identification and paternity tests. Yi et al. [115] used a similar objective amplification detection to perform genotyping test on 11 human papilloma virus (HPV) with high sensitivity (detection limit was 200 copies) and accuracy. Validity of the method was further confirmed with using clinical samples. All HPV types were recognized with a good compatibility degree of more than 90%, which proved the high feasibility and reliability of the established separation and detection method in the screening and prognosis of cervical cancer.

MS detection can rapidly analyze multiple components at the same time with high sensitivity. Moreover, the detectable biological species of MS cover small molecules, peptides, amino acid and protein, which have a broad size distribution. More importantly, unknown substances can be analyzed and speculated based on the mass-to-charge ratio obtained from MS results. Consequently, MS plays an increasingly important role in biological medicine research, such as investigations of signaling pathways and protein structures. Though MS has lots of merits, the direct employment of MS in cell analysis is of great challenge because cell products always have complex compositions and strong matrix interference, which leads to fail results. As a solution, effective separation units are frequently empolyed to purify or separate the test solution before MS detection.

As exhibited in Fig. 6.27, since the 1950s, the developed separation units have experienced gas chromatography, liquid chromatography, capillary electrophoresis and microfluidic chip in turn [116]. Among those, the microfluidic chip is much more particular than other column-based separation techniques, because microfluidic chips have superiorities in connecting effective separation units with cell experiments with high integration, online processability and low reagent consumption [117]. In chip-MS systems, the electrospray ionization mass spectrometry

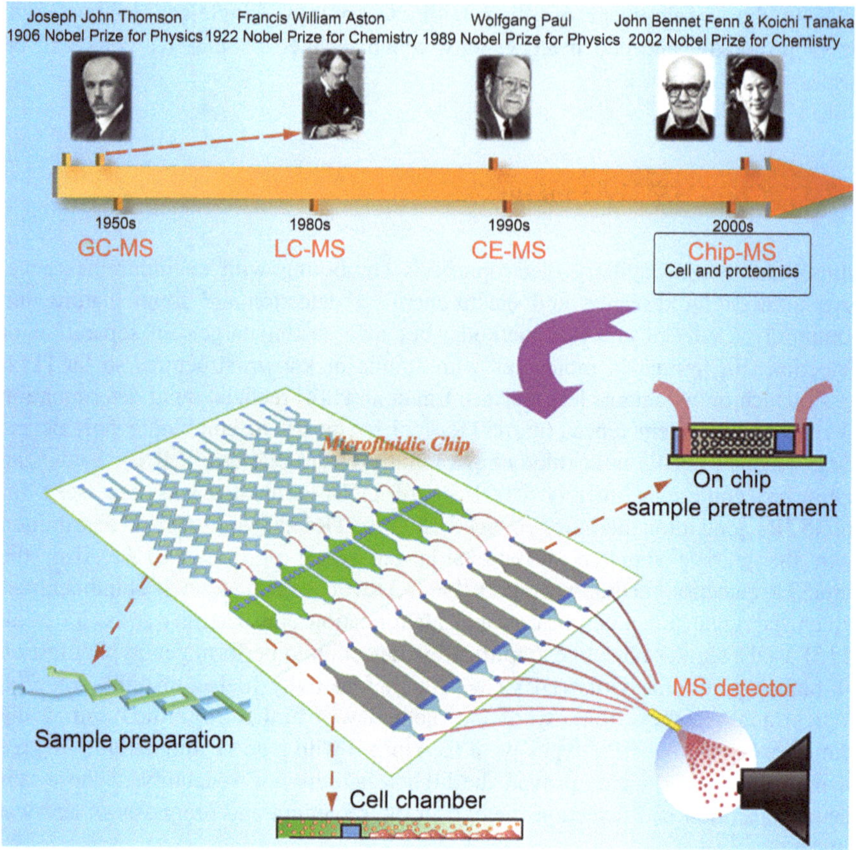

Fig. 6.27 The development of MS with coupling to different separation technology. The chip-MS system integrated sample preparation, cell chambers, on-chip sample pretreatment, and direct MS detection. (Reprinted with permission from Ref. [116])

(ESI-MS) is the most frequently used MS technique. On-chip sample pretreatments and the interface design of sample injection to MS are very important to obtain good detection results. Integrating monolithic columns or extracted materials are intensively utilized in sample pretreatment on chips [81, 82, 118] to accomplish desalination and target molecule enrichment of solutions with complex compositions, such as serum. In Gao's microfluidic device [81], SPE micro-column chambers were integrated on-chip, through which the rinsing and enrichment of target molecules were completed in fluids. They selected the methotrexate, an inhibitor of folate metabolism enzyme systems, as the drug model. HepG2 and Caco-2 cells were used as target cells to evaluate the effectiveness of the fabricated microfluidic chip on drug absorption. The results confirmed the feasibility of the fabricated platform in investigation of real-time pharmacokinetics of living cells. Moreover, to make the design and operation of sample pretreatment on chip simpler, droplet paper-spray ion source interfaces and inkjet toothpick-spray ion source techniques were developed [119]. In the droplet paper-spray, non-easily ionized matters were maintained on paper fibers in advance, because paper fibers are capable of improving the separation and purification of complex solutions simultaneously. Liu et al. [120] fabricated a multi-channel microfluidic platform and determined the association constant of a noncovalent protein-protein complex using the paper-based electrospray MS (Fig. 6.28a). The results proved that the platform is effective in protein analysis. In the same way, the inkjet toothpick-spray ion source technique also realized separating and purifying complex matrixes on toothpick wood fibers. Moreover, the ink-jet system can be applied in accurately generating picoliter droplets on toothpick tips to form sprays for MS analysis. Luo et al. [119] set up a testing platform consisting of the inkjet droplet generation and a toothpick electrospray ion source (Fig. 6.28b). In MS detection, theobromine was used as the internal standard in linear measurement of caffeine. Good linearity exhibited in range of 1–200 ppm. More importantly, due to the shorter duration of electrospray, the obtained peak intensity ratio of different peaks almost equaled to the peak area ratio, which makes the quantitative examination possible on this platform.

Matrix-assisted laser desorption/ionization mass spectrometry (MALDI-MS) has good salinity tolerance, which is very suitable for biological macromolecules detection, such as proteins and peptides. MALDI-MS is particularly suitable for single-cell and rasterization analysis due to its characteristic of rasterization detection. However, in MALDI-MS analysis, providing direct laser illumination eutectic laminas is necessary. Therefore, an open test area is required, which brings difficulties in connecting microfluidic chips to MALDI-MS. Zhang et al. [121] achieved direct, in situ MALDI-MS lipid analysis of mammalian cells on the platform shown in Fig. 6.29. First, they had mammalian cells directly cultured on indium tin oxide (ITO) coated glass. Then, a homogeneous eutectic lamina was prepared through electrospray ionization. The in situ determination of cell membrane lipid was carried out using a MALDI-MS. As a result, abundant information

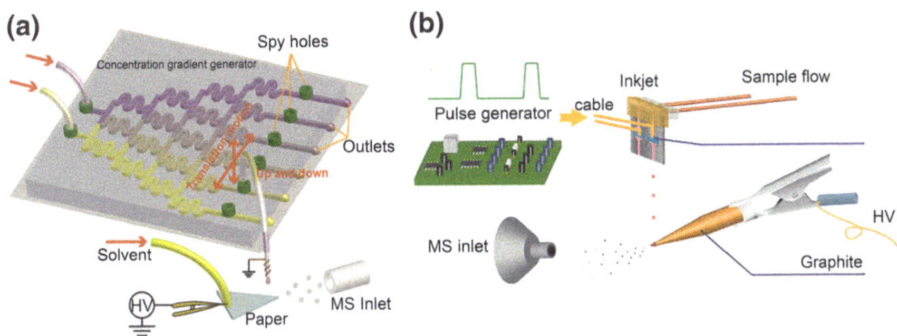

Fig. 6.28 Droplet spray ion source interfaces. **a** A droplet paper-spray ion source interface (Reprinted with permission from Ref. [119]) and **b** a inkjet toothpick-spray ion source technique in real-time monitoring of cells. (Reprinted with permission from Ref. [120])

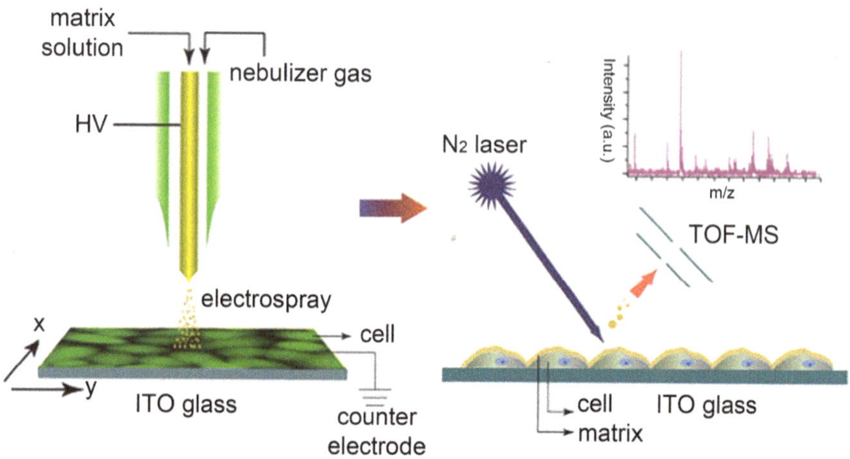

Fig. 6.29 Direct and in situ lipid analysis of mammalian cells on ITO glass by MALDI-MS. (Reprinted with permission from Ref. [121])

of the membrane lipid was obtained, which can be used as a relatively rough molecular mapping in cell identification.

6.8 Summary

The exploitation of microfluidic chips originally aims at developing portable chemical analysis equipments. So far, microfluidic chips are no longer just miniaturized analytical instruments, they are also being expected to realize biostructures or tissues constructions in vitro, and investigate biological targets, such as cells, proteins, and metabolites. Now, microfluidic chips are the most powerful tool for drug evaluation, diagnosis and therapy, biological research and fields based on cells, because they have tremendous advantages such as low reagent consumption, high throughput, easy integration, good designability, and the ability of mimicking complex microenvironment.

The principle materials of constructing microfluidics experienced the age of inorganic materials now have entered the organic era dominated by synthetic polymers. Performances of microfluidic devices in cell culture, cell observations and cell analysis can be improved by using adequate biomaterials. The hydrogels and smart materials have great potential to improve the biocompatibility and biological functionality of microfluidic devices. Meanwhile, making use of biomaterials and optimized chip design, 3D cell culture and single-cell analysis that can hardly be performed on traditional in vitro devices are realized on microfluidic chips, which have great potential in the future. In addition, the combination of microfluidic chips with MS detection efficiently extended the scope of testing targets and is capable of analyzing complex compositions, which is the future direction of on-chip cell analysis.

References

1. Paguirigan AL, Beebe DJ (2008) Microfluidics meet cell biology: bridging the gap by validation and application of microscale techniques for cell biological assays. BioEssays 30 (9):811–821. doi:10.1002/bies.20804
2. Terry SC, Jerman JH, Angell JB (1979) A gas chromatographic air analyzer fabricated on a silicon wafer. IEEE Trans Electron Dev 26(12):1880–1886. doi:10.1109/T-ED.1979.19791
3. Ozaydin-Ince G, Coclite AM, Gleason KK (2012) CVD of polymeric thin films: applications in sensors, biotechnology, microelectronics/organic electronics, microfluidics, MEMS, composites and membranes. Rep Prog Phys 75(1):16501. doi:10.1088/0034-4885/75/1/016501
4. Tanzi S, Østergaard PF, Matteucci M, Christiansen TL, Cech J, Marie R, Taboryski R (2012) Fabrication of combined-scale nano-and microfluidic polymer systems using a multilevel dry etching, electroplating and molding process. J Micromech Microeng 22(11):115008. doi:10.1088/0960-1317/22/11/115008
5. Qin D, Xia Y, Whitesides GM (2010) Soft lithography for micro-and nanoscale patterning. Nat Protoc 5(3):491–502. doi:10.1038/nprot.2009.234
6. Wang J, He Y, Xia H, Niu LG, Zhang R, Chen QD, Zhang YL, Li YF, Zeng SJ, Qin JH, Lin BC, Sun HB (2010) Embellishment of microfluidic devices via femtosecond laser micronanofabrication for chip functionalization. Lab Chip 10(15):1993–1996. doi:10.1039/c003264f
7. Zhang L, Wang W, Ju XJ, Xie R, Liu Z, Chu LY (2015) Fabrication of glass-based microfluidic devices with dry film photoresists as pattern transfer masks for wet etching. RSC Adv 5(8):5638–5646. doi:10.1039/C4RA15907A
8. Sun G, Zhao X, Kim CJC (2012) Fabrication of very-high-aspect-ratio microstructures in complex patterns by photoelectrochemical etching. J Microelectromech Syst 21(6):1504–1512. doi:10.1109/JMEMS.2012.2211574
9. You JB, Min KI, Lee B, Kim DP, Im SG (2013) A doubly cross-linked nano-adhesive for the reliable sealing of flexible microfluidic devices. Lab Chip 13(7):1266–1272. doi:10.1039/c2lc41266g
10. Dochow S, Beleites C, Henkel T, Mayer G, Albert J, Clement J, Krafft C, Popp J (2013) Quartz microfluidic chip for tumour cell identification by Raman spectroscopy in combination with optical traps. Anal Bioanal Chem 405(8):2743–2746. doi:10.1007/s00216-013-6726-3
11. Ou J, Glawdel T, Ren CL, Pawliszyn J (2009) Fabrication of a hybrid PDMS/SU-8/quartz microfluidic chip for enhancing UV absorption whole-channel imaging detection sensitivity and application for isoelectric focusing of proteins. Lab Chip 9(13):1926–1932. doi:10.1039/b821438g
12. van Midwoud PM, Janse A, Merema MT, Groothuis GM, Verpoorte E (2012) Comparison of biocompatibility and adsorption properties of different plastics for advanced microfluidic cell and tissue culture models. Anal Chem 84(9):3938–3944. doi:10.1021/ac300771z
13. Xiao Y, Zhang B, Liu H, Miklas JW, Gagliardi M, Pahnke A, Thavandiran N, Sun Y, Simmons C, Keller G, Radisic M (2014) Microfabricated perfusable cardiac biowire: a platform that mimics native cardiac bundle. Lab Chip 14(5):869–882. doi:10.1039/c3lc51123e

14. Tsougeni K, Petrou PS, Papageorgiou DP, Kakabakos SE, Tserepi A, Gogolides E (2012) Controlled protein adsorption on microfluidic channels with engineered roughness and wettability. Sens Actuators B Chem 161(1):216–222. doi:10.1016/j.snb.2011.10.022
15. Battle KN, Jackson JM, Witek MA, Hupert ML, Hunsucker SA, Armistead PM, Soper SA (2014) Solid-phase extraction and purification of membrane proteins using a UV-modified PMMA microfluidic bioaffinity μSPE device. Analyst 139(6):1355–1363. doi:10.1039/c3an02400h
16. Mogi K, Sugii Y, Yamamoto T, Fujii T (2014) Rapid fabrication technique of nano/microfluidic device with high mechanical stability utilizing two-step soft lithography. Sens Actuators B Chem 201:407–412. doi:10.1016/j.snb.2014.05.047
17. O'Connor RS, Hao X, Shen K, Bashour K, Akimova T, Hancock WW, Kam LC, Milone MC (2012) Substrate rigidity regulates human T cell activation and proliferation. J Immunol 189(3):1330–1339. doi:10.4049/jimmunol.1102757
18. Zhou J, Ellis AV, Voelcker NH (2010) Recent developments in PDMS surface modification for microfluidic devices. Electrophoresis 31(1):2–16. doi:10.1002/elps.200900475
19. Bartalena G, Loosli Y, Zambelli T, Snedeker JG (2012) Biomaterial surface modifications can dominate cell–substrate mechanics: the impact of PDMS plasma treatment on a quantitative assay of cell stiffness. Soft Matter 8:673. doi:10.1039/c1sm06250f
20. Zhang Y, Ren L, Tu Q, Wang X, Liu R, Li L, Wang JC, Liu W, Xu J, Wang J (2011) Fabrication of reversible poly(dimethylsiloxane) surfaces via host-guest chemistry and their repeated utilization in cardiac biomarker analysis. Anal Chem 83(24):9651–9659. doi:10.1021/ac202517x
21. Annabi N, Selimović Š, Acevedo Cox JP, Ribas J, Afshar Bakooshli M, Heintze D, Weiss AS, Cropek D, Khademhosseini A (2013) Hydrogel-coated microfluidic channels for cardiomyocyte culture. Lab Chip 13(18):3569–3577. doi:10.1039/c3lc50252j
22. Roy E, Galas JC, Veres T (2011) Thermoplastic elastomers for microfluidics: Towards a high-throughput fabrication method of multilayered microfluidic devices. Lab Chip 11 (18):3193–3196. doi:10.1039/c1lc20251k
23. Rolland JP, Hagberg EC, Denison GM, Carter KR, De Simone JM (2004) High-resolution soft lithography: enabling materials for nanotechnologies. Angew Chem Int Ed 43 (43):5796–5799. doi:10.1002/anie.200461122
24. Song L, Zhao J, Yuan XY (2014) Strengthening of hydrogels based on polysaccharide and polypeptide. Prog Chem 26:385–393
25. Kharkar PM, Kiick KL, Kloxin AM (2013) Designing degradable hydrogels for orthogonal control of cell microenvironments. Chem Soc Rev 42(17):7335–7372. doi:10.1039/c3cs60040h
26. Guan Y, Zhang Y (2013) Boronic acid-containing hydrogels: synthesis and their applications. Chem Soc Rev 42(20):8106–8121. doi:10.1039/c3cs60152h
27. Seiffert S, Weitz DA (2010) Controlled fabrication of polymer microgels by polymer-analogous gelation in droplet microfluidics. Soft Matter 6:3184. doi:10.1039/c0sm00071j
28. Kothapalli CR, van Veen E, de Valence S, Chung S, Zervantonakis IK, Gertler FB, Kamm RD (2011) A high-throughput microfluidic assay to study neurite response to growth factor gradients. Lab Chip 11(3):497–507. doi:10.1039/c0lc00240b
29. Kang DH, Kim HN, Kim P, Suh KY (2014) Poly(ethylene glycol) (PEG) microwells in microfluidics: Fabrication methods and applications. BioChip J 8(4):241–253. doi:10.1007/s13206-014-8401-y
30. Pei Y, Wang X, Huang W, Liu P, Zhang L (2013) Cellulose-based hydrogels with excellent microstructural replication ability and cytocompatibility for microfluidic devices. Cellulose 20(4):1897–1909. doi:10.1007/s10570-013-9930-6
31. Yu J, Ge L, Huang J, Wang S, Ge S (2011) Microfluidic paper-based chemiluminescence biosensor for simultaneous determination of glucose and uric acid. Lab Chip 11(7):1286–1291. doi:10.1039/c0lc00524j

32. Liu F, Wang S, Zhang M, Wang Y, Ge S, Yu J, Yan M (2014) Aptamer based test stripe for ultrasensitive detection of mercury(II) using a phenylene-ethynylene reagent on nanoporous silver as a chemiluminescence reagent. Microchim Acta 181(5–6):663–670. doi:10.1007/s00604-014-1171-3
33. He Q, Ma C, Hu X, Chen H (2013) Method for fabrication of paper-based microfluidic devices by alkylsilane self-assembling and UV/O3-patterning. Anal Chem 85(3):1327–1331. doi:10.1021/ac303138x
34. Mu X, Zhang L, Chang S, Cui W, Zheng Z (2014) Multiplex microfluidic paper-based immunoassay for the diagnosis of hepatitis C virus infection. Anal Chem 86(11):5338–5344. doi:10.1021/ac500247f
35. Thuo MM, Martinez RV, Lan W, Liu X, Barber J (2014) Fabrication of low-cost paper-based microfluidic devices by embossing or cut-and-stack methods. Chem Mater 26 (14):4230–4237. doi:10.1021/cm501596s
36. Larsen EK, Larsen NB (2013) One-step polymer surface modification for minimizing drug, protein, and DNA adsorption in microanalytical systems. Lab Chip 13(4):669–675. doi:10.1039/C2LC40750G
37. Wu D, Zhao B, Dai Z, Qin J, Lin B (2006) Grafting epoxy-modified hydrophilic polymers onto poly(dimethylsiloxane) microfluidic chip to resist nonspecific protein adsorption. Lab Chip 6(7):942–947. doi:10.1039/b600765a
38. Vu TT, Fouet M, Gue AM, Sudor J (2014) A new and easy surface functionalization technnology for monitoring wettability in heterogeneous nano-and microfluidic devices. Sens Actuators B Chem 196:64–70. doi:10.1016/j.snb.2014.01.085
39. Bodas D, Khan-Malek C (2007) Hydrophilization and hydrophobic recovery of PDMS by oxygen plasma and chemical treatment-An SEM investigation. S Sens Actuators B Chem 123(1):368–373. doi:10.1016/j.snb.2006.08.037
40. Kuddannaya S, Chuah Y (2013) Surface chemical modification of poly (dimethylsiloxane) for the enhanced adhesion and proliferation of mesenchymal stem cells. ACS Appl Mater Interfaces 5(19):9777–9784. doi:10.1021/am402903e
41. Sun T, Wang G, Feng L, Liu B, Ma Y, Jiang L, Zhu D (2004) Reversible Switching between Superhydrophilicity and Superhydrophobicity. Angew Chem Int Ed 116(3):361–364. doi:10.1002/anie.200352565
42. Jung YD, Khan M, Park SY (2014) Fabrication of temperature- and pH-sensitive liquid-crystal droplets with PNIPAM-b-LCP and SDS coatings by microfluidics. J Mater Chem B 2(30):4922–4928. doi:10.1039/C4TB00476K
43. Sun T, Qing G (2011) Biomimetic smart interface materials for biological applications. Adv Mater 23(12):H57–H77. doi:10.1002/adma.201004326
44. Sun T, Song W, Jiang L (2005) Control over the responsive wettability of poly (N-isopropylacrylamide) film in a large extent by introducing an irresponsive molecule. Chem Commun 13:1723–1725. doi:10.1039/b417670g
45. Schmidt S, Zeiser M, Hellweg T, Duschl C, Fery A, Möhwald H (2010) Adhesion and mechanical properties of PNIPAM microgel films and their potential use as switchable cell culture substrates. Adv Funct Mater 20:3235–3243. doi:10.1002/adfm.201000730
46. Wylie RG, Ahsan S, Aizawa Y, Maxwell KL, Morshead CM, Shoichet MS (2011) Spatially controlled simultaneous patterning of multiple growth factors in three-dimensional hydrogels. Nat Mater 10(10):799–806. doi:10.1038/nmat3101
47. Caron MM, Emans PJ, Coolsen MM, Voss L, Surtel DA, Cremers A, van Rhijn LW, Welting TJ (2012) Redifferentiation of dedifferentiated human articular chondrocytes: Comparison of 2D and 3D cultures. Osteoarthritis Cartilage 20(10):1170–1178. doi:10.1016/j.joca.2012.06.016
48. Yoshii Y, Waki A, Yoshida K, Kakezuka A, Kobayashi M, Namiki H, Kuroda Y, Kiyono Y, Yoshii H, Furukawa T, Asai T, Okazawa H, Gelovani JG, Fujibayashi Y (2011) The use of nanoimprinted scaffolds as 3D culture models to facilitate spontaneous tumor cell migration and well-regulated spheroid formation. Biomaterials 32(26):6052–6058. doi:10.1016/j.biomaterials.2011.04.076

49. Kang E, Choi YY, Jun Y, Chung BG, Lee SH (2010) Development of a multi-layer microfluidic array chip to culture and replate uniform-sized embryoid bodies without manual cell retrieval. Lab Chip 10(20):2651–2654. doi:10.1039/c0lc00005a
50. Wu J, Chen Q, Liu W, Zhang Y, Lin JM (2012) Cytotoxicity of quantum dots assay on a microfluidic 3D-culture device based on modeling diffusion process between blood vessels and tissues. Lab Chip 12(18):3474–3480. doi:10.1039/c2lc40502d
51. Chen Q, Wu J, Zhuang Q, Lin X, Zhang J, Lin JM (2013) Microfluidic isolation of highly pure embryonic stem cells using feeder-separated co-culture system. Sci Rep 3:2433. doi:10.1038/srep02433
52. Leipzig ND, Wylie RG, Kim H, Shoichet MS (2011) Differentiation of neural stem cells in three-dimensional growth factor-immobilized chitosan hydrogel scaffolds. Biomaterials 32 (1):57–64. doi:10.1016/j.biomaterials.2010.09.031
53. Sapir Y, Kryukov O, Cohen S (2011) Integration of multiple cell-matrix interactions into alginate scaffolds for promoting cardiac tissue regeneration. Biomaterials 32(7):1838–1847. doi:10.1016/j.biomaterials.2010.11.008
54. Mack PJ, Zhang Y, Chung S, Vickerman V, Kamm RD, García-Cardeña G (2009) Biomechanical regulation of endothelium-dependent events critical for adaptive remodeling. J Biol Chem 284(13):8412–8420. doi:10.1074/jbc.M804524200
55. Chung S, Sudo R, Zervantonakis IK, Rimchala T, Kamm RD (2009) Surface-treatment-induced three-dimensional capillary morphogenesis in a microfluidic platform. Adv Mater 21(47):4863–4867. doi:10.1002/adma.200901727
56. Shin Y, Han S, Jeon JS, Yamamoto K, Zervantonakis IK, Sudo R, Kamm RD, Chung S (2012) Microfluidic assay for simultaneous culture of multiple cell types on surfaces or within hydrogels. Nat Protoc 7(7):1247–1259. doi:10.1038/nprot.2012.051
57. Kim C, Chung S, Yuchun L, Kim M-C, Chan JKY, Asada HH, Kamm RD (2012) In vitro angiogenesis assay for the study of cell-encapsulation therapy. Lab Chip 12(16):2942–2950. doi:10.1039/C2LC40182G
58. Klemm D, Kramer F, Moritz S, Lindström T, Ankerfors M, Gray D, Dorris A (2011) Nanocelluloses: a new family of nature-based materials. Angew Chem Int Ed 50(24):5438–5466. doi:10.1002/anie.201001273
59. Gea S, Reynolds CT, Roohpour N, Wirjosentono B, Soykeabkaew N, Bilotti E, Peijs T (2011) Investigation into the structural, morphological, mechanical and thermal behaviour of bacterial cellulose after a two-step purification process. Bioresour Technol 102(19):9105–9110. doi:10.1016/j.biortech.2011.04.077
60. Nwe N, Furuike T, Tamura H (2010) Selection of a biopolymer based on attachment, morphology and proliferation of fibroblast NIH/3T3 cells for the development of a biodegradable tissue regeneration template: Alginate, bacterial cellulose and gelatin. Process Biochem 45(4):457–466. doi:10.1016/j.procbio.2009.11.002
61. Fu L, Zhang J, Yang G (2013) Present status and applications of bacterial cellulose-based materials for skin tissue repair. Carbohydr Polym 92(2):1432–1442. doi:10.1016/j.carbpol.2012.10.071
62. Isogai A, Saito T, Fukuzumi H (2011) TEMPO-oxidized cellulose nanofibers. Nanoscale 3 (1):71–85. doi:10.1039/c0nr00583e
63. Lin N, Dufresne A (2014) Nanocellulose in biomedicine: current status and future prospect. Eur Polym J 59:302–325. doi:10.1016/j.eurpolymj.2014.07.025
64. Bhattacharya M, Malinen MM, Lauren P, Lou YR, Kuisma SW, Kanninen L, Lille M, Corlu A, Guguen-Guillouzo C, Ikkala O, Laukkanen A, Urtti A, Yliperttula M (2012) Nanofibrillar cellulose hydrogel promotes three-dimensional liver cell culture. J Control Release 164(3):291–298. doi:10.1016/j.jconrel.2012.06.039
65. Torres-Rendon JG, Femmer T, De Laporte L, Tigges T, Rahimi K, Gremse F, Zafarnia S, Lederle W, Ifuku S, Wessling M, Hardy JG, Walther A (2015) Bioactive gyroid scaffolds formed by sacrificial templating of nanocellulose and nanochitin hydrogels as instructive platforms for biomimetic tissue engineering. Adv Mater 27(19):2989–2995. doi:10.1002/adma.201405873

66. Mo X, Li Q, Yi Lui LW, Zheng B, Kang CH, Nugraha B, Yue Z, Jia RR, Fu HX, Choudhury D, Arooz T, Yan J, Lim CT, Shen S, Hong Tan C, Yu H (2010) Rapid construction of mechanically-confined multi-cellular structures using dendrimeric intercellular linker. Biomaterials 31(29):7455–7467. doi:10.1016/j.biomaterials.2010.06.020
67. Luo Y, Wang C, Hossain M, Qiao Y, Ma L, An J, Su M (2012) Three-dimensional microtissue assay for high-throughput cytotoxicity of nanoparticles. Anal Chem 84 (15):6731–6738. doi:10.1021/ac301191j
68. Chandrasekaran S, Giang UB, King MR, DeLouise LA (2011) Microenvironment induced spheroid to sheeting transition of immortalized human keratinocytes (HaCaT) cultured in microbubbles formed in polydimethylsiloxane. Biomaterials 32(29):7159–7168. doi:10.1016/j.biomaterials.2011.06.013
69. Liu T, Winter M, Thierry B (2014) Quasi-spherical microwells on superhydrophobic substrates for long term culture of multicellular spheroids and high throughput assays. Biomaterials 35(23):6060–6068. doi:10.1016/j.biomaterials.2014.04.047
70. Hong S, Pan Q, Lee LP (2012) Single-cell level co-culture platform for intercellular communication. Integr Biol 4(4):374–380. doi:10.1039/c2ib00166g
71. Henrich D, Wilhelm K, Warzecha J, Frank J, Barker J, Marzi I, Seebach C (2013) Human endothelial-like differentiated precursor cells maintain their endothelial characteristics when cocultured with mesenchymal stem cell and seeded onto human cancellous bone. Mediators Inflamm 2013:364591. doi:10.1155/2013/364591
72. Huh D, Matthews BD, Mammoto A, Hsin HY, Ingber DE (2010) Reconstituting organ-level lung functions on a chip. Science 328(5986):1662–1668. doi:10.1126/science.1188302
73. Agarwal A, Goss JA, Cho A, McCain ML, Parker KK (2013) Microfluidic heart on a chip for higher throughput pharmacological studies. Lab Chip 13(18):3599–3608. doi:10.1039/C3LC50350J
74. Booth R, Kim H (2012) Characterization of a microfluidic in vitro model of the blood-brain barrier (μBBB). Lab Chip 12(10):1784–1792. doi:10.1039/c2lc40094d
75. Alessandri K, Sarangi BR, Gurchenkov VV, Sinha B, Kießling TR, Fetler L, Rico F, Scheuring S, Lamaze C, Simon A, Geraldo S, Vignjevic D, Doméjean H, Rolland L, Funfak A, Bibette J, Bremond N, Nassoy P (2013) Cellular capsules as a tool for multicellular spheroid production and for investigating the mechanics of tumor progression in vitro. Proc Natl Acad Sci U S A 110(37):14843–14848. doi:10.1073/pnas.1309482110
76. Onoe H, Okitsu T, Itou A, Kato-Negishi M, Gojo R, Kiriya D, Sato K, Miura S, Iwanaga S, Kuribayashi-Shigetomi K, Matsunaga YT, Shimoyama Y, Takeuchi S (2013) Metre-long cell-laden microfibres exhibit tissue morphologies and functions. Nat Mater 12(6):584–590. doi:10.1038/nmat3606
77. Lee SA, No DY, Kang E, Ju J, Kim DS, Lee SH (2013) Spheroid-based three-dimensional liver-on-a-chip to investigate hepatocyte-hepatic stellate cell interactions and flow effects. Lab Chip 13(18):3529–3537. doi:10.1039/c3lc50197c
78. Gao D, Liu H, Jiang Y, Lin JM, Gao D, Liu H, Jiang Y (2012) Recent developments in microfluidic devices for in vitro cell culture for cell-biology research. Trends Analyt Chem 35:150–164. doi:10.1016/j.trac.2012.02.008
79. Jeon NL, Dertinger SKW, Chiu DT, Choi IS, Stroock AD, Whitesides GM (2000) Generation of solution and surface gradients using microfluidic systems. Langmuir 16 (22):8311–8316. doi:10.1021/la000600b
80. Toh AGG, Wang ZP, Yang C, Nguyen NT (2014) Engineering microfluidic concentration gradient generators for biological applications. Microfluid Nanofluid 16(1–2):1–18. doi:10.1007/s10404-013-1236-3
81. Gao D, Li H, Wang N, Lin JM (2012) Evaluation of the absorption of methotrexate on cells and its cytotoxicity assay by using an integrated microfluidic device coupled to a mass spectrometer. Anal Chem 84(21):9230–9237. doi:10.1021/ac301966c
82. Chen Q, Wu J, Zhang Y, Lin JM (2012) Qualitative and quantitative analysis of tumor cell metabolism via stable isotope labeling assisted microfluidic chip electrospray ionization mass spectrometry. Anal Chem 84(3):1695–1701. doi:10.1021/ac300003k

83. Scherber C, Aranyosi AJ, Kulemann B, Thayer SP, Toner M, Iliopoulos O, Irimia D (2012) Epithelial cell guidance by self-generated EGF gradients. Integr Biol 4(3):259–269. doi:10.1039/c2ib00106c
84. Byrne MB, Leslie MT, Gaskins HR, Kenis PJA (2014) Methods to study the tumor microenvironment under controlled oxygen conditions. Trends Biotechnol 32(11):556–563. doi:10.1016/j.tibtech.2014.09.006
85. Wang L, Liu W, Wang Y, Wang J, Tu Q, Liu R, Wang J (2013) Construction of oxygen and chemical concentration gradients in a single microfluidic device for studying tumor cell-drug interactions in a dynamic hypoxia microenvironment. Lab Chip 13(4):695–705. doi:10.1039/c2lc40661f
86. Lin X, Chen Q, Liu W, Zhang J, Wang S, Lin Z, Lin JM (2015) Oxygen-induced cell migration and on-line monitoring biomarkers modulation of cervical cancers on a microfluidic system. Sci Rep 5:9643. doi:10.1038/srep09643
87. Rosso F, Giordano A, Barbarisi M, Barbarisi A (2004) From cell-ECM interactions to tissue engineering. J Cell Physiol 199(2):174–180. doi:10.1002/jcp.10471
88. Wu J, Chen Q, Liu W, Lin JM (2013) A simple and versatile microfluidic cell density gradient generator for quantum dot cytotoxicity assay. Lab Chip 13(10):1948–1954. doi:10.1039/c3lc00041a
89. Wu C, Asokan SB, Berginski ME, Haynes EM, Sharpless NE, Griffith JD, Gomez SM, Bear JE (2012) Arp2/3 is critical for lamellipodia and response to extracellular matrix cues but is dispensable for chemotaxis. Cell 148(5):973–987. doi:10.1016/j.cell.2011.12.034
90. Sun YS, Peng SW, Cheng JY (2012) In vitro electrical-stimulated wound-healing chip for studying electric field-assisted wound-healing process. Biomicrofluidics 6(3):034117. doi:10.1063/1.4750486
91. Tong Z, Cheung LS, Stebe KJ, Konstantopoulos K (2012) Selectin-mediated adhesion in shear flow using micropatterned substrates: multiple-bond interactions govern the critical length for cell binding. Integr Biol 4(8):847–856. doi:10.1039/c2ib20036h
92. Przybyla LM, Voldman J (2012) Attenuation of extrinsic signaling reveals the importance of matrix remodeling on maintenance of embryonic stem cell self-renewal. Proc Natl Acad Sci U S A 109(3):835–840. doi:10.1073/pnas.1103100109
93. Moledina F, Clarke G, Oskooei A, Onishi K, Günther A, Zandstra PW (2012) Predictive microfluidic control of regulatory ligand trajectories in individual pluripotent cells. Proc Natl Acad Sci U S A 109(9):3264–3269. doi:10.1073/pnas.1111478109
94. Lin HK, Zheng S, Williams AJ, Balic M, Groshen S, Scher HI, Fleisher M, Stadler W, Datar RH, Tai YC, Cote RJ (2010) Portable filter-based microdevice for detection and characterization of circulating tumor cells. Clin Cancer Res 16(20):5011–5018. doi:10.1158/1078-0432.CCR-10-1105
95. Gao D, Liu J, Bin Wei H, Li HF, Guo GS, Lin JM (2010) A microfluidic approach for anticancer drug analysis based on hydrogel encapsulated tumor cells. Anal Chim Acta 665(1):7–14. doi:10.1016/j.aca.2010.03.015
96. Biju V (2014) Chemical modifications and bioconjugate reactions of nanomaterials for sensing, imaging, drug delivery and therapy. Chem Soc Rev 43(3):744–764. doi:10.1039/C3CS60273G
97. Li Q, Liu L, Liu JW, Jiang JH, Yu RQ, Chu X (2014) Nanomaterial-based fluorescent probes for live-cell imaging. TrAC—Trends Anal Chem 58:130–144. doi:10.1016/j.trac.2014.03.007
98. Lee JW, Lee S, Jang S, Han KY, Kim Y, Hyun J, Kim SK, Lee Y (2013) Preparation of non-aggregated fluorescent nanodiamonds (FNDs) by non-covalent coating with a block copolymer and proteins for enhancement of intracellular uptake. Mol BioSyst 9(5):1004–1011. doi:10.1039/C2MB25431J
99. Hua X, Zhou Z, Yuan L, Liu S (2013) Selective collection and detection of MCF-7 breast cancer cells using aptamer-functionalized magnetic beads and quantum dots based nano-bio-probes. Anal Chim Acta 788:135–140. doi:10.1016/j.aca.2013.06.001

100. Yang D, Dai Y, Ma P, Kang X, Cheng Z, Li C, Lin J (2013) One-step synthesis of small-sized and water-soluble NaREF4 upconversion nanoparticles for in vitro cell imaging and drug delivery. Chem Eur J 19(8):2685–2694. doi:10.1002/chem.201203634
101. Zheng C, Zheng M, Gong P, Jia D, Zhang P, Shi B, Sheng Z, Ma Y, Cai L (2012) Indocyanine green-loaded biodegradable tumor targeting nanoprobes for in vitro and in vivo imaging. Biomaterials 33(22):5603–5609. doi:10.1016/j.biomaterials.2012.04.044
102. Gao D, Liu H, Lin J-M, Wang Y, Jiang Y (2013) Characterization of drug permeability in Caco-2 monolayers by mass spectrometry on a membrane-based microfluidic device. Lab Chip 13(5):978–985. doi:10.1039/c2lc41215b
103. Lin L, Chen H, Wei H, Wang F, Lin JM (2011) On-chip sample pretreatment using a porous polymer monolithic column for solid-phase microextraction and chemiluminescence determination of catechins in green tea. Analyst 136(20):4260–4267. doi:10.1039/c1an15530j
104. Lin L, Gao Z, Wei H, Li H, Wang F, Lin JM (2011) Fabrication of a gel particle array in a microfluidic device for bioassays of protein and glucose in human urine samples. Biomicrofluidics 5(3):034112–3411210. doi:10.1063/1.3623412
105. Yu Z, Huang Y (2015) Sample preparation for single cell sequencing on integrated micro-fluidic devices. Sci Sin Chim 45(11):1090–1101. doi:10.1360/N032015-00100
106. Chen Q, Wu J, Zhang Y, Lin Z, Lin JM (2012) Targeted isolation and analysis of single tumor cells with aptamer-encoded microwell array on microfluidic device. Lab Chip 12 (24):5180–5185. doi:10.1039/c2lc40858a
107. Liu C, Liu J, Gao D, Ding M, Lin JM (2010) Fabrication of microwell arrays based on two-dimensional ordered polystyrene microspheres for high-throughput single-cell analysis. Anal Chem 82(22):9418–9424. doi:10.1021/ac102094r
108. Wei H, Chueh B, Wu H, Hall EW, Li C, Schirhagl R, Lin JM, Zare RN (2011) Particle sorting using a porous membrane in a microfluidic device. Lab Chip 11(2):238–245. doi:10.1039/C0LC00121J
109. Wu J, Li H, Chen Q, Lin X, Liu W, Lin J-M (2014) Statistical single-cell analysis of cell cycle-dependent quantum dot cytotoxicity and cellular uptake using a microfluidic system. RSC Adv 4(47):24929–24934. doi:10.1039/c4ra01665c
110. Mach AJ, Kim JH, Arshi A, Hur SC, Di Carlo D (2011) Automated cellular sample preparation using a Centrifuge-on-a-Chip. Lab Chip 11(17):2827–2834. doi:10.1039/C1LC20330D
111. Wang BL, Ghaderi A, Zhou H, Agresti J, Weitz DA, Fink GR, Stephanopoulos G (2014) Microfluidic high-throughput culturing of single cells for selection based on extracellular metabolite production or consumption. Nat Biotechonl 32(5):473–478. doi:10.1038/nbt.2857
112. Lin X, Leung KH, Lin L, Lin L, Lin S, Leung CH, Ma DL, Lin JM (2016) Determination of cell metabolite VEGF165 and dynamic analysis of protein-DNA interactions by combination of microfluidic technique and luminescent switch-on probe. Biosens Bioelectron 79:41–47. doi:10.1016/j.bios.2015.11.089
113. Su WT, Feng K, Qin JH (2015) Progress of Microfluidics-based immunoassays in detection of cardiac markers. Chin J Anal Chem 43(10):1490–1498. doi:10.11895/j.issn.0253-3820.150440
114. Lin X, Wu J, Li H, Wang Z, Lin JM (2013) Determination of mini-short tandem repeat (miniSTR) loci by using the combination of polymerase chain reaction (PCR) and microchip electrophoresis. Talanta 114:131–137. doi:10.1016/j.talanta.2013.04.012
115. Yi L, Xu X, Lin X, Li H, Ma Y, Lin JM (2014) High-throughput and automatic typing via human papillomavirus identification map for cervical cancer screening and prognosis. Analyst 139(13):3330–3335. doi:10.1039/c4an00329b
116. He X, Chen Q, Zhang Y, Lin JM (2014) Recent advances in microchip-mass spectrometry for biological analysis. Trends Anal Chem 53:84–97. doi:10.1016/j.trac.2013.09.013

117. Li H, Zhang Y, Lin JM (2014) Recent advances in coupling techniques of microfluidic device-mass spectrometry for cell analysis. Sci Sin Chim 44(5):777–783. doi:10.1360/N032014-00007
118. Gao D, Wei H, Guo G-S, Lin JM (2010) Microfluidic cell culture and metabolism detection with electrospray ionization quadrupole time-of-flight mass spectrometer. Anal Chem 82 (13):5679–5685. doi:10.1021/ac101370p
119. Luo C, Ma Y, Li H, Chen F, Uchiyama K, Lin JM (2013) Generation of picoliter droplets of liquid for electrospray ionization with piezoelectric inkjet. J Mass Spectrom 48(3):321–328. doi:10.1002/jms.3159
120. Liu W, Chen Q, Lin X, Lin JM (2015) Online multi-channel microfluidic chip-mass spectrometry and its application for quantifying noncovalent protein–protein interactions. Analyst 140(5):1551–1554. doi:10.1039/c4an02370f
121. Zhang Y, Li H, Ma Y, Lin J (2014) Lipid profiling of mammalian cells with in situ matrix-assisted laser desorption ionization-mass spectrometry. Sci Chin Chem 57(3):442–446. doi:10.1007/s11426-013-4960-3

Chapter 7
Droplet-Based Microfluidic Technology for Cell Analysis

Junming Wang and Jin-Ming Lin

Abstract The mainstream of microfluidic chip research has transformed from the platform construction and method development to a wide range of applications. As a typical multiphase micro-functional unit, the droplet can be used as an independent microreactor with a volume ranging from pL to nL and formation rate up to thousands of droplets per second. It has the advantages of restricted diffusion, accelerated mixing, high heat transfer, effective mass transfer and so on. Droplet-based microfluidic technology has been emerged as a powerful tool to carry out high throughput screening and droplet manipulation research. Thus cell analysis, especially single cell analysis, and cell manipulation in droplets are easy and effective to be implemented with the assistance of a series of analytical methods. In this Chapter, we have briefly introduced an overview of droplet generation and corresponding principle by microfluidic chip. Besides, some usual methods coupled with droplet analysis have also been presented such as fluorescence analysis, mass spectrometry, capillary electrophoresis and others. Then we have mainly discussed the progress of droplets in cell analysis of recent decades. Finally, a summary and possible predication of droplet-based microfluidic chip are made at the end of this chapter.

Keywords Microfluidics · Droplet · Single cell analysis · Cell manipulation

7.1 Introduction

The microfluidic technology, also known as Lab Chip, or Micro-Total-Analysis Systems (μTAS), has developed rapidly in last decades and became core tools of research in a broad range of fields, such as biology, medicine, pharmacy, tissue engineering, material science and so on [1–3]. Especially in cell biological studies,

J. Wang · J.-M. Lin (✉)
Department of Chemistry, Tsinghua University, Beijing 100084
People's Republic of China
e-mail: jmlin@mail.tsinghua.edu.cn

J.-M. Lin (ed.), *Cell Analysis on Microfluidics*, Integrated Analytical Systems,
https://doi.org/10.1007/978-981-10-5394-8_7

microfluidic technology can revolutionize and replace the traditional methods of utilizing multi-well plates and microscopy slides for the observation and analysis of cell cultures in vitro. The main reasons of such dramatical superiority are from giant advantages of microfluidic technology in cell research [4–7], including: (1) the sizes of microfluidic channels are usually range from 10 to 100 μm, which is almost commensurate with cell size and shape, making it benefit for the accurate manipulation and analysis; (2) the multidimensional mesh structure of microfluidic chips has formed sealed micro-environment, similar with cellular environment in vivo; (3) the requirement of cells and other reagents can be significantly reduced as the tiny space in microfluidic chips, thus saving much costs and sample consumption; (4) microfluidic chip can integrate diverse technologies and laborious research operations, such as sample preparation, introduction, cell culture, reaction, sorting, separation and detection, into a piece of intact chip, contributing to integration and automation of practical application in the future; (5) it can get abundant biological information via high-throughput assay; (6) the primary materials constructing microfluidic chips, such as PDMS (Polydimethylsiloxane), have good biocompatibility, good permeability and transparence, convenient for cell culture, observation and study; (7) high surface to volume ratios can accelerate energy transfer and often reduce reaction time, providing effective cell research environment. All of these inherent and extrinsic superiority make microfluidic technology expand fast and widely applied.

Droplet-based microfluidics, as one subcategory of microfluidics, is a newly developed technology for precise and accurate manipulation of many or single individual small volume droplet [8]. It not only possesses the advantages of microfluidic technology above mentioned, but also overcomes some shortcomings of conventional limitation, such as the restricted liquid mixing in the channel due to laminar flow driven pressure, cross contamination of samples easily happened in narrow space and so on [9]. At the same time, this technology also has a higher specific surface area, higher mass and heat transfer efficiency, shorter diffusion distance, less reagent consumption and other merits. Two immiscible phases, such as water and oil, are utilized to produce discrete tiny droplets of femoliter to microliter volumes, with a high speed up to 20,000 per second [10]. One phase is named as continuous phase (also mobile phase), and the other is dispersed phase. Their roles and compositions can be exchanged depending on purpose of study. Unlike other microfluidic continuous-flow systems, droplet-based microfluidic technology creates a clear separation space between droplets, thus to isolate and obtain smaller, easier to control, independent but intact rooms for cell research. Since a large number of micro-chambers can be produced in a very short time, parallel processing and experiments can be easily achieved, which provides the basis for obtaining large amounts of data. Hence, droplet-based microfluidic technology is also more superior than the common microfluidic technology in high-throughput study and analysis. As mentioned reasons, droplet-based microfluidic technology is widely used in enzyme kinetics research and analysis, protein crystallization, nanoparticle and molecular synthesis, single cell analysis and so on (Fig. 7.1). In the whole process of droplet-based microfluidic technology, a series

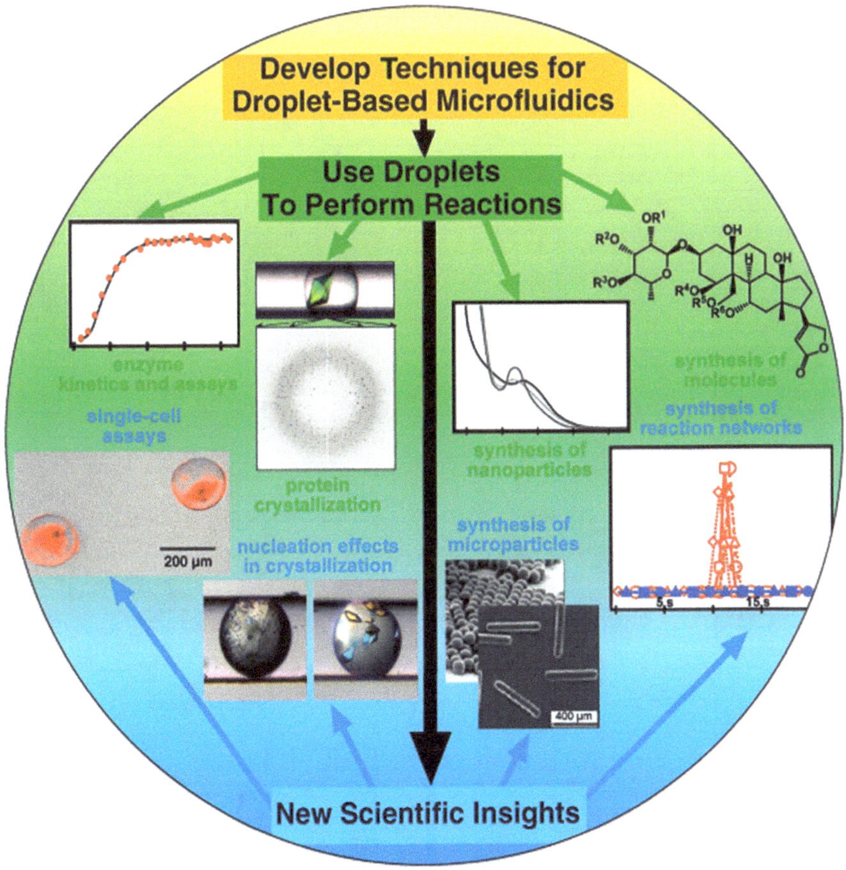

Fig. 7.1 Some main applications of droplet-based microfluidic technology (Reprinted with permission from Ref. [11])

of functions such as micro-droplet generation, transport, split, fusion, sorting, positioning, capture and control of two-phase behavior, can be realized easily and combined with other analytical detection techniques for further study (Fig. 7.2) [13–15]. Therefore, it has been developed rapidly in recent years and become one potential trend of microfluidic technology.

Of course, some shortcomings not beneficial for cell culture and research also exist in droplet-based microfluidic technology. For example, the interior of a microdroplet can provide a really suitable physical and chemical environment, but an interfacial tension is created at its surface due to direct contacting with the continuous phase, which may have an adverse effect on cell growth [16]. Thus, Baroud et al. separated the continuous phase from the microchannel to reduce the side effect of interfacial tension on the droplet interior [17]. Another aspect is the small amount of consumption and use of the reagents and cells in microfluidic

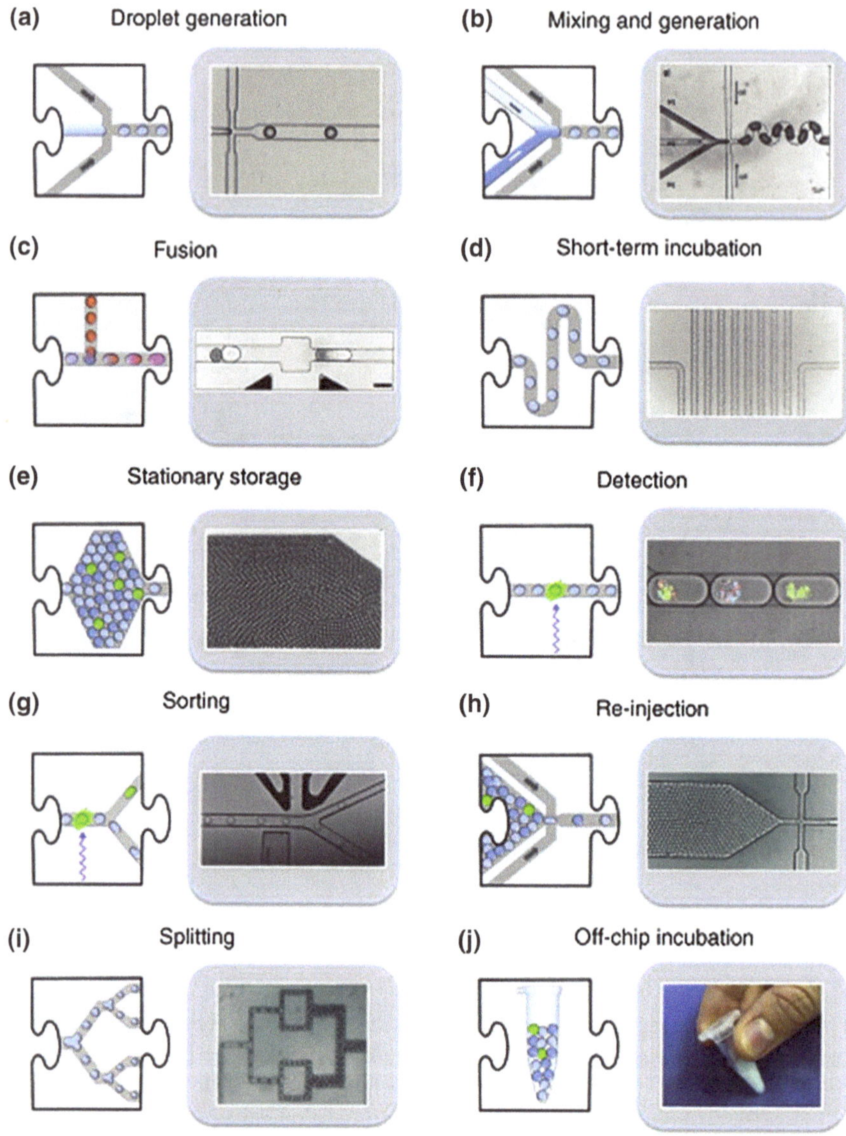

Fig. 7.2 A whole of unit operations for droplet microfluidics. **a–j** droplet generation; mixing and generation; fusion; short-term incubation; stationary storage; detection; sorting; re-injection; splitting and off-chip incubation (Reprinted with permission from Ref. [12])

droplet technology, which reversely bothers the detection of certain trace substances and needs high technical requirements compared with traditional methods. It desires for appropriate technology of analysis and detection to make further progress and development to satisfy the demands. So far, droplet-based microfluidic technology has already been combined with a variety of technologies, such as electric field, surface acoustic wave, for cells sorting and screening, as well as some detection technologies such as fluorescence, mass spectrometry, raman spectroscopy, electrophoresis, PCR (polymerase chain reaction), and electrical detection methods, for the detection and analysis of cells and related substances. Schmid et al. combined surface acoustic wave with micro-droplet technology to achieve sequential droplet sorting at high speeds up to 3000 per second without large shear forces production [18]. Mazutis et al. severally injected cells and green fluorescent protein-conjugated antibodies from two wells of a Y-shaped structure to produce droplets of packaging cells and antibodies. As a process of culture and fluorescent screening, they have detailed a binding assay for detecting antibodies secreted from single mouse hybridoma cell [19]. Hatch et al. utilized the combination of droplet microfluidics with PCR technology to manufacture a continuous, high throughput, automated portable devices for realizing multichannel and simultaneous fluorescence imaging [20]. With the constant development and progress of these means and the emergence of new technologies, droplet-based microfluidic technology will play a greater role in cell analysis and achieve more effective analysis through mutual combination.

Taking into account of above favorable and unfavorable factors, droplet-based microfluidic technology have mainly focused on cell manipulation, metabolite analysis and single cell analysis in cell research at present. The ability to generate discrete droplets via droplet-based microfluidic technology provides a new strategy for the studies of cell-cell interactions as employment of different designs in microfluidic chip, or formation of three-dimensional tumor spheroids by using some special materials such as gel. Sabhachandani et al. have reported a method to obtain three-dimensional spherical tumor of MCF-7 (human breast cancer cell) and co-cultured with HS-5 lactococcus lactis by individual injecting cells contained alginate solution and Ca^{2+} solution into the channel from both sides of the T-shaped structure chip, thus to produce the cell-loaded monodispersed calcium alginate droplets [21]. It's capable to produce single cell droplets quickly and efficiently by proper adjustment of flow rate and concentration as the high flux and small volume in droplets generation. Brouzes et al. conducted droplets fused via utilizing multiple droplets generated microfluidic structures, allowing single cells to be cultivated for several days in different drug environments, then to implement the cytotoxicity assays and analysis using the same method [22]. Droplet-based microfluidic technology has special and dramatical advantages of simple, fast, cheap, efficient and no aids of aptamers and antibodies in single cell generation when compared with other methods, so most of the studies have focused on single cell analysis in the field of cell research.

Hence, we will mainly review the development and progress of droplet-based microfluidic technology in cell analysis over the years, to summarize its advantages and shortcomings, and to look forward to its potentials and prospects in future. This chapter is mainly introduced in the following sections: principles of droplet generation, methods for droplet generation, analytical methods in droplet microfluidic, single cell analysis and cell manipulation.

7.2 Principles of Droplet Generation

It's very simple to generate continuous droplets via use of microfluidic chips today after years of study and development. The earliest droplet generation dates back to 1950s, when de Fonbrune sprayed water droplets into mineral oil to produce stable droplets in 1947, and some similar approaches are still in use nowadays [23]. Subsequent Lederberg and Nossal have successfully encapsulated single cells with droplets through modified methods and proved the rule that a cell would only produce an antibody [24]. The most common methods for generating continuous droplets today are based on the use of geometrical structures such as V-shaped crossings or T-shaped crossings, to severally inject two mutually immiscible liquids at high speed, producing droplets of high flux. The basic principles of these methods are almost similar. As soon as the two immiscible liquids come into contact, the interface are formed. One of the liquid phases would be self-dispersal into discrete droplets and surrounded by the other liquid phase due to the role of surface tension [25]. It mainly depends on the liquid phases, flow velocity and the surface energy of the channel to decide the affiliation of monodisperse droplet. For example, the aqueous phase become dispersed phase when oil and water are used for two liquid phases and typically hydrophobic PDMS as channel. Whereas, the oil phase may change into dispersed phase if the hydrophobic channel is hydrophilic modification [26].

The droplets also need to meet some specific characters in practical applications although easy formation, just as droplets for cell encapsulation, including whether the size of the droplet is appropriate and uniform, the shape meets the demand, the composition and distribution are the identical, and the monodispersity reaches the standard [27]. Some droplets, also named smart droplets, can specifically receive external stimulation, such as temperature, pH, UV irradiation, magnetic field and electric field, and respond in an active manner, by combining smart or intelligent materials (Fig. 7.3). The formation and stability of droplets are often the reflection of the balance among inertia force, viscous resistance, shear force, buoyancy force and interfacial tension. In addition, particularly high surface area to volume ratios caused by the limitation of small sizes of the microchannel and droplet, and the boundary effect in the channel make the behavior of the droplet peculiarly complicated. Some parameters are commonly used for describing the microfluidic kinetic of droplet formation, such as surface tension, Reynolds number, Weber number, Bond number, capillary number, surfactant effect and so on [25].

7.2.1 *Surface Tension σ*

The surface tension σ is a kind of tension acting on the surface due to the imbalance of molecular attraction and quantized in units of newton per meter (N/m) or joules per square meter (J/m^2). The surface tension can greatly influence the droplet

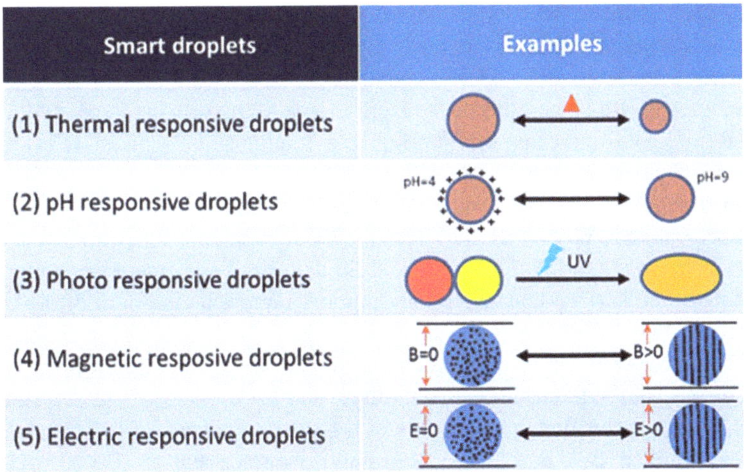

Fig. 7.3 Explanation and examples of five smart droplets. *1* droplet size decreases with increasing temperature; *2* droplets are positively charged at pH 4 and becoming neutral at pH 9; *3* two droplets are fused together under the exposure to UV light; *4* and *5* are magnetically induced particles and electrically induced particles (Reprinted with permission from Ref. [30])

behavior because the interface area is totally surrounded by the continuous phase, leading to large interfacial area and dramatically strong force. So the droplets will decrease the interface area as much as possible to reduce the surface energy in the presence of surface tension. In theory, the droplet should be the size of circular when its surface energy performs minimal. But the droplets in the microfluidic chips have been found to be tend towards a certain oval shape rather than totally circular in the actual process, possible the cause of small size of the channel and the boundary effects [28].The curvature of the interface can lead to pressure jump, named Laplace pressure (ΔP), between the outside and inside of the interface as the deformation of the droplet interface curvature. One formula is used to describe the Laplace pressure, as $\Delta P = \sigma\left(\frac{1}{R_1} + \frac{1}{R_2}\right)$, where R_1 and R_2 are the radius of curvature at different locations in droplet interface, and have inverse proportion to Laplace pressure. The interfaces of droplets have elasticity under the effect of ΔP. This property is used to carry out the droplet control and manipulation in recent years, such as droplet capture, merger and dilution [29].

7.2.2 Dimensionless Numbers

The dimensionless number is an important parameter for discrimination of different fluids, main including Reynolds number, Weber number, Bond number and capillary number.

(1) Reynolds number (Re), is a dimensionless number to characterize fluid flow through a chip channel. $Re = \rho Ul/\mu$, where ρ, U, μ and l represent mass density, characteristic velocity, dynamic viscosity and characteristic length, respectively. Re is the ratio of the inertia force to the viscous force, and a value of $Re \leq 2300$ indicates that the liquid is a laminar flow under selected conditions. In general, the Re of fluid in the microfluidic channel is small, which means negligible inertial force and the regular fluid condition [31].
(2) Weber number (We), is used to describe the movement of the fluid. $We = \rho U^2 l/\sigma$, where ρ, U and σ represent the fluid density, the characteristic velocity of the flow and surface tension. Weber number represents the ratio of the inertial force to the surface tension. The droplet is likely to be disintegrated if the Weber number is larger than a certain critical value when droplets are formed by droplet-based microfluidic technology, and the droplet will be deformed but not broken if the Weber number is at a moderate value [32].
(3) Bond number (Bo), is the ratio of gravity to surface tension. $Bo = \Delta\rho g l^2/\sigma$, g and Δρ represent acceleration of gravity and density difference between the two liquids, respectively. Bo is usually much smaller than 1 at the microscale, meaning that the effect of gravity can be ignored.
(4) Capillary number (Ca), also named as the interfacial tension number, is defined as the ratio of viscous stress to surface tension. $Ca = \mu_c U/\sigma$, where μ_c is the fluid viscosity, and U is the velocity, and σ also represents the surface tension. A low value of Ca means that the surface tension are dominant over viscous stress, and the droplets perform spherical to reduce the surface area and energy. While a high value of Ca appears opposite meaning that the viscous stress plays a dominant role, and the droplets will become deformation by the flow and present asymmetric structure [33].

From the definition of above, we can discover that the Reynolds number, Weber number and Bond number have small effect on droplet generation in most cases, while the surface tension and capillary number own a dominate role on the micrometer scale of microfluidic technology. Furthermore, other factors have also contribute their efforts, such as the hydrophilic of channel, boundary effects and so on. It is worth noting that the design of the microchannel still plays an important role in the generation of droplets.

7.3 Methods for Droplet Generation

There are many methods have been developed for droplet formation, the core of droplet-based microfluidic technology, after extensive and years of study. In general, these methods can be categorized into two groups, passive and active approaches. The passive methods are mainly based on different microfluidic geometries, such as T-junction, flow focussing structure and co-flowing structure, utilizing the principle of hydrodynamics for droplet formation [34, 35]. This method is simple, efficient and fast, so it is widely studied and used, but the flexibility of droplet is relatively poor as limited by channel geometry [36]. The active approaches are primarily driven through the control of external field to generate droplets, such as pneumatic drive, optical drive, thermal drive, electric field drive and other methods [30]. People can control the size and frequency of the generated droplets well by this method without totally dependence of chip structure, but the velocity of droplet production is very slow. We will mainly focus on the passive approach since it's the most common configuration of droplets formation.

7.3.1 T-junction

T-junction are the most commonly used microfluidic geometry for droplet generation (Fig. 7.4a). Thorsen firstly used T-junction to generate droplets in 2001 [37]. In this configuration, the main channel, injected with the mobile phase, and the side channel, introduced with the dispersed phase, are intersected at focused zone in a perpendicular way to create droplets, under the effect of surface tension and shear stress. At the same time, two asymmetric forces, upstream water pressure of the side channel and the pressure of the mobile phase in the main channel, together with the surface tension, can cut off continuous dispersed phase to single and independent droplets [38]. A brief summary of the droplet formation process is dropping, spraying, and deformation. The frequency and size of generated droplet can be effectively regulated and controlled by adjustment of the flow rate, channel size and liquid viscosity. For example, increasing channel width of the downstream can effectively accelerate the rate of droplets formation and improve the stability of droplets. And multi-wrapped droplets can be achieved through reasonable use of multiple T-junction (Fig. 7.4b) [39]. However, cells encapsulated in droplets will be greater effected due to the asymmetry force in droplet generation process, so T-junction is most commonly used in other aspects but not cell analysis.

7.3.2 Flow Focusing Structure

In flow focussing structures, three channels, one main channel and two symmetric side channels, are focused on a narrow region connecting the downstream channel (Fig. 7.4a). In brief, the dispersed phase is injected into the side channels, while the continuous phase is introduced into main channel, which approach the confluence at the straight line in the downstream channel. The advantage of this design is that the dispersed phase in focused zone will only bear the driving force on the main channel and force from other directions would be counteracted as the symmetrical structure of side channels, leading to less affected and more stable droplets. It can

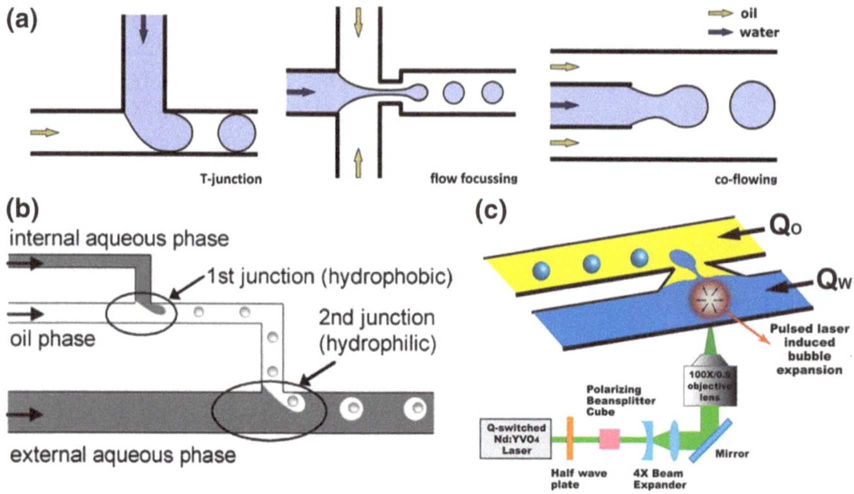

Fig. 7.4 Common methods of generating droplet. **a** The two most commonly used methods of passive generation droplets are the T-junction and the flow focussing, and the less commonly used co-flowing method (Reprinted with permission from Ref. [40]). **b** A double T-junction cross structure to obtain droplets of oil/water/oil structure (Reprinted with permission from Ref. [39]). **c** Pulsed laser-generated holes promote droplets on demand (Reprinted with permission from Ref. [41])

be seen that the size of the droplet is related to the velocity ratio of the two phases. The larger the flow rate of the mobile phase is when compared with the dispersed phase, the smaller and faster the generated droplet is. In addition, the size of the focusing region, the viscosity of the liquid can also influence the formation and size of droplets. Similarly, the application of multiple flow focussing structures is designed to obtain multilayered droplets, which are often used for cell co-culture in cell analysis [42]. However, some small satellites droplets are often produced accompanied with droplets generation in this process, which greatly reduces the monodispersity of the whole system. So some control methods, such as screening systems, adjusting the flow rate, selecting the appropriate two phases and designing appropriate focused zone, have been studied to control the non-uniformity [43].

T-junction and flow focussing structure are the basic structures for droplet formation in droplet-based microfluidic technology. Different designs are usually implemented in the actual process to meet the needs of different research. Song et al. vertically connected the downstream channel in flow focussing structures with another main channel to form a second T-junction, thereby achieving the purpose of mixing three reagents in one droplet [44]. Adamson et al. integrated the T-junction with the dendritic bifurcation channels to generate multiple droplet arrays [45]. Both methods are simple, fast and easy to be achieved, so they have been widely used in droplet based microfluidic technology.

7.3.3 Co-flowing Structure

The co-flowing structure has utilized a drive pipe to create interface between two co-flowing immiscible fluids (Fig. 7.4a). That is to say, the dispersed phase flows into the inner channel while the outer channel is mobile phase. One serious disadvantage of this structure is that the dispersed phase has suffered strong viscous shear force of continuous phase for its roomy diameter. As the continuous phase-touched droplets march forward, their surface tension force have counteracted with drag force, thus to pinch the droplet off from the continuous phase. Compared with other two methods, this approach is complex and difficult to a certain extent, and strong uniform forces in the process of generation make it unusually applied for cell research.

7.3.4 Other Methods

Though some methods to generate droplets, such as air pressure driving, light driving, heat driving and electric field driving, are not convenient, fast and high flux as the former methods, they also attract a lot of research due to the stable size and controllable frequency of the droplets [46]. The basic principle of these methods is that the dispersed phase is pinched into the mobile phase to form droplets by the power of regular, controlled, continuous pulse pressure, which is applied to the dispersed phase near the mobile phase. Park et al. irradiated pulsed laser light into the dispersed phase to produce small voids, which can extrude the dispersed phase into the mobile phase, and then form droplets of small volumes (Fig. 7.4c) [41]. Collins et al. imposed the surface acoustic waves to the dispersed phase and utilized the generated sound to force the dispersed phase into the mobile phase, forming droplets under the action of shear stress and surface tension. The size and frequency of droplets could be adjusted and controlled by the frequency and amplitude of the surface acoustic waves [47]. He et al. applied a high-voltage pulse to the dispersed phase so as to move the two-phase interface towards the direction of potential. The dispersed phase could be incised into single droplet with a minimal volume of fL or pL in this process [48]. The droplets formed by these methods have been widely used in mass spectrometry analysis, protein detection, DNA analysis, particle encapsulation and other fields, but not in cell analysis, especially cell manipulation. One reason is that some extra field driving, such as electric field driving, maybe have bad impact on cell or biological sample by produced high voltage. Besides, the frequency of droplet generation is low, and the homogeneity of the wrapped cells is poor by using such methods, which is far inferior to T-junction and flow focussing structure. Therefore, these technologies need to be further improved and optimized.

7.4 Analytical Methods in Droplet Microfluidic

The analytical methods in droplet microfluidic, especially cell analysis, have a higher standards whether off-line analysis or online analysis, due to the high flux of generated droplets, the small size of the droplets, a small amount of cells or even single cell usually contained in one droplet, together with the different size and frequency of the droplets in different experiments. Now the commonly used methods are fluorescence, mass spectrometry, electrochemical detection, capillary electrophoresis and so on.

7.4.1 Fluorescence

Fluorescence detection is the most common and effective method in microfluidic technology. This method usually requires an assistance of fluorescence reagents and fluorescence microscopy. It has the characteristics of high specificity, sensitivity and accuracy, and can be used for analysis of the enzyme activity, proteins, DNA, small molecules, even single molecules and individual cells (Fig. 7.5a) [19, 49–51]. For example, Chen et al. were able to detect enzyme activity for a long time by encapsulating enzyme proteins in fluorescent substrates contained droplets [52]. Harris et al. utilized microfluidic and fluorescence techniques to successfully detect single-molecule nucleotide sequences without amplification of the sequence [53]. Gu et al. have constructed a multifunctional droplet manipulation platform, which could achieve many functions of efficient single cell encapsulation, single cell enzyme activity analysis, and single cell DNA purification with the aid of fluorescence microscopy (Fig. 7.5c) [55]. In these analysis, quantitative analysis of fluorescence intensities and high-resolution images via the use of high-resolution microscopy are often required to obtain more complete and meaningful data. However, the effective analysis of the droplets need to be conducted in the condition of slow generation velocity or idle state, or after being collected and captured since the droplets are generated very quickly, because production frequency of droplets are usually much faster than frame number of a charge-coupled camera (Fig. 7.5b) [54, 56]. Shi et al. attached a droplet trapping array behind the T-junction, capturing the formed droplets in a specific small space and then analyzing them [57]. Fidalgo et al. reduced the flow rates of generated droplets and left the fluorescent droplets and water droplets alternately flowing past the fluorescence detection region, allowing for selective separation of the fluorescent droplets [58]. Thousands of droplets can be monitored simultaneously for long periods of time using fluorescence detection because of the small size of droplets, so a large amount of statistical data and parallel results can be obtained in an experiment. Baret et al. could directly detect the bacteria of overexpressing β-galactosidase in a large number of Escherichia coli encapsulated droplets, and compare their enzyme activity using fluorescence detection [59].

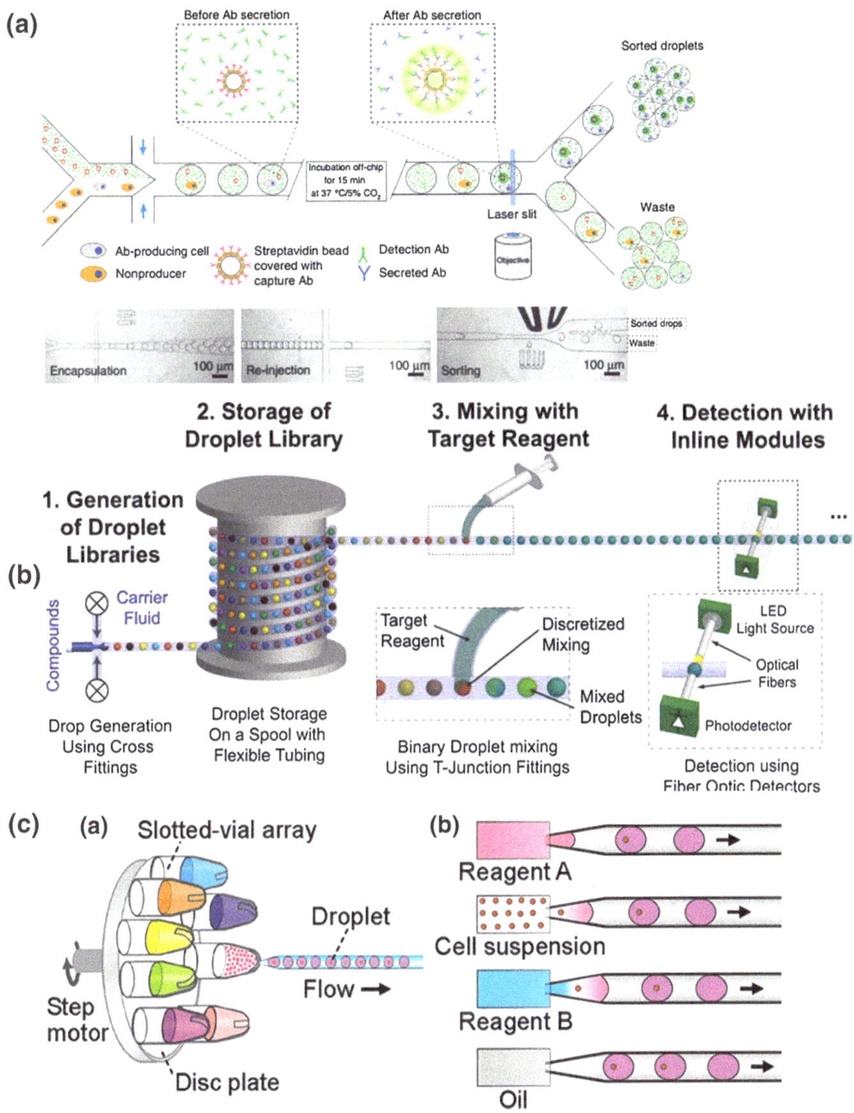

Fig. 7.5 Several common methods for combining droplets with fluorescence detection. **a** Direct mixing with a Y-shaped channel (Reprinted with permission from Ref. [19]). **b** Addition of target reagents by droplets fusion (Reprinted with permission from Ref. [54]). **c** The different reagents were gradually added to obtain blending droplets (Reprinted with permission from Ref. [55])

Fluorescence detection method can also provide the information of spatial distribution of the material in the droplet, such as protein secretion, cell migration, droplet breakage, and connection between droplets [60]. For example, Khorshidi et al. continuously observed and analyzed the dynamic behavior of a large number of droplets containing HEK293T single cells up to 11 h using automated microscopy and image analysis [61]. Chokkalingam et al. encapsulated microspheres of capturing secretions on T-cell membranes, and then incubated, cooled, gelled the single-cell droplets, finally detecting the secreted cytokines by flow cytometry after adding fluorescent antibodies [62]. Courtois et al. have studied the time-dependent release of molecules from droplets by observing the fluorescence of the fluorescent derivatives in the droplets [63]. The emergence of fluorescence lifetime imaging microscopy, has achieved observation in spatial distribution and ultra-short time interval imaging as low as 1 microsecond, providing possible for ultra-sensitive and ultra-fast detection [64]. Solvas has observed the mixing behavior of two droplets in microsecond time resolution by using a fluorescence lifetime imaging microscope [65]. Laser-induced fluorescence microscopy can be realized for analysis and detection of highly sensitive and high-throughput, making it possible for a large number of droplets detection, such as single-cell analysis, PCR detection, cell screening and so on [51, 66]. For example, Agrest et al. have implemented ultra-high throughput cell sorting to study biochemical reactions and protein evolution with the assistance of laser-induced fluorescence microscopy [67]. Therefore, the visualization, localization and quantification of fluorescence detection make it well suited for droplet analysis and cell analysis.

7.4.2 Mass Spectrometry (MS)

Many study have investigated the association of microfluidic chip with mass spectrometry as it is a label-free method and can provide the structure information of substance compared with fluorescence detection methods [68, 69]. Lin's group has carried out a series of study about combination of microfluidic chip with mass spectrometry and made great progress by developing micro-solid phase extraction cartridge for sample pretreatment and constructing MS connector. This combination of microfluidic chip with MS has been widely used for cell analysis, cell co-culture, drug metabolism and cancer therapy research. Chen et al. directly linked microfluidic chip to the mass spectrometry via attaching a micro-solid phase extraction cartridge to the end of the microfluidic structure for removal of salt in the cell culture medium [70]. However, combination of the droplet microfluidic with mass spectrometry is more difficult, mainly as the followings: (1) The mobile phase surrounding the droplets tends to interfere the detection of mass spectrometry and dilute the target content in the droplet; (2) The cells and target molecules are very rare as small volume of droplets, thus requiring high sensitivity of the mass spectrometer; (3) The cell culture medium can't be directly accessed into mass spectrometry for containing much salts when cells are contained in the droplets; (4) High flux of the droplets generation also limits its combination with mass spectrometry [71–73].

The first thought of researchers is to separate the droplet from the mobile phase, in order to reduce the interference of mobile phase when facing these difficulties [77]. Smith et al. collected and stored the surfactant-stabilized droplets, and then re-injected into the microfluidic chip attached to the mass spectrometer to perform high-flux droplet analysis [78]. Fidalgo et al. have constructed a two-phase interface of water-oil in a chip, each leading to different channels, and then pushed the generated droplets into the water phase from the oil phase by using the electric field (Fig. 7.6a) [68]. The droplets were sprayed into mass spectrometry after moving into the fused silica nozzle. Su et al. drove the generated droplets onto the glass to form arrays, and then pipetted them into the capillary that connecting with the mass spectrometer for detection [79]. Similarly, Küster et al. formed droplets microarray in an indium tin oxide-coated glass, which were then detected by matrix assisted laser desorption ionization (MALDI) [80]. Of course, some people also used clever designs to omit these steps [79, 81]. Gasilova et al. punched an observation hole in the top of the main channel and connected with the mass spectrometer inlet, and then sprayed the droplets of flowing through the observation port by using the high pressure, thus accompanying little of mobile phase and minimum effect on the MS analysis (Fig. 7.6b). In addition, the droplets have been generated without mobile phase or just chosen as sampling device due to the existence of the above difficulties (Fig. 7.6c, d). For example, Chen et al. generated single-cell droplets without mobile phase by utilizing piezoelectric ceramics to squeeze cell-containing organic solvents. So the droplets could be directly dropped onto a high-pressure needle and sprayed into mass spectrometry for obtaining information [75]. Lee and Zhang et al. used droplets as agent to extract individual cell and then for mass spectrometry analysis by laser or electric spraying [82, 83]. Liu et al. employed a

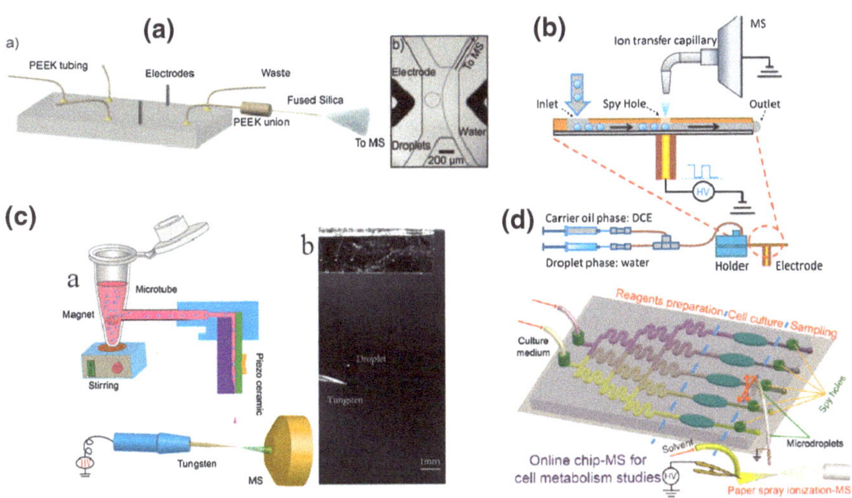

Fig. 7.6 Some usually ways of combination droplet microfluidics with mass spectrometry. **a** The droplets were separated from the mobile phase by an electric field (Reprinted with permission from Ref. [68]). **b** The droplets are directly sprayed into the mass spectrum from the channel using high voltage pulses (Reprinted with permission from Ref. [74]). **c** Single cell droplets were directly produced by piezoelectric ceramics to drip onto the high pressure needle (Reprinted with permission from Ref. [75]). **d** Capillary sampling and generation of droplets combined with multi-channel microfluidic chip and paper spray mass spectrometry (Reprinted with permission from Ref. [76])

small diameter capillary to inhale the cell culture medium from multi-channel microfluidic chip by capillary action, and then got stable droplets with height difference into the paper spray mass spectrometry to obtain metabolite signal [76]. Even so, many difficulties and problems still remain unresolved. Therefore, the combination of droplet microfluidic technology and mass spectrometry also need more in-depth research and more technical support.

7.4.3 Electrochemical Detection and Capillary Electrophoresis

Electrochemical detection is a simple, rapid, effective, high sensitivity and low cost detection method, and has been mainly used to measure the electrical signal changes, for detection of the droplet size, frequency, speed and conductivity [84–86]. Luo et al. have firstly measured droplet size and ion concentration with a detection limit of 20 μM by utilizing different electrochemical signals as the different impedance in water and oil [87]. Moiseeva et al. also successfully detect the droplet chains based on thin-film electrodes and the difference of two-phase impedance [85]. In addition, the droplet size, speed and frequency can be well analyzed since changes in capacitance when droplets flow into the specific detection area. Liu and Elbuken have successfully measured the capacitance change of the droplet stream by different arrangement of the electrodes, and then analyzed these properties of the droplets [88, 89].

Capillary electrophoresis is a simple, fast and efficient separation technique, which is often used in conjunction with other detection methods. If the components of the droplet are complex or easily interfered with each other, capillary electrophoresis can be used for separation to obtain a good analytical result before entering the detector. Edgar et al. were able to simultaneously analyze the various substances contained in the droplets via droplets separation of capillary electrophoresis and detection by using laser-induced fluorescence [90]. As the low efficiency of capillary electrophoresis in microfluidic chip, Wang et al. modified the surface of the chip to get high efficiency of capillary electrophoresis, and successfully detect the neurotransmitters in the brain of rats by such method [91].

7.4.4 Other Methods

Except for the above methods, other usually methods are Raman spectroscopy, chemiluminescence, absorption spectrophotometry, nuclear magnetic resonance and so on. Cecchini et al. have constructed an ultrafast surface-enhanced Raman scattering system for the detection and analysis of droplets, with highest rate up to 32 droplets per second. [92]. Trivedi et al. allowed quick and efficient screening of droplet libraries by combining droplet generation with absorbance detection [49]. Shen et al. integrated the three reagents required in chemiluminescence into one droplet to complete luminescence and detection by means of liquid-liquid extraction after trapping droplets on both sides of the main channel [93].

There are still many problems in the analysis of droplets when they are applied to cell analysis although many detection methods have been carried out. Fluorescence, capillary electrophoresis, and chemiluminescence need assistant of additional reagents to complete the detection. Mass spectrometry, Raman spectroscopy, nuclear magnetic resonance, and absorption spectrophotometry, require complex pre-treatment, or still seriously interfere the detection owing to matrix effect, or have relatively low detection limit. Therefore, further technical progress, improvement and the establishment of new methods have broad prospects and requirement in droplet microfluidic technology.

7.5 Single Cell Analysis

Tumor is a disease that seriously threatens the health of human beings nowadays, and the heterogeneity of tumor cells is one of the important reasons for its easy relapse and difficult treatment [94]. As the traditional analysis is often the average results of abundant cells, some key information is likely to be hidden. So the analysis of individual tumor cells shows great meaning and has been widely concerned and explored for better mastering and understanding the occurrence, development, transferring and treatment of tumors [95, 96]. Droplet microfluidic technology has an intrinsic advantage in single cell analysis because droplets can provide a limited environment with easy accessibility by regulation of cell loading patterns. Besides, it's also convenient to carry out analysis and detection of cellular secretion and factors because of long survival of cells in droplets and metabolism in a narrow space [97]. Meanwhile, the droplet also is a moving chamber that can be high-throughput screening according to the property of the droplets and cells, which is contributing to observation and analysis of follow-up [98].

7.5.1 Encapsulation of Single Cells

One of the key points in single cell analysis is the separation of single cells from a large population of cells. There are two main ways for single cells encapsulation by using droplet microfluidic technology: passive and active methods [10]. Passive encapsulation is a way that actuates cells into droplets by adjusting the flow rate, cell concentration and channel width (Fig. 7.7a). This method is convenient, simple and high throughput, but the number of cells in each droplet is variable and presents a pattern of Poisson distribution, even no cells in considerable number of droplets [99, 100]. The active encapsulation is the utilization of external force such as sound, electricity, light, magnetic to wrap cells into droplets based on microfluidic chip (Fig. 7.7b). This approach is more accurate and produces higher proportion of single cell droplets than passive method, but with shortcomings of low throughput and efficiency [101, 102].

Passive encapsulation is the most common method of generating single-cell droplets today, and a large number of studies have used this method. The main process is that the cells are suspended to a certain concentration in the dispersed phase and incised into single droplet to complete single cell separation in the confluence zone by flowing phase on the basis of typical T-junction or flow focussing structure [105]. The droplets produced in this way don't have uniform number of cells in each droplet with a presentation of Poisson distribution as the spatial distribution of the cells and the moment of reaching the focusing region are random, thus greatly reducing its research significance [106]. The most common approach to solve this problem is the sorting of droplets, which will be described in

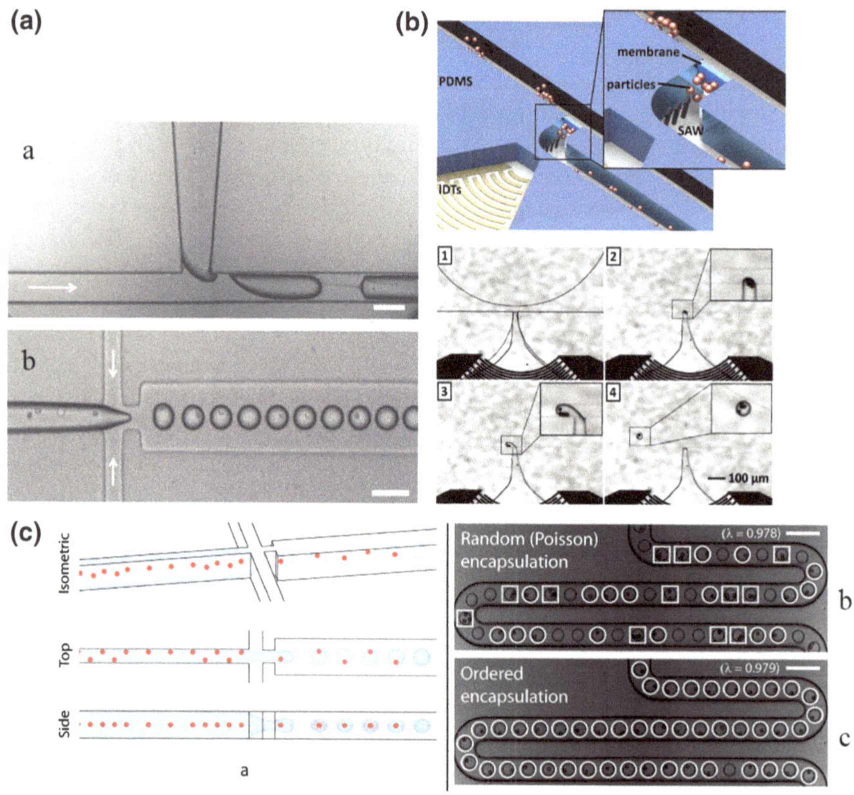

Fig. 7.7 Common methods of encapsulating single cell into one droplet. **a** Basic T-junction and flow focussing structures were used to encapsulate single cells (Reprinted with permission from Ref. [103]). **b** Surface acoustic wave was used to actively wrap a single cell (Reprinted with permission from Ref. [104]). **c** Regulating cells for stable spacing by high aspect ratio channels to obtain high occupancy rates for single cell droplets (Reprinted with permission from Ref. [108])

detail later. Others have come up with some sensible methods, such as regulation the rates of cell arriving at the focal region and droplet production to the same, making one cell in one droplet [107]. On that account, Edd et al. engendered a more stable spacing between cells by passing through a high-aspect-ratio channel, and kept the frequency of arriving at the focusing region with the same as droplets, resulting in a significant increase in the ratio of single cell droplets (Fig. 7.7c) [108]. The realization of this method is based on a series of hydrodynamics, such as hydrodynamic repulsive effect, repellent effect and parabolic shear gradient, which make the cells arranged orderly in long, wide and high, respectively, and easily controlled by reasonable regulation. Kemna et al. used a helical channel to align the cells in good order of horizontal and vertical, and most of the droplets only contained a single cell by a straightforward confluence with flowing phase in flow focussing structure. [102]. This method needs to arrange appropriate conditions of the cells in advance, so its application is limited as the heterogeneity of different cells during the process and it's difficult to balance the flow rate.

Active encapsulation refers to droplets generation with the assistance of external force, for the main advantage of accurate single-cell droplets production. Such as sound, electricity, light, magnetic force can be applied near the interface of the two phases and drive the cells into the focus area to control the generation of single cell droplets [104, 109–111]. He et al. immediately applied a high voltage pulse to produce a single cell droplet once the cells reached the two-phase interface with optical capture of the cell site [112]. Schoendube et al. used two electrodes to precisely determine the position and velocity of the droplets and installed a piezoelectric actuator in the focal region. After the electrode-measured signal was transferred to the piezoelectric actuator in real time, the force released by the piezoelectric actuator perfectly contacted with cells, generating single-cell droplets [113]. Another method was that the surface acoustic wave was used to act on the cell-containing dispersed phase to generate a thin liquid bridge, and then the mobile phase divided the bridge into droplets, but the number of cells is also uncertain by this method [101]. It is evident that the generation of single-cell droplets by active encapsulation results in very low rates due to the need of precise control of the cell's location for exerting external forces, even in comparison with passive encapsulation. However, single-cell analysis often require a large number of samples for statistics and comparison. Um et al. constructed a microwell array containing mesh-grid for simultaneous merging and storage of single-cell droplets in a high-throughput manner [114]. This way is still not fast as passive method and also an off-line method, so how to effectively improve the rate of single cell droplets is an important problem to be faced in active methods.

7.5.2 Droplet Sorting

As mentioned above, the number of cells in a droplet is often not deterministic, thus making subsequent experiments much more difficult. The main purpose of sorting is to select high-purity single cell droplets or other required droplets. For example, Wu et al. screened fluorescent-labeled lymphoma cells at high-speed by using laser pulse excitation, up to 45,000 cells per second (Fig. 7.8c) [115]. Another goal of sorting is to reduce the work of analysis in follow-up. A large number of cell-free or multiple cells-contained droplets can be filtered out by sorting, greatly reducing the subsequent experimental workload. Sorting is mainly based on the different characters among the target droplets with other droplets, such as physical size, magnetic, mechanical properties or optical properties and so on [109, 111, 116]. Jing et al. instituted a small column array after flow focussing structure and sorted cells-contained droplets by adjusting the spacing of the column, due to different sizes of the cell droplet and empty droplets (Fig. 7.8b) [116]. Chen et al. covered cells with magnetic beads to obtain magnetic droplets after cells encapsulation, which were then separated from all the droplets depended on the attraction of an electromagnet. Nam et al. used a standing surface acoustic wave to encapsulate cells into droplets and separate single cell droplet based on density (Fig. 7.8a) [117]. There are two ways of sorting: active and passive sorting, according to the different modes of action [118].

Passive sorting is a way that makes full use of microfluidic chip structure and the physical properties of droplets to sort, without external field or force [119]. Size screening of micro-column arrays in passive method is a common approach as the

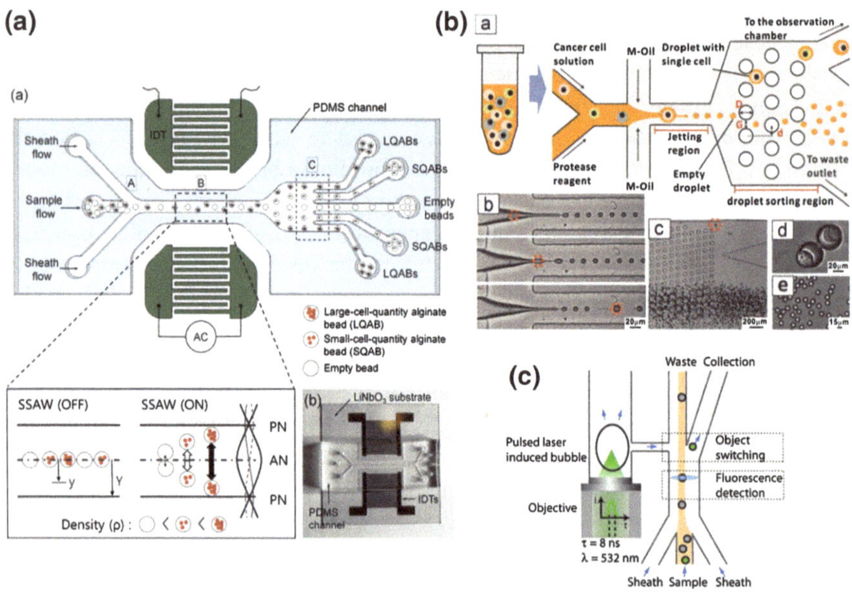

Fig. 7.8 Active cell droplet sorting based on microfluidic chip. **a** Using surface acoustic wave to sort out the cells without cells, containing a small number of cells and many cells of the droplets and collecting them (Reprinted with permission from Ref. [117]). **b** Size screening by micro-column arrays (Reprinted with permission from Ref. [116]). **c** Fluorescence detection and high purity pulsed laser screening (Reprinted with permission from Ref. [115])

presence of cells in the droplets changes the size of the droplets [120]. Joensson et al. set up a micro-column array to let droplets move towards different directions according to their size in order to allow droplets for determinate lateral movement. Some sorting methods are also on the basis of the dynamic effect of liquid [121]. Chabert et al. rendered the droplets lateral movement and spatial dispersion by using shear flow and compression flow of liquid, respectively, to achieve the purpose of self-sorting [122]. In addition, methods making use of channels with selective size and different fluid forces are also commonly used [123, 124]. Passive sorting doesn't need much equipment and is relatively simple to operate. Nevertheless, it is less used in practical application because it owns low efficiency and large error, and requires a complete understanding of every parameters in droplets for being likely to redesign the chip once any changed.

Active sorting, based on the droplet itself or the external cause of the differences in the physical or chemical properties, utilizes technology to find these difference and screens droplets by exerting external force [125]. Active methods provide the most flexible method for droplet sorting, and have employed most measurable parameters of droplets. The most common, efficient and accurate method is depended on the optical properties of the droplets, especially the fluorescence intensity [126]. Cell expression proteins, secretory proteins, surface proteins, and other molecules can be fluorescent labeled for sorting with a property of the high specificity, sensitivity, and easy detection [59, 127]. For example, Wu et al. sorted generated single-cell droplets at 1000 Hz by using fluorescent-labeled cells and laser sorting, obtaining droplets containing single cells up to 98% [128]. Joensson et al. used this principle and the enzyme signal amplification technique to analyze low-level surface markers in individual cells and to sort cells with higher marker content [129]. Fluorescence-activated dielectrophoresis force, as one of the commonly used driving forces, can be combined with fluorescence detection, and move the charged droplets in the electric field to achieve the purpose of droplet sorting. Agresti et al. sorted droplets based on fluorescence-induced electrophoretic equipment, realizing to screen 10^8 droplets in 10 h [67]. In addition, the electric field force, magnetic force, surface acoustic wave and other forces are also often used as the driving force for active sorting [103, 111, 130–132]. Chen et al. labeled cells with magnetic particles and then sorted out droplets with cells with the assist of magnetic fields [126]. The more the cells are contained in the droplet, the farer the droplets will move to both sides of the channel by surface acoustic wave. Franke et al. sorted the droplets in a wide channel to select out droplets containing no cells, a few cells and multiple cells by using this principle without prior labelling and cell encapsulation into liquid droplet [131]. Active sorting can greatly increase sorting efficiency and speed with high accuracy, based on not only whether the droplets contain cells, but also the characters of the cells in the droplet [22, 62]. At the same time, active sorting is less dependent on the structure of microfluidic chip than passive sorting, and still works on a wide range of droplet sizes, but usually requires additional reagents or equipment, which is one of the reasons for its higher sensitivity [18, 133].

7.5.3 Protein Analysis

The heterogeneity of cancer cells has reflected in different proteins expression of different cells, but single-cell protein analysis has been always a major challenge in cell research because of the small amount of protein in a single cell [134, 135]. While conventional methods such as flow cytometry can carry out single-cell protein analysis to a certain extent, but its practical significance is greatly weakened because of its damage to cells, and analysis without real-time [136, 137]. The single cell in the droplet is not only high-throughput, long-term survival, easy to detect, but also limited to a tiny space and not easily damaged, so it's considered as an ideal place for single cell protein analysis in recent years [10, 138].

The protein of cells mainly includes three types: surface protein, secretory protein and internal protein [139]. Because proteins can't amplify as DNA by polymerase chain reaction (PCR), the basic principle for trace protein detection is to amplify signals like enzyme-linked immunosorbent assays (ELISA) [60, 140]. Surface proteins and secretory proteins are easily captured by the addition of antibodies, so the basic idea is to collect protein by the antibody or aptamer, and then enhance signal through the other end of the antibody (Fig. 7.9a). Kony et al. connected antibodies with avidin bridge, which was linked with special DNA chain after capturing the cell surface protein. The DNA chain can conduct amplification of the signal through the DNA rolling chain amplification for protein detection [141]. In the detection of secreted proteins, nanoparticles modified with specific antibody were usually wrapped in the cell surface to perform enrichment and signal amplification in order to get better results as the protein will disperse into whole parts of the droplet [142]. Konry et al. used the microfluidic droplet system again to analyze the amount of surface protein CD86 and secreted protein IL-6 in single dendritic cells stimulated with drugs and co-cultured with T cells, demonstrating the feasibility of this method [145]. In addition, the feature that protease can specificly decompose corresponding substrate has also been used to detetct secreted protease by analysis of production (Fig. 7.9b). Antibodies are hard to enter the internal of cells due to the presence of cell membranes, so the detection of internal proteins is more difficult [146]. Now it's mainly implemented by extracting proteins from the cell internal after cell lysis or allowing cells to express fluorescent protein by transfection, and then to detect, except for some specific dyes that can entering the interior of the cell to label protein (Fig. 7.9c) [144]. For example, Huebner et al. have successfully detected droplet size, cell occupancy and fluorescence expression of a single cell through combination of T-junction, bending channel, laser induced fluorescence microscopy and transfected cells [51].

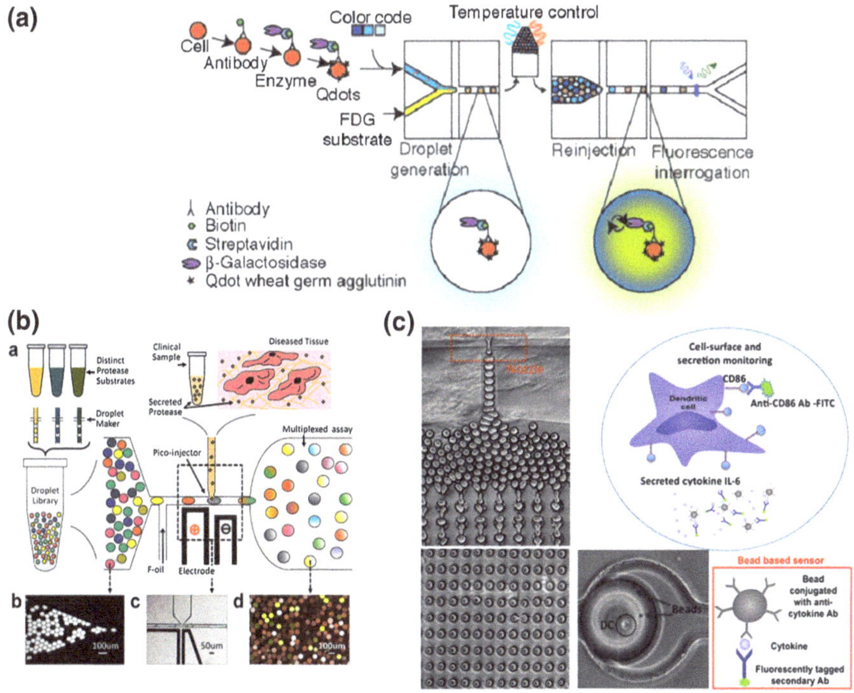

Fig. 7.9 Analysis of single cell proteins in microdroplets. **a** Surface proteins of single cell were analyzed by enzyme—amplified signal method (Reprinted with permission from Ref. [143]). **b** Simultaneous determination of a variety of cell secretory protein activity analysis (Reprinted with permission from Ref. [60]). **c** Cell surface proteins and secretory proteins were simultaneously monitored in single cell droplets (Reprinted with permission from Ref. [144])

7.5.4 Single Cell PCR

PCR, the abbreviation of polymerase chain reaction, is one of the most commonly used nucleic acid amplification means. High-throughput DNA sequencing method has been applied by combination droplet microfluidic technology with PCR in recent years [11]. Droplets provide a small, independent space, and a good environment for controllable and efficient replication reaction, reducing pollution of moving process in the traditional method [140]. In single-cell PCR, diluted cells or samples are dispersed into a large number of droplets, each with only a single or even no target DNA sequence, so a large number of parallel experiments can be carried out simultaneously, thus greatly increasing the throughout [150]. Moreover, template nucleic acid molecules are only located in droplets without fusion into the mobile phase in droplet microfluidic PCR, avoiding nucleic acid recombination with multiple cells in a conventional method, and not generating short, chimeric and other artificial products [151, 152]. These factors make droplet PCR very suitable for single-cell and single-molecule nucleic acid amplification. For example, Zhong et al. used multiple real-time droplet PCR to break the barrier of the traditional PCR, in which a color can only indicate one target [153].

Currently, there are two main approaches available to perform microfluidic PCR in droplet: digital microfluidic PCR and continuous-flow microfluidic PCR. Digital microfluidic PCR is performed by placing reagents and amplification reactions in fixed or semi-fixed droplets for accurate flow control, droplet manipulation, and rapid detection [154, 155]. Pekin et al. screened the mutated genes in a 20,000-fold normal gene and accurate quantified by using digital microfluidic droplet PCR [156]. While the continuous-flow microfluidic PCR is different as droplets are often in continual motion in microfluidic devices (Fig. 7.10c), which makes the operation great challenge and is less use but higher throughout and suitable for continuous amplification and detection [149, 157]. Mohr et al. integrated cell lysis, complementary DNA synthesis and gene amplification into the chip by manipulating the droplets using the magnetic field dynamics after adding cells and magnetic beads into the droplets [158]. Microfluidic PCR in droplet can orderly process a large number of droplets in a short time with a higher throughout. Difficulties in single-cell PCR mainly exist in three processes: cell lysis, DNA extraction, and PCR. The lysate after cell lysis can strongly inhibit the PCR reaction reversely, and it will also interfere with the PCR results if the DNA extraction is incomplete or contains impurities. To solve these problems, Novak et al. used agarose as the dispersed phase to encapsulate single cells and released the DNA from the cells with sodium dodecyl sulfate and proteinase K, in which impurities could be washed away, and finally reacted with the PCR reagents to proceed amplification and analysis (Fig. 7.10a) [147]. Some by-products or undesired reagents can be removed in this way, but it is not suitable for simultaneous cell staining and gene analysis due to the presence of the gel. Eastburn et al. have constructed a microfluidic platform to perform the whole PCR process by diluting the droplets containing the cell lysate with a large water droplet to reduce the inhibitory effect, and then separating parts of the droplets for amplification and analysis after adding the PCR reagents (Fig. 7.10b) [148]. In the PCR process, two types of thermal cycling are often used: on-chip and off-chip. The off-chip thermal cycling is simple, convenient, efficient, and capable of handling a large number of droplets simultaneously, but the off-chip thermal cycling is more convenient for automating [159]. Kim et al. performed droplets PCR reaction with the aid of laser for heating, obtaining good results because the laser can be well controlled to heat the droplets only and produce thermal reactions very quickly [160].

7.6 Cell Manipulation

Though the size of droplet is small, it can be much larger than a single cell, so droplets can be used not only for single-cell analysis, but also for manipulation of multiple cells [161]. When many cells are encapsulated in droplets, the number of cells in the droplet obeys no longer a Poisson distribution, but rather a Gaussian

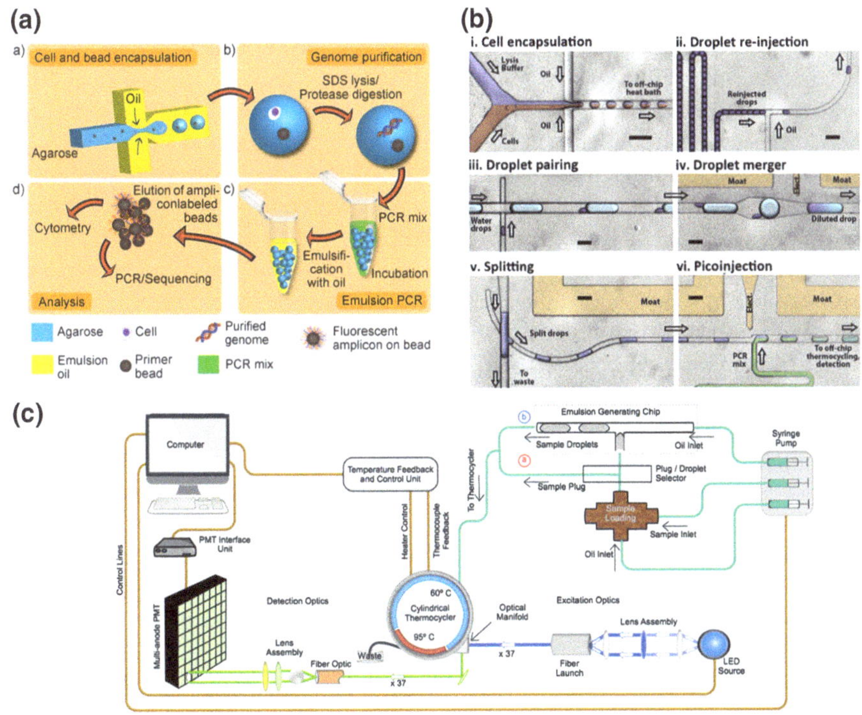

Fig. 7.10 Droplet microfluidic chip for single-cell PCR. **a** Single cell PCR analysis system based on gel droplets (Reprinted with permission from Ref. [147]). **b** A common water-in-oil droplet platform was used for single-cell PCR analysis (Reprinted with permission from Ref. [148]). **c** Architecture of a complete real-time PCR device (Reprinted with permission from Ref. [149])

distribution [39]. For droplet manipulation of cells, it is important not just by droplet generation and sorting, but by a series of other operations: fusion, splitting, and mixing [27, 162]. Some special cell manipulations can be achieved through droplets operations, such as cell three-dimensional (3D) culture and cell co-culture.

Compared with conventional 2D culture system, 3D cell culture more accords with the condition of cells in the organism, which makes people understand the cell differentiation, tissue production and function more deeply and is meaningful to study cells and tissues as important components of body [163]. Cells culture in gel is considered to be the most commonly used method for 3D cell culture as the gel has good biocompatibility, provides good support for cells growth and is convenient for the addition of cytokines [164]. Meanwhile, gel can perform the conversion between the liquid and semi-solid state under certain conditions, which is exploited to carry out micro-droplet 3D culture. Headen and colleagues wrapped human bone marrow mesenchymal stem cells with RGD (Arg-Gly-Asp peptide) functionally modified liquid polyethylene glycol gel, which would become coagulation state under the action of triethanolamine, and then cells microspheres were successfully formed after cultivation (Fig. 7.11a) [165]. The formation of gel droplets by droplet microfluidic technology can not only accurately control the size, shape and dispersibility of the gel, but also precisely regulate the chemical composition of the gel. Now there are two ways for formation of microgel droplets. One is to firstly generate droplets with small molecule monomers, and then polymerize these monomers to form micro-gel (Fig. 7.11b) [166]. Such way is convenient to operate and easy to implement with closely linked two-step operation, but difficult to control the degree of reaction and morphology of droplets. The other way serves prepolymer as disperse phase to form droplets, and then to develop micro-gel via cross-linking reaction by adding cross-linking agent. The difficulties in this approach are that the prepolymer is a non-Newtonian fluid and owns different fluid properties contrast with the mobile phase, while its advantage is getting more controllable microgels [167]. The latter method is more suitable for cell culture, such as commonly used alginate, agarose, carrageenan, gelatin, and so on [168, 169]. For example, Siltanen et al. formed stem cells encapsulated hydrogels in a way of pre-polymerisation and then cross-linking, and have found that the formed stem cell spheres expressed higher content of endoderm markers [170]. Now three kinds of cross-linking methods are commonly used: chemical crosslinking, temperature crosslinking and ion crosslinking. Kumacheva et al. encapsulated cells and formed microgels with an agarose precursor capable of gelatinization by thermal inversion [171]. Liu et al. formed calcium ion droplets and alginate droplets, respectively, and then fused two droplets to produce microgels to achieve cell encapsulation and 3D culture as the property that alginate turns into gelation in presence of calcium ions [166]. The 3D cell culture in droplet provides an effective to study drug screening, cell differentiation and phenotypic change. Du and colleges accomplished drug screening of a large number of cell microspheres formed by microfluidic droplet array just with a small amount of drugs [172].

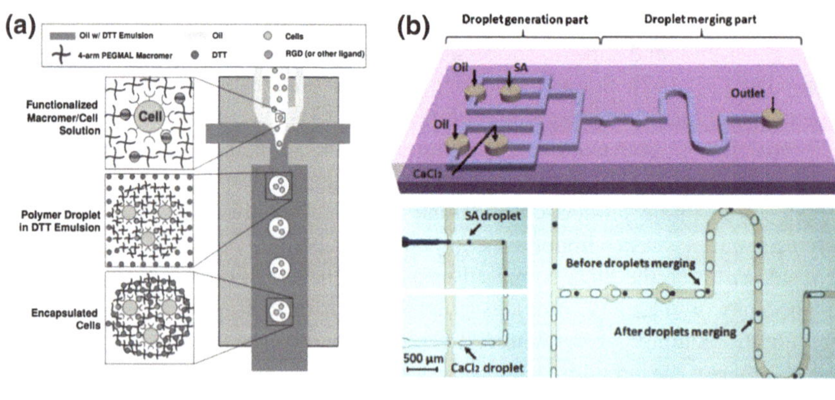

Fig. 7.11 Cells were wrapped in biocompatible gel for 3D cell culture. **a** The microfluidic device was used to generate droplets with macromonomers encapsulating the cells and then were polymerized to form gel [165]. **b** The droplets containing two components of synthetizing gel are formed and the gel was formed by droplet fusion [166]. (Reprinted with permissions from Refs. [165] and [166])

Except for 3D culture, cell co-culture is also an important direction for biological research as the cells in vivo are not isolated, but closely contacts with other cells through directly or indirectly communication. Since the droplets are born in sequence and can be operated with fusion and separation between each other, Tumarkin et al. mixed two cell lines contained in agarose with a Y-channel, and formed droplets through a T-junction, and finally developed into micro-gel by the temperature cross-linking, in order to get the purpose of two cells co-culture [175]. Yu et al. fused two types of droplets pathway containing E. coli and phage, respectively, by applying the same method for the detection of phage content in water after culture for a certain time (Fig. 7.12a). Chen et al. have developed a new core-shell structure consisting of human hepatocellular carcinoma cells and NIH-3T3 cells, respectively, by wrapping the water droplets with gels and analyzed the secretion of albumin and urea in single culture and co-culture (Fig. 7.12b). In addition, Zhang et al. printed arrays of two cell with accurate and controllable interval by using Inkjet to study intercellular drug metabolism (Fig. 7.12c). Liu and colleges applied high-voltage pulse into three metal nozzles which were next to each other, so that the generated droplets got fused together in the process of spray for cell co-culture [176]. In contrast with other methods, cells co-culture in droplet enables co-culture between a few cells, or even single cells and high-throughput study, and provides a more precise control of droplet behavior. Furthermore, cell patterning [177], the study of cell growth [178], and live cell monitoring [179], through union of droplets and nanoparticles are also widely used in droplet microfluidic technology.

7.7 Summary and Prospect

The droplet-based microfluidic technology inherits the advantages of microfluidic technology and further performs some characters of high-throughput, low consumption, and higher specific surface area, so it has been developed rapidly in recent years and has become a hotspot and powerful tool in the fields of chemical and biological research. It mainly depends on T-junction, flow focussing structures or external forces, to produce monodisperse and high-flux droplets with a minimal volume of fL or pL. This process is easy to operate and provides tiny and independent space for single or small number of cells. The generated cell droplets are charactered with a variety of analytical methods, such as fluorescence, mass spectrometry, electrical detection and chemiluminescence, to fulfill aim of study the cell states, phenotypes, metabolism, genes and so on. The single-cell analysis, 3D cell culture and cell co-culture are well controlled by the design of the microfluidic chip structure and some pluripotent operations of fusion, splitting, sorting and mixing, which offers a possible direction in single cell level research and analysis.

For future droplet-based microfluidic technology, we expect that it will allow better integration with various detection methods for real-time assay of cell metabolism and other markers, or incorporation with new biocompatible materials to

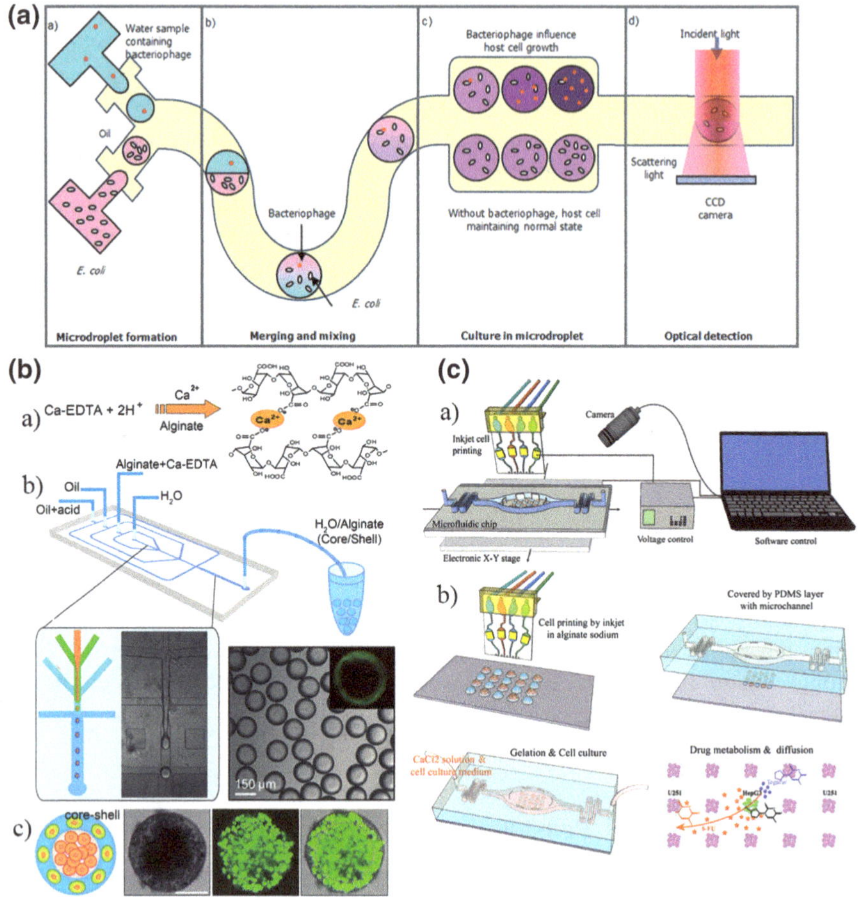

Fig. 7.12 Some common methods for cell co-culture using droplet microfluidic technology. *A* Cell co-culture was performed using droplet fusion. **a** droplet generation; **b** droplet fusion; **c** culture in droplets; **d** optical detection (Reprinted with permission from Ref. [173]). *B* core-shell structure for co-culture. **a** calcium ion-induced alginate crosslinked network; **b** microfluidic chip and partial enlarged view of core-shell structure; **c** hepatocytes located in the nucleus and fibroblasts located in shell (Reprinted with permission from Ref. [42]). *C* Inkjet printing was used to perform cell co-culture in droplet. **a** schematic of complete system; **b** an inkjet print, integrated chip, and drug metabolism process (Reprinted with permission from Ref. [174])

establish new methods and systems for cell research. Since droplets are high-throughput, some unexpected conclusion may be obtained by combination experimental results with big data analysis. One possible field of future droplet microfluidics may be the development of portable devices that apply some simple and sensitive detection techniques such as electrical or chemiluminescent detection for diagnosis of diseases. Even with few applications, active methods of generating droplets can overcome many disadvantages of passive methods in encapsulating cells and will promote progress in study of single cell and cell manipulation if further developed. In addition, efficient and extensive drug screening, development of new isothermal amplification methods and other applications also have broad prospects. We believe that droplet-based microfluidic technology will play a more important role in cell research and analysis, tumor therapy, disease diagnosis in the near future.

References

1. Woodruff K, Maerkl SJ (2016) A high-throughput microfluidic platform for mammalian cell transfection and culturing. Sci Rep 6:23937. doi:10.1038/srep23937
2. Vidi PA, Maleki T, Ochoa M, Wang L, Clark SM, Leary JF, Lelievre SA (2014) Disease-on-a-chip: mimicry of tumor growth in mammary ducts. Lab Chip 14(1):172–177. doi:10.1039/c3lc50819f
3. Huh D, Matthews BD, Mammoto A, Montoya-Zavala M, Hsin HY, Ingber DE (2010) Reconstituting organ-level lung functions on a chip. Science 328(5986):1662–1668. doi:10.1126/science.1188302
4. Chen Q, Wu J, Zhuang Q, Lin X, Zhang J, Lin JM (2013) Microfluidic isolation of highly pure embryonic stem cells using feeder-separated co-culture system. Sci Rep 3:2433. doi:10.1038/srep02433
5. Bergstrom G, Nilsson K, Mandenius CF, Robinson ND (2014) Macroporous microcarriers for introducing cells into a microfluidic chip. Lab Chip 14(18):3502–3504. doi:10.1039/c4lc00693c
6. Deng Y, Zhang Y, Sun S, Wang Z, Wang M, Yu B, Czajkowsky DM, Liu B, Li Y, Wei W, Shi Q (2014) An integrated microfluidic chip system for single-cell secretion profiling of rare circulating tumor cells. Sci Rep 4:7499. doi:10.1038/srep07499
7. Shapiro OH, Kramarsky-Winter E, Gavish AR, Stocker R, Vardi A (2016) A coral-on-a-chip microfluidic platform enabling live-imaging microscopy of reef-building corals. Nat Commun 7:10860. doi:10.1038/ncomms10860
8. Whitesides GM (2006) The origins and the future of microfluidics. Nature 442(7101):368–373. doi:10.1038/nature05058
9. Niu XZ, Zhang B, Marszalek RT, Ces O, Edel JB, Klug DR, deMello AJ (2009) Droplet-based compartmentalization of chemically separated components in two-dimensional separations. Chem Commun 41:6159–6161. doi:10.1039/b918100h
10. Niu X, deMello AJ (2012) Building droplet-based microfluidic systems for biological analysis. Biochem Soc Trans 40(4):615–623. doi:10.1042/BST20120005
11. Song H, Chen DL, Ismagilov RF (2006) Reactions in droplets in microfluidic channels. Angew Chem Int Edit 45(44):7336–7356. doi:10.1002/anie.200601554
12. Kintses B, van Vliet LD, Devenish SR, Hollfelder F (2010) Microfluidic droplets: new integrated workflows for biological experiments. Curr Opin Chem Biol 14(5):548–555. doi:10.1016/j.cbpa.2010.08.013
13. Boybay MS, Jiao A, Glawdel T, Ren CL (2013) Microwave sensing and heating of individual droplets in microfluidic devices. Lab Chip 13(19):3840–3846. doi:10.1039/c3lc50418b
14. Srisa-Art M, Bonzani IC, Williams A, Stevens MM, deMello AJ, Edel JB (2009) Identification of rare progenitor cells from human periosteal tissue using droplet microfluidics. Analyst 134(11):2239–2245. doi:10.1039/b910472k
15. Wang T, Zhu H, Zhuo J, Zhu Z, Papakonstantinou P, Lubarsky G, Lin J, Li M (2013) Biosensor based on ultrasmall MoS_2 nanoparticles for electrochemical detection of H_2O_2

released by cells at the nanomolar level. Anal Chem 85(21):10289–10295. doi:10.1021/ac402114c
16. Schaerli Y, Hollfelder F (2009) The potential of microfluidic water-in-oil droplets in experimental biology. Mol BioSyst 5(12):1392–1404. doi:10.1039/b907578j
17. Baroud CN, de Saint Vincent MR, Delville JP (2007) An optical toolbox for total control of droplet microfluidics. Lab Chip 7(8):1029–1033. doi:10.1039/b702472j
18. Schmid L, Weitz DA, Franke T (2014) Sorting drops and cells with acoustics: acoustic microfluidic fluorescence-activated cell sorter. Lab Chip 14(19):3710–3718. doi:10.1039/c4lc00588k
19. Mazutis L, Gilbert J, Ung WL, Weitz DA, Griffiths AD, Heyman JA (2013) Single-cell analysis and sorting using droplet-based microfluidics. Nat Protoc 8(5):870–891. doi:10.1038/nprot.2013.046
20. Hatch AC, Fisher JS, Tovar AR, Hsieh AT, Lin R, Pentoney SL, Yang DL, Lee AP (2011) 1-Million droplet array with wide-field fluorescence imaging for digital PCR. Lab Chip 11(22):3838–3845. doi:10.1039/c1lc20561g
21. Sabhachandani P, Motwani V, Cohen N, Sarkar S, Torchilin V, Konry T (2016) Generation and functional assessment of 3D multicellular spheroids in droplet based microfluidics platform. Lab Chip 16(3):497–505. doi:10.1039/c5lc01139f
22. Brouzes E, Medkova M, Savenelli N, Marran D, Twardowski M, Hutchison JB, Rothberg JM, Link DR, Perrimon N, Samuels ML (2009) Droplet microfluidic technology for single-cell high-throughput screening. Proc Natl Acad Sci U S A 106(34):14195–14200. doi:10.1073/pnas.0903542106
23. Theberge AB, Courtois F, Schaerli Y, Fischlechner M, Abell C, Hollfelder F, Huck WT (2010) Microdroplets in microfluidics: an evolving platform for discoveries in chemistry and biology. Angew Chem Int Edit 49(34):5846–5868. doi:10.1002/anie.200906653
24. Nossal GJ, Lederberg J (1958) Antibody production by single cells. Nature 181(4620):1419–1420
25. Baroud CN, Gallaire F, Dangla R (2010) Dynamics of microfluidic droplets. Lab Chip 10(16):2032. doi:10.1039/c001191f
26. Shen F, Li Y, Liu Z-M, Cao R-T, Wang G-R (2015) Advances in micro-droplets coalescence using microfluidics. Chin J Anal Chem 43(12):1942–1954. doi:10.1016/s1872-2040(15)60886-6
27. Yang C-G, Xu Z-R, Wang J-H (2010) Manipulation of droplets in microfluidic systems. TrAC Trends Anal Chem 29(2):141–157. doi:10.1016/j.trac.2009.11.002
28. Niu X, Gulati S, Edel JB, deMello AJ (2008) Pillar-induced droplet merging in microfluidic circuits. Lab Chip 8(11):1837–1841. doi:10.1039/b813325e
29. Bayley H, Cronin B, Heron A, Holden MA, Hwang WL, Syeda R, Thompson J, Wallace M (2008) Droplet interface bilayers. Mol BioSyst 4(12):1191–1208. doi:10.1039/b808893d
30. Wu J, Wen W, Sheng P (2012) Smart electroresponsive droplets in microfluidics. Soft Matter 8(46):11589. doi:10.1039/c2sm26286j
31. Movahednejad E, Ommi F, Hosseinalipour SM (2010) Prediction of droplet size and velocity distribution in droplet formation region of liquid spray. Entropy 12(6):1484–1498. doi:10.3390/e12061484
32. Gu H, Duits MH, Mugele F (2011) Droplets formation and merging in two-phase flow microfluidics. Int J Mol Sci 12(4):2572–2597. doi:10.3390/ijms12042572
33. Zhao C-X, Middelberg APJ (2011) Two-phase microfluidic flows. Chem Eng Sci 66(7):1394–1411. doi:10.1016/j.ces.2010.08.038
34. Cao J, Köhler JM (2015) Droplet-based microfluidics for microtoxicological studies. Eng Life Sci 15(3):306–317. doi:10.1002/elsc.201400074
35. Zec H, Shin DJ, Wang TH (2014) Novel droplet platforms for the detection of disease biomarkers. Expert Rev Mol Diag 14(7):787–801. doi:10.1586/14737159.2014.945437
36. Fleury J-B, Schiller UD, Thutupalli S, Gompper G, Seemann R (2014) Mode coupling of phonons in a dense one-dimensional microfluidic crystal. New J Phys 16(6):063029. doi:10.1088/1367-2630/16/6/063029

37. Thorsen T, Roberts RW, Arnold FH, Quake SR (2001) Dynamic pattern formation in a vesicle-generating microfluidic device. Phys Rev Lett 86(18):4163–4166. doi:10.1103/PhysRevLett.86.4163
38. Basova EY, Foret F (2015) Droplet microfluidics in (bio)chemical analysis. Analyst 140 (1):22–38. doi:10.1039/c4an01209g
39. Okushima S, Nisisako T, Torii T, Higuchi T (2004) Controlled production of monodisperse double emulsions by two-step droplet breakup in microfluidic devices. Langmuir 20 (23):9905–9908. doi:10.1021/la0480336
40. Collins DJ, Neild A, deMello A, Liu AQ, Ai Y (2015) The Poisson distribution and beyond: methods for microfluidic droplet production and single cell encapsulation. Lab Chip 15 (17):3439–3459. doi:10.1039/c5lc00614g
41. Park SY, Wu TH, Chen Y, Teitell MA, Chiou PY (2011) High-speed droplet generation on demand driven by pulse laser-induced cavitation. Lab Chip 11(6):1010–1012. doi:10.1039/c0lc00555j
42. Chen Q, Utech S, Chen D, Prodanovic R, Lin JM, Weitz DA (2016) Controlled assembly of heterotypic cells in a core-shell scaffold: organ in a droplet. Lab Chip 16(8):1346–1349. doi:10.1039/c6lc00231e
43. Tan YC, Lee AP (2005) Microfluidic separation of satellite droplets as the basis of a monodispersed micron and submicron emulsification system. Lab Chip 5(10):1178–1183. doi:10.1039/b504497a
44. Song H, Tice JD, Ismagilov RF (2003) A microfluidic system for controlling reaction networks in time. Angew Chem Int Edit 115(7):792–796. doi:10.1002/anie.200390203
45. Adamson DN, Mustafi D, Zhang JX, Zheng B, Ismagilov RF (2006) Production of arrays of chemically distinct nanolitre plugs via repeated splitting in microfluidic devices. Lab Chip 6 (9):1178–1186. doi:10.1039/b604993a
46. Xiao Z, Niu M, Zhang B (2012) Droplet microfluidics based microseparation systems. J Sep Sci 35(10–11):1284–1293. doi:10.1002/jssc.201200115
47. Collins DJ, Alan T, Helmerson K, Neild A (2013) Surface acoustic waves for on-demand production of picoliter droplets and particle encapsulation. Lab Chip 13(16):3225–3231. doi:10.1039/c3lc50372k
48. He M, Kuo JS, Chiu DT (2005) Electro-generation of single femtoliter- and picoliter-volume aqueous droplets in microfluidic systems. Appl Phys Lett 87(3):031916. doi:10.1063/1.1997280
49. Huebner A, Srisa-Art M, Holt D, Abell C, Hollfelder F, deMello AJ, Edel JB (2007) Quantitative detection of protein expression in single cells using droplet microfluidics. Chem Commun 12:1218–1220. doi:10.1039/b618570c
50. Kapanidis AN, Strick T (2009) Biology, one molecule at a time. Trends Biochem Sci 34 (5):234–243. doi:10.1016/j.tibs.2009.01.008
51. Casadevall iSX, Niu X, Leeper K, Cho S, Chang SI, Edel JB, deMello AJ (2011) Fluorescence detection methods for microfluidic droplet platforms. JoVE-J Vis Exp 58. doi:10.3791/3437
52. Chen CH, Sarkar A, Song YA, Miller MA, Kim SJ, Griffith LG, Lauffenburger DA, Han J (2011) Enhancing protease activity assay in droplet-based microfluidics using a biomolecule concentrator. J Am Chem Soc 133(27):10368–10371. doi:10.1021/ja2036628
53. Harris TD, Buzby PR, Babcock H, Beer E, Bowers J, Braslavsky I, Causey M, Colonell J, Dimeo J, Efcavitch JW, Giladi E, Gill J, Healy J, Jarosz M, Lapen D, Moulton K, Quake SR, Steinmann K, Thayer E, Tyurina A, Ward R, Weiss H, Xie Z (2008) Single-molecule DNA sequencing of a viral genome. Science 320(5872):106–109. doi:10.1126/science.1150427
54. Trivedi V, Doshi A, Kurup GK, Ereifej E, Vandevord PJ, Basu AS (2010) A modular approach for the generation, storage, mixing, and detection of droplet libraries for high throughput screening. Lab Chip 10(18):2433–2442. doi:10.1039/c004768f
55. Gu SQ, Zhang YX, Zhu Y, Du WB, Yao B, Fang Q (2011) Multifunctional picoliter droplet manipulation platform and its application in single cell analysis. Anal Chem 83(19):7570–7576. doi:10.1021/ac201678g

56. Huebner A, Bratton D, Whyte G, Yang M, Demello AJ, Abell C, Hollfelder F (2009) Static microdroplet arrays: a microfluidic device for droplet trapping, incubation and release for enzymatic and cell-based assays. Lab Chip 9(5):692–698. doi:10.1039/b813709a
57. Shi W, Qin J, Ye N, Lin B (2008) Droplet-based microfluidic system for individual Caenorhabditis elegans assay. Lab Chip 8(9):1432–1435. doi:10.1039/b808753a
58. Fidalgo LM, Whyte G, Bratton D, Kaminski CF, Abell C, Huck WT (2008) From microdroplets to microfluidics: selective emulsion separation in microfluidic devices. Angew Chem Int Edit 47(11):2042–2045. doi:10.1002/anie.200704903
59. Baret JC, Miller OJ, Taly V, Ryckelynck M, El-Harrak A, Frenz L, Rick C, Samuels ML, Hutchison JB, Agresti JJ, Link DR, Weitz DA, Griffiths AD (2009) Fluorescence-activated droplet sorting (FADS): efficient microfluidic cell sorting based on enzymatic activity. Lab Chip 9(13):1850–1858. doi:10.1039/b902504a
60. Chen CH, Miller MA, Sarkar A, Beste MT, Isaacson KB, Lauffenburger DA, Griffith LG, Han J (2013) Multiplexed protease activity assay for low-volume clinical samples using droplet-based microfluidics and its application to endometriosis. J Am Chem Soc 135 (5):1645–1648. doi:10.1021/ja307866z
61. Khorshidi MA, Rajeswari PK, Wahlby C, Joensson HN, Andersson Svahn H (2014) Automated analysis of dynamic behavior of single cells in picoliter droplets. Lab Chip 14 (5):931–937. doi:10.1039/c3lc51136g
62. Chokkalingam V, Tel J, Wimmers F, Liu X, Semenov S, Thiele J, Figdor CG, Huck WT (2013) Probing cellular heterogeneity in cytokine-secreting immune cells using droplet-based microfluidics. Lab Chip 13(24):4740–4744. doi:10.1039/c3lc50945a
63. Courtois F, Olguin LF, Whyte G et al (2009) Controlling the retention of small molecules in emulsion microdroplets for use in cell-based assays. Anal Chem 81(8):3008–3016. doi:10.1021/ac802658n
64. Farkas DL, Nicolau DV, Leif RC, Song YS, Won YJ, Lee S-H, Kim DY (2014) Fluorescence lifetime imaging of lipids during 3T3-L1 cell differentiation. Langmuir 8947:89471W. doi:10.1117/12.2036160
65. Casadevall iSX, Srisaart M, Demello AJ, Edel JB (2010) Mapping of Fluidic Mixing in Microdroplets with 1 μs Time Resolution Using Fluorescence Lifetime Imaging. Anal Chem 82(9):3950–3956. doi:10.1021/ac100055g
66. Mazutis L, Araghi AF, Miller OJ, Baret JC, Frenz L, Janoshazi A, Taly V, Miller BJ, Hutchison JB, Link D, Griffiths AD, Ryckelynck M (2009) Droplet-based microfluidic systems for high-throughput single DNA molecule isothermal amplification and analysis. Anal Chem 81(12):4813
67. Agresti JJ, Antipov E, Abate AR, Ahn K, Rowat AC, Baret JC, Marquez M, Klibanov AM, Griffiths AD, Weitz DA (2010) Ultrahigh-throughput screening in drop-based microfluidics for directed evolution. Proc Natl Acad Sci U S A 107(9):4004–4009. doi:10.1073/pnas.0910781107
68. Fidalgo LM, Whyte G, Ruotolo BT, Benesch JL, Stengel F, Abell C, Robinson CV, Huck WT (2009) Coupling microdroplet microreactors with mass spectrometry: reading the contents of single droplets online. Angew Chem Int Edit 48(20):3665–3668. doi:10.1002/anie.200806103
69. Wang XL, Zhu Y, Fang Q (2014) Coupling liquid chromatography/mass spectrometry detection with microfluidic droplet array for label-free enzyme inhibition assay. Analyst 139 (1):191–197. doi:10.1039/c3an01917a
70. Chen Q, Wu J, Zhang Y, Lin JM (2012) Qualitative and quantitative analysis of tumor cell metabolism via stable isotope labeling assisted microfluidic chip electrospray ionization mass spectrometry. Anal Chem 84(3):1695–1701. doi:10.1021/ac300003k
71. Sun S, Slaney TR, Kennedy RT (2012) Label free screening of enzyme inhibitors at femtomole scale using segmented flow electrospray ionization mass spectrometry. Anal Chem 84(13):5794–5800. doi:10.1021/ac3011389

72. Kuster SK, Pabst M, Jefimovs K, Zenobi R, Dittrich PS (2014) High-resolution droplet-based fractionation of nano-LC separations onto microarrays for MALDI-MS analysis. Anal Chem 86(10):4848–4855. doi:10.1021/ac4041982
73. Luo C, Ma Y, Li H, Chen F, Uchiyama K, Lin JM (2013) Generation of picoliter droplets of liquid for electrospray ionization with piezoelectric inkjet. J Mass Spectrom 48(3):321–328. doi:10.1002/jms.3159
74. Gasilova N, Yu Q, Qiao L, Girault HH (2014) On-chip spyhole mass spectrometry for droplet-based microfluidics. Angew Chem Int Edit 53(17):4408–4412. doi:10.1002/anie.201310795
75. Chen F, Lin L, Zhang J, He Z, Uchiyama K, Lin JM (2016) Single-cell analysis using drop-on-demand inkjet printing and probe electrospray ionization mass spectrometry. Anal Chem 88(8):4354–4360. doi:10.1021/acs.analchem.5b04749
76. Liu W, Lin JM (2016) Online monitoring of lactate efflux by multi-channel microfluidic chip-mass spectrometry for rapid drug evaluation. ACS Sens 1:344–347. doi:10.1021/acssensors.5b00221
77. Kelly RT, Page JS, Marginean I, Tang K, Smith RD (2009) Dilution-free analysis from picoliter droplets by nano-electrospray ionization mass spectrometry. Angew Chem Int Edit 48(37):6832–6835. doi:10.1002/anie.200902501
78. Smith CA, Li X, Mize TH, Sharpe TD, Graziani EI, Abell C, Huck WT (2013) Sensitive, high throughput detection of proteins in individual, surfactant-stabilized picoliter droplets using nanoelectrospray ionization mass spectrometry. Anal Chem 85(8):3812–3816. doi:10.1021/ac400453t
79. Su Y, Zhu Y, Fang Q (2013) A multifunctional microfluidic droplet-array chip for analysis by electrospray ionization mass spectrometry. Lab Chip 13(10):1876–1882. doi:10.1039/c3lc00063j
80. Kuster SK, Fagerer SR, Verboket PE, Eyer K, Jefimovs K, Zenobi R, Dittrich PS (2013) Interfacing droplet microfluidics with matrix-assisted laser desorption/ionization mass spectrometry: label-free content analysis of single droplets. Anal Chem 85(3):1285–1289. doi:10.1021/ac3033189
81. Verboket PE, Borovinskaya O, Meyer N, Gunther D, Dittrich PS (2014) A new microfluidics-based droplet dispenser for ICPMS. Anal Chem 86(12):6012–6018. doi:10.1021/ac501149a
82. Lee JK, Jansson ET, Nam HG, Zare RN (2016) High-resolution live-cell imaging and analysis by laser desorption/ionization droplet delivery mass spectrometry. Anal Chem 88(10):5453–5461. doi:10.1021/acs.analchem.6b00881
83. Zhang XC, Wei ZW, Gong XY, Si XY, Zhao YY, Yang CD, Zhang SC, Zhang XR (2016) Integrated droplet-based microextraction with ESI-MS for removal of matrix interference in single-cell analysis. Sci Rep 6:24730. doi:10.1038/srep24730
84. Zhu Y, Fang Q (2013) Analytical detection techniques for droplet microfluidics—a review. Anal Chim Acta 787:24–35. doi:10.1016/j.aca.2013.04.064
85. Moiseeva EV, Fletcher AA, Harnett CK (2011) Thin-film electrode based droplet detection for microfluidic systems. Sens Actuat B-Ch 155(1):408–414. doi:10.1016/j.snb.2010.11.028
86. Cahill BP, Land R, Nacke T, Min M, Beckmann D (2011) Contactless sensing of the conductivity of aqueous droplets in segmented flow. Sens Actuat B-Ch 159(1):286–293. doi:10.1016/j.snb.2011.07.006
87. Luo C, Yang X, Fu Q, Sun M, Ouyang Q, Chen Y, Ji H (2006) Picoliter-volume aqueous droplets in oil: electrochemical detection and yeast cell electroporation. Electrophoresis 27(10):1977–1983. doi:10.1002/elps.200500665
88. Liu S, Gu Y, Le Roux RB, Matthews SM, Bratton D, Yunus K, Fisher AC, Huck WT (2008) The electrochemical detection of droplets in microfluidic devices. Lab Chip 8(11):1937–1942. doi:10.1039/b809744e
89. Elbuken C, Glawdel T, Chan D, Ren CL (2011) Detection of microdroplet size and speed using capacitive sensors. Sens Actuat A-Ph 171(2):55–62. doi:10.1016/j.sna.2011.07.007

90. Edgar JS, Pabbati CP, Lorenz RM et al (2006) Capillary electrophoresis separation in the presence of an immiscible boundary for droplet analysis. Anal Chem 78(19):6948. doi:10.1021/ac0613131
91. Wang M, Roman GT, Perry ML et al (2009) Microfluidic chip for high efficiency electrophoretic analysis of segmented flow from a microdialysis probe and in vivo chemical monitoring. Anal Chem 81(21):9072. doi:10.1021/ac901731v
92. Cecchini MP, Hong J, Lim C, Choo J, Albrecht T, Demello AJ, Edel JB (2011) Ultrafast surface enhanced resonance Raman scattering detection in droplet-based microfluidic systems. Anal Chem 83(8):3076–3081. doi:10.1021/ac103329b
93. Shen H, Fang Q, Fang ZL (2006) A microfluidic chip based sequential injection system with trapped droplet liquid-liquid extraction and chemiluminescence detection. Lab Chip 6 (10):1387–1389. doi:10.1039/b605332g
94. Buettner F, Natarajan KN, Casale FP, Proserpio V, Scialdone A, Theis FJ, Teichmann SA, Marioni JC, Stegle O (2015) Computational analysis of cell-to-cell heterogeneity in single-cell RNA-sequencing data reveals hidden subpopulations of cells. Nat Biotechnol 33 (2):155–160. doi:10.1038/nbt.3102
95. Schubert C (2011) Single-cell analysis: The deepest differences. Nature 480(7375):133. doi:10.1038/480133a
96. Hsu PP, Sabatini DM (2008) Cancer cell metabolism: Warburg and beyond. Cell 134 (5):703–707. doi:10.1016/j.cell.2008.08.021
97. Lim B, Reddy V, Hu X, Kim K, Jadhav M, Abedini-Nassab R, Noh YW, Lim YT, Yellen BB, Kim C (2014) Magnetophoretic circuits for digital control of single particles and cells. Nat Commun 5:3846. doi:10.1038/ncomms4846
98. Spiller DG, Wood CD, Rand DA, White MR (2010) Measurement of single-cell dynamics. Nature 465(7299):736–745. doi:10.1038/nature09232
99. Kang A, Park J, Ju J, Jeong GS, Lee SH (2014) Cell encapsulation via microtechnologies. Biomaterials 35(9):2651–2663. doi:10.1016/j.biomaterials.2013.12.073
100. Li JL, Day D, Gu M (2010) Design of a compact microfludic device for controllable cell distribution. Lab Chip 10(22):3054–3057. doi:10.1039/c0lc00090f
101. Moon D, Im DJ, Lee S, Kang IS (2014) A novel approach for drop-on-demand and particle encapsulation based on liquid bridge breakup. Exp Therm Fluid Sci 53:251–258. doi:10.1016/j.expthermflusci.2013.12.016
102. Kemna EW, Schoeman RM, Wolbers F, Vermes I, Weitz DA, van den Berg A (2012) High-yield cell ordering and deterministic cell-in-droplet encapsulation using Dean flow in a curved microchannel. Lab Chip 12(16):2881–2887. doi:10.1039/c2lc00013j
103. Hong J, deMello AJ, Jayasinghe SN (2010) Bio-electrospraying and droplet-based microfluidics: control of cell numbers within living residues. Biomed Mater 5(2):21001. doi:10.1088/1748-6041/5/2/021001
104. Collins DJ, Alan T, Neild A (2014) The particle valve: on-demand particle trapping, filtering, and release from a microfabricated polydimethylsiloxane membrane using surface acoustic waves. Appl Phys Lett 105(3):033509. doi:10.1063/1.4891424
105. Amini H, Lee W, Di Carlo D (2014) Inertial microfluidic physics. Lab Chip 14(15):2739–2761. doi:10.1039/c4lc00128a
106. Lagus TP, Edd JF (2013) A review of the theory, methods and recent applications of high-throughput single-cell droplet microfluidics. J Phys D Appl Phys 46(11):114005. doi:10.1088/0022-3727/46/11/114005
107. Morimoto Y, Tan WH, Tsuda Y, Takeuchi S (2009) Monodisperse semi-permeable microcapsules for continuous observation of cells. Lab Chip 9(15):2217–2223. doi:10.1039/b900035f
108. Edd JF, Di CD, Humphry KJ et al (2008) Controlled encapsulation of single-cells into monodisperse picolitre drops. Lab Chip 8(8):1262. doi:10.1039/B805456H
109. Brouzes E, Kruse T, Kimmerling R, Strey HH (2015) Rapid and continuous magnetic separation in droplet microfluidic devices. Lab Chip 15(3):908–919. doi:10.1039/c4lc01327a

110. Morimoto Y, Tan WH, Takeuchi S (2009) Three-dimensional axisymmetric flow-focusing device using stereolithography. Biomed Microdevic 11(2):369–377. doi:10.1007/s10544-008-9243-y
111. Verbruggen B, Tóth T, Cornaglia M, Puers R, Gijs MAM, Lammertyn J (2014) Separation of magnetic microparticles in segmented flow using asymmetric splitting regimes. Microfluid Nanofluid 18(1):91–102. doi:10.1007/s10404-014-1409-8
112. He M, Edgar JS, Jeffries GDM et al (2009) Selective encapsulation of single cells and subcellular organelles into picoliter- and femtoliter-volume droplets. Anal Chem 77(6):1539. doi:10.1021/ac0480850
113. Schoendube J, Wright D, Zengerle R, Koltay P (2015) Single-cell printing based on impedance detection. Biomicrofluidics 9(1):014117. doi:10.1063/1.4907896
114. Um E, Rha E, Choi SL, Lee SG, Park JK (2012) Mesh-integrated microdroplet array for simultaneous merging and storage of single-cell droplets. Lab Chip 12(9):1594. doi:10.1039/C2LC21266H
115. Wu TH, Chen Y, Park SY, Hong J, Teslaa T, Zhong JF, Di Carlo D, Teitell MA, Chiou PY (2012) Pulsed laser triggered high speed microfluidic fluorescence activated cell sorter. Lab Chip 12(7):1378–1383. doi:10.1039/c2lc21084c
116. Jing T, Ramji R, Warkiani ME, Han J, Lim CT, Chen CH (2015) Jetting microfluidics with size-sorting capability for single-cell protease detection. Biosens Bioelectron 66:19–23. doi:10.1016/j.bios.2014.11.001
117. Nam J, Lim H, Kim C, Yoon Kang J, Shin S (2012) Density-dependent separation of encapsulated cells in a microfluidic channel by using a standing surface acoustic wave. Biomicrofluidics 6(2):24120–2412010. doi:10.1063/1.4718719
118. Chen A, Byvank T, Chang WJ, Bharde A, Vieira G, Miller BL, Chalmers JJ, Bashir R, Sooryakumar R (2013) On-chip magnetic separation and encapsulation of cells in droplets. Lab Chip 13(6):1172–1181. doi:10.1039/c2lc41201b
119. McGrath J, Jimenez M, Bridle H (2014) Deterministic lateral displacement for particle separation: a review. Lab Chip 14(21):4139–4158. doi:10.1039/c4lc00939h
120. Tan Y-C, Ho YL, Lee AP (2007) Microfluidic sorting of droplets by size. Microfluid Nanofluid 4(4):343–348. doi:10.1007/s10404-007-0184-1
121. Joensson HN, Uhlen M, Svahn HA (2011) Droplet size based separation by deterministic lateral displacement-separating droplets by cell–induced shrinking. Lab Chip 11(7):1305–1310. doi:10.1039/c0lc00688b
122. Chabert M, Viovy JL (2008) Microfluidic high-throughput encapsulation and hydrodynamic self-sorting of single cells. Proc Natl Acad Sci U S A 105(9):3191–3196. doi:10.1073/pnas.0708321105
123. Landenberger B, Hofemann H, Wadle S, Rohrbach A (2012) Microfluidic sorting of arbitrary cells with dynamic optical tweezers. Lab Chip 12(17):3177–3183. doi:10.1039/c2lc21099a
124. Shim JU, Ranasinghe RT, Smith CA et al (2013) Ultrarapid generation of femtoliter microfluidic droplets for single-molecule-counting immunoassays. ACS Nano 7(7):5955–5964. doi:10.1021/nn401661d
125. Levison HF, Morbidelli A (2003) The formation of the Kuiper belt by the outward transport of bodies during Neptune's migration. Nature 426(6965):419–421. doi:10.1038/nature02120
126. Chen Y, Wu TH, Kung YC, Teitell MA, Chiou PY (2013) 3D pulsed laser-triggered high-speed microfluidic fluorescence-activated cell sorter. Analyst 138(24):7308–7315. doi:10.1039/c3an01266b
127. Huang LR, Edward CC, Robert HA, James CS (2004) Continuous particle separation through deterministic lateral displacement. Science 14(304):987–990. doi:10.1126/science.1094567
128. Wu L, Chen P, Dong Y, Feng X, Liu BF (2013) Encapsulation of single cells on a microfluidic device integrating droplet generation with fluorescence-activated droplet sorting. Biomed Microdevic 15(3):553–560. doi:10.1007/s10544-013-9754-z

129. Joensson HN, Samuels ML, Brouzes ER et al (2009) Detection and analysis of low-abundance cell-surface biomarkers using enzymatic amplification in microfluidic droplets. Angew Chem Int Edit 48(14):2518–2521. doi:10.1002/anie.200804326
130. Link DR, Grasland-Mongrain E, Duri A, Sarrazin F, Cheng Z, Cristobal G, Marquez M, Weitz DA (2006) Electric control of droplets in microfluidic devices. Angew Chem Int Edit 45(16):2556–2560. doi:10.1002/anie.200503540
131. Franke T, Braunmuller S, Schmid L, Wixforth A, Weitz DA (2010) Surface acoustic wave actuated cell sorting (SAWACS). Lab Chip 10(6):789–794. doi:10.1039/b915522h
132. Eastburn DJ, Sciambi A, Abate AR (2014) Identification and genetic analysis of cancer cells with PCR-activated cell sorting. Nucleic Acids Res 42(16):e128. doi:10.1093/nar/gku606
133. Kemna EW, Segerink LI, Wolbers F et al (2013) Label-free, high-throughput, electrical detection of cells in droplets. Analyst 138(16):4585–4592. doi:10.1039/C3AN00569K
134. Yu D, Gustafson WC, Han C, Lafaye C, Noirclerc-Savoye M, Ge WP, Thayer DA, Huang H, Kornberg TB, Royant A, Jan LY, Jan YN, Weiss WA, Shu X (2014) An improved monomeric infrared fluorescent protein for neuronal and tumour brain imaging. Nat Commun 5:3626. doi:10.1038/ncomms4626
135. Cesinaro AM, Natoli C, Grassadonia A, Tinari N, Iacobelli S, Trentini GP (2002) Expression of the 90 K tumor-associated protein in benign and malignant melanocytic lesions. J Invest Dermatol NLM 119(1):187–190. doi:10.1046/j.1523-1747.2002.17642.x
136. Ramdzan YM, Polling S, Chia CP, Ng IH, Ormsby AR, Croft NP, Purcell AW, Bogoyevitch MA, Ng DC, Gleeson PA, Hatters DM (2012) Tracking protein aggregation and mislocalization in cells with flow cytometry. Nat Methods 9(5):467–470. doi:10.1038/nmeth.1930
137. Franci C, Takkunen M, Dave N, Alameda F, Gomez S, Rodriguez R, Escriva M, Montserrat-Sentis B, Baro T, Garrido M, Bonilla F, Virtanen I, Garcia de Herreros A (2006) Expression of Snail protein in tumor-stroma interface. Oncogene 25(37):5134–5144. doi:10.1038/sj.onc.1209519
138. Guo MT, Rotem A, Heyman JA, Weitz DA (2012) Droplet microfluidics for high-throughput biological assays. Lab Chip 12(12):2146–2155. doi:10.1039/c2lc21147e
139. Kang D-K, Monsur Ali M, Zhang K, Pone EJ, Zhao W (2014) Droplet microfluidics for single-molecule and single-cell analysis in cancer research, diagnosis and therapy. TrAC-Trend Anal Chem 58:145–153. doi:10.1016/j.trac.2014.03.006
140. Hummer D, Kurth F, Naredi-Rainer N, Dittrich PS (2016) Single cells in confined volumes: microchambers and microdroplets. Lab Chip 16(3):447–458. doi:10.1039/c5lc01314c
141. Konry T, Smolina I, Yarmush JM, Irimia D, Yarmush ML (2011) Ultrasensitive detection of low-abundance surface-marker protein using isothermal rolling circle amplification in a microfluidic nanoliter platform. Small 7(3):395–400. doi:10.1002/smll.201001620
142. Easley CJ, Rocheleau JV, Head WS, Piston DW (2009) Quantitative measurement of zinc secretion from pancreatic islets with high temporal resolution using droplet-based microfluidics. Anal Chem 81(21):9086–9095. doi:10.1021/ac9017692
143. Joensson HN, Samuels ML, Brouzes ER, Medkova M, Uhlen M, Link DR, Andersson-Svahn H (2009) Detection and analysis of low-abundance cell-surface biomarkers using enzymatic amplification in microfluidic droplets. Angew Chem Int Edit 48(14):2518–2521. doi:10.1002/anie.200804326
144. Konry T, Golberg A, Yarmush M (2013) Live single cell functional phenotyping in droplet nano-liter reactors. Sci Rep 3:3179. doi:10.1038/srep03179
145. Konry T, Dominguez-Villar M, Baecher-Allan C, Hafler DA, Yarmush ML (2011) Droplet-based microfluidic platforms for single T cell secretion analysis of IL-10 cytokine. Biosens Bioelectron 26(5):2707–2710. doi:10.1016/j.bios.2010.09.006
146. Wu N, Zhu Y, Brown S, Oakeshott J, Peat TS, Surjadi R, Easton C, Leech PW, Sexton BA (2009) A PMMA microfluidic droplet platform for in vitro protein expression using crude E. coli S30 extract. Lab Chip 9(23):3391–3398. doi:10.1039/b911581a

147. Novak R, Zeng Y, Shuga J, Venugopalan G, Fletcher DA, Smith MT, Mathies RA (2011) Single-cell multiplex gene detection and sequencing with microfluidically generated agarose emulsions. Angew Chem Int Edit 50(2):390–395. doi:10.1002/anie.201006089
148. Eastburn DJ, Sciambi A, Abate AR (2013) Ultrahigh-throughput mammalian single-cell reverse-transcriptase polymerase chain reaction in microfluidic drops. Anal Chem 85 (16):8016–8021. doi:10.1021/ac402057q
149. Hatch AC, Ray T, Lintecum K, Youngbull C (2014) Continuous flow real-time PCR device using multi-channel fluorescence excitation and detection. Lab Chip 14(3):562–568. doi:10.1039/c3lc51236c
150. Zhang Y, Ozdemir P (2009) Microfluidic DNA amplification—a review. Anal Chim Acta 638(2):115–125. doi:10.1016/j.aca.2009.02.038
151. Zhang Y, Jiang HR (2016) A review on continuous-flow microfluidic PCR in droplets: advances, challenges and future. Anal Chim Acta 914:7–16. doi:10.1016/j.aca.2016.02.006
152. Curcio M, Roeraade J (2003) Continuous segmented-flow polymerase chain reaction for high-throughput miniaturized DNA amplification. Anal Chem 75(1):1–7. doi:10.1021/ac0204146
153. Zhong Q, Bhattacharya S, Kotsopoulos S, Olson J, Taly V, Griffiths AD, Link DR, Larson JW (2011) Multiplex digital PCR: breaking the one target per color barrier of quantitative PCR. Lab Chip 11(13):2167–2174. doi:10.1039/c1lc20126c
154. Wang P, Jing F, Li G, Wu Z, Cheng Z, Zhang J, Zhang H, Jia C, Jin Q, Mao H, Zhao J (2015) Absolute quantification of lung cancer related microRNA by droplet digital PCR. Biosens Bioelectron 74:836–842. doi:10.1016/j.bios.2015.07.048
155. Leman M, Abouakil F, Griffiths AD, Tabeling P (2015) Droplet-based microfluidics at the femtolitre scale. Lab Chip 15(3):753–765. doi:10.1039/c4lc01122h
156. Pekin D, Skhiri Y, Baret JC, Le Corre D, Mazutis L, Salem CB, Millot F, El Harrak A, Hutchison JB, Larson JW, Link DR, Laurent-Puig P, Griffiths AD, Taly V (2011) Quantitative and sensitive detection of rare mutations using droplet-based microfluidics. Lab Chip 11(13):2156–2166. doi:10.1039/c1lc20128j
157. Hartung R, Brosing A, Sczcepankiewicz G, Liebert U, Hafner N, Durst M, Felbel J, Lassner D, Kohler JM (2009) Application of an asymmetric helical tube reactor for fast identification of gene transcripts of pathogenic viruses by micro flow-through PCR. Biomed Microdevic 11(3):685–692. doi:10.1007/s10544-008-9280-6
158. Mohr S, Zhang YH, Macaskill A, Day PJR, Barber RW, Goddard NJ, Emerson DR, Fielden PR (2007) Numerical and experimental study of a droplet-based PCR chip. Microfluid Nanofluid 3(5):611–621. doi:10.1007/s10404-007-0153-8
159. Leng X, Zhang W, Wang C, Cui L, Yang CJ (2010) Agarose droplet microfluidics for highly parallel and efficient single molecule emulsion PCR. Lab Chip 10(21):2841–2843. doi:10.1039/c0lc00145g
160. Kim H, Dixit S, Green CJ et al (2009) Nanodroplet real-time PCR system with laser assisted heating. Opt Express 17(1):218–227. doi:10.1364/OE.17.000218
161. Yun H, Kim K, Lee WG (2013) Cell manipulation in microfluidics. Biofabrication 5 (2):022001. doi:10.1088/1758-5082/5/2/022001
162. Ding Y, Qiu F, Casadevall iSX, Chiu F, Nelson B, deMello A (2016) Microfluidic-based droplet and cell manipulations using artificial bacterial flagella. Micromachines 7(2):25. doi:10.3390/mi7020025
163. Huh D, Hamilton GA, Ingber DE (2011) From 3D cell culture to organs-on-chips. Trends Cell Biol 21(12):745–754. doi:10.1016/j.tcb.2011.09.005
164. Koehler KR, Mikosz AM, Molosh AI, Patel D, Hashino E (2013) Generation of inner ear sensory epithelia from pluripotent stem cells in 3D culture. Nature 500(7461):217–221. doi:10.1038/nature12298
165. Headen DM, Aubry G, Lu H, Garcia AJ (2014) Microfluidic-based generation of size-controlled, biofunctionalized synthetic polymer microgels for cell encapsulation. Adv Mater 26(19):3003–3008. doi:10.1002/adma.201304880

166. Liu K, Deng Y, Zhang N, Li S, Ding H, Guo F, Liu W, Guo S, Zhao X-Z (2012) Generation of disk-like hydrogel beads for cell encapsulation and manipulation using a droplet-based microfluidic device. Microfluid Nanofluid 13(5):761–767. doi:10.1007/s10404-012-0998-3
167. Seiffert S (2011) Functional microgels tailored by droplet-based microfluidics. Macromol Rapid Commun 32(20):1600–1609. doi:10.1002/marc.201100342
168. Sakakihara S, Araki S, Iino R, Noji H (2010) A single-molecule enzymatic assay in a directly accessible femtoliter droplet array. Lab Chip 10(24):3355–3362. doi:10.1039/c0lc00062k
169. Tumarkin E, Kumacheva E (2009) Microfluidic generation of microgels from synthetic and natural polymers. Chem Soc Rev 38(8):2161–2168. doi:10.1039/b809915b
170. Siltanen C, Yaghoobi M, Haque A, You J, Lowen J, Soleimani M, Revzin A (2016) Microfluidic fabrication of bioactive microgels for rapid formation and enhanced differentiation of stem cell spheroids. Acta Biomater 34:125–132. doi:10.1016/j.actbio.2016.01.012
171. Kumachev A, Greener J, Tumarkin E, Eiser E, Zandstra PW, Kumacheva E (2011) High-throughput generation of hydrogel microbeads with varying elasticity for cell encapsulation. Biomaterials 32(6):1477–1483. doi:10.1016/j.biomaterials.2010.10.033
172. Du GS, Pan JZ, Zhao SP, Zhu Y, den Toonder JM, Fang Q (2013) Cell-based drug combination screening with a microfluidic droplet array system. Anal Chem 85(14):6740–6747. doi:10.1021/ac400688f
173. Yu JQ, Huang W, Chin LK, Lei L, Lin ZP, Ser W, Chen H, Ayi TC, Yap PH, Chen CH, Liu AQ (2014) Droplet optofluidic imaging for lambda-bacteriophage detection via co-culture with host cell Escherichia coli. Lab Chip 14(18):3519–3524. doi:10.1039/c4lc00042k
174. Zhang J, Chen F, He Z, Ma Y, Uchiyama K, Lin JM (2016) A novel approach for precisely controlled multiple cell patterning in microfluidic chips by inkjet printing and the detection of drug metabolism and diffusion. Analyst 141(10):2940–2947. doi:10.1039/c6an00395h
175. Tumarkin E, Tzadu L, Csaszar E, Seo M, Zhang H, Lee A, Peerani R, Purpura K, Zandstra PW, Kumacheva E (2011) High-throughput combinatorial cell co-culture using microfluidics. Integr Biol 3(6):653–662. doi:10.1039/c1ib00002k
176. Liu Z, Shum HC (2013) Fabrication of uniform multi-compartment particles using microfludic electrospray technology for cell co-culture studies. Biomicrofluidics 7(4):44117. doi:10.1063/1.4817769
177. Peng Lee C, Hsin Chen Y, Hang Wei Z (2013) Fabrication of hexagonally packed cell culture substrates using droplet formation in a T-shaped microfluidic junction. Biomicrofluidics 7(1):14101. doi:10.1063/1.4774315
178. Chen H, Sun J, Wolvetang E, Cooper-White J (2015) High-throughput, deterministic single cell trapping and long-term clonal cell culture in microfluidic devices. Lab Chip 15(4):1072–1083. doi:10.1039/c4lc01176g
179. Taylor RJ, Falconnet D, Niemisto A, Ramsey SA, Prinz S, Shmulevich I, Galitski T, Hansen CL (2009) Dynamic analysis of MAPK signaling using a high-throughput microfluidic single-cell imaging platform. Proc Natl Acad Sci U S A 106(10):3758–3763. doi:10.1073/pnas.0813416106

Chapter 8
Single Cell Analysis on Microfluidic

Qiushi Huang and Jin-Ming Lin

Abstract Under the background of the limitation of the world's resources and to saving cost in each field, microfluidic devices was born in chemical biology analysis. As technique developed, the target of microfluidic devices changes from cell population into single cell analysis gradually. In this chapter, the works of single cell analysis on microfluidic are reviewed by steps of tissue dissociation, cell sorting, single cell isolation, single cell lysis and single cell analysis. Several pioneering works are listed to show the prospect in this study field.

Keyword Microfluidics · Single cell analysis · Tissue dissociation
Cell sorting · Single cell isolation · Single cell lysis

8.1 Introduction

8.1.1 Lab on a Chip

In recent years, the increasing of world's population leads to the rise of life, food, clothing, medical care and other costs. Due to the limitation of the world's resources, we must save costs in each aspect. Under this background, a new type of laboratory equipment—lab on a chip, arises at this historic moment. The birth of the lab on a chip helps to reduce the cost of the analysis in the medical test [1].

Lab on a chip also called nano/micro fluidic system is a tiny sized chip with multi-function feature. Most of these equipments have the characteristics of disposable use, low production cost, portable operation. They can also save time and cost of experiments, as well as reducing the consumption of sample in test [2].

Lab on a chip is usually a part of a series of microelectronic mechanical system, which is called "micro total analysis system" (μTAS). The device is usually applied

Q. Huang · J.-M. Lin (✉)
Department of Chemistry, Tsinghua University, Beijing 100084,
People's Republic of China
e-mail: jmlin@mail.tsinghua.edu.cn

J.-M. Lin (ed.), *Cell Analysis on Microfluidics*, Integrated Analytical Systems,
https://doi.org/10.1007/978-981-10-5394-8_8

in medical diagnosis, such as detection of infectious diseases, large molecules and biological molecular analysis, whose advantage is to reduce the overall testing costs.

8.1.2 Single Cell Analysis and Microfluidic

Most eukaryotic cells come to be a single individual life. In the process of evolution, they are able to cooperate closely and are associated with each other to form tissues, organs and even the whole individual life. Single cell analysis, however, usually only focuses on individual cells in the tissues or organs [3]. Even under the condition of the same culture, homologous cells show differences at the single-cell level with morphology and gene expression level, as well as the differences in growth characteristics [4–8].

In addition, single cell analysis plays an important role in explaining the mechanism of cellular life process. When we study single cells, trying to understand how they play a role in biological systems, we can see the development of living systems, the growth of life or specialization and the evolution of the single cells and the whole organisms. If we can understand life phenomena at the molecular level, we may be able to understand the nuances of cells induced biological phenomena, such as learning and memory or understanding tiny changes in cells caused by diseases or the dysfunction of life, such as cancer.

In order to develop new treatment interventions aimed at a cellular level, understanding the biological heterogeneity and transport channel becomes particularly important. In these studies, rare but important cell types will be applied, such as stem cells or progenitor cells. Therefore, a wide and varied single-cell analysis research methods are developed to analyze their morphological and physiological characteristics, and to obtain a quantitative biological molecules, gene, transcription, and chemical information, even in a simulated condition of a body monitoring dynamic changes of the cells. At the same time, in the organisms with complete functions, a single cell can be marked and tracked, finally be analyzed in cells with no interference of the environment.

At present, the challenge of the single cell analysis is that the size of a single cell is too small to deal with. However, in order to obtain sufficient and effective data statistics, the experiment is presented in high flux. At the same time the cost also need to be decreased to allow more laboratory which can be economically affordable [3]. Single cell analysis comes out to be molecular level measurement of transcription, translation, regulation and internal signals under the condition of single cell. The goal is to analyze from a single cell level to comprehensively understand cell population [9]. This method allows researchers to uncover the relationship between the individual cells and tissues, organs and even living organisms. Single cell analysis develops rapidly in the past ten years, a large number of related research works have been published [3, 10, 11].

Under this background, due to the study of single cell heterogeneity and randomness, researches have changed from analysis of a large number of cells to the single cell analysis [12, 13]. These cell populations' random variation may be due to the influence of the inner noise. These effects include the differences between results of transcription, translation of different cells, such as ion, mRNA and protein. These components show differences by different factors such as reaction rate and the random molecular collision.

Considering the flexibility and dynamic characteristics of the cell membrane, reaction and molecular collisions occur randomly. Therefore, it is not reasonable assuming that all cells in a cell population in a certain moment are exactly the same. Only large number of experiments in single cell analysis can reveal the heterogeneity and provide statistical data to establish the model. In the process of research and technology development, the scale of sample pretreatment in analysis has been decreased, which makes the sample pretreatment easier and more convenient. Therefore, microfluidic systems have been developed and applied to a wide range of single cell analysis [14, 15]. Since the remarkable microfluidic instrumental role played in the single cell analysis, we can expect that the sustainable development of microfluidic analysis techniques will promote the development of single cell analysis.

8.2 Tissue Dissociation

8.2.1 Conventional Approaches of Tissue Dissociation

The first step in single cell analysis is to obtain the cells. In order to make the data from the single cell analysis representative, the acquired cells and the environment of their sources are very important. Normally, cells obtained by biopsy are well representative of their native microenvironment. In order to obtain the suspended cells from the excised living tissue, the three-dimensional structure of the extracellular matrix and the interconnected cells must be disrupted. The traditional way is utilizing enzymes to deal with intact biological tissue, such as the use of collagenase digestion of extracellular matrix proteins. When the intercellular adhesive junction is destroyed, the intact biological tissue is dissociated into a cell suspension via slow mechanical agitation [16–18].

A target of tissue dissociation is to obtain as many viable cells as possible for the subsequent experimental processing preparation. Several parameters may affect the results of tissue dissociation, such as animal species, tissue type, enzyme type, chemical concentration, treatment time, temperature, agitation method, and etc. These controllable parameters are empirically determined by trial and error in order to maximize cell yield and viability.

Although the development of a general scheme for tissue dissociation is a difficult task, there are still a number of research literatures dealing with the dissociation of various mammalian tissues, Such as rat brain [19], heart and intestinal [20].

Tissue dissociation is a very important step for cell analysis. This is because once the problem arises in the pre-processing time, the corresponding error will have a negative impact on subsequent experimental measurements.

8.2.2 Tissue Dissociation on Microfluidic

Traditional tissue culture operations rely on providing the cells with nutrients and renewal media at a fixed point in time, thereby maintaining nutrient and waste concentrations throughout the life of the cell stable. In order to create a better environment to simulate the flow of tissue in vivo, Hattersley et al. [21] designed a microfluidic device for on chip perfusion culture, analysis and dissociation of intact rat liver tissue, as shown in Fig. 8.1a.

The feasibility of the tissue dissociation experiment on the chip was analyzed. Experimental procedures are as followed: Ca^{2+} ions were removed by perfusing immobilized tissue using ethylene glycol tetraacetic acid (EGTA) and then perfused with EBS (balanced salt solution) to remove EGTA to prevent the inhibition of collagenase. Next, the collagenase was perfused to digest the tissue. Finally, the cold dispersing buffer was rapidly perfused at a high flow rate (500 μL/min). Scattered cells can be collected directly into the centrifuge tube at the output of the chip. This method

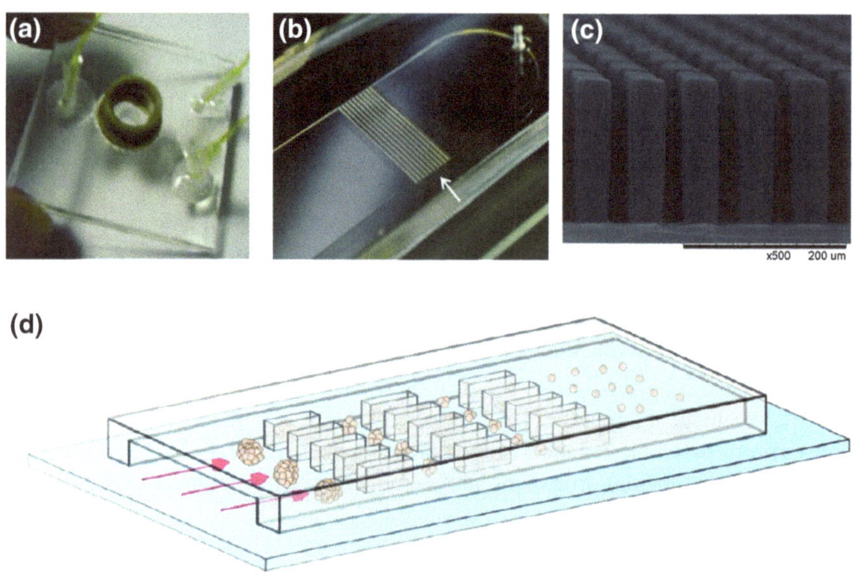

Fig. 8.1 **a** The whole tissue can be cultured and dissociated from the chip. Reprinted with permission from Ref [21]. **b** Single channel "μ-CDC" chip fabricated by soft lithography and PDMS. Reprinted with permission from Ref [22]. **c** Scanning electron microscopy, tissue can be mechanically separated into individual cells by microcolumns. Reprinted with permission from Ref [22]. **d** Schematic diagram of tissue dissociation procedure. Reprinted with permission from Ref [22]

of tissue dissociation uses microfluidic chips to reduce the exposure of tissue or cells to non-sterile environment, thereby reducing the risk of contamination. However, due to the long dissociation time, it is not suitable for gene expression analysis. In order to strictly separate single cells from tissues, Lin et al. [22] designed a microfluidic chip for cell separation called "μ-CDC". The chip consists of a number of micro pillars which are 50 micron wide and 167 micron high with a pitch of 20 microns, as shown in the Figs. 8.1b,c. In this work, the neurospheres of mouse brain tumors DC115 and KT98 were dissociated into single cells after passing through a designed microcolumn array with a syringe pump at a rate of 3–15 ml/min. The yield of single cells obtained by the μ-CDC chip was between 91 and 95%, and the yield of single cell viabilities was between 80 and 85%. This method is also advantageous in dissociating tissue from the conventional grinding process, but the rate of separation is relatively slow due to inherent defects in the mechanical separation.

8.3 Cell Sorting

8.3.1 Conventional Approaches of Cell Sorting

After dissociation of tissues, a mixed cell suspension containing many different types of cells is obtained. However, scientists are usually interested in the target cell among a large population of cells in the study. Thus, the following step in single cell analysis is to separate the mixed cell suspension into a single cell suspension. The most common cell sorting method is flow cytometry (FACS), which relies on a fluorescent dye that binds specifically to a specific protein on cell surface. When the cell suspension flows through the laser beam, the fluorescent dye is excited to emit light and the results are recorded by the detector. Droplets with a single cell give a specific color based on the fluorescent dye, which is then transferred to the corresponding container by an electric field.

Another option for flow cytometry is the use of magnetic fields to isolate single cells with specific magnetic tags, a process known as magnetic activated cell sorting (MACS). Although the two flow cytometry methods (FACS and MACS) are widely used today, both methods have limitations in single cell analysis that limit the universality of their use, such as expensive instrument costs, complex cell surface markers, increased sample pretreatment time and reduced cell viability, as well as sample contamination risk. These limitations facilitate the development of microfluidic cell sorting platforms, which will be discussed in detail in the next section.

8.3.2 Cell Sorting on Microfluidic

The emergence of microfluidic cell sorting techniques has overcome some of the limitations of flow cytometry and magnetic activated cell sorting. Compared to conventional cell sorting methods, microfluidic devices are inherently more advantageous for cell sorting. The miniaturization of the device allows the reagent consumption in the experiment to be reduced while the portability is improved. The manufacturing laboratory tool can use standard micro-fabrication system and soft lithography technology to reduce the production cost greatly. The experiment simplifies the operation of the instrument in the chip, and also reducing the risk of contamination.

However, microfluidic cell sorting requires further development in order to replace conventional cell sorting techniques. The major drawback of microfluidic systems is the low throughput compared to commercial flow cytometry. Secondly, the soft lithography technology by the manufacture of microfluidic devices in the experiment easily lead to cell adhesion and blockage, affecting its service life. In this field, there are a lot of good work published [23–27]. In this section, we will discuss the work involved in the sample preparation of single cell analysis.

8.3.2.1 Electrophoresis and Dielectrophoresis

Most cells exhibit negative electronegativity on their surface at the neutral pH, therefore, the suspended cells move toward the positive electrode in a constant electric field [28]. As the different types of cells have different charges or sizes, the electric field force and resistance are different, resulting in differences in their movement state, thus cell was separated. This separation mode is known as electrophoresis (EP) [29]. In general, the separation of multiphase cell suspensions using electrophoresis is poorly performed because the specificity of the different cell types in electrophoretic migration is not very strong. Compared to conventional electrophoresis, dielectrophoresis (DEP) has a higher frequency of cell sorting due to its higher dielectric properties for different cell types. Such as dielectrophoretic field-flow fractionation [30], via this method, mixed cell suspensions flow through one channel and produce an up-going DEP force by the electrodes at the bottom of the channel to balance cell downward gravity. As shown in Fig. 8.2a, each cell of the same type is stabilized at a specific height in the channel, depending on the parabolic velocity and flow rate [29].

Compared with electrophoresis, dielectrophoresis has certain specificity in cell sorting. In the DEP method, the use of alternating current can avoid the electrochemical reaction on the electrode, and the possibility of damage to the cells is minimized. The use of dielectrophoresis is a relatively common method for sorting live and dead cells. In addition, the method can also be used as a marker-free counting of blood cell. Wei et al. [31] used electrophoresis to sort cells by electroporation, and maintains the viability of the cells. The chip is divided into three functional regions, including the electroporation segment (a), the concentration segment (b), and electrophoresis separation section (c) as shown in Fig. 8.2b. Green and red dots represent viable and non-viable cells, respectively. Figure 8.2b (d) shows that viable cells and non-viable cells are subjected to different forces and corresponding cellular motions in the sorting segment. This work makes the sorting efficiency and cell activity in cell sorting achieve a better balance.

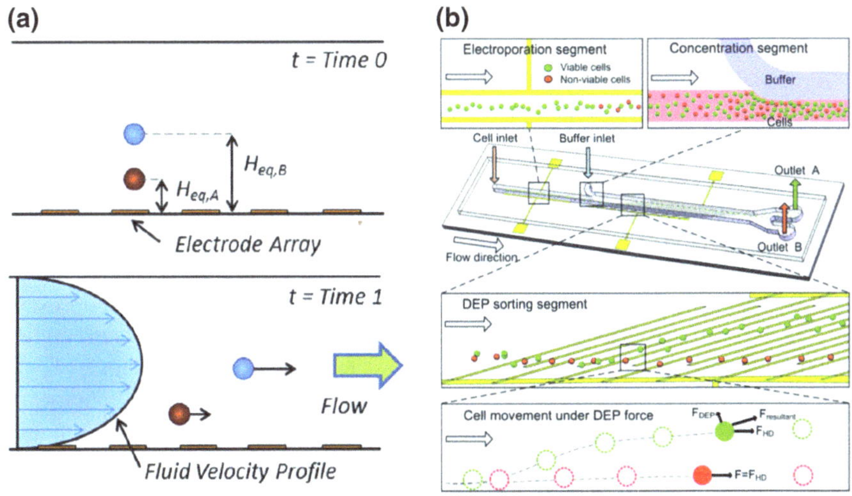

Fig. 8.2 **a** Dielectrophoretic field-flow cell sorting. Reprinted with permission from Ref [23]. **b** Schematic diagram of cell flow through an electroporation microchip. Reprinted with permission from Ref [31]

8.3.2.2 Electro-Osmotic Flow

Similar to the segregation mechanism of electrophoresis and dielectrophoresis, the cell sorting method based on electro-osmotic flow also applies the e field, but the experimental phenomena are different. Electro-osmotic flow (EOF) refers to fluid movement induced solved ions movement in the electric field, while the suspended particles associated with ion mobility caused the secondary fluid movement. Dittrich et al. [32] demonstrated a cell sorting microchip that uses a pump to drive primary flow to produce electro-osmotic flow. This method utilizes pressure-driven to achieve a fast and stable flow rate, allowing cell sorting to be carried out with high throughput. The advantage of electro-osmotic flow in the microfluidic device is to accurately control the flow of different channels. However, electro-osmotic flow requires the fabrication of chip electrodes, increasing the complexity of chip fabrication. In addition, cell viability may be reduced by exposure to an electric field. Nevertheless, this method allows precise control of cell sorting based on cell volume.

8.3.2.3 Acoustic

Acoustic-based cell sorting has recently become a new cell sorting method because it has no influence on cell viability. The key to the acoustic-based cell sorting model is cellular motions responding to the pressure generated by acoustic wave. According to the different types of sound waves, the sorting method can be divided into a number of types, including standing waves, standing surface acoustic waves and traveling waves.

By matching the wavelength of the acoustic wave with the size of the microfluidic channel, discrete standing waves are produced, resulting in two different regions in the chip channel in the direction of the acoustic wave. The first zone has no fluctuating pressure called the node; the second zone is called the anti-node, as the fluctuating pressure changes between the maximum and minimum values, as shown in Fig. 8.3a.

When the particles flow through the channel, due to different acoustic contrast factors on the standing wave they produce different responses. These responses are affected by the cell density and the compression coefficient of the surrounding medium. Cells with a positive acoustic contrast factor will migrate to nodes in the standing wave and cells with negative acoustic contrast factors will migrate toward the antinodes in the standing waves, allowing the cells to be separated by different outlets. Grenvall et al. [33] fabricated a microchip with a two-dimensional acoustic focusing region that separated the cells to five different outlets according to the size of the cells, as shown in Fig. 8.4. This device was used to sort white blood cells of high purity and viability. In this work, however, the sorting throughput is not very high, which can only reach to 150 cells/second.

Standing surface acoustic wave (SSAW) is a standing wave formed by the interdigital transducer (IDT) at the bottom of the microfluidic channel. The interdigital transducers are mounted on a microfluidic chip in the form of a piezoelectric substrate. The acoustic wave in the fluid changes from a transverse wave to a longitudinal wave due to the action of the substrate, and a pressure node is generated. These acoustic waves generated by the cross-shaped electrodes separate the cells into different streamlines and are finally separated through different outlets, as shown in Fig. 8.3b. Because the standing surface acoustic wave can move objects

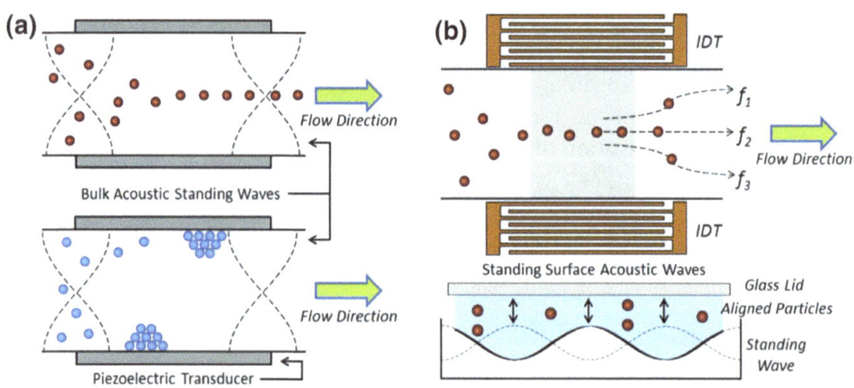

Fig. 8.3 **a** The cell sorting is performed by dispersing standing waves; **b** The cell sorting is performed by standing surface acoustic wave. Reprinted with permission from Ref [23]

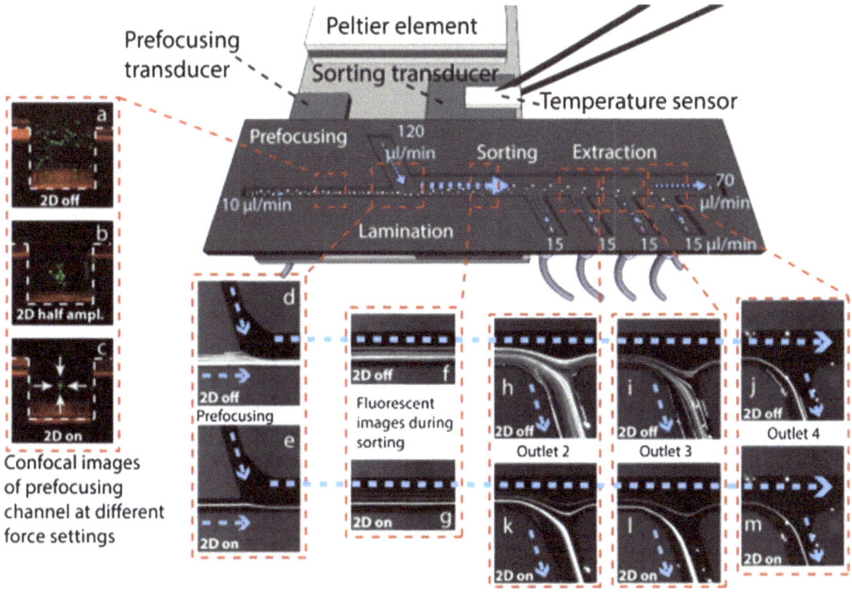

Fig. 8.4 Schematic diagram of cell sorting chip with two-dimensional acoustically focused region. Reprinted with permission from Ref [33]

into different deflection according to different contrast factors in the fluid, it is considered to be a more flexible method for sorting small and hard-to-isolate cells compared to bulk standing waves. Li et al. [34] designed a chip device based on standing surface acoustic waves that can sort droplets in the oil phase, and finally into five different outlets, as shown in Fig. 8.5a. This device can be used for genomics and transcriptomics upstream component sorting experiments.

Most of the acoustic-based cell sorting is dependent on standing waves, but there are still some outstanding publications using the traveling acoustic wave for cell sorting. The standing wave usually needs to match the wavelength of the acoustic wave with the width of the microfluidic channel, so the frequency of the acoustic wave is limited and the sorting rate can't be increased. The traveling wave has a good solution to this problem because there is no wavelength limitation. Schmid et al. [35] designed a microfluidic chip that uses a fluorescence-induced traveling wave to sort cells into three channels. In this work, the cell sorting rate was increased by a factor of 10 compared with the use of a standing surface acoustic wave [34].

Similar to electrically-based cell sorting, the acoustic method often requires the integration of a sensor in the chip. This makes the chip manufacturing, operation become more complex and increase the cost. Furthermore, bulk standing waves are not suitable for sorting subpopulations of cell types.

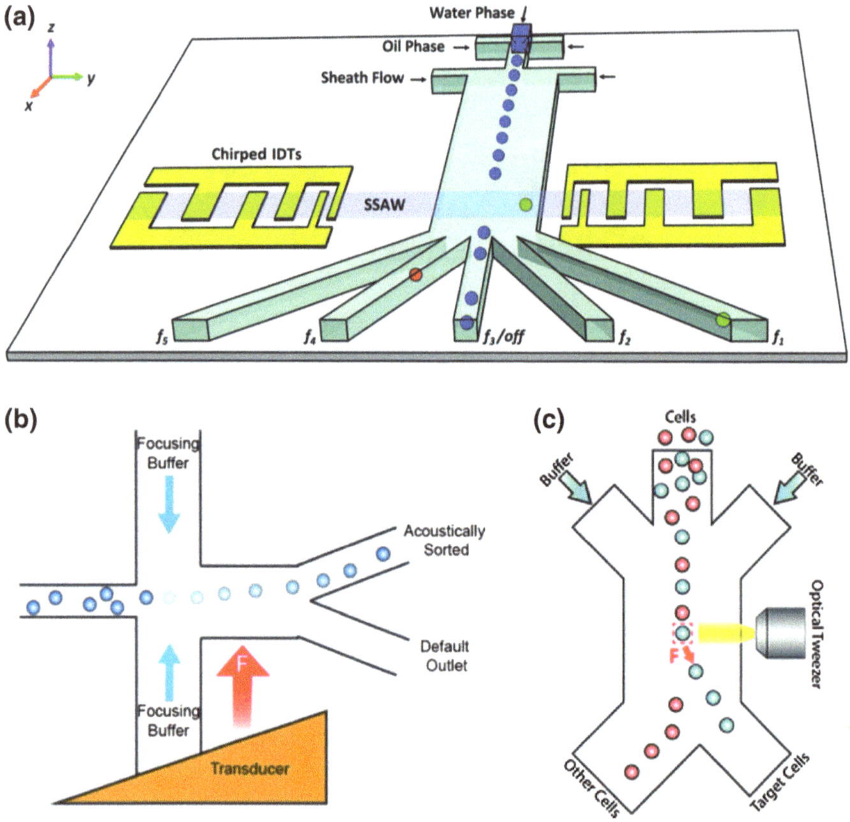

Fig. 8.5 **a** Schematic diagram of a device for sorting droplets based on standing surface acoustic wave. Reprinted with permission from Ref [34]. **b** Schematic diagram of cell sorting by surface acoustic wave; **c** Schematic diagram of optical-based cell sorting. Reprinted with permission from Ref [9]

8.3.2.4 Optical

The use of light to move the particles can be traced back to 45 years ago when focused laser was found to move the particles in the liquid [36]. Similar to the electro-osmotic conductivity in the electrodeposition, the movement of particles in the fluid during the optical drive is dependent on the difference in refractive index between itself and the surrounding fluid. When the refractive index of particles is higher than the surrounding fluid, the particles will migrate to the area where the light intensity is the highest, and vice versa to the area where the light intensity is the lowest. In both cases, the particles will move in the direction of light propagation [37, 38]. Since the principle of optical manipulation has been discovered, it has been applied to cell sorting and cell manipulation in microfluidic chips, the schematic of which is shown in Fig. 8.5c.

Unlike other sorting methods, optical-based cell sorting methods have minimal damage to cell viability. Bragheri et al. [39] reported a fluorescence-induced cell sorting device. In their work, however, due to the delay of software communication, this device can only sort 100 cells per minute. Low throughput experiments limited the application prospects of the device. In fact, all microfluidic chips have a drawback of low throughput compared to conventional flow cytometry. To solve this problem, Chen et al. [40] reported a microfluidic device had a sorting capacity of 23,000 cells/second, with a purity of 90% and a purity of 45% at 45,000 cells/second. The device utilizes a tightly focused laser to induce cavitation bubbles in the fluid, providing a mechanical force for cell deflection, as shown in Fig. 8.6a. The key innovation in achieving this high separation rate is in a three-dimensional multi-layer device structure. But there are two drawbacks, one is the construction complexity of the three-dimensional microfluidic channel, and the other is the dilution of the sample.

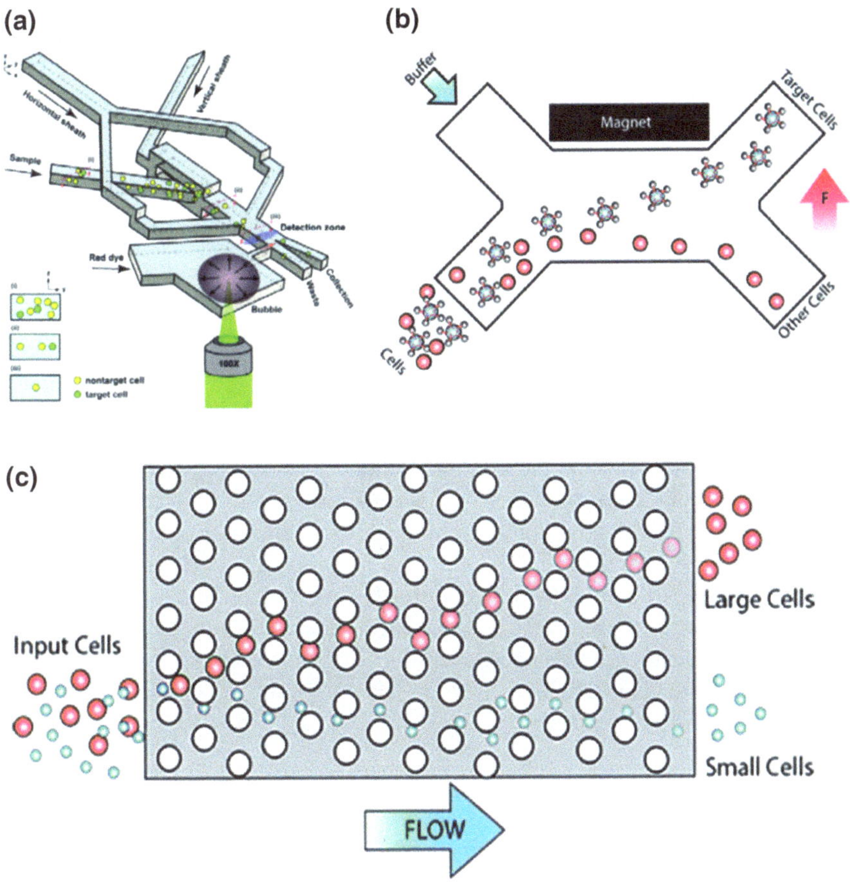

Fig. 8.6 **a** Schematic of laser cell sorting microfluidic device with high throughput. Reprinted with permission from Ref [40]; **b** Schematic diagram of cell sorting by magnetic field. Reprinted with permission from Ref [9]; **c** Schematic diagram of cell sorting based on deterministic lateral displacement. Reprinted with permission from Ref [9]

8.3.2.5 Other Approaches

In addition to the above methods, there are other cell sorting methods integrated in the microfluidic chip. When cells are modified by cell-specific antibodies with magnetic particles, these cells can be separated in a microfluidic chip by a magnetic field, as shown in Fig. 8.6b. Compared with the cell sorting through the electric field, the simplicity of this method has great advantages, not only avoiding the chemical reaction between electrode and the electrolyte but also affecting cell viability.

Deterministic lateral displacement method uses a column array to direct the cells to produce different motions in the fluid depending on their sizes. Cells smaller than the critical radius move in the direction of primary fluid flow. Cells above the critical radius are deflected in the fluid by the designed array, as shown in Fig. 8.6c. The separable critical particle size depends on the design of the different column arrays in the chip [41]. In addition to the size of the cells, similarly, shape and deformability can also be used for this type of cell sorting [42].

In addition to the above classification methods, there are inertia [43], filtration [44] and other methods integrated in the microfluidic chip, which will not be presented as a key contents in this section. Readers interested can further study from the relevant references.

8.3.3 Outlook for Cell Sorting on Microfluidic

Microfluidic cell sorting is now a mature field and has been established prior to single cell analysis. In the discussed work, most are examples of microfluidic cell sorting. In this area, flow cytometry is still a very useful tool. High throughput, simple operation, short analysis time, high purity and other advantages make it widely used so far. At the same time, the shortcomings of flow cytometry, such as large sample demand, low recovery rate and high cost promote the development of microfluidic systems which can solve these problems. At present, microfluidic systems need to be improved in terms of cell sorting because of their low sorting throughput. Limited experimental throughput have limited their application. In addition, due to channel blockage and cell adhesion problems, microfluidic chip's low recovery rate is still a big problem. In the future, perhaps the microfluidic system can find a place in the aseptic sorting market, but this requires more experiments and system validation. It is believed that the future will continue to develop microfluidic system to replace the traditional method and commercialization.

8.4 Single Cell Isolation

8.4.1 Conventional Approaches of Single Cell Isolation

After obtaining a cell suspension by cell sorting, the next step in single cell analysis is to isolate individual cells from a discrete population for analysis. In actual single cell analysis, individual cell immobilization is not sufficient and the cells must be distinguished in a specific way so that the lysates of one cell do not contaminate the other. Traditionally, flow cytometry can be used to sort cells and locate individual cells into a single well for single cell analysis [45]. Flow cytometry based on cell isolation has high throughput, automation and other advantages. The drawback is the high cost and demand for a large number of samples as discussed above. In addition, an open experimental environment increases the risk of introducing contamination. Nevertheless, flow cytometry is still a major commercial technique in cell analysis.

The micromanipulation technique is achieved by manual selection and transfer of individual cells to the containers needed for the next experiment. This can be done simply by mechanical transfer through a micropipe. A particular advantage of micromanipulation is that cell concentration is not required because the cells can be visually recognized directly prior to manipulation. Single cell isolation can be ensured by manual identification and selection under a microscope. In addition, the micromanipulation is ideal for materials with small sample sizes. However, there are also two unavoidable disadvantages existing that the low throughput of micromanipulation and the risk of introducing contaminants in the open experimental system.

Optical tweezers, as a new micromanipulation technique, are easy to isolate single cells. This principle is the same as that of the previously mentioned principle of optical-based cell sorting, and the particles are moved at the center of the focal point of light by the momentum of refracting and scattering photons. Single cell isolation byoptical tweezers was also proved available previously [46, 47].

Compared to the conventional mechanical micromanipulation, the use of optical tweezers for single cell isolation decrease the risk of sample contamination due to the closed operating environment, and consume little sample which is similar to other micromanipulation technology.

The most challenging part of single cell analysis is the isolation of individual cells. Under normal circumstances, high throughput and low cost are the two inconsistent conditions. Single cell separation must remain effective for small samples. In addition, due to the small amount sample needed, the effect of sample contamination will be amplified, so the possibility of contamination of the sample must be minimized.

8.4.2 Single Cell Isolation on Microfluidic

Microfluidic devices offer many advantages in single cell isolation. Each single cell compartment is miniaturized to reduce lysate dilution, which is critical for the analysis of low levels of biomolecules, such as ribonucleic acid and other biological molecules in the cell which can be as low as 0.01–2.5 picograms/single cell [48]. The volume of reagent consumed by the microfluidic system is reduced, thereby reducing the cost of the experiment. The isolated single cells can be tightly aligned by microscopic visualization. In addition, most microfluidic systems are fully automated, closed systems, reducing the risk of sample contamination. In this section, we discuss the single cell isolation method in microscopically visible analysis, which focuses on the design of single cell isolation methods.

8.4.2.1 Valves

On chip valves offer an easily conceptualized system for single cell isolation. The valve can be used to control the flow of fluid in the channel, but also to control the flow direction. Quake et al. [49, 50] popularized the commonly used microfluidic valve structure, thus called the Quake valve, as shown in Fig. 8.7a.

On chip valve systems are easily conceptualized. Most chip valves are double-layer pneumatic valves or Quake valves. The bottom of the chip is the channel through which the cells, buffers, etc. flow, and the top of the chip places a large channel perpendicularly intersecting the underlying channel. When the large channel is inflated, the bottom channel is forced to close (Fig. 8.7b). In addition to the characteristics of the double-layer architecture, these chip valve devices are usually computer-controlled. These two features make the equipment's manufacturing costs raise and the operation become more complex. Finally, single cells may be injected into the microfluidic system via pipettes or syringes, but it is necessary for the experimenter to confirm single cell isolation by microscopy, which greatly limits the throughput of single cell isolation.

8.4.2.2 Dielectrophoretic

The use of dielectrophoresis to isolate single cells has gained a great success in its application. In fact, the DEP system from a silica biosystem (Fig. 8.8a) is commercially a mature microfluidic single cell isolation system. The system includes an individual cartridge and analysis platform. The cartridge is an individually controlled array of electrodes that capture a single cell through a dielectrophoretic cages. After capturing the cells, the captured cells are visually recognized and

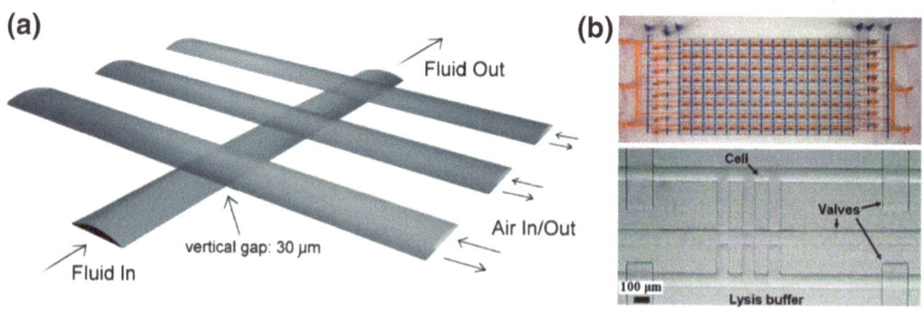

Fig. 8.7 **a** Schematic diagram of a chip with three Quake valves. Reprinted with permission from Ref [49]. **b** Plane diagram of the Quake valve. Reprinted with permission from Ref [88]

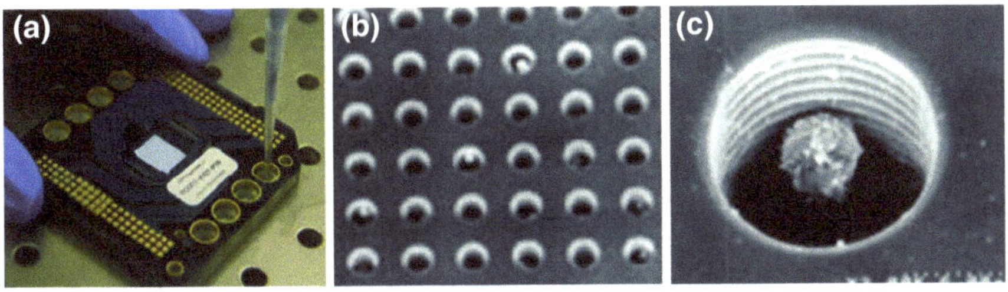

Fig. 8.8 **a** Diagram of dielectrophoresis array chip. Reprinted with permission from Ref [89]. **b** An experimental of single cells isolated by a porous array. Reprinted with permission from Ref [90]. **c** Microporous capture single cells. Reprinted with permission from Ref [90]

manipulated into other electrophoretic cages or separated from the chip. Carpenter et al. [51] isolated the single neuroblastoma cells from the pathogens by the dielectrophoresis array for genome sequencing. In this approach, the experimenter moved the captured single cells into the microtubes after the capture and recognition of the cells. The manual operation limits the experimental throughput. In addition, repeated measurements are necessary steps to distinguish cell heterogeneity between experimental noise, thus increasing the cost of the experiment.

8.4.2.3 Microwells

Microwells provide a simple way to isolate single cells by using physical boundaries, as physically separated by a multiwell plate, as shown in Fig. 8.8b. The key difference between microporous and conventional porous plates is that they have a smaller size and volume. In a microporous array, cells are subjected to gravitational attraction into microwells, and other cells outside the microwells are washed away. In this way, a sufficiently large array of wells can be fabricated to assure the capture of enough single cells for analysis. The advantage of the microwell method is that the device is simple and easy to operate, but one drawback of this method is that a single cell occupies only a part of the microwell, and therefore more cells may be present in one microwell. However, the efficiency of single cell capture can be optimized by adjusting the size, shape, and concentration of the cell suspension. Since the system has been in a mixed state, the lysates of the cells may mix. Thus this method is more suitable for single cell studies under microscopic conditions than for molecular analysis.

8.4.2.4 Hydrodynamic

Single cell isolation may also come out through passive hydrodynamic mechanisms. This method is very attractive to researchers because it does not require complex experimental systems and operations. There are two subclasses in the hydrodynamic approach, including cell capture in the circulating vortex, and the using of physical barriers.

A single cell can be trapped in the vortex center using four vortices generated by cyclic oscillations of the sound waves around a cylinder [52]. This effect is referred to as "hydrodynamic tweezers", as shown in Fig. 8.9a. The cell lysates are not isolated in this method, so the method is primarily oriented toward microscopic analysis.

The second type of single cell isolation through hydrodynamic is the use of physical barriers such as U-shaped cups to mechanically separate single cells on the chip, as shown in Fig. 8.9b. The designed barrier should have a cross-section that allows fluid to flow through unoccupied traps, increasing the chance of capturing

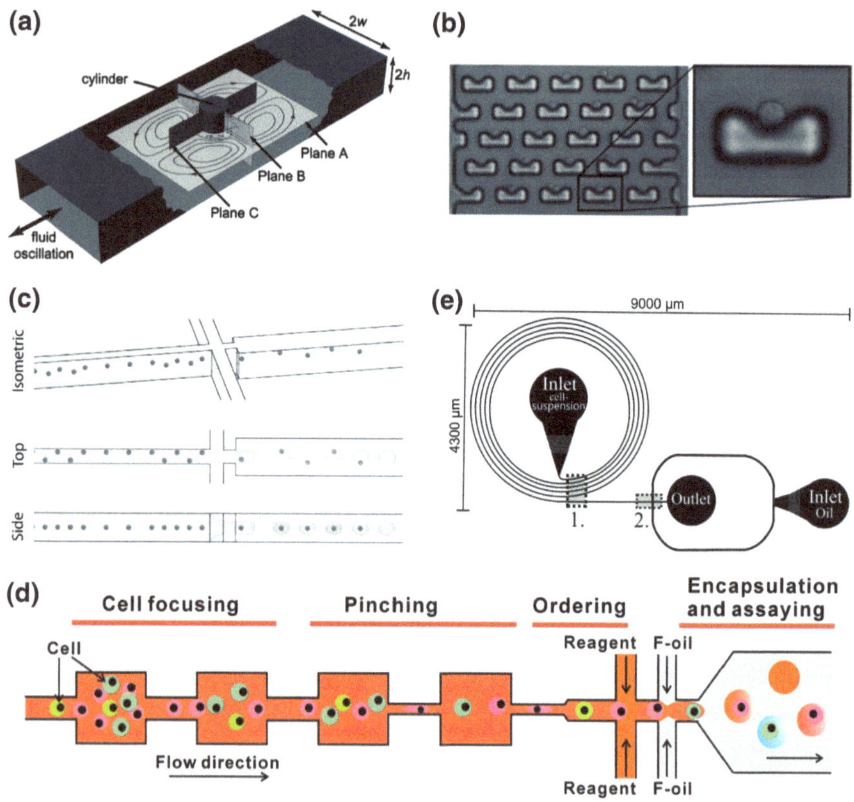

Fig. 8.9 **a** Schematic diagram of single cells isolation via vortex. Reprinted with permission from Ref. [52]. **b** Single cell isolation via cup-shaped physical barriers. Reprinted with permission from Ref. [91]. **c** Schematic diagram of single cell isolation based on droplets. Reprinted with permission from Ref. [92]. **d** Schematic diagram of single cell isolation of lung cancer based on droplet. Reprinted with permission from Ref. [56]. **e** Single cells were isolated by droplet formed via helical channels. Reprinted with permission from Ref. [57]

single cells. Recently, Guan et al. [53] reported that a chip captured single cells with a hydrodynamic mechanism, which achieved 100% capture efficiency.

The method of single cell isolation via hydrodynamics mechanism discussed above has the advantage of high efficiency by varying the size and shape of the obstacles. However, almost all cells captured by this mechanism are used for culture and transient imaging analysis, instead of single cell lysate analysis.

8.4.2.5 Droplets

Cell isolation via droplets has become a very popular method in single cell analysis, whose principle is shown in Fig. 8.9c. The mono-dispersed droplets are generated by two immiscible fluids flowing in the same direction [54]. This method is particularly useful for molecular analysis of single cell lysates since each individual droplet is isolated from one another and can function as a separate chemical reactor.

Single cell isolation via droplets requires less complexity than the other methods, without a two-layer valve system or an integrated electrode system. Droplet generation and cell encapsulation occur in continuous fluid flow, allowing separation to have a high throughput, typically up to several kilohertz. Droplets can be reduced to a minimum of picoliter levels, reducing the dilution analyte. Single cell isolation based on droplets is generally consistent with the Poisson distribution, which means that more than half of the droplets do not contain cells [55].

At present, there have been many studies reported improve the separation efficiency of single cells. For example, Ramji et al. [56] invented a chip device with an initial platform for isolating lung cancer cells, as shown in Fig. 8.9d. These cells are separated by a series of equally spaced channels by expansion and contraction and maintain a uniform distance. The individual cells are then combined with a reagent and encapsulated in a droplet. The cell separation efficiency of this experiment can reach 55%, only 25% of all droplets are empty. Kemna et al. [57] uses long helical microchannels to isolate single cells, and cells are sorted through long channels in order to achieve a final cell encapsulation efficiency of 77%, as shown in Fig. 8.9e. Increasing the proportion of individual cells in a droplet can be achieved by pre-enrichment or post-screening steps. In fact, the simple integration of the droplet formation with the other steps allows the method to be used more broadly.

In single cell analysis, the droplet method has become a powerful tool. Single cell molecular analysis such as PCR and fluorescence-based protein analysis have been carried out by micromation and sorting in droplets of picogram order. Articles published on single cell analysis and droplet microfluidic methods have doubled in the past five years. Therefore, we can expect considerable growth in this area of research in the future [58, 59].

8.5 Single Cell Lysis

8.5.1 Conventional Approaches of Single Cell Lysis

Once the single cells required for the analysis are isolated and the biomolecules to be studied are determined, the next step is to quantify the target. Transmembrane proteins are usually quantified by fluorescent antibody probes, in contrast to biomolecules in cells that are usually analyzed by extraction. In order to extract intracellular molecules, it is necessary to disrupt the plasma membrane of the cells. Conventional methods of destroying plasma membranes can be divided into two categories: detergent-free and detergent-based.

Methods of detergent-free include mechanical agitation, temperature cycles and sonication. Cell lysis by mechanical agitation can be accomplished by rotating impeller blades, mortar and pestle [60] or high speed agitator [61]. However, these methods are not suitable for single cell lysis, but are suitable for multi-cell suspensions or samples with larger sizes. Sonication is through the immersion liquid piezoelectric transducers producing high frequency sound waves to destroy cell membrane. Cheow et al. [62] lysed single cells of liver cancer using a nanoliter well immersed in an ultrasonic cleaner. However, sonication requires 50s to lyse a single cell, during which time sufficient heat is generated to denature the protein in the cell. However, excess heat can be alleviated by means of short pulses and ice-water baths [63]. The freezing-thawing cycle can also be used to lyse cells. The ice crystals formed during the freezing process can break down the cells during thawing. Kim et al. [64] obtained a good result by the method above to lyse single lymphocytes for DNA extraction. However, in order to obtain a good extraction rate, the freezing-thawing cycle is often repeated for a number of times, making the process lengthy.

Cell lysis via detergent-based method may be a more gentle and faster way, and can be miniaturized to single cell lysis. Detergents break cell membranes by dissolving proteins on the cell membrane and interfering with the interaction of phospholipids with proteins. Many types of detergents are currently available, as long as they are amphipathic. Detergent selection is usually empirical, however, there are some general rules. Nonionic or zwitterionic detergents are better able to prevent protein denaturation than ionic detergents. Han et al. [65] tested the ability of various detergent and concentrations in single cell lysis. A non-ionic detergent can dissolve the cell membrane within 10 min at 2% concentration, while maintaining the integrity of the nucleus within 1 h.

In general, the goal of cell lysis is to minimize the changes in the native structure and expression of intracellular molecules, which means that lysis must be both gentle and rapid. The traditional approach usually involves some problems in miniaturization to single cell lysis, while microfluidic technology provides a new approach to cell lysis, which is particularly well suited for lysis of single cells, as discussed in the next section.

8.5.2 *Single Cell Lysis on Microfluidic*

Microfluidic systems provide an ideal platform for single cell lysis. Since the chip can be fabricated with specific geometries and precise size, which avoids interference from other cells in single cell lysis. The length of the microfluidic chip is matched on a single cell to minimize dilution to increase detection sensitivity. Most of the chips are optically transparent, allowing visualization. There are many ways of microfluidic single cell lysis, which will be discussed in the next section. Each has its own advantages. Depending on the experimental influencing factors, selecting the appropriate technique is crucial to the experiment results.

8.5.2.1 Mechanical

Mechanical cell lysis uses mechanical forces to penetrate cell membranes, and mechanical forces can be induced by shear, compression, collisions, and other methods. Hoefemann et al. [66] demonstrated that single cells can be lysed in the course of continuous microfluidic flow. The cells are broken down by pass through an integrated heater, a foam produced by the heater, which presses the cells down to the channel walls, as shown in Fig. 8.10. Single cell lysis can achieve 100% efficiency in less than 20 ms. However, the above experiments did not perform single cell molecular analysis, and therefore the diffusion and separation of cell lysates should also be considered when using this method for single cell molecular analysis.

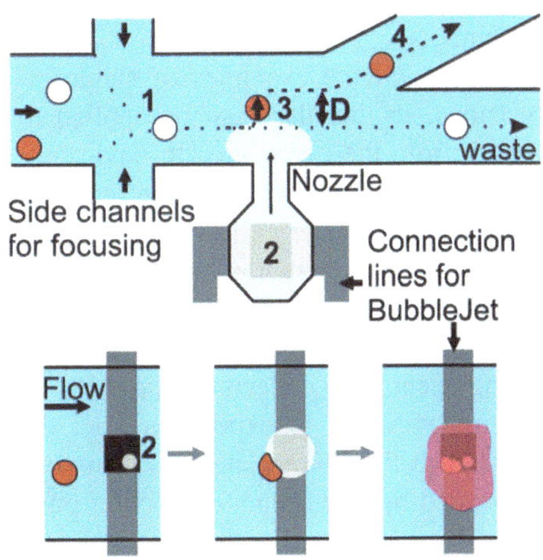

Fig. 8.10 Schematic diagram of single cell lysis by mechanical force in a microfluidic chip. Reprinted with permission from Ref [66]

8.5.2.2 Thermal

Thermal Cell lysis relies on heat-induced cell membrane protein denaturation, which opens the cell. Heating method does not bring other impurities in the system for contamination. However, many intracellular macromolecules such as proteins are very sensitive to temperature. Thermal cell lysis methods are rarely used for protein analysis. Gong et al. [67] loaded 6086 single cells in wells of the chip and lysed the cells by heating at 50 °C for 40 min. The lysed product was used to amplify the complementary DNA product. As for single cell molecular analysis, the limitation of thermal method is the lysis time. Since thermal cracking typically takes a few minutes, this method is not suitable for measuring signals that occur within a few seconds in the cell. Thermal methods are therefore most often used to measure events such as gene expression.

8.5.2.3 Chemical

Since chemical cell lysis has been used for a large number of cellular assays, it is also a popular technique for single cell lysis in unicellular assays. The rate of chemical cell lysis is affected by many factors, such as the detergent or enzyme used, the concentration and the contact efficiency. The first two parameters to a certain extent, limit the study range of biomolecules. Ionic detergents such as sodium dodecyl sulfate can dissolve cells quickly, but also denature proteins. The fluid flow at the microscale usually occurs in the viscous state, and the molecular transfer and mixing are mainly in the form of diffusion. So the process of cell lysis will be relatively long. To overcome this limitation, peristaltic pump, syringe pump, or a reagent is injected manually through a pipette to facilitate mixing of the reagents.

So far, all the examples of single cell lysis discussed are based on cell suspensions, so that they can be applied to microfluidic systems. As a result, cells detach from their own growth environment, making it impossible to correlate the measured data with information about their natural environment. To address this limitation, Sarkar et al. [68] developed a microfluidic probe that selectively lyses single cells of adherent cultures and collects their lysates while minimizing dilution of the sample. The device produces an effusion chemical agent at the tip and slowly flows into the cell, as shown in Fig. 8.11. The tip of the probe, while providing the lysing agent, also collects the product of cell lysis and transfers it to the analytical instrument. Prior to experiment, the device was placed over a target cell in a standard cell culture plate by means of a micromanipulator. The lysing agent is controlled by two separate injection pumps. Although this method has a low lysis throughput and requires manual control during experiments, it is unique in its ability to analyze single cell protein activity in adherent cell cultures.

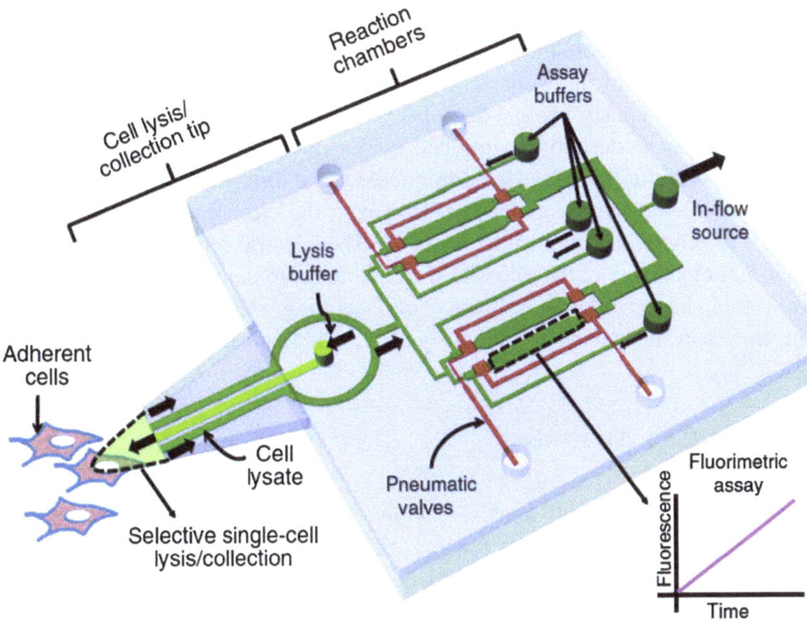

Fig. 8.11 Schematic diagram of a microfluidic probe. Reprinted with permission from Ref [68]

8.5.2.4 Electrical

The plasma membrane of the cell is composed of a lipid bilayer, which protects the cell from the external environment. When lipid bilayers are exposed to an electric field, molecular reforming and thermal phase transformation occurs and new pores are formed. If the electric field is relatively mild and the exposure time is short, the pore formation process is reversible [69]. This phenomenon, known as electroporation, is commonly used to load therapeutic and genetic material into cells. When the cells are exposed to a strong electric field or long-term exposure, the pore will become permanent. The osmotic pressure between the cytoplasm and the surrounding medium becomes unbalanced and the cells undergo swelling and rupture. One obvious advantage of cell lysis by electrical dissociation is that the strength of the electric field can be quickly adjusted to avoid denaturation of the target biomolecule. Therefore, the electrolytic dissociation method has the property of selectively degrading the cell membrane. Mellors et al. [70] built a chip for single cell analysis of erythrocyte proteins via capillary electrophoresis and electrospray ionization mass spectrometry. All flows on the chip were driven via electro-osmotic pumping. An injected cell suspension traveled to a T intersection where an increase in the electric potential (4 kV) lyses cells. The authors found single cell lysing events occurring at a rate of 0.2 cells/s. Kim et al. [71] integrated microwell compartmentalization, DEP immobilization, and electrical cell lysis all on one chip. The group reported 95% of 3600 wells were loaded with single cells due to DEP facilitated trapping. Following DEP immobilization, reagent exchange was done rapidly in 30 s via pressure driven flow, without perturbing the cell positioning. To confine lysates, the wells are physically closed by pressing a PDMS membrane on the top of the wells. The group reported that 100% of trapped cells were lysed simultaneously via a series of 30 V electrical pulses.

Electrical dissociation allows rapid cell lysis without destroying the target biomolecules required for the assay. In addition, electrical dissociation can easily avoid contamination of the sample under test. However, the electrical method has some limitations in single cell molecular analysis. The electrical method requires the integration of electrodes in the microfluidic chip and also the corresponding control system, which has significant influence on operating and preparation costs. At present, the main method of single cell lysis is still chemical method.

8.6 Single Cell Analysis

The final step is single cell analysis, which is to detect and characterize the molecules obtained from a series of steps - tissue dissociation, cell sorting, single cell isolation, and single cell lysis. Finally obtain the required cell information. The commonly used single cell analysis methods include fluorescence imaging, electrochemical analysis and mass spectrometry.

8.6.1 Single Cell Fluorescence Imaging

During a long time, fluorescence imaging is the most commonly used single cell analysis method, which plays a vital role in single cell analysis [72–74]. Single cell fluorescence imaging has a unique advantage in the intracellular molecule analysis. It provides images with high contrast and chemical-specific property. Although fluorescence imaging has the advantages described above, it requires that the target molecule is able to generate fluorescence. For molecules that do not fluoresce themselves, it is necessary to chemically label or label the target molecules with fluorescent functional groups that can emit light, which limits the use of fluorescence imaging to some extent. The development of new fluorescent labeling method has been a hotspot in fluorescence imaging [75–78]. In addition, the labeling of metabolite molecules themselves have a certain impact on the nature, especially small molecule metabolites. Due to spectral overlap, single cell fluoroscopy is also difficult to simultaneously image multiple metabolites.

8.6.2 Single Cell Electrochemical Analysis

The method of electrochemical analysis can detect the metabolites in single cells quantitatively in real time. The electrochemical method in neuroscience research, for example, Adams researched on neurotransmitters in the 1970s pioneered this approach [79]. At present, electrochemical methods in single cell analysis is no longer limited to the study of neurotransmitters, extending to the cells of various small molecules or macromolecules [80–82]. However, similar to the fluorescence imaging method, the electrochemical method is also difficult to achieve multi-component simultaneous detection.

8.6.3 Single Cell Mass Spectrometry

Mass spectrometry is a good method for single cell analysis. It has high sensitivity, the ability to detect multiple substances simultaneously and the ability to structure the molecules of interest. At present, a variety of ionization methods have been developed for desorption/ionization of different types of samples, including electrospray/nano-electrospray ionization (ESI/Nano-ESI), laser ablation/laser desorption ionization (LA/LDI), and secondary ionization mass spectrometry (SIMS). These ionization methods can ionize a variety of substances, such as proteins, peptides, esters, and small molecule metabolites [83]. Since unicellular mass spectrometry is unlabeled and capable of simultaneous analysis of multiple components, it has received increasing attention in recent years in single cell research [84–87].

8.7 Future Outlook

Single cell analysis is an emerging and rapidly expanding field. Therefore, the emergence of new technologies has made it difficult to predict the future of concern and application. However, it is undeniable that innovative microfluidic technology has made great contributions to the field of single cell analysis and will continue to play an important role. In the future, more microfluidic devices for tissue dissociation are likely to emerge. However, microfluidic techniques for tissue dissociation are still not as mature as the techniques used for cell sorting, since traditional tissue dissociation methods are the cheapest and most effective. In contrast, cell sorting is still driven by economic factors. Often the rare cell populations that need isolation are enriched by cell sorting. This step requires a simple, low-cost and high-throughput cell sorting system, which also are the advantage of microfluidic devices. Therefore, we expect the work in the field of microfluidic cell sorting to grow continually.

Modern single cell analysis techniques are particularly exciting. Technological advances in amplification, sequencing and microfluidics allow us to detect the fundamental unit of life—cell in an unprecedented way. Single cell analysis can have lasting promotion on research of cancer, immunology, developmental biology, stem cells, and so on. Researchers are now investigating single cells by transferring cells from the microenvironment where they grow to the analysis buffer. Thus, the effect of the cell's native environment is completely stripped, and the process of cell analysis is also irreversible. It is generally believed that the growth space of the cell environment has a significant meaning and influence, which also promoted the microfluidic chip in simulating the biological environment at the same time.

Microfluidic devices have greater potential for achieving single cell multi-parameter measurements and have higher test speeds compared to conventional single cell device. However, the sensitivity and specificity of microfluidic devices still need to be improved. In the future, the development of microfluidic technology will bring more applications, making the field of single cell analysis move forward.

References

1. Valizadeh A, Khosroushahi AY (2015) Single-cell analysis based on Lab Chip fluidic system. Anal Methods 7(20):8524–8533. doi:10.1039/c5ay02029h
2. Wang X, Chen S, Kong M, Wang Z, Costa KD, Li RA, Sun D (2011) Enhanced cell sorting and manipulation with combined optical tweezer and microfluidic chip technologies. Lab Chip 11(21):3656–3662. doi:10.1039/c1lc20653b
3. Galler K, Brautigam K, Grosse C, Popp J, Neugebauer U (2014) Making a big thing of a small cell–recent advances in single cell analysis. Analyst 139(6):1237–1273. doi:10.1039/c3an01939j
4. Di Carlo DLL (2006) Dynamic single-cell analysis for quantitative biology. Anal Chem 78 (23):7918–7925. doi:10.1021/ac060541s
5. Newman JR, Ghaemmaghami S, Ihmels J, Breslow DK, Noble M, DeRisi JL, Weissman JS (2006) Single-cell proteomic analysis of S. cerevisiae reveals the architecture of biological noise. Nature 441(7095):840–846. doi:10.1038/nature04785
6. Kaern M, Elston TC, Blake WJ, Collins JJ (2005) Stochasticity in gene expression: from theories to phenotypes. Nat Rev Genet 6(6):451–464. doi:10.1038/nrg1615
7. Colman-Lerner A, Gordon A, Serra E, Chin T, Resnekov O, Endy D, Pesce CG, Brent R (2005) Regulated cell-to-cell variation in a cell-fate decision system. Nature 437(7059):699–706. doi:10.1038/nature03998
8. Strovas TJ, Sauter LM, Guo X, Lidstrom ME (2007) Cell-to-cell heterogeneity in growth rate and gene expression in Methylobacterium extorquens AM1. J Bacteriol 189(19):7127–7133. doi:10.1128/JB.00746-07
9. Hosic S, Murthy SK, Koppes AN (2016) Microfluidic sample preparation for single cell analysis. Anal Chem 88(1):354–380. doi:10.1021/acs.analchem.5b04077
10. Blainey PC, Quake SR (2014) Erratum: dissecting genomic diversity, one cell at a time. Nat Methods 11(3):349. doi:10.1038/nmeth0314-349b
11. Liang J, Cai W, Sun Z (2014) Single-cell sequencing technologies: current and future. J Genet Genomics 41(10):513–528. doi:10.1016/j.jgg.2014.09.005
12. Tan DW, Jensen KB, Trotter MW, Connelly JT, Broad S, Watt FM (2013) Single-cell gene expression profiling reveals functional heterogeneity of undifferentiated human epidermal cells. Development 140(7):1433–1444. doi:10.1242/dev.087551
13. MarinovGKW BA, McCue K, Schroth GP, Gertz J, MyersRMW BJ (2013) From single-cell to cell-pool transcriptomes: stochasticity in gene expression and RNA splicing. Genome Res 24(3):496. doi:10.1101/gr.161034.113
14. Lecault V, White AK, Singhal A, Hansen CL (2012) Microfluidic single cell analysis: from promise to practice. Curr Opin Chem Biol 16(3–4):381–390. doi:10.1016/j.cbpa.2012.03.022

15. Yin H, Marshall D (2012) Microfluidics for single cell analysis. Curr Opin Biotech 23 (1):110–119. doi:10.1016/j.copbio.2011.11.002
16. Robin JD, Wright WE, Zou Y, Cossette SC, Lawlor MW, Gussoni E (2015) Isolation and immortalization of patient-derived cell lines from muscle biopsy for disease modeling. J Vis Exp 95:e52307. doi:10.3791/52307
17. Diana K, Philip W, Veronika K, Holger J, Heinz Günther J, Süleyman E (2011) Vascular wall-resident CD44+ multipotent stem cells give rise to pericytes and smooth muscle cells and contribute to new vessel maturation. PLoS ONE 6(5):e20540. doi:10.1371/journal.pone.0020540
18. Uchida N, Buck DW, He D, Reitsma MJ, Masek M, Phan TV, Tsukamoto AS, Gage FH, Weissman IL (2001) Direct isolation of human central nervous system stem cells. P Natl Acad Sci 97(26):14720–14725. doi:10.1073/pnas.97.26.14720
19. Welser-Alves JV, Boroujerdi A, Milner R (2014) Isolation and culture of primary mouse brain endothelial cells. Methods Mol Biol 1135:345–356. doi:10.1007/978-1-4939-0320-7_28
20. Geem D, Medinacontreras O, Kim W, Huang CS, Denning TL (2011) Isolation and characterization of dendritic cells and macrophages from the mouse intestine. J Vis Exp 54 (63):51–54. doi:10.3791/4040
21. Hattersley SM, Dyer CE, Greenman J, Haswell SJ (2008) Development of a microfluidic device for the maintenance and interrogation of viable tissue biopsies. Lab Chip 8(11):1842–1846. doi:10.1039/b809345h
22. Lin CH, Lee DC, Chang HC, Chiu IM, Hsu CH (2013) Single-cell enzyme-free dissociation of neurospheres using a microfluidic chip. Anal Chem 85(24):11920–11928. doi:10.1021/ac402724b
23. Th SC, Reyes CD, López GP (2011) Microfluidic cell sorting: a review of the advances in the separation of cells from debulking to rare cell isolation. Lab Chip 15(5):2689–2697(2689). doi:10.1039/c4lc01246a
24. Thompson AM, Paguirigan AL, Kreutz JE, Radich JP, Chiu DT (2014) Microfluidics for single-cell genetic analysis. Lab Chip 14(17):3135–3142. doi:10.1039/c4lc00175c
25. Sajeesh P, Sen AK (2014) Particle separation and sorting in microfluidic devices: a review. Microfluid Nanofluid 17(1):1–52. doi:10.1007/s10404-013-1291-9
26. Chen Y, Li P, Huang PH, Xie Y, Mai JD, Wang L, Nguyen NT, Huang TJ (2014) Rare cell isolation and analysis in microfluidics. Lab Chip 14(4):626–645. doi:10.1039/c3lc90136j
27. Autebert J, Coudert B, Bidard FC, Pierga JY, Descroix S, Malaquin L, Viovy JL (2012) Microfluidic: An innovative tool for efficient cell sorting. Methods 57(3):297–307. doi:10.1016/j.ymeth.2012.07.002
28. Mehrishi JN, Bauer J (2002) Electrophoresis of cells and the biological relevance of surface charge. Electrophoresis 23(13):1984–1994
29. Voldman J (2006) Electrical forces for microscale cell manipulation. Annu Rev Biomed Eng 8(8):425–454. doi:10.1146/annurev.bioeng.8.061505.095739
30. Shim S, Stemkehale K, Tsimberidou AM, Noshari J, Anderson TE, Gascoyne PR (2012) Antibody-independent isolation of circulating tumor cells by continuous-flow dielectrophoresis. Biomicrofluidics 7(1):32–50. doi:10.1063/1.4774304
31. Wei Z, Li X, Zhao D, Yan H, Hu Z, Liang Z, Li Z (2014) Flow-through cell electroporation microchip integrating dielectrophoretic viable cell sorting. Anal Chem 86(20):10215–10222. doi:10.1021/ac502294e
32. Dittrich PS, Schwille P (2003) An integrated microfluidic system for reaction, high-sensitivity detection, and sorting of fluorescent cells and particles. Anal Chem 75(21):5767–5774. doi:10.1021/ac034568c
33. Grenvall C, Magnusson C, Lilja H, Laurell T (2015) Concurrent isolation of lymphocytes and granulocytes using prefocused free flow acoustophoresis. Anal Chem 87(11):5596–5604. doi:10.1021/acs.analchem.5b00370
34. Li S, Ding X, Guo F, Chen Y, Lapsley MI, Lin SCS, Wang L, Mccoy JP, Cameron CE, Huang TJ (2013) An on-chip, multichannel droplet sorter using standing surface acoustic waves (SSAW). Anal Chem 85(11):5468–5474. doi:10.1021/ac400548d

35. Schmid L, Weitz DA, Franke T (2014) Sorting drops and cells with acoustics: acoustic microfluidic fluorescence-activated cell sorter. Lab Chip 14(19):3710–3718. doi:10.1039/c4lc00588k
36. Ashkin A (1970) Acceleration and trapping of particles by radiation pressure. Phys Rev Lett 24(4):156–159. doi:10.1103/PhysRevLett.24.156
37. Joná A, Zemánek P (2008) Light at work: The use of optical forces for particle manipulation, sorting, and analysis. Electrophoresis 29(24):4813–4851. doi:10.1002/elps.200800484
38. Moffitt JR, Chemla YR, Smith SB, Bustamante C (2008) Recent advances in optical tweezers. Annu Rev Biochem 77(1):205–228. doi:10.1146/annurev.biochem.77.043007.090225
39. Bragheri F, Minzioni P, Martinez VR, Bellini N, Paiè P, Mondello C, Ramponi R, Cristiani I, Osellame R (2012) Optofluidic integrated cell sorter fabricated by femtosecond lasers. Lab Chip 12(19):3779–3784. doi:10.1039/c2lc40705a
40. Chen Y, Wu TH, Kung YC, Teitell MA, Chiou PY (2013) 3D pulsed laser-triggered high-speed microfluidic fluorescence-activated cell sorter. Analyst 138(24):7308–7315. doi:10.1039/c3an01266b
41. Inglis DW, Davis JA, Austin RH, Sturm JC (2006) Critical particle size for fractionation by deterministic lateral displacement. Lab Chip 6(5):655–658. doi:10.1039/b515371a
42. Beech JP, Holm SH, Adolfsson K, Tegenfeldt JO (2012) Sorting cells by size, shape and deformability. Lab Chip 12(6):1048–1051. doi:10.1039/c2lc21083e
43. Warkiani ME, Guan G, Luan KB, Lee WC, Bhagat AA, Chaudhuri PK, Tan DS, Lim WT, Lee SC, Chen PC (2014) Slanted spiral microfluidics for the ultra-fast, label-free isolation of circulating tumor cells. Lab Chip 14(1):128–137. doi:10.1039/c3lc50617g
44. Mcfaul SM, Lin BK, Ma H (2012) Cell separation based on size and deformability using microfluidic funnel ratchets. Lab Chip 12(13):2369–2376. doi:10.1039/c2lc21045b
45. Jaitin DA, Kenigsberg E, Kerenshaul H, Elefant N, Paul F, Zaretsky I, Mildner A, Cohen N, Jung S, Tanay A (2014) Massively parallel single-cell RNA-seq for marker-free decomposition of tissues into cell types. Science 343(6172):776–779. doi:10.1126/science.1247651
46. Marshall IPG, Blainey PC, Spormann AM, Quake SR (2012) A single-cell genome for Thiovulum sp. Appl Environ Microb 78(24):8555–8563. doi:10.1128/AEM.02314-12
47. Pamp SJ, Harrington ED, Quake SR, Relman DA, Blainey PC (2012) Single-cell sequencing provides clues about the host interactions of segmented filamentous bacteria (SFB). Genome Res 22(6):1107–1119. doi:10.1101/gr.131482.111
48. Livesey FJ (2003) Strategies for microarray analysis of limiting amounts of RNA. Brief Funct Genomics 2(1):31–36. doi:10.1093/bfgp/2.1.31
49. Unger MA, Chou HP, Thorsen T, Scherer A, Quake SR (2000) Monolithic microfabricated valves and pumps by multilayer soft lithography. Science 288(5463):113–116. doi:10.1126/science.288.5463.113
50. Thorsen T, Maerkl SJ, Quake SR (2002) Microfluidic large-scale integration. Science 298(5593):580–584. doi:10.1126/science.1076996
51. Carpenter EL, Rader JA, Ruden J, Rappaport EF, Hunter KN, Hallberg PL, Krytska K, O'Dwyer PJ, Mosse YP (2014) Dielectrophoretic capture and genetic analysis of single neuroblastoma tumor cells. Front Oncol 4:201. doi:10.3389/fonc.2014.00201
52. Lutz BR, Chen J, Schwartz DT (2005) Microscopic steady streaming eddies created around short cylinders in a channel: flow visualization and Stokes layer scaling. Phys Fluids 17(17):779–797. doi:10.1063/1.1824137
53. Guan A, Shenoy A, Smith R, Li Z (2015) Streamline based design guideline for deterministic microfluidic hydrodynamic single cell traps. Biomicrofluidics 9(2):2010–2030. doi:10.1063/1.4914469
54. Utada AS, Lorenceau E, Link DR, Kaplan PD, Stone HA, Weitz DA (2005) Monodisperse double emulsions generated from a microcapillary device. Science 308(5721):537–541. doi:10.1126/science.1109164
55. Lim SW, Tran TM, Abate AR (2015) PCR-activated cell sorting for cultivation-free enrichment and sequencing of rare microbes. PLoS ONE 10(1):e0113549–e0113549. doi:10.1371/journal.pone.0113549

56. Ramji R, Wang M, Bhagat AA, Tan SWD, Thakor NV, Teck LC, Chen CH (2014) Single cell kinase signaling assay using pinched flow coupled droplet microfluidics. Biomicrofluidics 8 (3):034104. doi:10.1063/1.4878635
57. Kemna EW, Schoeman RM, Wolbers F, Vermes I, Weitz DA, Van Den Berg A (2012) High-yield cell ordering and deterministic cell-in-droplet encapsulation using Dean flow in a curved microchannel. Lab Chip 12(16):2881–2887. doi:10.1039/c2lc00013j
58. Tran TM, Lan F, Thompson CS, Abate AR (2013) From tubes to drops: droplet-based microfluidics for ultrahigh-throughput biology. J Phys D Appl Phys 46(11):114004–114020 (114017). doi:10.1088/0022-3727/46/11/114004
59. Guo MT, Rotem A, Heyman JA, Weitz DA (2012) Droplet microfluidics for high-throughput biological assays. Lab Chip 12(12):2146–2155. doi:10.1039/c2lc21147e
60. Zhou J, Bruns MA, Tiedje JM (1996) DNA recovery from soils of diverse composition. Appl Environ Microb 62(2):316–322
61. Middelberg APJ (1995) Process-scale disruption of microorganisms. Biotechnol Adv 13 (3):491–551
62. Cheow LF, Sarkar A, Kolitz S, Lauffenburger D, Han J (2014) Detecting kinase activities from single cell lysateusing concentration-enhancedmobility shift assay. Anal Chem 86 (15):7455–7462. doi:10.1021/ac502185v
63. Zhang H, Jin W (2004) Determination of different forms of human interferon-γ in single natural killer cells by capillary electrophoresis with on-capillary immunoreaction and laser-induced fluore scence detection. Electrophoresis 25(7–8):1090–1095. doi:10.1002/elps.200305803
64. Kim SA, Yoon JA, Kang MJ (2009) An efficient and reliable DNA extraction method for preimplantation genetic diagnosis: a comparison of allele drop out and amplification rates using different single cell lysis methods. Fertil Steril 92(2):S236. doi:10.1016/j.fertnstert.2008.06.033
65. Lin H, Zi X, Garmire LX, Yu W, Weissman SM, Pan X, Rong F (2014) Co-detection and sequencing of genes and transcripts from the same single cells facilitated by a microfluidics platform. Sci Rep 4(6204):6485. doi:10.1038/srep06485
66. Hoefemann H, Wadle S, Bakhtina N, Kondrashov V, Wangler N, Zengerle R (2012) Sorting and lysis of single cells by BubbleJet technology. Sensor Actuat B-Chem 168(12):442–445. doi:10.1016/j.snb.2012.04.005
67. Gong Y, Ogunniyi AO, Love JC (2010) Massively parallel detection of gene expression in single cells using subnanolitre wells. Lab Chip 10(18):2334–2337. doi:10.1039/c004847j
68. Sarkar A, Kolitz S, Lauffenburger DA, Han J (2014) Microfluidic probe for single-cell lysis and analysis in adherent tissue culture. Nat Commun 5(3):3421. doi:10.1038/ncomms4421
69. Tsong TY (1991) Electroporation of cell membranes. Biophys J 60(2):297–306. doi:10.1016/S0006-3495(91)82054-9
70. Mellors JS, Jorabchi K, Smith LM, Ramsey JM (2010) Integrated microfluidic device for automated single cell analysis using electrophoretic separation and electrospray ionization mass spectrometry. Anal Chem 82(3):967–973. doi:10.1021/ac902218y
71. Kim SH, Yamamoto T, Fourmy D, Fujii T (2011) Electroactive microwell arrays for highly efficient single-cell trapping and analysis. Small 7(22):3239–3247. doi: 10.1002/smll.201101028
72. Lenn T, Leake MC (2012) Experimental approaches for addressing fundamental biological questions in living, functioning cells with single molecule precision. Open Biol 2(6):120090. doi:10.1098/rsob.120090
73. Schmolze DB, Standley C, Fogarty KE, Fischer AH (2011) Advances in microscopy techniques. Arch Pathol Lab Med 135(135):255–263. doi:10.1043/1543-2165-135.2.255
74. Padilla-Parra S, Tramier M (2012) FRET microscopy in the living cell: different approaches, strengths and weaknesses. BioEssays 34(34):369–376. doi:10.1002/bies.201100086
75. Yao S, Belfield KD (2012) Two-photon fluorescent probes for bioimaging. Eur J Org Chem 17:3199–3217. doi:10.1002/ejoc.201200281

76. Dmitriev RI, Papkovsky DB (2012) Optical probes and techniques for O2 measurement in live cells and tissue. Cell Mol Life Sci 69(12):2025–2039. doi:10.1007/s00018-011-0914-0
77. Cottet-Rousselle C, Ronot X, Leverve X, Mayol JF (2011) Cytometric assessment of mitochondria using fluorescent probes. Cytom Part A 79A(6):405–425. doi:10.1002/cyto.a.21061
78. Wombacher R, Cornish VW (2011) Chemical tags: Applications in live cell fluorescence imaging. J Biophotonics 4(6):391–402. doi:10.1002/jbio.201100018
79. Adams RN (1976) Probing brain chemistry with electroanalytical techniques. Anal Chem 48 (48):1126A–1138A
80. Trouillon R, Svensson MI, Berglund EC, Cans AS, Ewing AG (2012) Highlights of selected recent electrochemical measurements in living systems. Electrochim Acta 84(12):84–95. doi:10.1016/j.electacta.2012.04.164
81. Cecala C, Sweedler JV (2012) Sampling techniques for single-cell electrophoresis. Analyst 137(13):2922–2929. doi:10.1039/c2an16211c
82. Lama RD, Charlson K, Anantharam A, Hashemi P (2012) Ultrafast detection and quantification of brain signaling molecules with carbon fiber microelectrodes. Anal Chem 84(19):8096–8101. doi:10.1021/ac301670h
83. Lei Z, Huhman DV, Sumner LW (2011) Mass spectrometry strategies in metabolomics. J Biol Chem 286(29):25435–25442. doi:10.1074/jbc.R111.238691
84. Svatoš A (2011) Single-cell metabolomics comes of age: new developments in mass spectrometry profiling and imaging. Anal Chem 83(13):5037–5044. doi:10.1021/ac2003592
85. Tsuyama N, Mizuno H, Masujima T (2011) Mass spectrometry for cellular and tissue analyses in a very small region. Anal Sci 27(2):163–170
86. Tsuyama N, Mizuno H, Masujima T (2012) Molecular and functional analysis of cellular phenomena using single-cell mass spectrometry. Biol Pharm Bull 35(9):1425–1431
87. Wu B, Becker JS (2011) Imaging of elements and molecules in biological tissues and cells in the low-micrometer and nanometer range. Int J Mass Spectrom 307(1):112–122. doi:10.1016/j.ijms.2011.01.019
88. Shi Q, Qin L, Wei W, Geng F, Fan R, Shin YS, Guo D, Hood L, Mischel PS, Heath JR (2012) Single-cell proteomic chip for profiling intracellular signaling pathways in single tumor cells. P Natl Acad Sci USA 109(2):419–424. doi:10.1073/pnas.1110865109
89. Fabbri F, Carloni S, Zoli W, Ulivi P, Gallerani G, Fici P, Chiadini E, Passardi A, Frassineti GL, Ragazzini A (2013) Detection and recovery of circulating colon cancer cells using a dielectrophoresis-based device: KRAS mutation status in pure CTCs. Cancer Lett 335 (1):225–231. doi:10.1016/j.canlet.2013.02.015
90. Lindström S, Andersson-Svahn H (2011) Miniaturization of biological assays—overview on microwell devices for single-cell analyses. Biochim Biophys Acta 1810(3):308–316. doi:10.1016/j.bbagen.2010.04.009
91. Di CD, Wu LY, Lee LP (2006) Dynamic single cell culture array. Lab Chip 6(11):1445–1449. doi:10.1039/b605937f
92. Edd JF, Di CD, Humphry KJ, Köster S, Irimia D, Weitz DA, Toner M (2008) Controlled encapsulation of single-cells into monodisperse pic olitre drops. Lab Chip 8(8):1262–1264. doi:10.1039/b805456h

Chapter 9
Microfluidics-Mass Spectrometry for Cell Analysis

Ling Lin and Jin-Ming Lin

Abstract Microchip-mass spectrometry has been emerging as excellent analytical tool in the field of complex biological samples analysis. Microchip can play an important role, such as cell culture and sample preparation steps prior to mass spectral identification, which are beneficial from its ability to handle small sample quantities with the potential for high-throughput parallel analysis. Recent progress in chip-mass spectrometry including various approaches that combined microchip devices with electrospray ionization (ESI) and matrix-assisted laser desorption/ionization (MALDI) is described. Then the main applications of chip-mass spectrometry in proteomics and cell analysis during the recent years are reviewed. We also outlook their implications in the future. The goal of this chapter is to guide new and interested researchers to the important areas and strive toward full integration of microchip and mass spectrometry within biological research field.

Keywords Microfluidics · Mass spectrometry · Sample pretreatment
Cell analysis

9.1 Introduction

Mass spectrometry (MS), is the most used analytical tool today due to its high sensitivity and selectivity [1–3]. It has been developed as a powerful tool for the qualitative and quantitative detection of analytes ranging from inorganic elements to small organic molecules to macromolecules, such as proteins and nucleic acids.

L. Lin
National Center for Nanoscience and Technology,
Chinese Academy of Sciences, Beijing 100190, People's Republic of China
e-mail: linling@nanoctr.cn

J.-M. Lin (✉)
Department of Chemistry, Tsinghua University, Beijing 100084,
People's Republic of China
e-mail: jmlin@mail.tsinghua.edu.cn

J.-M. Lin (ed.), *Cell Analysis on Microfluidics*, Integrated Analytical Systems,
https://doi.org/10.1007/978-981-10-5394-8_9

MS has been extensively applied to the study of drug cytotoxicity [4], cell—cell interaction [5, 6], and cellular metabolism [7]. MS has become increasingly important as an analytical tool for bioanalysis because of its sensitivity. MS allows the ionization of intact molecules to obtain a highly accurate molecular weight, making identification of molecules easier. As MS technology advanced, the advent of two "soft" ionization methods, namely the electrospray ionization (ESI) [8] and the matrix-assisted laser desorption/ionization (MALDI) [9, 10] revolutionized MS analysis. ESI produces highly charged ions directly from a liquid, thus facilitating online coupling of chromatographic separations to the mass spectrometer. However, the presence of high concentrations of salts in the samples can lead to severe ion suppression which makes analysis impossible. MALDI has a high tolerance to buffers and salts, but MALDI spectra tend to be very noisy in the mass range below 500 daltons, due to the presence of matrix ions. Therefore coupling mass spectrometry to separation technology was proposed. Despite many advantages in mass spectrometry, the direct application of MS detection to real complex sample is greatly limited. Here, gas chromatography (GC)-MS, liquid chromatography (LC)-MS and capillary electrophoresis (CE)-MS based on column separation have made greatly contribution for the separation and detection of molecules. Mass spectrometry based technology with higher throughput, more rapid and convenient analysis, and lower sample consumption are still in great demand to meet the new challenges in life science.

Miniaturized total analysis systems have attracted an increasing interest in analytical chemistry [11, 12]. There has been great progress in microfluidic devices, which have realized miniaturization of chemical analysis and integration of various chemical processes. This research field is called "micro total analysis systems (μ-TAS)" or "Lab-on-a-chip" [13–16]. Small fluidic channels, with sizes from tens to hundreds of μm which are fabricated using micro fabrication technology on a substrate of several cm have been utilized as a space in which reaction and analytical processes occur. The device is often called a "microchip." By controlling the size, fluidics, surface properties, temperature of the micro scale channel, various chemical and analytical processes have been integrated and miniaturized. As a result, many advantages such as short analysis time, low consumption of sample and reagent materials, small waste volumes, effective reactions due to the large specific interfacial area, and small space requirements, have been achieved in this field. This strategy is effective not only from an industrial viewpoint but also widely applicable in various chemical fields, such as organic chemistry, physical chemistry, and cellular biochemistry. In the technology used for biological analysis, several microfluidic devices have been reported as using optical imaging [17], electrophoresis [18], and chemiluminescence [19, 20]. But these methods were limited in many aspects, for example, they lack the information for the characterization of chemical structure. An alternative strategy is to connect the microchip with mass spectrometer for signal analysis.

From 1997, Karger [21] and Ramsey [22] developed the microchip-mass spectrometry in modern biological analysis. A lot of microfluidic chip-mass spectrometry studies on the interface and applications were also reported [23, 24]. A successful research has been carried for peptide separation and identification by coupling a microchip with electrospray ionization quadrupole time-of-flight mass spectrometer [25, 26]. After these works, series dynamic analytical methods for cell culture and metabolism detection [27–29], cell-cell co-culture and their information exchange [30–32] have been developed, and realized their potential as an integrated platform for cellular biology analysis [27].

In our previous works, we developed integrated microfluidic devices containing multiple functional parts to directly couple with mass spectrometry through a silica-fused capillary for cells microenvironments analysis [4, 28–30]. Since 2008 we presents a newly analysis methods for cells study coupling microfluidic devices to ESI-MS and MALDI-MS [25, 31]. Attracted considerable research in recent years showing this method enormous potential for application in the biochemical analysis. That includes cell metabolism [27, 32–34], drug absorption [35, 36], drug permeability [37], cell–cell interactions [38–41], and cell secretion [6, 42]. Here, drug metabolism or cell secretion studies related functional parts, including cell culture, metabolism generation or cell secretion, sample pretreatment, and detection, were all integrated into the microfluidic devices. Compared with conventional methods, our methods offered several advantages, such as short analysis time (less than 10 min), low sample and reagent consumption (less than 100 mL). In order to detect rare samples, we also developed micro-sampling droplet combined with by paper spray ionization for online chemical monitoring of cell culture [7]. Our established platforms have the potential for drug screening and the diagnosis of specific disease with high throughput and low cost.

This chapter will provide a critical survey of recent advances and future trends of microchip-MS in biological sample analysis, with a special focus on various functional sample preparation and separation units integrated on microfluidic chips and the recent interface developments between chip and mass spectrometry. Then the biological applications of chip-MS, including proteomics and cell analysis, are summarized and discussed.

9.2 Mass Spectrometer Interface

Due to the recent development in microfluidic technologies, miniaturized microfluidic ion sources have received considerable attention in coupling with MS, and have gained a great achievement [32]. The interface between microfluidic chip and MS detector is a major challenge in the development of chip-MS technique. ESI can directly couple to microfluidic chips. The interface based on MALDI is typically performed off-line through deposition of the chip effluent onto a MALDI target or by directly ionizing from the chip.

9.2.1 ESI Interface

The major challenges for coupling of microfluidics to ESI-MS focus on the development of stable and effective interfaces. The existing chip-ESI MS interfaces are mainly the following three approaches: (1) spraying directly from the edge of a chip; (2) spraying from a chip-inserted fused-silica chip; (3) spraying from a microfluidic chip-integrated ESI-tip.

Several approaches for integrating a single electrospray emitter with different materials have been developed. For example, a robust PDMS microchip ESI-MS interface which containing an auxiliary channel to provide electrical contact was developed. Lin's group developed integrated microfluidic devices containing multiple functional parts to directly couple with ESI-Q-TOF MS through a silica-fused capillary for biochemical analysis, which proposed a new tool for the cell analysis and opened a new research area in the life science. As shown in Fig. 9.1, drug metabolism or cell secretion studies related functional parts, including cell culture, metabolism generation or cell secretion, sample pretreatment, and detection, were all integrated into the microfluidic device. Compared with conventional methods, our methods offered several advantages, such as short analysis time (less than 10 min), low sample and reagent consumption (less than 100 μL). They established platform has the potential for drug screening and the diagnosis of specific disease with high throughput and low cost. Besides, Sun et al. [43] proposed a novel PDMS membrane-based microfluidic emitter for nano ESI-MS (Fig. 9.2a). This emitters were cutting into ~100-μm-thick, and stable electrosprays could be achieved at low potentials (~2 kV) and low flow rates (10 nL/min). This membrane-based ESI emitter enabled convenient integration with on-chip separations (e.g., microchip CE) for MS detection and multilayer PDMS microfluidic devices (e.g., pneumatic valve-integrated microchips). The

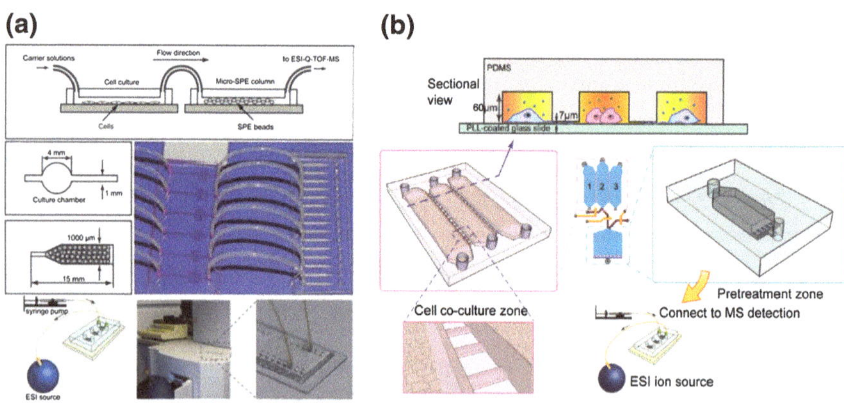

Fig. 9.1 PDMS microchip with MS cell analysis. **a** Cell culture and sample pretreatment prior to ESI-Q-TOF-MS. **b** Schematic illustration of chip-MS platform for cell signaling analysis. (Reprinted with permission from Ref. [33, 39].)

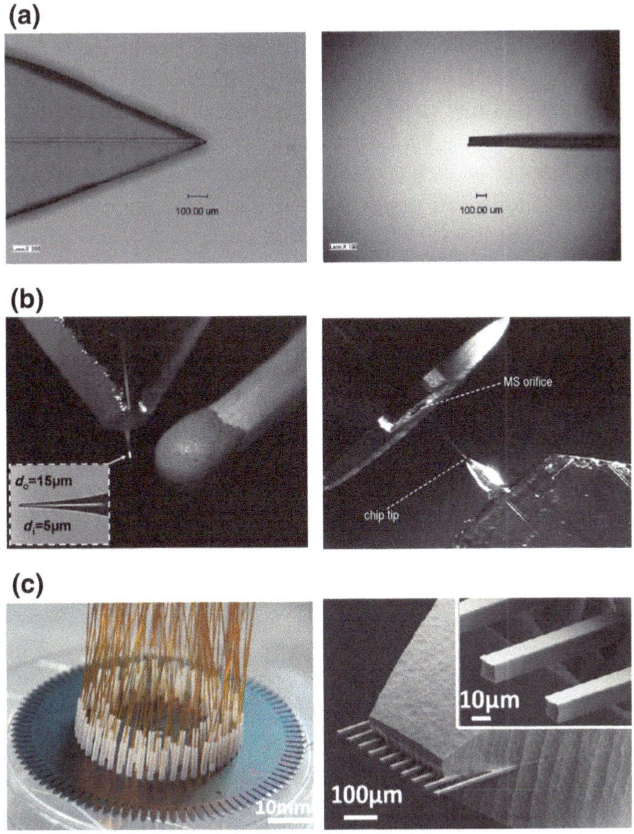

Fig. 9.2 Microchip ES-MS interface. **a** Photographs of a PDMS-membrane-based microfluidic ESI emitter from *top view* and *side view* [43]. **b** Visualization of the integration nanospray emitter for electrospray process [45]. **c** Microchip dual ESI device with analyte emitter active and reference emitter active [53]. (Reprinted with permission from Ref. [43, 45, 53])

integration of electrospray emitters on glass-based microfluidic chips is also of greatly interest due to important as the interface for chip to mass spectrometer [44]. Hoffman et al. [45] developed a glass-based Chip-MS platform with a monolithically integrated nanospray tip (Fig. 9.2b). The nanospray emitter was manufactured by computer numerical control (CNC) milling followed by heating and pulling. The electrospray performance of the integrated glass emitter was found to be comparable to commercially available nanospray needles. To improve the stability of the electrospray, the same research group optimized the manufacturing process including CNC milling and HF etching to generate a well-controlled opening of the fused pulled tips. These tips are of very small emitter dimensions (only a few microns) without dead-volume.

Recently, Single cell genomics has become a new frontier [46]. Proof-of-principle experiments have been performed to characterize peptides and metabolites in single cells using MALDI-MS [47] and ESI-MS [48]. However, limited sample processing strategies and low coverage of proteome and metabolome hindered their further developments. Thus, efficient sample processing in a single cell, high-throughput capabilities, as well as the improvement in detection sensitivity are urgently needed in single cell omics studies. The integration of multiple electrosprays simultaneously is a powerful method to improve MS sensitivity. Sequential multiple electrosprays using multichannel [49], multiple nozzles [50, 51], and gated multi-inlets [52] can realize high-throughput. Here, Mao et al. [53] reported a silicon-based monolithic multinozzle emitter array (MEA), it opens up the possibility of a fully integrated microfluidic system for ultrahigh-sensitivity and ultrahigh-throughput proteomics and metabolomics analysis (Fig. 9.2c). Digital microfluidic (DMF) device usually coupled with MALDI in an offline manner, however Baker et al. [54] realized the online coupling of digital microfluidic devices with electrospray ionization using an eductor, which consisted of a transfer capillary, a standard ESI needle, and a tapered gas nozzle. It has a good potential for rapid, versatile and high-throughput microfluidic analysis.

9.2.2 MALDI Interface

MALDI is another common soft ionization method, which was initially proposed in late 1980s [9, 10]. MALDI generally uses for large molecules analysis, such as proteins, peptides, and other biomolecules [43, 55]. Recently, high-density microarrays on microchip were directly adapted to the MALDI-MS analysis. Ability to obtain thousands of spectra from tissues or other samples by MALDI-MS makes it an optimal technology for high-throughput cell analysis and drug research [56–59]. On-line interfaces for MALDI-MS can be realized by delivering the matrix and analyte into the mass spectrometer mainly by the following three strategies, such as aerosol particles, capillaries, or mechanical means. However, the online coupling is challenge because a MALDI target is under vacuum while the operations on microfluidics are at atmospheric pressure. Microfluidic-based MALDI analysis is usually accomplished by dropping, spraying, or spotting the chip effluent onto a sample target and adding matrix. Spotting is the most widely used method for sample deposition, in which precise volume control of the droplets was very important. The most direct method for sample deposition is spotting with a robotic target spotter. The matrix and samples are online mixed before deposition or spotted onto a matrix-coated target. Using this approach, we developed inkjet automated single cells and matrices printing system for single cell MALDI analysis [60]. Single or several cells were introduced and printed onto ITO glass substrate by inkjet technology and phospholipids were detected by MALDI-MS cell analysis platform (Fig. 9.3a). A gold nanoparticles modified porous silicon chip based surface assisted laser desorption/ionization mass spectrometry (SALDI-MS) was

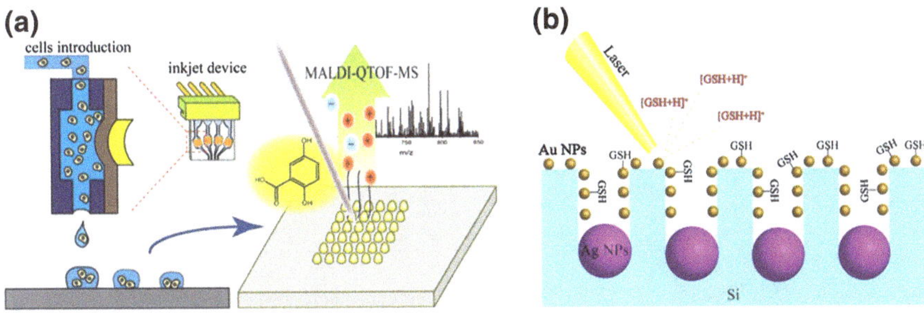

Fig. 9.3 Microchip MALDI interface. **a** Inkjet automated single cells and matrices printing system for MALDI cell analysis [60]. **b** Gold nanoparticles modified porous silicon chip for MALDI glutathione (GSH) in cells [61]. (Reprinted with permission from Ref. [60, 61])

developed to capture and analyze glutathione (GSH) in cells [61]. The silicon chip was array patterned for high throughput SALDI-MS detection and showed great potential for more efficient analysis of small thiol biomarkers in complex biological samples (Fig. 9.3b).

9.3 On-Chip Sample Pretreatment

The integration of sample pretreatment within microfluidic devices has become an important research area. In biological analysis, the analytes of interest are commonly at low abundance and mixed with complex interferences. Considering the limited volume of analysis system, pretreatment is roughly divided as preconcentration and separation. On-chip sample preconcentration and separation are especially needed during chip-MS analysis.

9.3.1 Sample Preconcentration

Before analytes are prepared on MS targets, samples usually need to go through preconcentration. Hence, preconcentration is necessary for effective MS analysis. For example, Lin group realized the extraction of cell metabolites from culture medium via microdialysis (Fig. 9.4a) [7]. A homemade microdialysis hollow fiber was immersed into cell culture medium, and as dialysate flowed into the fiber, glucose from culture media diffused through dialysis interface. The other end of the fiber came to paper spray ionization for MS detection. With the continuously flowing dialysate and generating droplets, on-line monitoring at glucose level in cell culture medium was achieved with the shortest time interval of 1.5 s. This work demonstrated a potential platform with label free MS detection for the monitoring of cellular culture system.

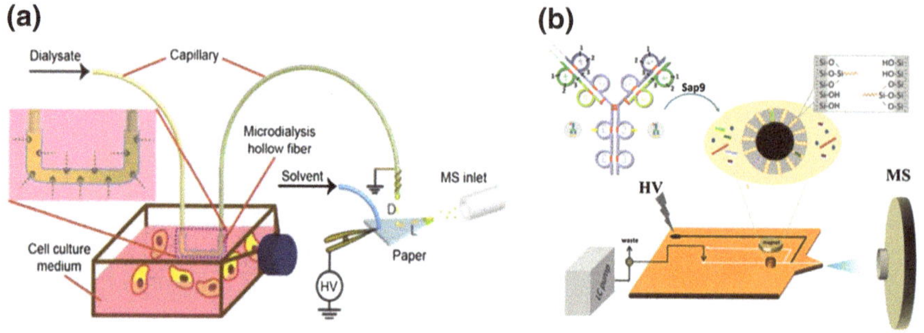

Fig. 9.4 On-Chip sample preconcentration. **a** Microdialysis to extract glucose for paper spray ionization from cell culture medium [7]. **b** Schematic representation of the experimental setup. For the electrospray induction, high voltage (HV) is applied to the SPE-GEMS/MS microchip emitter, placed in front of the ion transfer capillary inlet of MS instrument. Meanwhile, for the fractionation of the analyte, enriched on the magnetic SPE sorbent, the eluent flow and composition are controlled by analytical LC pump and microsplitter valve [65]. (Reprinted with permission from Ref. [7, 65])

Novel paper spray cartridge with an integrated solid phase extraction (SPE) column was described by Zhang and Manicke [62]. The cartridge performs extraction and pre-concentration, as well as sample ionization by paper spray, from complex samples such as plasma. The cartridge allows for selective enrichment of target molecules from larger sample volumes and removal of the matrix, which significantly improved the signal intensity of target compounds in plasma samples by paper spray ionization.

Furthermore Hu's group proposed a chip-based array magnetic solid phase microextraction (MSPME) system and on-line combined it with ICP-MS via microflow concentric nebulization for the determination of trace heavy metals in cells [63]. Wheeler group introduced a method to couple solid-phase microextraction (SPME) with HPLC-MS using digital microfluidics (DMF) [64]. In this technique, analytes in sample solution are extracted into a SPME fiber coating off-line, and then eluted into a droplet of desorption solvent on a digital microfluidic device, prior to analysis by HPLC-MS. The low volumes inherent to DMF allow for pre-concentration of analytes prior to analysis. The new SPME-DMF-HPLC-MS method was applied to the quantification of pg/mL-level free steroid hormones in urine. Girault group introduced an on-chip solid-phase extraction–gradient elution–tandem mass spectrometry (SPE-GEMS/MS), as shown in Fig. 9.4b [65]. This technique combines in a microfluidic format online sample preconcentration/purification on SPE sorbent with further fractionation and MS/MS analysis. Mikkonen et al. [66] used isoelectric focusing (IEF) to enrich samples in an open channel. After preconcentration, matrix was applied to the analytes by electrospray matrix deposition, resulting in effective crystallization in the channel for subsequent MALDI-MS detection. Then, they developed a novel method for preconcentration and purification of the Alzheimer's disease related amyloid beta (Aβ) peptides by IEF in microchannels combined with their analysis by micropillar-matrix-assisted laser desorption ionization-time-of-flight-mass spectrometry (MALDI-TOF-MS) [67]. Yang and co-workers presented a microfluidic aptamer-based biosensor for detection of low-molecular-weight biomarkers in patient samples by using computer simulation and performing MALDI-MS analysis directly from the open microchannel [68]. Using a microfluidic device that integrates aptamer-based specific analyte extraction, isocratic elution, and detection by matrix-assisted laser desorption/ionization time-of-flight (MALDI-TOF) mass spectrometry demonstrated rapid, sensitive and label-free detection of arginine vasopressin (AVP) in human plasma ultrafiltrate. This integrated on-chip sample processing enables the quantitative detection of low-abundance AVP by MALDI-TOF mass spectrometry in a rapid and label-free manner.

9.3.2 Sample Separation

On-chip separation technique was also developed on a chip-MS platform to achieve multi-component, high throughput analysis. Here, Ramsey's group developed a microfabricated fluidic device for the automated real-time analysis of individual cells using capillary electrophoresis (CE) and electrospray ionization-mass spectrometry (ESI-MS) [69]. The microfluidic structure incorporates a means for rapid lysis of single cells within a free solution electrophoresis channel, where cellular constituents were separated, and an integrated electrospray emitter for ionization of separated components. The eluent was characterized using mass spectrometry. The dissociated heme group and the α and β subunits of hemoglobin from individual erythrocytes were detected as cells continuously flowed through the device. The average analysis throughput was approximately 12 cells per minute, demonstrating the potential of this method for high-throughput single cell analysis. Moreover, they utilized capillary electrophoresis–mass spectrometry (CE–MS) in an integrated microfluidic platform to analyze an intact, lysine-linked antibody drug conjugate (ADC) in order to assess post translational modifications and drug load variants [70].

Feasibility of multichannel parallel separation for sensitivity improvement in interfacing microchip electrophoresis hyphenated with inductively coupled plasma mass spectrometry (MCE-ICP-MS) was reported by Wang group [71]. By using 2–20 array lanes for parallel separation, the sensitivity of the MCE-ICP-MS system was proportionally improved by 2–20 folds. There are other work s concerning the application of MCE in metabolites probing with combination of electrochemical [72] or ESI-MS detection [73]. The sensitivity and quantitative ability of CE can be further strengthened via modification of sample introduction manner such as inkjet direct sampling (Fig. 9.5) [74]. About rapid liquid phase lithography prototyping creates the multifunctional device economically. Belder's group presented a microfluidic chips containing porous polymer monolithic columns as a means to facilitate chemical transformations as well as both downstream chromatographic separation and mass spectrometric analysis [75].

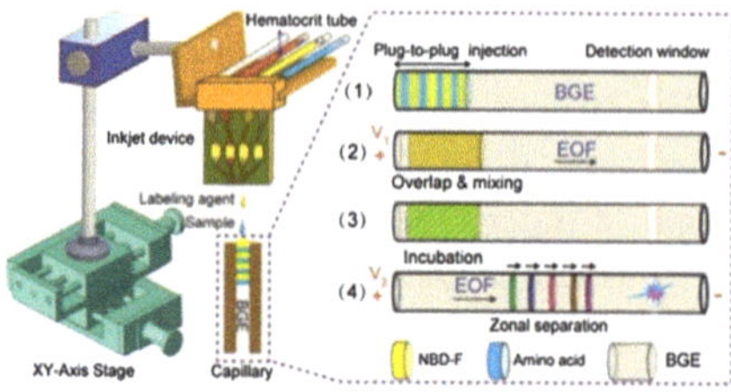

Fig. 9.5 On-Chip sample separation. Illustration of ink jet sample introduction to CE and NBD-F derivatization of amino acids by drop sample introduction [74] (Reprinted with permission from Ref. [74])

9.4 Analytical Application

9.4.1 Proteomics

Proteomics has generated great research interest in elucidating cell function, to investigate the relationship between diseases state and clinical diagnostics [76]. Currently, one of the most successful analytical strategies of proteomics is 2D gel electrophoresis separation, in-gel digestion by proteases, and MS detection and analysis. However, many necessary sample-preparation steps in this process remain time consuming and labor intensive. Microfluidic systems have been proposed as the means of automating sequential sample-pretreatment steps and increasing sample throughput in parallel proteomics analysis [77]. Multiple processing steps can be seamlessly integrated on a single chip or multiple chips to allow fully automated sample processing, which can address the labor-intensity issues associated with proteomics analyses. Protein identification forms the basis for understanding protein functions and their roles in biological networks. Microfluidic-based approaches have been developed to perform various tasks that are essential for protein identification, such as enzymatic digestion [78–82], separation [83–85], and sample infusion [86]. Integrated microfluidic systems capable of performing multiple functions have also been reported for protein identification [87].

Wilson group developed a means to integrate hydrogen/deuterium exchange (HDX) labeling upstream of this system to study allosteric effects in enzyme inhibition [88]. As shown in Fig. 9.6a, a variable position micromixer was used to adjust the reaction time of the beta lactamase enzyme, TEM-1, with a covalent inhibitor clavulanate. This system was successfully applied to characterize structural changes and dynamics that occur upon acylation of TEM-1 by clavulanate. Both rapid structural changes at the active site and subtle changes in active-remote regions were identified.

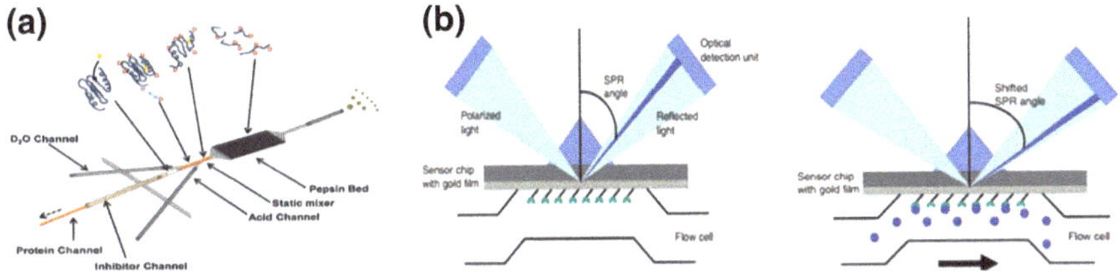

Fig. 9.6 Chip-MS devices for proteomic applications. **a** Schematic of the device used for HDX pulse labeling [88]. **b** The principle of surface plasmon resonance (SPR) [89]. (Reprinted with permission from Ref. [88, 89])

MS-based investigation of protein–protein interactions usually involves the enrichment of a bait protein and its interacting partners followed by separation, digestion, and MS detection. Microfluidic-based approaches can greatly facilitate the identification of protein complexes and their related dynamic pathways. Recent developments of surface plasmon resonance (SPR) and MS have provided the feasibility of better integration that could be dedicated to the identification and characterization of protein–protein interactions [89]. Madeira et al. discoveedr protein–protein interactions using the surface plasmon resonance technique coupled to mass spectrometry (MS), as shown in Fig. 9.6b. A peptide or a protein is immobilized on a sensor chip and then exposed to brain extracts injected through the surface of the chip by a microfluidic system. The interactions between the immobilized ligand and the extracts can be monitored in real time.

9.4.2 *Metabolomics*

Cell metabolite analysis is of great interest to analytical chemistry and clinical medicine society, and some metabolites having been identified as important indicators of major diseases such as cancer and HIV. Recently, our group has published a series of works employing micro-SPE columns integrated to cell culture chip for establishment of drug cellular testing platform, study of cell-cell communication and drug metabolism. The device contained three components: One part composed of drug and culture medium injection entrance; the second part was used as cell culture and observation chamber for drug-induced cell apoptosis; the third part was composed of array of SPE micro columns for sample desalination. For examples, Gao et al. introduced a microfluidic platform to test drug permeability of intestine and coupled mass spectrometer for online curcumin detection [90]. Cell culture as well as sample desalting and extraction was integrated in the same chip by operating two independent channels separated by a semipermeable polycarbonate (PC) membrane, and integrating SPE micro-columns. Permeation of curcumin on microfluidic platform was characterized by HPLC and electrospray ionization quadrupole time-of-flight mass spectrometer (ESI-Q-TOF MS). Gao et al.

developed an integrated chip-MS platform for high-throughput drug screening with an online ESI-Q-TOF MS. By using the technology combination, characterization of drug absorption and evaluation of cytotoxicity could be simultaneously realized (Fig. 9.7a) [35]. Chen et al. reported a microfluidic chip integrated ESI-Q-TOF MS platform based on a stable isotope labeling technology for qualitative and quantitative analysis of cell metabolism under antitumor drug genistein treatment [27]. They also reported a cell-compatible paper chip for in situ sensing of live cell components by PSI-MS, which allowed profiling the cellular lipids and quantitative measurement of drug metabolism with minimum sample pretreatment (Fig. 9.7b) [91]. Wei et al. fabricated a cell co-culture chip in which the secreted proteins were qualitatively and semi-quantitatively determined by a directly coupled mass spectrometer (Fig. 9.1b) [39]. PC12 cells and GH3 cells were co-cultured under various conditions to simulate the nervous system regulation of different organs. A SPE column was integrated to remove salts from the cells secretion prior to MS detection. A three-layer PDMS microfluidic device was fabricated to integrate valves for avoiding contamination between the cells co-culture zone and the pretreatment zone. The metabolites could be directly detected online with an ESI-Q-TOF MS after micro-SPE pretreatment. Zhang et al. constructed an in vitro liver model in a microchip to imitate and detect prodrug capecitabine (CAP) metabolism (Fig. 9.7c) [6]. CAP was metabolized into active metabolite in HepG2

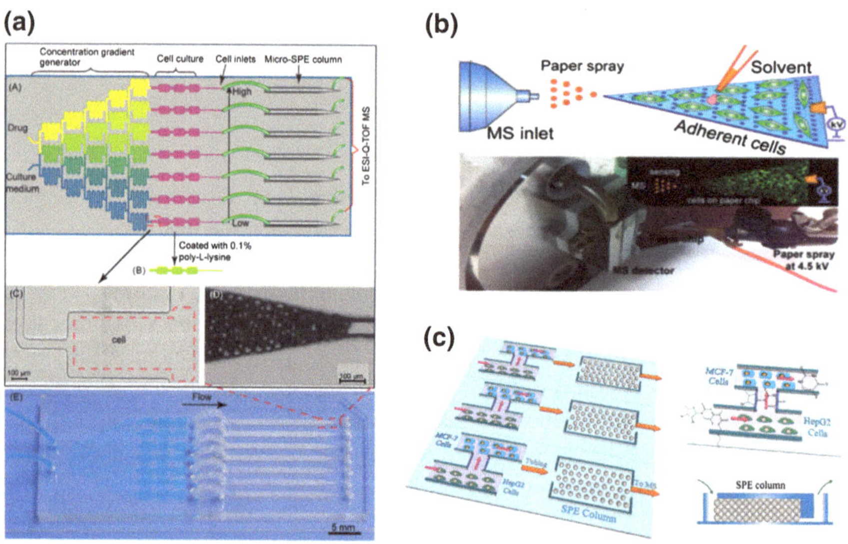

Fig. 9.7 Cell metabolite analysis on chip-MS platform. **a** A schematic drawing shows the upstream concentration gradient generator, the downstream cell cultivation modules, and sample pretreatment module prior to ESI-Q-TOF MS detection. [35] **b** Schematic diagram of cell cultures on paper-based chip-MS for cell lipid analysis. [91] **c** Schematic of cell co-culture and on-chip preconcentration process on microchip. [5]. (Reprinted with permission from Ref. [5, 35, 91])

cell and then transformed into final effective drug in tumor cells. SPE microcolumns were integrated into this device as pretreatment units prior to ESI-Q-TOF MS.

9.4.3 Glycomics

Protein glycosylation plays an important role in various biological processes, such as modification of protein function, regulation of protein–protein interactions, and control of turnover rates of proteins. Glycosylation is one of the most common post-translational protein modifications, having major roles in various biological processes [92–97]. Quantification is an important part of a glycomics study as in recent years; glycans have been identified as potential biomarkers for a number of diseases, including cancers, immune deficiencies, and cardiovascular diseases [98–101]. Accordingly, quantitation reliability is critical for analysis; meanwhile high-throughput and highly efficient quantitative methods are in high demand because of the vast number of samples required for quantitative glycomic analyses of clinical samples.

Glycans can be labeled with chromophores and detected and quantified by fluorescence spectroscopy. However, spectroscopy-based quantitation methods rely greatly on internal or external standards, and quantitation reliability decreases with increasing sample complexity. In recent years, mass spectrometry (MS) has become a dominant analytical instrument for glycomic research because of the ability to offer glycan identification, structure elucidation, and compositional quantitation through high resolution MS and triple quadrupole MS operated in selected reaction monitoring (SRM) or multiple reaction monitoring (MRM) modes [102, 103]. For the overall analysis of glycans or oligosaccharides, the competitive ionization in the ion source of a mass spectrometer is a significant barrier to observation of many trace-level analytes. The use of a separation technique prior to MS analysis is perhaps the most efficient approach to minimize the adverse effects from competitive ionization. The most reported separation methods are graphitized carbon chromatography [104], hydrophilic interaction chromatography, high performance anion-exchange chromatography, chip-based reverse-phase LC-MS. Here, Lebrilla's group first used an Agilent HPLC-Chip/Time-of-Flight MS system for the separation of serum glycans, and then online detection by nano ESIMS [105]. The microfluidic chip consisted of an enrichment column, an analytical column packed both with porous graphitized carbon as the stationary phase, and an integrated nano-ESI emitter. By analyzing serum glycans in two groups of prostate cancer patients with different prognoses, they found a number of significant differences in N-glycan abundances. Lebrilla's group has demonstrated the application of Chip-MS to glycomics by profiling N-glycans from human serum [106] and human milk glycome [107–110]. They have achieved [105] the comprehensive glycan profiling of human serum with the capability of isomer separation and quantitation by Chip LC-MS. In this study, a microfluid ic chip pa cked with graphiti zed carbon was used to chroma tographically separate the glycans from the serum sampl es of two groups of prostate cancer patients. More than 300 N-glycan species (including isomeric structures) were identified, corresponding to over 100 N-glycan compositions.

9.4.4 Single Cell Analysis

Nowadays, numerous methods have been developed to disperse and confine single cells in discrete compartments for further mass spectrometry (MS)-coupled analysis. The coupling of a microfluidic chip with MS has been emerging as a powerful analytical tool in cell-based assays [29, 111–113].

A stable isotope labeling assisted microfluidic chip-electrospray ionization MS platform was developed to qualitatively and quantitatively analyze cell metabolism [27]. Proteins secreted by co-cultured cells were qualitatively and semi-quantitatively determined by MS which was coupled with an integrated microfluidic device [39]. An electrochemical chip was coupled online to MS or LC–MS to generate phase I and phase II drug metabolites and to demonstrate protein modification by reactive metabolites. This new screening method gave the potential of this electrochemical chip as a complementary tool for a variety of drug metabolism studies in the early stages of drug discovery [113].

Recently, Liu's group [114] developed a microfluidic platform deploying the double nano-electrode cell lysis technique for automated analysis of single cells with mass spectrometric detection. The proposed microfluidic chip features integration of a cell-sized high voltage zone for quick single cell lysis, a microfluidic channel for electrophoretic separation, and a nanoelectrospray emitter for ionization in MS detection. Built upon this platform, a microchip electrophoresis-mass spectrometric method (MCE-MS) has been developed for automated single cell analysis. In the method, cell introduction, cell lysis, and MCE-MS separation are computer controlled and integrated as a cycle into consecutive assays. Analysis of large numbers of individual PC-12 neuronal cells (both intact and exposed to 25 mM KCl) was carried out to determine intracellular levels of dopamine (DA) and glutamic acid (Glu). To conduct the studies of cell metabolomics at single-cell resolution, MALDI-MS was used as an analytical detector on microplate chip. Amantonico et al. [47] validated a single-cell mass spectrometric method to evaluate the metabolic levels in the populations of unicellular organisms (Fig. 9.8a). Several cellular components (ADP, ATP, GTP, and UDP-Glucose) in single-cells of Closterium acerosum was easily analyzed by negative-mode MALDI-MS. Recently, Our group [115] reported a single-cell lipid profiling platform combining an inkjet nozzle cell manipulator with probe electrospray ionization mass spectrometry (PESI-MS) (Fig. 9.8b). Through inkjet sampling of a cell suspension, droplets with single cells were generated, precisely dripped onto a tungsten-made electrospray ionization needle, and immediately sprayed under a high-voltage electric field. Lipid fingerprints of single cells were obtained by a mass spectrometry (MS) detector.

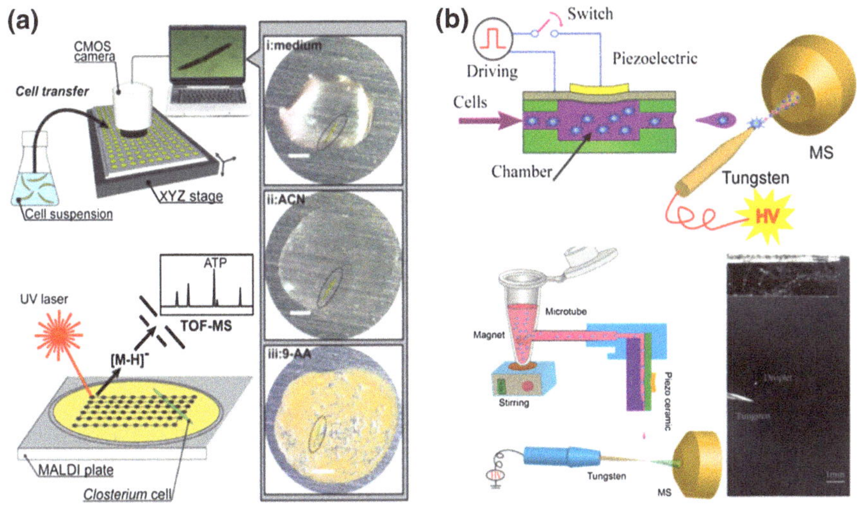

Fig. 9.8 Single cell analysis using Chip-MS. **a** Single-cell metabolites detection using MALDI-MS [47]. **b** Single-cell lipid profiling platform combining an inkjet nozzle cell manipulator with probe electrospray ionization mass spectrometry (PESI-MS) [113]. (Reprinted with permission from Ref. [47, 113])

9.5 Conclusion and Future Perspective

Coupling microfluidic chips to MS takes advantages of both technologies, addressing issues in conventional analytical methods such as efficiency, sensitivity and throughput. In this chapter, we documented the advances in coupling microfluidic chips to MS in the past decade, covering both innovations in Chip-MS interfaces and their applications in biological research and clinical diagnosis. ESI interfaces have evolved from off-the-edge spraying, inserted fused-silica capillary emitters to fully integrated micro fabricated emitters. There are remarkable advantages in integrated emitters such as easy fabrication and zero dead volume, and the number of corresponding publications has been rapidly increasing. In addition to a large number of innovations in Chip-MS interfaces, a wide range of applications has also been reported, including proteomics, glycemic, metabolic studies, cell analysis, and clinical diagnosis. Integration of multiple functions has been a major trend in coupling microfluidic chips with MS for automated sample preparation and MS detection with high sensitivity and throughput. Highly-automated, fully-integrated chip-MS would revolutionize the way that we do chemical and biological analysis today.

References

1. Bings NH, Skinner CD, Wang C, Colyer CL, Harrison DJ et al (2000) Coupling electrospray mass spectrometry to microfluidic devices with low dead volume connections. In: Harrison DJ, van den Berg, Albert (eds), Micro Total Analysis Systems '98, Springer, Dordrecht 141–144
2. Astorga-Wells J, Jornvall H, Bergman T (2003) A microfluidic electrocapture device in sample preparation for protein analysis by MALDI mass spectrometry. Anal Chem 75:5213–5219
3. Grimm RL, Beauchamp JL (2003) Field-induced droplet ionization mass spectrometry. J Phys Chem B 107:14161–14163
4. Gao D, Li H, Wang N, Lin JM (2012) Evaluation of the absorption of methotrexate on cells and its cytotoxicity assay by using an integrated microfluidic device coupled to a mass spectrometer. Anal Chem 84:9230–9237
5. Ong TH, Kissick DJ, Jansson ET, Com TJ, Romanova EV et al (2015) Classification of large cellular populations and discovery of rare cells using single cell matrix-assisted laser desorption/ionization time-of-flight mass spectrometry. Anal Chem 87:7036–7042
6. Zhang J, Wu J, Li HF, Chen QS, Lin JM (2015) An in vitro liver model on microfluidic device for analysis of capecitabine metabolite using mass spectrometer as detecto. Biosens Bioelectron 68:322–328
7. Liu W, Wang NJ, Lin XX, Ma Y, Lin JM (2014) Interfacing microsampling droplets and mass spectrometry by paper spray ionization for online chemical monitoring of cell culture. Anal Chem 86:7128–7134
8. Fenn JB, Mann M, Meng CK, Wong SF, Whitehouse CM (1989) Electrospray ionization for mass spectrometry of large biomolecules. Science 246:64–71
9. Karas M, Hillenkamp F (1988) Laser desorption ionization of proteins with molecular massesexceeding 10,000 daltons. Anal Chem 60:2299–2301
10. Tanaka K, Waki H, Ido Y, Akita S, Yoshida Y et al (1988) Protein and polymer analyses up to m/z 100 000 by laser ionization time-of-flight mass spectrometry. Rapid Commun Mass Sp 2:151–153
11. Janasek D, Franzke J, Manz A (2006) Scaling and the design of miniaturized chemical-analysis systems. Nature 442:374–380
12. Ríos A, Escarpa A, González MC, Crevillén AG (2006) Challenges of analytical microsystems. TrAC Trend Anal Chem 25:467–479
13. Reyes DR, Iossifidis D, Auroux PA, Manz A (2002) Micro total analysis systems. 1. introduction, theory, and technology. Anal Chem 74:2623–2636
14. Auroux PA, Iossifidis D, Reyes DR, Manz A (2002) Micro total analysis systems. 2. analytical standard operations and applications. Anal Chem 74:2637–2652
15. Vilkner T, Janasek D, Manz A (2004) Micro total analysis systems. 3. Recent Developments. Anal Chem 76:3373–3385
16. Dittrich P, Tachikawa K, Manz A (2006) Micro total analysis systems. Latest Advancements and Trends. Anal Chem 78:3887–3907
17. Liu W, Dechev N, Foulds IG, Burke R, Parameswaran A et al (2009) A novel permalloy based magnetic single cell micro array. Lab Chip 9:2381–2390
18. Taniguchi Y, Choi PJ, Li GW, Chen H, Babu M et al (2010) Quantifying E. coli Proteome and Transcriptome with Single-Molecule Sensitivity in Single Cells. Science 329:533–538
19. Su R, Lin JM, Qu F, Gao Y, Chen Z et al (2004) Capillary electrophoresis microchip coupled with on-line chemiluminescence detection. Anal Chim Acta 508:11–15
20. Su R, Lin JM, Uchiyama K, Yamada M (2004) Integration of a flow-type chemiluminescence detector on a glass electrophoresis chip. Talanta 64:1024–1029
21. Koster S, Verpoorte E (2007) A decade of microfluidic analysis coupled with electrospray mass spectrometry: An overview. Lab Chip 7:1394–1412
22. Whitesides GM (2006) The origins and the future of microfluidics. Nature 442:368–373

23. Lee J, Soper SA, Murray KK (2009) Microfluidic chips for mass spectrometry-based proteomics. J Mass Spectrom 44:579–593
24. Meyvantsson I, Beebe DJ (2008) Cell culture models in microfluidic systems. Annu Rev Anal Chem 1:423–449
25. Li H, Liu J, Cai Z, Lin JM (2008) Coupling a microchip with electrospray ionization quadrupole time-of-flight mass spectrometer for peptide separation and identification. Electrophoresis 29:1889–1894
26. Zheng Y, Li H, Guo Z, Lin JM, Cai Z (2007) Chip-based CE coupled to a quadrupole TOF mass spectrometer for the analysis of a glycopeptide. Electrophoresis 28:1305–1311
27. Chen Q, Wu J, Zhang YD, Lin JM (2012) Qualitative and quantitative analysis of tumor cell metabolism via stable isotope labeling assisted microfluidic chip electrospray ionization mass spectrometry. Anal Chem 84:1695–1701
28. Gao D, Liu H, Jiang Y, Lin JM (2013) Recent advances in microfluidics combined with mass spectrometry: technologies and applications. Lab Chip 13:3309–3322
29. Li H, Zhang Y, Lin JM (2014) Recent advances in coupling techniques of microfluidic device-mass spectrometry for cell analysis. Scientia Sinica Chimica 44(5):777–783
30. Jie M, Mao S, Li H, Lin JM (2017) Multi-channel microfluidic chip-mass spectrometry platform for cell analysis. Chin Chem Lett. doi:10.1016/j.cclet.2017.05.024
31. Wei H, Li H, Lin JM A microfluidic device combined with ESI-Q-TOF mass spectrometer applied to analyze herbicides on a single C_{30} bead. J Chromatogr A 1216: 9134–9142
32. Mao S, Gao D, Liu W, Wei H, Lin JM (2012) Imitation of drug metabolism in human liver and cytotoxicity assay using a microfluidic device coupled to mass spectrometric detection. Lab Chip 12(1):219–226
33. Gao D, Wei H, Guo GS, Lin JM (2010) Microfluidic cell culture and metabolism detection with electrospray ionization quadrupole time-of-flight mass spectrometer. Anal Chem 82:5679–5685
34. Zhang J, Chen F, He Z, Ma Y, Uchiyama K, Lin JM (2016) A novel approach for precisely controlled multiple cell patterning in microfluidic chip by inkjet printing and the detection of drug metabolism and diffusion. Analyst 141:2940–2947
35. Gao D, Li H, Wang N, Lin JM (2012) Evaluation of the absorption of methotrexate on cells and its cytotoxicity assay by using an integrated microfluidic device coupled to mass spectrometer. Anal Chem 84(21):9230–9237
36. Liu W, Lin JM (2016) Online monitoring of lactate efflux by multi-channel microfluidic chip-mass spectrometry for rapid drug evaluation. ACS Sensor 1(4):344–347
37. Gao D, Liu H, Lin JM, Wang Y, Jiang Y (2013) Characterization of drug permeability in Caco-2 monolayers by mass spectrometry on a membrane-based microfluidic device. Lab Chip 13:978–985
38. Mao S, Zhang J, Li H, Lin JM (2013) A novel strategy for high throughput signaling molecule detection to study cell–to–cell communication. Anal Chem 85(2):868–876
39. Wei H, Li H, Mao S, Lin JM (2011) Cell signaling analysis by mass spectrometry under co-culture conditions on an integrated microfluidic device. Anal Chem 83(24):9306–9313
40. Wu J, Jie M, Dong X, Qi H, Lin JM (2016) Multi-channel cell co-culture for drug development based on glass microfluidic chip-mass spectrometry coupled platform. Rapid Commun Mass. SP 30:80–86
41. Jie M, Li H, Lin L, Zhang J, Lin JM (2016) Integrated microfluidic system for cell co-culture and simulation of drug metabolism. RSC Adv. 6:54564–54572
42. Wei H, Li H, Gao D, Lin JM (2010) Multi-channel microfluidic devices combined with electrospray ionization quadrupole time-of-flight mass spectrometry applied to the monitoring of glutamate release from neuronal cells. Analyst 135:2043–2050
43. Sun X, Kelly RT, Tang K, Smith RD (2011) Membrane-based emitter for coupling microfluidics with ultrasensitive nanoelectrospray ionization-mass spectrometry. Anal Chem 83:5797–5803

44. Yue GE, Roper MG, Jeffery ED, Easley CJ, Balchunas C et al (2005) Glass microfluidic devices with thin membrane voltage junctions for electrospray mass spectrometry. Lab Chip 5:619–627
45. Hoffmann P, Haeusig U, Schulze P, Belder D (2007) Microfluidic glass chips with an integrated nanospray emitter for coupling to a mass spectrometer. Angew Chem Int Edit 46:4913–4916
46. Wang D, Bodovitz S (2010) Single cell analysis: The new frontier in 'omics'. Trends Biotechnol 28:281–290
47. Amantonico A, Urban PL, Fagerer SR, Balabin RM, Zenobi R (2010) Single-cell MALDI-MS as an analytical tool for studying intrapopulation metabolic heterogeneity of unicellular organisms. Anal Chem 82:7394–7400
48. Lapainis T, Rubakhin SS, Sweedler JV (2009) Capillary electrophoresis with electrospray ionization mass spectrometric detection for single-cell metabolomics. Anal Chem 81:5858–5864
49. Xue Q, Foret F, Dunayevskiy YM, Zavracky PM, McGruer NE et al (1997) Multichannel microchip electrospray mass spectrometry. Anal Chem 69:426–430
50. Schultz GA, Corso TN, Prosser SJ, Zhang S (2000) A fully integrated monolithic microchip electrospray device for mass spectrometry. Anal Chem 72:4058–4063
51. Kim W, Guo M, Yang P, Wang D (2007) Microfabricated monolithic multinozzle emitters for nanoelectrospray mass spectrometry. Anal Chem 79:3703–3707
52. Moini M, Jiang L, Bootwala S (2011) High-throughput analysis using gated multi-inlet mass spectrometry. Rapid Commun Mass Sp 25:789–794
53. Mao P, Wang HT, Yang PD, Wang DJ (2011) Multinozzle emitter arrays for nanoelectrospray mass spectrometry. Anal Chem 83:6082–6089
54. Baker CA, Roper MG (2012) Online coupling of digital microfluidic devices with mass spectrometry detection using an eductor with electrospray ionization. Anal Chem 84:2955–2960
55. Ericson C, Phung QT, Horn DM, Peters EC, Fitchett JR et al (2003) An automated noncontact deposition interface for liquid chromatography matrix-assisted laser desorption/ionization mass spectrometry. Anal Chem 75:2309–2315
56. Lee J, Soper SA, Murray KK (2009) Development of an efficient on-chip digestion system for protein analysis using MALDI-TOF MS. Analyst 134:2426–2433
57. Xie W, Gao D, Jin F, Jiang Y, Liu H (2015) Study of phospholipids in single cells using an integrated microfluidic device combined with matrix-assisted laser desorption/ionization mass spectrometry. Anal Chem 87:7052–7059
58. Lazar IM, Kabulski JL (2013) Microfluidic LC device with orthogonal sample extraction for on-chip MALDI-MS detection. Lab Chip 13:2055–2065
59. Yang M, Nelson R, Ros A (2016) Toward analysis of proteins in single cells: a quantitative approach employing isobaric tags with MALDI mass spectrometry realized with a microfluidic platform. Anal Chem 88:6672–6679
60. Korenaga A, Chen F, Li H, Uchiyama K, Lin JM (2016) Inkjet automated single cells and matrices printing system for matrix-assisted laser desorption/ionization mass spectrometry. Talanta 162:474–478
61. Wu J, Jie M, Li H, He Z, Wang S (2017) Gold nanoparticles modified porous silicon chip for SALDI-MS determination of glutathione in cells. Talanta 168:222–229
62. Zhang C, Manicke NE (2015) Development of a paper spray mass spectrometry cartridge with integrated solid phase extraction for bioanalysis. Anal Chem 87:6212–6219
63. Wang H, Wu Z, Chen B, He M, Hu B (2015) Chip-based array magnetic solid phase microextraction on-line coupled with inductively coupled plasma mass spectrometry for the determination of trace heavy metals in cells. Analyst 140:5619–5626
64. Choi K, Boyacı E, Kim J, Seale B, Barrera-Arbelaez L et al (2016) A digital microfluidic interface between solid-phase microextraction and liquid chromatography–mass spectrometry. J Chromatogr A 1444:1–7

65. Gasilova N, Srzentić K, Qiao L, Liu B, Beck A et al (2016) On-chip mesoporous functionalized magnetic microspheres for protein sequencing by extended bottom-up mass spectrometry. Anal Chem 88:1775–1784
66. Mikkonen S, Rokhas MK, Jacksén J, Emmer Å (2012) Sample preconcentration in open microchannels combined with MALDI-MS. Electrophoresis 33:3343–3350
67. Mikkonen S, Jacksén J, Roeraade J, Thormann W, Emmer Å (2016) Microfluidic isoelectric focusing of amyloid beta peptides followed by micropillar-matrix-assisted laser desorption ionization-mass spectrometry. Anal Chem 88:10044–10051
68. Yang J, Zhu J, Pei R, Oliver JA, Landry DW et al (2016) Integrated microfluidic aptasensor for mass spectrometric detection of vasopressin in human plasma ultrafiltrate. Anal Methods 8:5190–5196
69. Mellors JS, Jorabchi K, Smith LM, Ramsey JM (2010) Integrated microfluidic device for automated single cell analysis using electrophoretic separation and electrospray ionization mass spectrometry. Anal Chem 82:967–973
70. Redman EA, Mellors JS, Starkey JA, Ramsey JM (2016) Characterization of intact antibody drug conjugate variants using microfluidic capillary electrophoresis–mass spectrometry. Anal Chem 88:2220–2226
71. Cheng H, Liu J, Xu Z, Wang Y, Ye M (2016) Improving sensitivity for microchip electrophoresis interfaced with inductively coupled plasma mass spectrometry using parallel multichannel separation. J Chromatogr A 1461:198–204
72. Saylor RA, Reid EA, Lunte SM (2015) Microchip electrophoresis with electrochemical detection for the determination of analytes in the dopamine metabolic pathway. Electrophoresis 36:1912–1919
73. Nordman N, Sikanen T, Moilanen ME, Aura S, Kotiaho T (2011) Rapid and sensitive drug metabolism studies by SU-8 microchip capillary electrophoresis-electrospray ionization mass spectrometry. J Chromatogr A 1218:739–745
74. Chen FM, Rang Y, Weng Y, Lin LY, Zeng HL et al (2015) Drop-by-drop chemical reaction and sample introduction for capillary electrophoresis. Analyst 140:3953–3959
75. Dietze C, Schulze S, Ohla S, Gilmore K, Seeberger PH et al (2016) Integrated on-chip mass spectrometry reaction monitoring in microfluidic devices containing porous polymer monolithic columns. Analyst 141:5412–5416
76. Haynes PA, Roberts TH (2007) Subcellular shotgun proteomics in plants: looking beyond the usual suspects. Proteomics 7:2963–2975
77. Chao TC, Hansmeier N (2013) Microfluidic devices for high-throughput proteome analyses. Proteomics 13:467–479
78. Liuni P, Rob T, Wilson DJ (2010) A microfluidic reactor for rapid, low pressure proteolysis with on-chip electrospray ionization. Rapid Commun Mass Sp 24:315–320
79. Fan HZ, Bao HM, Zhang LY, Chen G (2011) Immobilization of trypsin on poly (urea-formaldehyde)-coated fiberglass cores in microchip for highly efficient proteolysis. Proteomics 11:3420–3423
80. Ji J, Nie L, Qiao L, Li Y, Guo L et al (2012) Proteolysis in microfluidic droplets: An approach to interface protein separation and peptide mass spectrometry. Lab Chip 12:2625–2629
81. Lee J, Soper SA, Murray KK (2011) A solid-phase bioreactor with continuous sample deposition for matrix-assisted laser desorption/ ionization time-of-flight mass spectrometry. Rapid Commun Mass Sp 25:693–699
82. Min Q, Zhang X, Wu R, Zou H, Zhu JJ (2011) A novel magnetic mesoporous silica packed S-shaped microfluidic reactor for online proteolysis of low-MW proteome. Chem Commun 47:10725–10727
83. Fritzsche S, Hoffmann P, Belder D (2010) Chip electrophoresis with mass spectrometric detection in record speed. Lab Chip 10:1227–1230
84. Nordman N, Sikanen T, Aura S, Tuomikoski S, Vuorensola K et al (2010) Feasibility of SU-8-based capillary electrophoresis-electrospray ionization mass spectrometry microfluidic chips for the analysis of human cell lysates. Electrophoresis 31:3745–3753

85. Chambers AG, Mellors JS, Henley WH, Ramsey JM (2011) Monolithic integration of two-dimensional liquid chromatography-capillary electrophoresis and electrospray ionization on a microfluidic device. Anal Chem 83:842–849
86. Sun X, Kelly RT, Tang K, Smith RD (2010) Ultrasensitive nanoelectrospray ionization-mass spectrometry using poly(dimethylsiloxane) microchips with monolithically integrated emitters. Analyst 135:2296–2302
87. Wang C, Jemere AB, Harrison DJ (2010) Multifunctional protein processing chip with integrated digestion, solid-phase extraction, separation and electrospray. Electrophoresis 31:3703–3710
88. Rob T, Gill PK, Golemi-Kotra D, Wilson DJ (2013) An electrospray ms-coupled microfluidic device for sub-second hydrogen/deuterium exchange pulselabelling reveals allosteric effects in enzyme inhibition. Lab Chip 13:2528–2532
89. Madeira A, Ohman E, Nilsson A, Sjogren B, Andren PE et al (2009) Coupling surface plasmon resonance to mass spectrometry to discover novel protein-protein interactions. Nature Protoc 4:1023–1037
90. Gao D, Liu H, Lin JM, Wang Y, Jiang Y (2012) Characterization of drug permeability in Caco-2 monolayers by mass spectrometry on a membrane-based microfluidic device. Lab Chip 13:978–985
91. Chen Q, He Z, Liu W, Lin X, Wu J, Li H, Lin JM (2015) Engineering cell-compatible paper chips for cell culturing, drug screening, and mass spectrometric sensing. Adv. Healthc. Mater. 4:2291–2296
92. Helenius A, Aebi M (2001) Intracellular functions of N-linked glycans. Science 291:2364–2369
93. Rudd PM, Woods RJ, Wormald MR, Opdenakker G, Downing AK et al (1995) The effects of variable glycosylation on the functional activities of ribonuclease, plasminogen and tissue-plasminogen activator. Biochim Biophys Acta 1248:1–10
94. Rudd PM, Wormald MR, Stanfield RL, Huang M, Mattsson N et al (1999) Roles for glycosylation of cell surface receptors involved in cellular immune recognition. J Mol Biol 293:351–366
95. Varki A (1993) Biological roles of oligosaccharides - all of the theories are correct. Glycobiology 3:97–130
96. Iida J, Meijne AML, Knutson JR, Furcht LT, McCarthy JB (1996) Cell surface chondroitin sulfate proteoglycans in tumor cell adhesion, motility and invasion. Semin Cancer Biol 7:155–162
97. Reitter JN, Means RE, Desrosiers RC, Reitter JN, Means RE et al (1998) A role for carbohydrates in immune evasion in AIDS. Nat Med 4:679–684
98. Dennis JW, Granovsky M, Warren CE (1999) Protein glycosylation in development and disease. Bioessays News Rev Mol Cell Dev Biol 21:412–421
99. Lowe JB, Marth JD (2003) A genetic approach to mammalian glycan function. Annu Rev Biochem 72:643–691
100. Mechref Y, Hu Y, Garcia A, Zhou S, Desantos-Garcia JL et al (2012) Defining putative glycan cancer biomarkers by MS. Bioanalysis 4:2457–2469
101. Mechref Y, Hu Y, Garcia A, Hussein A (2012) Identifying cancer biomarkers by mass spectrometry-based glycomics. Electrophoresis 33:1755–1767
102. Tao SJ, Huang YN, Boyes BE, Orlando R (2014) Liquid chromatography-selected reaction monitoring (LC-SRM) approach for the separation and quantitation of sialylated N-glycans linkage isomers. Anal Chem 86:10584–10590
103. Zhou SY, Hu YL, DeSantos-Garcia JL, Mechref Y (2015) Quantitation of permethylated n-glycans through multiple-reaction monitoring (MRM) LC-MS/MS. J Am Soc Mass Spectr 26:596–603
104. Dallas DC, Martin WF, Strum JS, Zivkovic AM, Smilowitz JT et al (2011) N-Linked Glycan Profiling of Mature Human Milk by High-Performance Microfluidic Chip Liquid Chromatography Time-of-Flight Tandem Mass Spectrometry. J Agr Food Chem 59:4255–4263

105. Hua S, An HJ, Ozcan S, Ro GS, Soares S et al (2011) Comprehensive native glycan profiling with isomer separation and quantitation for the discovery of cancer biomarkers. Analyst 136:3663–3671
106. Chu CS, Ninonuevo MR, Clowers BH, Perkins PD, An HJ et al (2009) Profile of native Nlinked glycan structures from human serum using high performance liquid chromatography on a microfluidic chip and time-of-flight mass spectrometry. Proteomics 9:1939–1951
107. Nwosu CC, Aldredge DL, Lee H, Lerno LA, Zivkovic AM et al (2012) Comparison of the human and bovine milk Nglycome via high-performance microfluidic chip liquid chromatography and tandem mass spectrometry. J Proteome Res 11:2912–2924
108. Tao N, Wu S, Kim J, An HJ, Hinde K et al (2011) Evolutionary glycomics: Characterization of milk oligosaccharides in primates. J Proteome Res 10:1548–1557
109. Leoz MLAD, Gaerlan SC, Strum JS, Dimapasoc LM, Mirmiran M et al (2012) Lacto-N-tetraose, fucosylation, and secretor status are highly variable in human milk oligosaccharides from women delivering preterm. J Proteome Res 11:4662–4672
110. Totten SM, Zivkovic AM, Wu S, Ngyuen U, Freeman SL et al (2012) Comprehensive profiles of human milk oligosaccharides yield highly sensitive and specific markers for determining secretor status in lactating mothers. J Proteome Res 11:6124–6133
111. Zhou H, Castro-Perez J, Lassman ME, Thomas T, Li W et al (2013) Measurement of apo(a) kinetics in humansubjects using a microfluidic device with tandem massspectrometry. Rapid Commun Mass Sp 27:1294–1302
112. Gallagher R, Dillon L, Grimsley A, Murphy J, Samuelsson K et al (2014) The application of a new microfluidic device for the simultaneous identification and quantitation of midazolam metabolites obtained from a single micro-litre of chimeric mice blood. Rapid Commun Mass Sp 28:1293–1302
113. Brink FTGVD, Büter L, Odijk M, Olthuis W, Karst U et al (2015) Mass spectrometric detection of short-lived drug metabolites generated in an electrochemical microfluidic chip. Anal Chem 87:1527–1535
114. Li X, Zhao S, Hu H, Liu YM (2016) A microchip electrophoresis-mass spectrometric platform with double cell lysis nano-electrodes for automated single cell analysis. J Chromatogr A 1451:156–163
115. Chen F, Lin L, Zhang J, He Z, Uchiyama K et al (2016) Single-cell analysis using drop-on-demand inkjet printing and probe electrospray ionization mass spectrometry. Anal Chem 88:4354–4360

Chapter 10
Biochemical Analysis Techniques Integrated on Microfluidic Chips and Their Applications

Jing Wu and Jin-Ming Lin

Abstract Biochemical analysis plays an essential role in understanding mechanism of life activities and giving biological insights into life process. Microfluidic technology explores new paradigms in biochemical analysis field because of its good incorporation with existing analytical techniques. In this chapter, we summarize the typical analytical methods integrated onto microfluidic platforms. Optical, electrical, magnetic and acoustic techniques have been highlighted. Applications of these microfluidic analytical methods into genetic and protein analysis also have been discussed. At last, the challenges and future directions about microfluidics-based biochemical analysis development have been remarked.

Keywords Biochemical analysis · Optical detector · Electronic manipulation
Magnetic operation · Surface acoustic wave · Genetic analysis
Protein analysis

10.1 Introduction

In vivo, cells live in a microenvironment with multiple biological and chemical cues varying in space and time. These biochemical signals can be divided into soluble and insoluble signaling molecules [1]. Soluble signaling molecules contain chemokines [2], cytokines [3], hormones [4] and growth factors secreted by neighbor or distant cells, or even by cell itself [5]. Insoluble signaling molecules refer to components of the extracellular matrix (ECM) and membrane-bound proteins on neighbor cell surfaces. These soluble and insoluble signaling molecules

J. Wu
School of Science, China University of Geosciences (Beijing),
Beijing 100083, People's Republic of China
e-mail: wujing@cugb.edu.cn

J.-M. Lin (✉)
Department of Chemistry, Tsinghua University, Beijing 100084, China
e-mail: jmlin@mail.tsinghua.edu.cn

J.-M. Lin (ed.), *Cell Analysis on Microfluidics*, Integrated Analytical Systems,
https://doi.org/10.1007/978-981-10-5394-8_10

regulate cell physiology, phenotype and fate. Meanwhile, cells exert some signals or alter ECM to change their surrounding environment [6, 7]. Monitoring the biochemical interactions between cells and their microenvironment can predict cell behavior and give some indications on bio-analytical chemistry researches.

Cells typically reside in a three-dimensional (3D) environment with specific biochemical and mechanical features. Cell-based biochemical analysis taken under the conditions recreating intracellular microenvironment can give insights into the biochemical process and govern cell behavior [8]. However, conventional biochemical experiments always are conducted in two-dimensional (2D) culture conditions which poorly mimic the in vivo cellular environment [9, 10]. Fortunately, microfluidic technology provides a more physiologically relevant context for cell-based analysis and presents cells with controllable and reproducible biochemical cues [11–15]. 3D cell culture is facilely realized on microfluidic chips and retains native tissue-specific functions of cells as well as recapitulates tissue-tissue interfaces, spatio-temporal biochemical gradients [16–18]. Artificial organ models even are reconstructed on microfluidic platforms to reconstitute the complex organ structures and the critical organ functions [19–23]. With high compatibility, microchips can integrate with many biochemical analysis techniques showing great promise for basic biomedical and biochemical researches [24].

In this chapter, we summarize methods and techniques, such as optical detector, electrical manipulation, magnetic operation and acoustic wave, used in biochemical analysis on microfluidic chips and discuss the applications of these microtechnologies in biomacromolecular assays. Working principles of these microtechnologies are described in details and works aimed at gaining biological insights are emphasized. Finally, future prospects of biochemical analysis in microsystems to realize advanced studies are present.

10.2 Biochemical Analysis Techniques Integrated on Chips

10.2.1 Optical Detector

Optical detection of samples becomes common in microfluidic system due to the universal optical instruments in laboratories [25]. There are two modes for optical detection systems assembling onto microfluidic chips. The detectors propagating light in air before and after interaction with samples are categorized as "off-chip". In off-chip mode, the components of optical systems, such as light source, mirrors and detectors, are separated from the chip platform [26]. Conventional optical detection technologies, such as absorbance, fluorescence, infrared (IR) and surface plasmon resonance (SPR), have all been coupled with microchips through this mode. Absorbance and fluorescence are briefly discussed in following.

Absorption spectra arise from the capability of analytes to attenuate the incident electromagnetic radiation at various wavelengths. The shape and location of absorption spectra are related to sample structure and the intensities of absorption peak are proportional to sample concentration. Glucose consumption by live cells cultured on microfluidic chips always is measured through absorption detection. Schulz et al. [27, 28] implemented a fiber-coupled UV light source and a CCD spectrometer into the UV absorbance measurement system to quantitatively detected glucose concentration variation in live cell-cultured microfluidic reactors. They measured UV absorption intensity of NADH (λ_{ab} = 340 nm) which was correlated with the glucose concentration. Srinivasan et al. [29, 30] detected the optical absorbance of quinoneimine (λ_{ab} = 540 nm) which was a byproduct of glucose in Trinder's reaction by integrating an light-emitting diode (LED) and a photodiode into their detection system. Besides small molecular analytes, pathogenic bacteria also were reported to be detected by light absorption spectrometer. Jeon et al. [31] used magnetic nanoparticle clusters (MNCs) to capture *Escherichia coli* (*E. coli*) bacteria in milk and then injected them into a 3D-printed helical microchannel. The concentration of the *E. coli* bacteria was determined by light absorption intensity. The limit of detection of this method was 10 cfu mL^{-1} in buffer solution and 100 cfu mL^{-1} in milk.

Fluorescence is the optical emission of fluorophores after they absorb electromagnetic excitation radiation. As a high sensitive and high selective detection method for cellular and molecular sensing, fluorescence is a preferred choice to be coupled onto microdevices to obtain structural and functional information or to monitor cell behavior. An excitation part and an emission part comprise a whole optical setup of induced fluorescence. The detection window on microdevices is set between the two parts. Laser, LED and mercury or xenon arc lamp are common excitation sources used in off-chip fluorescence detection. Native chromophores inside the molecular structures or chemically linked fluorescent labels are the origin of fluorescence of biomacromolecules. Fluorescent proteins and synthetic dyes are universal fluorescent labels for microanalysis. Green fluorescent protein (GFP) has been used as fluorescent indicators to represent specific cellular dynamics and factors since it is discovered in 1994 [32]. The visualization of phagocytosis of *E. coli* on anopto-microfluidic device was facilitated by loading GFP expressing *E. coli* into the cell capture module arrays before murine macrophage introduction [33]. Fluorescent dyes, fluorochrome-labeled inhibitor of caspases (FLICA) and propidium iodide (PI) were used to discriminate viable, apoptotic and necrotic cells on a cell trap microfluidic device [34]. Fluorescein was tested as a drug model on a 3D-printed, diffusion-based and dynamic in vitro pharmacokinetic (PK) device to give PK profiles by a commercial plate reader ($\lambda_{ex}/\lambda_{em}$ 480/520 nm) [35]. A droplet-based microfluidic protocol was present for high-throughput fluorescence-activated analysis and sorting of single cells. Antibodies (Ab) secreted from single mouse hybridoma cells were detected by co-compartmentalizing single mouse hybridoma cells, a fluorescent probe and single beads coated with anti-mouse IgG Ab in droplets. The beads captured the secreted Ab and then bound to the probe to generate distinguishable fluorescence signal (Fig. 10.1) [36]. We also reported fluorescein sodium and fluorescein isothiocyanate conjugated to bovine serum albumin (FITC-BSA) were utilized as models

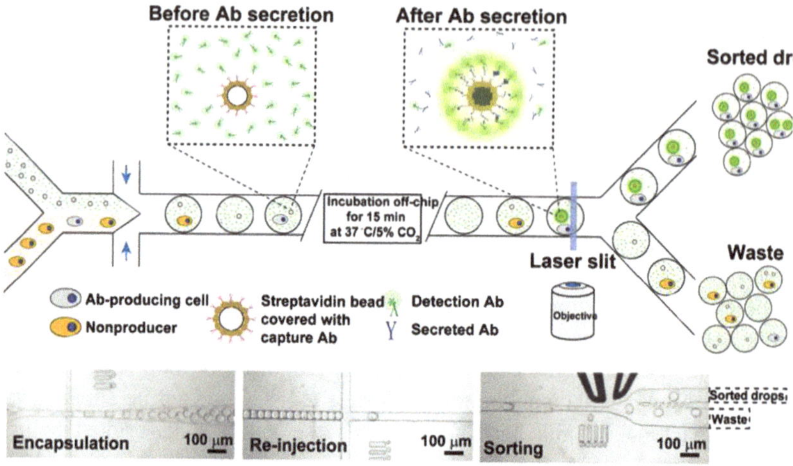

Fig. 10.1 Schematic principle of droplet-based microfluidic protocol for high-throughput fluorescence-activated analysis and sorting of single cells (Reprinted with permission from Ref. [36]. Copyright 2013 Nature America, Inc.)

to mimic the diffusion process between blood vessels and tissues. Fluorescent probes specific for cell viability (Calcein-AM/EthD-1) and apoptosis (Hoechst 33,342) analysis were used to demonstrate that cytotoxicity of quantum dots (QDs) were diffusion distance- and cell density-dependent [37, 38]. To improve sensitivity and specificity of fluorescent detection on microfluidic chips, novel fluorescent probes with high quantum yield have been devised. Nanomaterials are a representative new type of fluorescent probe. A thin layer of convalently bonded Ag@SiO_2 + FITC nanoparticles were patterned onto an embedded planar glass substrate to construct a fluorescent nanoparticle-based sensor for real-time, ratiometric pH imaging of biofilms. FITC chosen as a fluorescent pH indicator was motivated by the fact that its protonated and negatively charged forms have different quantum yields and maximum excitation wavelengths (λ_{ex} = 440 nm and 490 nm, respectively). This specific feature allowed quantitative pH measurements by ratioing images of emission signals recorded while exciting at each wavelength [39]. A gold nanoparticle-glutathione-fluorescein isothiocyanate (Au NP-GSH-FITC) probe coupled with on-line droplet-based microfluidic chip was developed as a novel fluorescent sensor for real-time and continuous monitoring the variation of bisulfide in microdialysis effluents. In the presence of bisulfide, the bonded GSH-FITC on Au NPs was replaced by bisulfide and released into the solution. The fluorescence of GSH-FITC was consequently recovered and the intensity increased with bisulfide concentration [40].

Another mode is the "on-chip" mode in which the optics are integrated onto the chips or fabricated with the fluidic functional units. "Optofluidic" is a prototype in which the optical property of the integrated optics can be tuned by fluidic technology. Valves and pumps were integrated onto a microfluidic mixing device to construct a single-molecule optofluidic analysis device. Titration of biomolecules was accurately accomplished with picoliter resolution on this platform [41]. Recently, a novel turbid optofluidic tissue phantom was fabricated to mimic the vascular perfusion and the turbid nature of cutaneous tissue. This phantom was composed of an interior hollow microfluidic structure which contained the geometry of an inlet, a river-like assay and an outlet. Representative optical parameters of human skin cutis, such as the absorption coefficient μ_a and the reduced scattering coefficient μ_s', were reflected on the multiple-layer phantom matrices which bridged the gap between tissue equivalent phantoms and Lab-on-Chip devices [42]. With the advance of manufacturing technologies, full integration of optical elements onto microfluidic chip is on the way.

10.2.2 Electronic Manipulation

Electrical analysis of biological cells has drawn wide interest because they are label-free and non-invasive. Due to the rich expertise on recording and processing of electronic signals and available technologies for miniaturization and integration, electrical analysis of biological samples on microfluidic devices shows prominent advantages [43]. Impedance spectroscopy, dielectrophoresis (DEP) and electrorheological fluid (ERF) are common electrical analytical methods integrated on microfluidic chip.

Impedance spectroscopy is a powerful tool for characterizing biological samples based on detecting cell membrane capacitance and resistance and cytoplasm conductivity and permittivity. Antimicrobial peptide (AMP) magainin I was immobilized onto gold microelectrode arrays to detect infectious agents using impedance spectroscopy. Detection limit of this method was revealed to be approximately 1 bacterium μL^{-1} which was a clinically useful detection range. A microfluidic sensor incorporating multiple alternating current electrical impedance and light scattering analysis was demonstrated to allow label-free discrimination of platelets, erythrocytes, monocytes, granulocytes and lymphocytes in whole blood samples. Two electrode pairs were combined for differential impedance detection and one-dimensional sheath flow focusing was implemented to analyze single particle at kHz count rates (Fig. 10.2a) [44]. Hydrodynamic trapping of single cells was combined with electrical impedance measurements within a microfluidic device to capture mouse embryonic stem cells (mESCs) in pre-defined locations and quantitatively analyzed the changes of electrical parameters of cells during differentiation [45].

DEP is a cell sorting method that utilizes nonuniform electric fields to apply various forces on cells with distinct dielectric properties to separate them. The electrical properties of cells, such as polarisability and conductivity, are relative to cellular composition, shape and size, thus, can be employed to differentiate cells. A long-range 3D DEP microfluidic platform with a V-shaped microchannel was

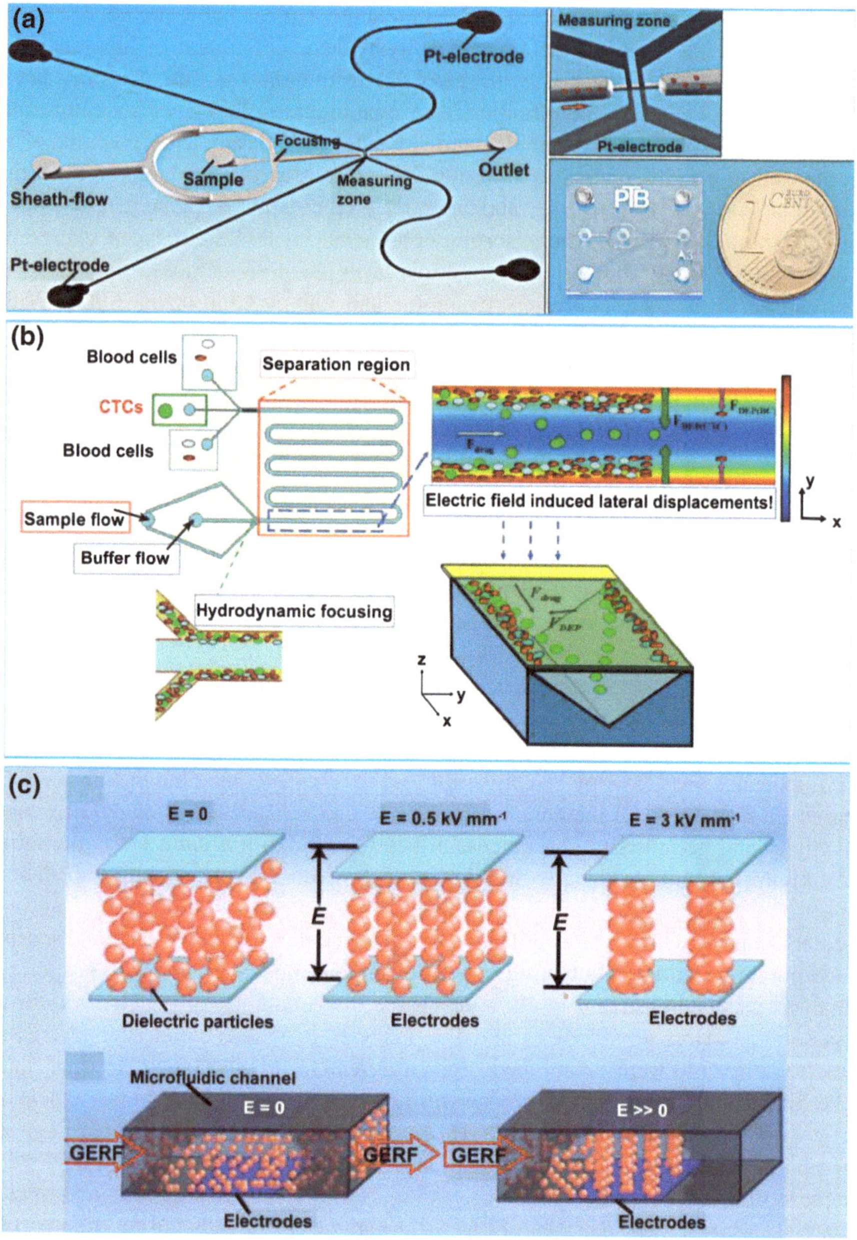

Fig. 10.2 Electrical manipulation on microfluidic chip. **a** A microfluidic sensor was designed for label-free flow cytometric cell differentiation combining electrical impedance and light scattering analysis (Reprinted with permission from Ref. [44]. Copyright The Royal Society of Chemistry 2016.). **b** A long-range 3D DEP microfluidic platform with a V-shaped microchannel was proposed to isolate rare cancer cells from diluted human blood (Reprinted with permission from Ref. [46]. Copyright The Royal Society of Chemistry 2015.). **c** Schematic working principle of GERF (Reprinted with permission from Ref. [52]. Copyright The Royal Society of Chemistry 2010.)

proposed to isolate rare cancer cells from diluted human blood (Fig. 10.2b). A hydrodynamic sheath flow was designed upstream in the chip to direct the suspension samples to the channel side walls forming a queen to allow DEP-induced lateral displacements. A sustained 3D lateral DEP particle force was generated to continuously sort particles into different downstream subchannels [46]. Cell-size trap-wells combined with high aspect ratio reaction-wells were fabricated on a chip to construct an electroactive double-well array. Trap-wells were used to deterministically trap single cells using DEP and reaction-wells were designed for confining the cell lysates extracted by lysing trapped single cells via electroporation [47]. Besides cell isolation, changes in cell state also could be monitored by DEP. By exploring the differences in the dielectric polarizabilities of the cells, the disease state of cells was discriminated by DEP [48]. We successfully detected human papillomavirus (HPV)16 E6/E7 mRNA by microchip electrophoresis (MCE). Nucleic acid sequence based amplification (NASBA), one-step and two-step real time-polymerase chain reaction (RT-PCR) were conducted in this approach. NASBA amplified HPV16 mRNA by isothermal amplification needing no complex and tedious operation. One-step RT-PCR simplified the amplification process and two-step RT-PCR enhanced sensitivity and resistance to interference. The MCE-based method realized high-throughput and rapid RNA assays [49]. Mutate and methylated genes of Bladder cancer cells also were identified by us based on MCE method [50].

ERF is a two-phase system composed of dielectric particles suspended in insulating oil [51]. Magnitude of the applied external electric field can alter the viscosity of ERF to make ERF solidify into an anisotropic solid during a short time. ERF keeps in liquid state when no electric field is applied. It begins to form chains upon appling electric field and grows into columns as the electric field increasing (Fig. 10.2c) [52]. The phase changing process is reversible. Thus, ERF can be used as a smart material offering potential applications of active-control clutches, dampers and valves on microfluidic devices. Wen et al. have made outstanding progresses in this field. They designed a microfluidic chip with switching and corresponding inverting functionalities by incorporating giant ERF (GERF) and two pairs of electrodes. One of electrodes was located in the GERF microchannel to control the flow status of GERF, while another one was set in other microchannel to detect the signal generated by passing-by droplet (signal droplet). The appearance of signal droplets controlled the switching of GERF from suspended state to flowing state or vice versa [52]. They also reported the first microfluid universal logic device by introducing GERF as the computing media to realize all 16 Boolean logic operations. GERF translated fluidic information into electronic and mechanic information, back and forth, making this microfluid universal logic gate fully compatible with electronic devices and to be building block of microfluidic processors [53].

10.2.3 Magnetic Operation

Magnetic operation mainly refers to manipulation via magnetic particles which possess apparent advantages that large surface-to-volume ratio, convenient bio-functionalization, easy actuation by magnetic fields and commercial availability. Application of magnetic particles into microfluidic devices becomes attractive on account of the following reasons: (1) the easy obtaining of magnetic particles and corresponding assay reagents serves as a solid basis for miniaturization of explorations; (2) magnetic manipulation is compatible with system miniaturization; (3) it allows diverse sophisticated fluids and molecular processing steps to be manipulated by external magnetic fields simplifying the construction of microfluidic devices. Now, key assay steps on microfluidic platforms, such as fluid mixing, analyte capture and detection, have been facilely manipulated by magnetic field via magnetic particles [54, 55].

Rapid fluid mixing is a necessary step for most point-of-care diagnostic devices. However, fluid mixing in microchannels is inefficient and slow due to the strong viscous forces induced by small dimensions of microchannels. As a result, the mixing of fluids has been an everlasting topic in the microfluidics field. Magnetic manipulation has been investigated to improve microfluidic mixing. The supra-particle structures of magnetic particles which are formed in a magnetic field because of dipole-dipole interaction can be used to stir fluids on microscale. Magnetically functionalized polydimethylsiloxane (PDMS)-based micropillar arrays were successfully implanted into a microchannel and used as micromixers simply controlled by on/off magnetic signal (Fig. 10.3a). Mixing efficiencies of the micropillar arrays with different densities were investigated and it was found that

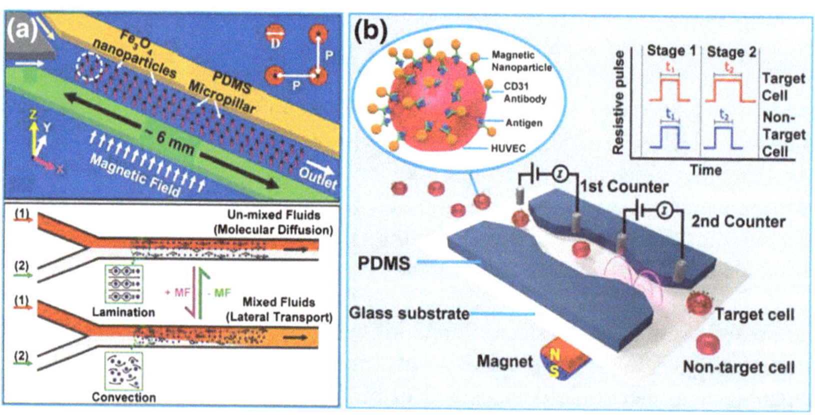

Fig. 10.3 Magnetic operation on microfluidic chip. **a** 3D schematic of the micromixer made of PDMS-based magnetic micropillars and working principle of the micromixer (Reprinted with permission from Ref. [56]. Copyright The Royal Society of Chemistry 2015.). **b** Schematic of the microfluidic magnetic particle assay for cell detection by coupling two identical micro Coulter counters (Reprinted with permission from Ref. [62]. Copyright 2016 American Chemical Society.)

higher density would render better mixing performance [56]. Magnetic nanoparticles were added into aqueous solutions to generate ferrofluid containing solutions. The aqueous fluid moved toward the position of an external magnet and emulsified at a junction with a larger width and height oil filled reservoir. Prior to emulsification, reagents and ferrofluids were introduced from two or more separate inlets and mixed driving by magnetic force [57].

The high surface-to-volume ratio and availability of many bio-functionalization options for magnetic particles give them promises as suitable tools for analyte capture and detection. Purification of nucleic acids is a prerequisite for downstream analysis, so nucleic acid capture is indispensable for microfluidic analysis. Transportation of magnetic purification particles between surface-adhering droplet arrays was utilized to facilitate nucleic acid extraction requiring no microfabrication facility. The process is quick and its repeatability and extraction efficiency are comparable to conventional tube-based operations [58]. After nucleic acid extraction, it can be analyzed on chip. Lee et al. extracted nucleic acid of bacteria *Helicobacter pylori* (*H. pylori*) by specific probe-conjugated magnetic particles. The target DNA fragments of the nucleic acid was amplified by single-nucleotide-polymorphism (SNP)-PCR and the PCR products were detected by optical method [59]. Cells as a type of bio-samples analyzed on microfluidic chips can be facilely captured by magnetic particles. Microfluidic magnetic cell separation was developed as a subfield of micro-magnetofluidics [60]. Multi-shaped nanomaterials such as multiwalled carbon nanotubes (MWCNT, tubes), graphene oxide (GO, planar) and Fe_3O_4 nanoparticles (MNPs, spheres) were simultaneously modified on a paper to create a biocompatible device for glucose analysis. Glucose was measured by bienzymatic colorimetric detection. The surface modification addressed the problems associated with the lack of homogeneity on color measurements. The limit of detection values were 62, 18 and 43 μM for paper-based devices modified with MWCNT, GO and MNPs, respectively [61]. Microfluidic magnetic particle assay for cell detection was developed as well. Zhe et al. [62] fabricated a cell detection device by coupling two identical micro Coulter counters. Antibody-functionalized magnetic particles were bound to specific antigens expressed on the target cells. The two micro Coulter counters were separated by a fluid chamber and a cylindrical magnet was externally applied to generate a high-gradient magnetic field (Fig. 10.3b). Transit time of magnetic particle-bonded target cells passing through the second counter was longer than that through the first counter because of the magnetic interaction between the magnetic beads and the magnetic field. While transit times of nontarget cells kept nearly the same when they passed through both counters. As a result, target cells were identified and their concentration was quantified depending on the transit time delay.

Magnetic manipulation enlarges the application of microfluidics. It is foreseen that the ongoing optimization of magnetic particles regarding their surface bio-functionalization, surface smoothness and size and magnetization uniformity will benefit microfluidic assays based on magnetic force.

10.2.4 Surface Acoustic Wave

Incorporating optics in microfluidics spurs the development of fluid-based optical elements with high flexibility [63–65]. Incorporating electrical techniques in microfluidics allows construction of digital microfluidics [66–68]. Incorporating magnetic in microfluidics enables highly efficient cell manipulation and separation [54, 69, 70]. Similarly, surface acoustic wave (SAW) technologies have been tried to be integrated onto microfluidic devices and bring prominent advantages that fast fluid actuation, high biocompatibility, cost-saving and compact devices and accessories, contact-free particle and label-free cell manipulation [71–73]. SAWs are produced by applying an appropriate electric field to a piezoelectric material and then propagate along the surface of an elastic material. Metallic interdigital transducers (IDTs) are indispensible for a typical SAW device and are set on the surface of a piezoelectric substrate to introduce the electric field. There exist many SAW modes while Rayleigh SAW is the most frequently used one on SAW-based microfluidic chip. Rayleigh SAW which is composed of a longitudinal and a vertically polarized shear component enables sensing of mass perturbation and elastic properties of a medium introduced on the wave's propagation path through coupling with media in contact with the wave propagation surface [74]. SAW-based microfluidics is categorized into two types: travelling SAW (TSAW) and standing SAW (SSAW).

TSAW is mainly used to manipulate micro-objects and actuate fluids depending on acoustic streaming force (ASF). The liquid's viscosity increases when a TSAW contacts a liquid causing part of the SAW to refract into the liquid as a longitudinal wave. The refracted longitudinal waves give a force in their propagation direction and generate flow within the confined liquid. The boundaries of the confined liquid reflect the actuated liquid and lead to internal streaming. SAW-induced ASF refers to such a non-linear phenomenon that transforms the SAW attenuation into a steady fluid flow [75]. Depending on TSAW, many microfluidic technologies, such as particle/cell sorting and concentrating, fluid mixing and translation, have been realized with easy [76, 77]. A detachable acoustofluidic system comprised of a reusable acoustic transducer and a disposable channel device was developed for noncontact continuous particle separation via TSAW. The channel device was placed onto the acoustic transducer and a microstructured pillar was used to couple acoustic waves into the fluid channel for noncontact particle manipulation (Fig. 10.4a). Particles with sizes of 10 and 15 μm were separated with high efficiency of 98% in a 49.5 MHz TSAW [78]. ASF coupled with an acoustic radiation force (ARF) enabled polystyrene (PS) microparticles aggregated in four distinct regimes inside a sessile droplet depending on the particle diameter, the SAW frequency, the acoustic wave field, the acoustic waves' attenuation length and the droplet volume. PS particles concentrated at the centre of the droplet formed the shape of a bead, around the periphery of the droplet formed the shape of a ring, at the side of the droplet formed the shape of an isolated island and close to the centre of the droplet formed the shape of a smaller ring [79]. Shilton et al. [80] used

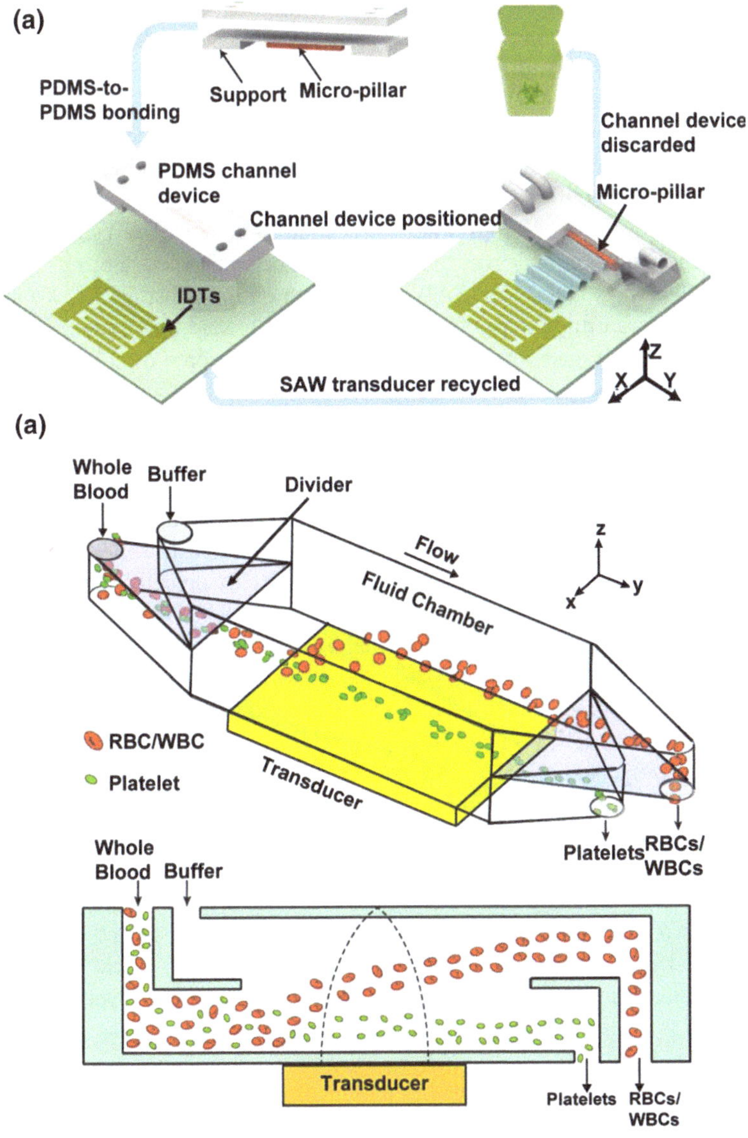

Fig. 10.4 SAW-based microfluidic devices. **a** A detachable acoustofluidic system for particle separation via TSAW (Reprinted with permission from Ref. [78]. Copyright 2016 American Chemical Society.). **b** Schematic of the SSAW-based platelet separation device and its side view (Reprinted with permission from Ref. [90]. Copyright The Royal Society of Chemistry 2016.)

TSAW with ultrahigh frequencies (1.1 GHz) to mix fluids in nanolitre droplets. Yeo et al. [81] used TSAW to drive intense chaotic streaming within a 100 μL chamber to mix fluids to enhance chemiluminescent detection.

Different from TSAW-based devices harnessing ASF, SSAW-based devices use the primary ARF acting on particles and cells via surrounding fluids. The principle of SSAW formation is similar to principle governing the interference of all waves. If a pair of identical IDTs fabricated on a piezoelectric material generates two TSAWs propagating toward each other, the SAWs' interference will result in a one-dimensional (1D) SSAW field. The amplitudes of the sound pressure field are not the same all through the 1D SSAW. The strongest and weakest amplitudes are nodes (minimum field) and anti-nodes (maximum field) of SSAWs at fixed locations on the substrate surface. The distance between neighboring nodes or anti-nodes is always half of the SAW wavelength on the substrate. If two pairs of identical IDTs are orthogonally fabricated on the substrate, two dimensional (2D) SSAW consequently generates [82]. SSAW can give a force on a suspended particle. The force is proportional to the square of the pressure amplitude p_0 and the volume of the particle V_p and inversely proportional to the wavelength λ (Eq. 1). Depending on the physical properties of the particle, it is moved either to a pressure node or to an anti-node by the force. In Eq. 2, ϕ represents the acoustic contrast factor which contains physical properties of the medium and the particle. Two particles with different signs of ϕ are moved by the ARF to different directions, thus, they are separated [83–85]. As a result, particles or cells can be actuated, sorted [86, 87] or focused in a flow stream under SSAW. Huang et al. have made great advances in SSAW-based microfluidics. They reported a SSAW-based microfluidic chip for cell and bead washing employing ARF generated in a SSAW field to actively extract cells or beads from their original medium. High-purity (>97%) white blood cells (WBCs) were prepared from lysed blood samples through cell washing [88]. Additionally, they successfully isolated circulating tumor cells (CTCs) from peripheral blood samples using title-angle SSAW. Device geometry and title angle were optimized to enhance cell throughput to be 20 times higher than previous reports [89]. Recently, they separated platelets from undiluted whole blood by acoustic separation on a microfluidic device. A thin reflective resonator in the device was fabricated by bonding a composite transducer onto the bottom of a stainless steel fluidic chamber. During processing, buffer and whole blood sample were introduced into two inlets at the upper and the lower level, respectively (Fig. 10.4b) [90].

$$F_r = -\left(\frac{\pi p_0^2 V_p \beta_f}{2\lambda}\right) \bullet \phi(\beta, \rho) \bullet \sin(2kx) \tag{10.1}$$

$$\phi = \frac{5\rho_p - 2\rho_f}{2\rho_p + \rho_f} - \frac{\beta_p}{\beta_f} \tag{10.2}$$

SAWs have demonstrated tremendous capability in microfluidic applications. SAW-based microfluidics exhibits a combination of advantages outstanding comparable techniques, such as non-invasive cell manipulation in native intracellular microenvironment. SAW microfluidic technologies show potential to be applied in the studies of stem cell differentiation, cell-to-cell communication and tissue engineering.

10.3 Applications

By integrating with various analytical techniques, microfluidic device as a powerful analytical method is revolutionizing many biochemical researches, especially genetic and protein analysis.

10.4 Genetic Analysis

The realization of rapid, sensitive, specific, quantitative and multiplexed genetic analysis could reframe diagnostic detections. The advent of microfluidics provides a promise for this aim by enabling the integration of sophisticated multistep biochemical assays in a monolithic, portable device while relatively small biological samples are needed [91].

The PCR technique is a process that amplifies individual DNA segments for functional analysis of genes in diseases. PCR was invented in 1984 by Kary Mullis and microfluidic PCR device was founded in 1993 by Watson et al. [92]. This simple chemical process is an important step existing in most genetic assays and has reformed medical, biological and forensic analysis and clinical diagnoses. Microscale PCR has been demonstrated to be more rapid because of the high surface-to-volume ratio of microchannels or microreactors. Fluorescence activated cell sorting (FACS) for individual single cell separation was combined with microfluidic technologies to implement single-cell PCR gene-expression analysis to dissect the cellular composition of primary human normal colon and colon cancer epithelia [93]. A fully integrated microfluidic device was present for performing high-precision RT-qPCR measurements of gene expression from hundreds of single cells. All steps of single-cell processing, such as cell capture, cell lysis, reverse transcription and quantitative PCR, were executed on this device (Fig. 10.5a). This work served as a foundation for the development of a variety of on-chip single-cell transcription analysis [94]. We also coupled PCR with MCE to analyze mini-short tandem repeat loci [95] and BCR-ABL fusion gene [96]. Recently, researchers have shifted their focus onto developing droplet-based microfluidic PCR as it can offer many advantages: (1) single-phase operation; (2) suitable for single-cell and single-molecule analysis; (3) has the most promising potential for system integration and automation [92, 97]. Microfabricated high-density droplets were used as picolitre reactors for DNA amplification to obtain an approximately 100-fold increase in throughput [98]. A simple T-junction microchannel was built on a microfluidic chip to generate numerous monodispersed droplets which served as microreactors for digital PCR [99]. Zhao et al. [100] developed a droplet digital

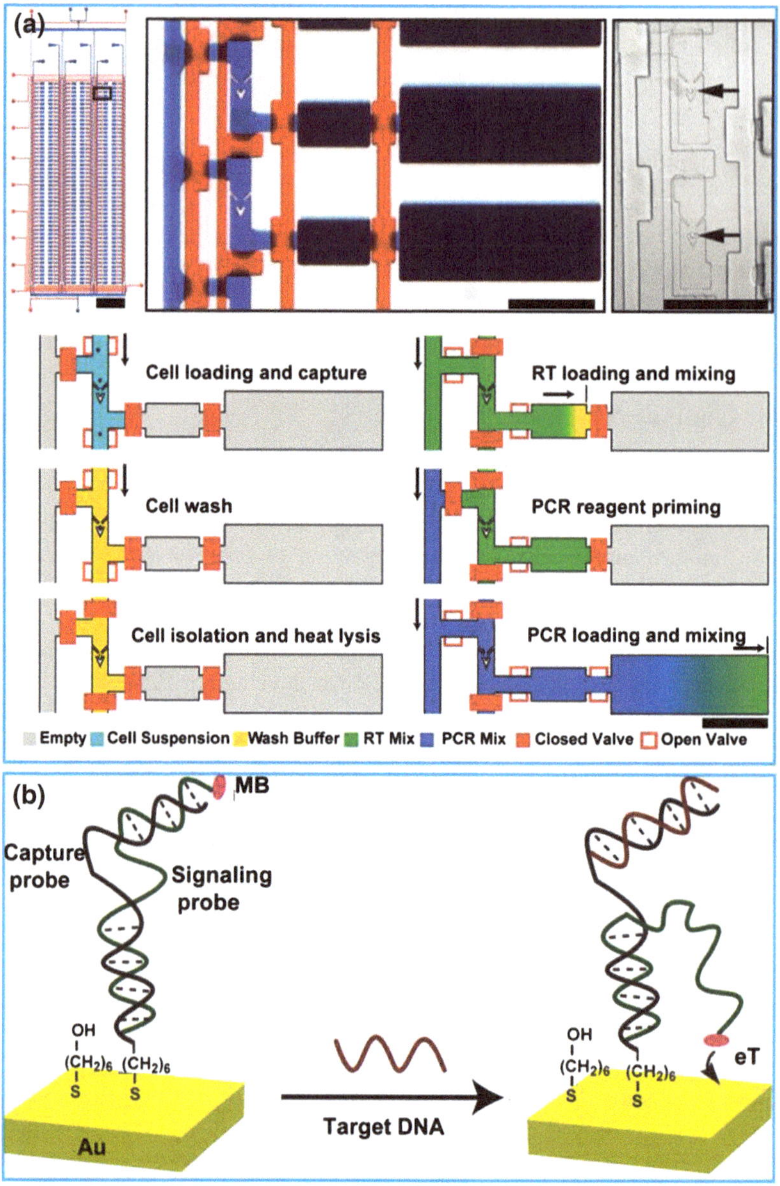

Fig. 10.5 Genetic analysis on microfluidic chip. **a** Schematic and optical images of design and operation of the microfluidic device for single-cell gene expression analysis. (Reprinted with permission from Ref. [94]. Copyright 2011 by the National Academy of Sciences of the USA.). **b** Schematic illustration of E-DNA sensor based on a target-induced strand displacement mechanism (Reprinted with permission from Ref. [106]. Copyright 2006 by the National Academy of Sciences of the USA.)

PCR system for quantification of lung cancer related microRNA. Two inlets on the chip were designed for oil and sample injection and highly uniform monodisperse water-in-oil emulsion droplets were subsequently generated at the cross section of the channel for detection and analysis. The droplet-based PCR with high throughput is ideal for exploration of genomic and cDNA libraries, therefore, greatly improves our understanding of disease pathophysiology and diagnosis.

Fluorescent assays are a frequently chosen detection method for nucleic acid analysis because of their convenient operation and robustness. Nanomaterials are optimum options for fluorescence enhancement. Au NP was reported to enhance fluorescence detection of tiny amount of nucleotides within microfluidic droplets in which single strand DNA (ssDNA) and relevant complementary strands DNA (cDNA) were contained. The droplets were generated through conventional flow-focusing manner and the hybridization between ssDNA and cDNA was realized in droplets flowing along a serpentine channel. Au NP efficiently amplified the fluorescence intensity and the detection limit was reduced to be 10 pM [101]. ssDNA-Cy3 probes on gold surface and a simple line-shape microchannel composed an optical DNA microfluidic sensor. The ssDNA-Cy3 probes exhibited enhanced fluorescence after binding with cDNA targets. This sensor also was capable for protein detection. The introduction of streptavidin resulted in fluorescence quenching because of the binding between hybridized DNA-biotin and streptavidin [102]. A GO-based paper chip was fabricated for fast, inexpensive and direct detection of nucleic acids based on fluorescence resonance energy transfer (FRET) assay. Cy3/FAM-labeled ssDNA was employed as the reporter, the GO was used as quencher and the glass fiber paper was served as data acquisition substrates [103].

Microfluidic electrochemical sensors have lowered the technology barrier for effective, point-of-care genetic detection because of no optical instrument requiring. A broad of electrochemistry-based biosensors monitored binding-induced physical changes in the rigidity of a redox-tagged probe DNA which has been site-specifically attached to an interrogating electrode [104]. Heeger et al. pointed out that DNA hybridization could be reagentlessly transducted into detectable electrochemical signal by means of a conformational change. They reported DNA hybridization induced a large conformational change in the electroactive, ferrocene-tagged DNA stem-loop structure which self-assembled onto a gold electrode. The electron-transfer tunneling distance between the electrode and the redoxable label was consequently altered and then changed electron transfer efficiency which was measured by cyclic voltammetry [105]. Additionally, they fabricated an electronic DNA (E-DNA) sensor based on a target-induced strand displacement mechanism. The E-DNA sensor was composed of a "capture probe" attached by its 5' terminus to a gold electrode and a 5' methylene blue (MB)-modified "signaling probe" that was complementary at both its 3' and 5' termini to the capture probe. Hybridization between target and signaling probes displaced the 5' end of the signaling probe generating a short, flexible single-stranded DNA element and producing up to a 7-fold increase in redox current (Fig. 10.5b). A detection limit of 400 fM was achieved because of the sensitive signal [106].

A 3×3 arrayed electrochemical sensors were fabricated for the analysis of DNA hybridization events. The design of this sensor involved a new dual layer microfluidic valved manipulation system which provided controlled and automated capabilities for high throughput analysis. Repeatability, accuracy and overall performance of the genetic sensor were improved because of this feature. The microfluidic sensor behaved semilogarithmic dose response and enabled a theoretical detection limit of 1 nM of complementary ssDNA target [107]. Microfluidic chip gives a promise as a broad genetic analysis platform.

10.5 Protein Analysis

Proteins are decisive in regulating cellular functions. They control cell division, internal and external signal transduction, and the expression state of the genome, in addition to a plethora of other functions. Thus, sensitive and scalable methods for protein characterization are of equal importance as genetic analysis for biochemical researches. Protein detection methods on microfluidic platforms can be roughly categorized into two kinds: fluorescent labeling methods and surface interaction-based methods.

FRET is a typical analytical method for on-chip protein analysis. FRET process occurs between a pair of donor and acceptor fluorophores when they are in close proximity. If the donor is excited, it can transfer energy to the close acceptor through radiationless ways, then, the acceptor emits a photon at a red shifted wavelength [108]. Xia et al. developed a series of micro/nanofluidic enrichment devices to detect proteins based on FRET mechanism. They assayed protease activity by a functionalized nanoprobe on a micro/nanofluidics preconcentrator combining with FRET detection method (Fig. 10.6a). The nanoprobe was formed by modifying AuNPs with dye labeled protein. The photoluminescence of donor dye was strongly quenched by AuNPs because of FRET. The dye labeled protein was cleaved from the AuNPs when the target protease was transported to the nanoprobes due to molecular recognition of enzyme to substrate. The fluorescence of dye was recovered as a consequence of the cleavage from AuNPs and enabled the protease activity assay [109]. They even detected trace quantities of protein depending on a similar method. In this approach, sample protein was adsorbed on AuNPs at first to occupy part of the AuNP surface. Then, dye-labeled protein was added to adsorb to the residual active sites of the AuNP surface. The unadsorbed dye-labeled protein remained in a free state in the system. Higher concentration of sample protein leaded to more free dye-labeled protein molecules remaining in the system generating more intense photoluminescence signal. The free dye-labeled protein molecules could be enriched by electric field and the photoluminescence signal was consequently amplified yielding an improved sensitivity [110]. During FRET process, the transfer between donor and acceptor is strongly dependent on the distance between them. As a result, the method can be used as a molecular ruler. Based on this principle, Yao et al. [111] revealed microsecond

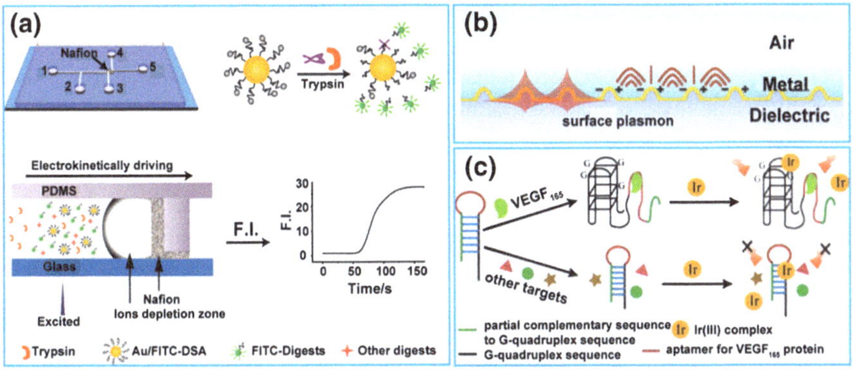

Fig. 10.6 Protein analysis on microfluidic chip. **a** Protein activity assay based on FRET mechanism combined with the micro/nanofluidicpreconcentration chip (Reprinted with permission from Ref. [109]. Copyright 2014 American Chemical Society.). **b** Schematic representation of the nanostructured metal thin film and the surface plasmon at the interface between the metal thin film and air (Reprinted with permission from Ref. [116]. Copyright 2016 Macmillan Publishers Limited, part of Springer Nature.). **c** Schematic illustration of VEGF165 assay based on G-quadruplex probe (Reprinted with permission from Ref. [119]. Copyright 2015 Elsevier Inc.)

protein folding events through detecting distance between donor-acceptor pair by combining the time-resolved FRET measurement and the ultrarapid hydrodynamic focusing microfluidic mixer. Intermolecular distance change was detected with temporal resolution at microsecond level and structural resolution at Angstrom level on the combined platform. The early folding states of Cytochrome c was investigated by monitoring the distance between the tryptophan-heme donor-acceptor pair. Protein-protein interaction and reaction progress were monitored in microfluidic droplets using FRET and confocal fluorescence lifetime microscope. A model system of Alexa Fluor 488 labeled biotinylated BSA, Alexa Fluor 594 labeled streptavidin and unlabeled chicken egg white avidin were detected in direct and competitive assay formats with nanomolar detection limits [112].

Surface plasmon resonance (SPR) is a well established detection method based on surface interactions. For SPR detection, a gold surface is necessary and derivatized with a protein layer in protein analysis. Binding of molecules to the gold layer induces a change in incident angle or wavelength which is dependent on the refractive index of the interface. As a result, mass transfer to the surface can be detected in real-time, allowing on-rates and off-rates to be measured. SPR integrated with microfluidics is well suitable for characterizing protein-protein interactions [113]. Nanohole-based SPR instrument quantified a broad range of antibody-ligand binding kinetics with equilibrium dissociation constants ranging from 200 pM to 40 nM. After optimizing surface modification protocols and antibody injection condtions, sequential SPR measurements were performed using a 12-channel microfluidic flow cell [114]. A titled fiber Bragg grating coated with a nanometer scale silver film was embedded inside a microfluidic channel to differentiate protein variation in urine samples of healthy, nephropatic and treated rats depending on SPR signals [115].

Nanomaterials and nanostructures are showing potential for integrating with microfluidic devices to characterize molecular interactions. A metal thin film with nanostructure was demonstrated to enhance transmission efficiency and sharp resonances of surface plasmon (Fig. 10.6b). The perpendicular transmission configuration and large-scale nanostructures provided a feasible platform for label-free SPR sensing of protein-protein interactions in microfluidic cells [116]. A method for on-chip bioorthogonal chemistry was developed and optimized to enable a variety of new SPR applications including on-chip assembly of functionalized nanoparticles which showed a preserved ability to interact with their target protein. Nanoparticle-protein interaction could be monitored in both orientations allowing experimental deconvolution of the intrinsic affinity of the small molecule for its target protein [117]. An antibody-modified roughness-controlled silicon nanowires (SiNWs) forest on chip was used to separate required protein analytes from complex biosamples because of the ultralarge binding surface area. The target proteins were released in a controlled liquid media and subsequently detected by supersensitive SiNW-based field-effect transistors (FETs) arrays fabricated on the same chip [118].

We also developed a series of microfluidic methods to identify proteins based on localized surface interactions. A novel G-quadruplex-selective luminescent iridium (III) complex was synthesized and applied as a probe for label-free quantitative analysis of $VEGF_{165}$ protein from cell metabolism as well as the kinetic study of the aptamer-protein interaction on microfluidic platform (Fig. 10.6c) [119]. $VEGF_{165}$ also was on-line qualitatively and semi-quantitatively analyzed by functional nucleic acid, hemin, 2,2'-azino-bis(3-ethylbenzothiazoline-6-sulphonic acid) (ABTS) and peroxide system. Cell co-culture and migration under different oxygen concentrations were investigated on the multifunctional microfluidic device [120]. Multiplex proteins from cell metabolism were assayed based on tunable aptamer and MCE. Electrophoretic mobility of proteins could be modulated by different lengths of aptamers allowing the proteins to be effectively separated in hydroxyethyl cellulose buffer with 1.00 mM Mg^{2+} [121]. Additionally, we developed an on-chip colorimetric assay method to detect thrombin from human plasma and serum using rolling circle amplification (RCA) and G-quadruplex DNAzyme. The microchannel was modified with aptamer to capture thrombin. A primer with a G-quadruplex circular template and a thrombin aptamer composed the secondary aptamer. Multiple G-quadruplex units were produced by RCA technology to improve the sensitivity. G-quadruplex DNAzyme catalyzed the H_2O_2-mediated oxidation of ABTS to generate colorimetric signal [122]. Besides that, we even developed a microfluidic chip-mass spectrometry (chip-MS) system to quantify noncovalent protein-protein interactions [123]. Both genetic and protein assays on microfluidic chip provides fundamental biological investigations to decipher mechanisms of some in vivo biochemical processes.

10.6 Conclusions and Outlooks

Microfluidics is advancing at a high speed and becoming a key miniaturized science and technology impacting biology, chemistry and engineering researches due to its flexible device design, excellent biomimetic microenvironment, high analysis speed and good compatibility with other microscale analytical methods. As a technology featuring microfabricated structures, microfluidics is good at manipulating fluids at microscale enabling faster reaction and detection. Micrometer sizes mean reduced reagent and sample consumption facilitating biochemical and clinic analysis. Multiple complex laboratory procedures, such as purification, separation and detection, have been miniaturized onto microfluidic devices with high throughput and sensitivity. Organ/tissue-on-chip recapitulating organ and tissue functions provides more comprehensive platform for advanced biochemical analysis.

Despite several advantages to date, there are some pending challenges needing to be addressed: (1) Non-specific adsorption of biomolecules within the microchannels critically affects the analytical performance of microfluidic chips. Although some strategies have been put forward to block the microchannels with blocking agents, a satisfactory solution still is on urgent demand. (2) The cost of integrating some components, such as electrodes, needs to be further reduced. (3) Large and multifunctional instruments, like as MS, nuclear magnetic resonance spectrometer, haven't been integrated onto the devices. Some preliminary trials that coupling these instruments with microchips have been attempted, however, fully-integrated chip-MS platforms need to be further developed. How to miniaturize these large instruments to make them adjustable to be coupled with chip is another problem need to be thought. With the progress of analytical methods and emerging of complementary technologies, such as smartphone-based rapid diagnostic test readers, next-generation of microfluidic system for biochemical analysis will be more multifunctional, fully-integrated and intelligent.

References

1. El-Ali J, Sorger PK, Jensen KF (2006) Cells on chips. Nature 442(7101):403–411. doi:10.1038/nature05063
2. Rosenkilde MM, Schwartz TW (2004) The chemokine system—a major regulator of angiogenesis in health and disease. APMIS 112(7–8):481–495. doi:10.1111/j.1600-0463.2004.apm11207-0808.x
3. Liu L, Ratner BD, Sage EH, Jiang S (2007) Endothelial cell migration on surface-density gradients of fibronectin, VEGF, or both proteins. Langmuir 23(22):11168–11173. doi:10.1021/la701435x
4. Forhead AJ, Fowden AL (2014) Thyroid hormones in fetal growth and prepartum maturation. J Endocrinol 221(3):R87–R103. doi:10.1530/joe-14-0025
5. Belloni AS, Albertin G, Forneris ML, Nussdorfer GG (2001) Proadrenomedullin-derived peptides as autocrine-paracrine regulators of cell growth. Histol Histopathol 16(4):1263–1274

6. Jamora C, Fuchs E (2002) Intercellular adhesion, signalling and the cytoskeleton. Nat Cell Biol 4(4):E101–E108. doi:10.1038/ncb0402-e101
7. Geiger B, Bershadsky A, Pankov R, Yamada KM (2001) Transmembrane extracellular matrix-cytoskeleton crosstalk. Nat Rev Mol Cell Bio 2(11):793–805. doi:10.1038/35099066
8. Fatehullah A, Tan SH, Barker N (2016) Organoids as an in vitro model of human development and disease. Nat Cell Biol 18(3):246–254
9. Li Y, Zhou Y, Wang H, Perrett S, Zhao Y, Tang Z, Nie G (2011) Chirality of glutathione surface coating affects the cytotoxicity of quantum dots. Angew. Chem Int Edit 50 (26):5860–5864. doi:10.1002/anie.201008206
10. Hoshino A, Fujioka K, Oku T, Suga M, Sasaki YF, Ohta T, Yasuhara M, Suzuki K, Yamamoto K (2004) Physicochemical properties and cellular toxicity of nanocrystal quantum dots depend on their surface modification. Nano Lett 4(11):2163–2169. doi:10.1021/nl048715d
11. Culbertson CT, Mickleburgh TG, Stewart-James SA, Sellens KA, Pressnall M (2014) Micro total analysis systems: Fundamental advances and biological applications. Anal Chem 86 (1, SI):95–118. doi:10.1021/ac403688g
12. Novo P, Dell'Aica M, Janasek D, Zahedi RP (2016) High spatial and temporal resolution cell manipulation techniques in microchannels. Analyst 141(6):1888–1905. doi:10.1039/c6an00027d
13. Patabadige DEW, Jia S, Sibbitts J, Sadeghi J, Sellens K, Culbertson CT (2016) Micro total analysis systems: Fundamental advances and applications. Anal Chem 88(1):320–338. doi:10.1021/acs.analchem.5b04310
14. Kovarik ML, Gach PC, Ornoff DM, Wang Y, Balowski J, Farrag L, Allbritton NL (2012) Micro total analysis systems for cell biology and biochemical assays. Anal Chem 84(2):516–540. doi:10.1021/ac202611x
15. Kovarik ML, Ornoff DM, Melvin AT, Dobes NC, Wang Y, Dickinson AJ, Gach PC, Shah PK, Allbritton NL (2013) Micro total analysis systems: Fundamental advances and applications in the laboratory, clinic, and field. Anal Chem 85(2, SI):451–472. doi:10.1021/ac3031543
16. Huh D, Hamilton GA, Ingber DE (2011) From 3D cell culture to organs-on-chips. Trends Cell Biol 21(12):745–754. doi:10.1016/j.tcb.2011.09.005
17. Sung KE, Beebe DJ (2014) Microfluidic 3D models of cancer. Adv Drug Deliver Rev 79–80:68–78. doi:10.1016/j.addr.2014.07.002
18. van Duinen V, Trietsch SJ, Joore J, Vulto P, Hankemeier T (2015) Microfluidic 3D cell culture: From tools to tissue models. Curr Opin Biotech 35:118–126. doi:10.1016/j.copbio.2015.05.002
19. Bersini S, Jeon JS, Dubini G, Arrigoni C, Chung S, Charest JL, Moretti M, Kamm RD (2014) A microfluidic 3D invitro model for specificity of breast cancer metastasis to bone. Biomaterials 35(8):2454–2461. doi:10.1016/j.biomaterials.2013.11.050
20. Unser AM, Mooney B, Corr DT, Tseng Y, Xie Y (2016) 3D brown adipogenesis to create "Brown-Fat-in-Microstrands". Biomaterials 75:123–134. doi:10.1016/j.biomaterials.2015.10.017
21. Maschmeyer I, Lorenz AK, Schimek K, Hasenberg T, Ramme AP, Hubner J, Lindner M, Drewell C, Bauer S, Thomas A, Sambo NS, Sonntag F, Lauster R, Marx U (2015) A four-organ-chip for interconnected long-term co-culture of human intestine, liver, skin and kidney equivalents. Lab Chip 15(12):2688–2699. doi:10.1039/c5lc00392j
22. Huh D, Matthews BD, Mammoto A, Montoya-Zavala M, Hsin HY, Ingber DE (2010) Reconstituting organ-level lung functions on a chip. Science 328(5986):1662–1668. doi:10.1126/science.1188302
23. Wilmer MJ, Ng CP, Lanz HL, Vulto P, Suter-Dick L, Masereeuw R (2016) Kidney-on-a-chip technology for drug-incuced nephrotoxiciy screening. Trends Biotechnol 34(2):156–170. doi:10.1016/j.tibtech.2015.11.001
24. Wu J, He Z, Chen Q, Lin J-M (2016) Biochemical analysis on microfluidic chips. TrAC. Trends Anal Chem 80:213–231. doi:10.1016/j.trac.2016.03.013

25. Choi J, Song H, Sung JH, Kim D, Kim K (2016) Microfluidic assay-based optical measurement techniques for cell analysis: A review of recent progress. Biosens Bioelectron 77:227–236. doi:10.1016/j.bios.2015.07.068
26. Noh J, Kim HC, Chung TD (2011) Biosensors in microfluidic chips.In Lin BC (ed) Microfluidics: Technologies and applications, Vol. 304, Topics in Current Chemistry, 1st edn. Springer, Berlin
27. Schulz CM, Ruzicka J (2002) Real-time determination of glucose consumption by live cells using a lab-on-valve system with an integrated microbioreactor. Analyst 127(10):1293–1298. doi:10.1039/b206907p
28. Schulz CM, Scampavia L, Ruzicka J (2002) Real-time monitoring of lactate extrusion and glucose consumption of cultured cells using a lab-on-valve system. Analyst 127(12):1583–1588. doi:10.1039/b209371p
29. Srinivasan V, Pamula VK, Fair RB (2004) Droplet-based microfluidic lab-on-a-chip for glucose detection. Anal Chim Acta 507(1):145–150. doi:10.1016/j.aca.2003.12.030
30. Srinivasan V, Pamula VK, Fair RB (2004) An integrated digital microfluidic lab-on-a-chip for clinical diagnostics on human physiological fluids. Lab Chip 4(4):310–315. doi:10.1039/b403341h
31. Lee W, Kwon D, Choi W, Jung GY, Au AK, Folch A, Jeon S (2015) 3D-Printed microfluidic device for the detection of pathogenic bacteria using size-based separation in helical channel with trapezoid Cross-Section. Sci Rep 5:7717. doi:10.1038/srep07717
32. Chalfie M, Tu Y, Euskirchen G, Ward WW, Prasher DC (1994) Green fluorescent protein as a marker for gene expression. Science 263(5148):802–805. doi:10.1126/science.8303295
33. Kijanka GS, Dimov IK, Burger R, Ducrée J (2015) Real-time monitoring of cell migration, phagocytosis and cell surface receptor dynamics using a novel, live-cell opto-microfluidic technique. Anal Chim Acta 872:95–99. doi:10.1016/j.aca.2014.12.035
34. Valero A, Merino F, Wolbers F, Luttge R, Vermes I, Andersson H, van den Berg A (2005) Apoptotic cell death dynamics of HL60 cells studied using a microfluidic cell trap device. Lab Chip 5(1):49–55. doi:10.1039/b415813j
35. Lockwood SY, Meisel JE, Monsma JFJ, Spence DM (2016) A Diffusion-Based and dynamic 3D-Printed device that enables parallel in vitro pharmacokinetic profiling of molecules. Anal Chem 88(3):1864–1870. doi:10.1021/acs.analchem.5b04270
36. Mazutis L, Gilbert J, Ung WL, Weitz DA, Griffiths AD, Heyman JA (2013) Single-cell analysis and sorting using droplet-based microfluidics. Nat Protoc 8(5):870–891. doi:10.1038/nprot.2013.046
37. Wu J, Chen Q, Liu W, Zhang Y, Lin J-M (2012) Cytotoxicity of quantum dots assay on a microfluidic 3D-culture device based on modeling diffusion process between blood vessels and tissues. Lab Chip 12(18):3474–3480. doi:10.1039/c2lc40502d
38. Wu J, Chen Q, Liu W, Lin J-M (2013) A simple and versatile microfluidic cell density gradient generator for quantum dot cytotoxicity assay. Lab Chip 13(10):1948–1954. doi:10.1039/c3lc00041a
39. Gashti MP, Asselin J, Barbeau J, Boudreau D, Greener J (2016) A microfluidic platform with pH imaging for chemical and hydrodynamic stimulation of intact oral biofilms. Lab Chip 16(8):1412–1419. doi:10.1039/c5lc01540e
40. Zhu X, Xu L, Wu T, Xu A, Zhao M, Liu S (2014) Continuous monitoring of bisulfide variation in microdialysis effluents by on-line droplet-based microfluidic fluorescent sensor. Biosens Bioelectron 55:438–445. doi:10.1016/j.bios.2013.12.056
41. Kim S, Streets AM, Lin RR, Quake SR, Weiss S, Majumdar DS (2011) High-throughput single-molecule optofluidic analysis. Nat Methods 8(3):242–245. doi:10.1038/nmeth.1569
42. Chen C, Ahmed M, Häfner T, Klämpfl F, Stelzle F, Schmidt M (2016) Fabrication of a turbid optofluidic phantom device with tunable μ_a and μ_s to simulate cutaneous vascular perfusion. Sci Rep 6:30567. doi:10.1038/srep30567
43. Valero A, Braschler T, Renaud P (2010) A unified approach to dielectric single cell analysis: impedance and dielectrophoretic force spectroscopy. Lab Chip 10(17):2216–2225. doi:10.1039/c003982a

44. Simon P, Frankowski M, Bock N, Neukammer J (2016) Label-free whole blood cell differentiation based on multiple frequency AC impedance and light scattering analysis in a micro flow cytometer. Lab Chip 16(12):2326–2338. doi:10.1039/c6lc00128a
45. Zhou Y, Basu S, Laue E, Seshia AA (2016) Single cell studies of mouse embryonic stem cell (mESC) differentiation by electrical impedance measurements in a microfluidic device. Biosens Bioelectron 81:249–258. doi:10.1016/j.bios.2016.02.069
46. Cheng I, Huang W, Chen T, Liu C, Lin Y, Su W (2015) Antibody-free isolation of rare cancer cells from blood based on 3D lateral dielectrophoresis. Lab Chip 15(14):2950–2959. doi:10.1039/c5lc00120j
47. Kim SH, Fujii T (2016) Efficient analysis of a small number of cancer cells at the single-cell level using an electroactive double-well array. Lab Chip 16(13):2440–2449. doi:10.1039/c6lc00241b
48. Adekanmbi EO, Srivastava SK (2016) Dielectrophoretic applications for disease diagnostics using lab-on-a-chip platforms. Lab Chip 16(12):2148–2167. doi:10.1039/c6lc00355a
49. Liu Q, Lin X, Lin L, Yi L, Li H, Lin J-M (2015) A comparative study of three different nucleic acid amplification techniques combined with microchip electrophoresis for HPV16 E6/E7 mRNA detection. Analyst 140(19):6736–6741. doi:10.1039/c5an00944h
50. Deng Y, Yi L, Lin X, Lin L, Li H, Lin J-M (2015) A non-invasive genomic diagnostic method for bladder cancer using size-based filtration and microchip electrophoresis. Talanta 144:136–144. doi:10.1016/j.talanta.2015.05.065
51. Sheng P, Wen W (2012) Electrorheological fluids: Mechanisms, dynamics, and microfluidics applications. Annu Rev Fluid Mech 44:143–174. doi:10.1146/annurev-fluid-120710-101024
52. Wang L, Zhang M, Li J, Gong X, Wen W (2010) Logic control of microfluidics with smart colloid. Lab Chip 10(21):2869–2874. doi:10.1039/c0lc00003e
53. Zhang M, Wang L, Wang X, Wu J, Li J, Gong X, Qin J, Li W, Wen W (2011) Microdroplet-based universal logic gates by electrorheological fluid. Soft Matter 7 (16):7493–7497. doi:10.1039/c1sm05687e
54. van Reenen A, de Jong AM, den Toonder JMJ, Prins MWJ (2014) Integrated lab-on-chip biosensing systems based on magnetic particle actuation—a comprehensive review. Lab Chip 14(12):1966–1986. doi:10.1039/c3lc51454d
55. Jamshaid T, Neto ETT, Eissa MM, Zine N, Kunita MH, El-Salhi AE, Elaissari A (2016) Magnetic particles: From preparation to lab-on-a-chip, biosensors, microsystems and microfluidics applications. TrAC. Trends Anal Chem 79:344–362. doi:10.1016/j.trac.2015.10.022
56. Zhou B, Xu W, Syed AA, Chau Y, Chen L, Chew B, Yassine O, Wu X, Gao Y, Zhang J, Xiao X, Kosel J, Zhang X, Yao Z, Wen W (2015) Design and fabrication of magnetically functionalized flexible micropillar arrays for rapid and controllable microfluidic mixing. Lab Chip 15(9):2125–2132. doi:10.1039/c5lc00173k
57. Kahkeshani S, Di Carlo D (2016) Drop formation using ferrofluids driven magnetically in a step emulsification device. Lab Chip 16(13):2474–2480. doi:10.1039/c6lc00645k
58. Shi X, Chen C, Gao W, Chao S, Meldrum DR (2015) Parallel RNA extraction using magnetic beads and a droplet array. Lab Chip 15(4):1059–1065. doi:10.1039/c4lc01111b
59. Chao C, Wang C, Che Y, Kao C, Wu J, Lee G (2016) An integrated microfluidic system for diagnosis of the resistance of Helicobacter pylori to quinolone-based antibiotics. Biosens Bioelectron 78:281–289. doi:10.1016/j.bios.2015.11.046
60. Hejazian M, Li W, Nguyen N (2015) Lab on a chip for continuous-flow magnetic cell separation. Lab Chip 15(4):959–970. doi:10.1039/c4lc01422g
61. Figueredo F, Garcia PT, Corton E, Coltro WKT (2016) Enhanced analytical performance of paper microfluidic devices by using Fe_3O_4 nanoparticles, MWCNT, and graphene oxide. ACS Appl Mater Inter 8(1):11–15. doi:10.1021/acsami.5b10027
62. Liu F, Pawan KC, Zhang G, Zhe J (2016) Microfluidic magnetic bead assay for cell detection. Anal Chem 88(1):711–717. doi:10.1021/acs.analchem.5b02716

63. Chen J, Kang Z, Wang G, Loo JFC, Kong SK, Ho H (2015) Optofluidic guiding, valving, switching and mixing based on plasmonic heating in a random gold nanoisland substrate. Lab Chip 15(11):2504–2512. doi:10.1039/c5lc00406c
64. Monat C, Domachuk P, Eggleton BJ (2007) Integrated optofluidics: A new river of light. Nat Photonics 1(2):106–114. doi:10.1038/nphoton.2006.96
65. Schmidt H, Hawkins AR (2011) The photonic integration of non-solid media using optofluidics. Nat Photonics 5(10):598–604. doi:10.1038/nphoton.2011.163
66. Zhu H, Lin X, Su Y, Dong H, Wu J (2015) Screen-printed microfluidic dielectrophoresis chip for cell separation. Biosens Bioelectron 63:371–378. doi:10.1016/j.bios.2014.07.072
67. Anand RK, Johnson ES, Chiu DT (2015) Negative dielectrophoretic capture and repulsion of single cells at a bipolar electrode: The impact of faradaic ion enrichment and depletion. J Am Chem Soc 137(2):776–783. doi:10.1021/ja5102689
68. Hai A, Shappir J, Spira ME (2010) In-cell recordings by extracellular microelectrodes. Nat Methods 7(3):200–202. doi:10.1038/nmeth.1420
69. Cornaglia M, Trouillon R, Tekin HC, Lehnert T, Gijs MAM (2014) Magnetic particle-scanning for ultrasensitive immunodetection On-Chip. Anal Chem 86(16):8213–8223. doi:10.1021/ac501568g
70. Zhu G, Nguyen N (2012) Rapid magnetofluidic mixing in a uniform magnetic field. Lab Chip 12(22):4772–4780. doi:10.1039/c2lc40818j
71. Li S, Guo F, Chen Y, Ding X, Li P, Wang L, Cameron CE, Huang TJ (2014) Standing surface acoustic wave based cell coculture. Anal Chem 86(19):9853–9859. doi:10.1021/ac502453z
72. Yeo LY, Friend JR (2014) Surface acoustic wave microfluidics. Annu Rev Fluid Mech 46 (1):379–406. doi:10.1146/annurev-fluid-010313-141418
73. Ding X, Li P, Lin SS, Stratton ZS, Nama N, Guo F, Slotcavage D, Mao X, Shi J, Costanzo F, Huang TJ (2013) Surface acoustic wave microfluidics. Lab Chip 13(18):3626–3649. doi:10.1039/c3lc50361e
74. Gedge M, Hill M (2012) Acoustofluidics 17: Theory and applications of surface acoustic wave devices for particle manipulation. Lab Chip 12(17):2998–3007. doi:10.1039/c2lc40565b
75. Wiklund M, Green R, Ohlin M (2012) Acoustofluidics 14: applications of acoustic streaming in microfluidic devices. Lab Chip 12(14):2438–2451. doi:10.1039/c2lc40203c
76. Destgeer G, Sung HJ (2015) Recent advances in microfluidic actuation and micro-object manipulation via surface acoustic waves. Lab Chip 15(13):2722–2738. doi:10.1039/c5lc00265f
77. Collins DJ, Neild A, Ai Y (2016) Highly focused high-frequency travelling surface acoustic waves (SAW) for rapid single-particle sorting. Lab Chip 16(3):471–479. doi:10.1039/c5lc01335f
78. Ma Z, Collins DJ, Ai Y (2016) Detachable acoustofluidic system for particle separation via a traveling surface acoustic wave. Anal Chem 88(10):5316–5323. doi:10.1021/acs.analchem.6b00605
79. Destgeer G, Cho H, Ha BH, Jung JH, Park J, Sung HJ (2016) Acoustofluidic particle manipulation inside a sessile droplet: four distinct regimes of particle concentration. Lab Chip 16(4):660–667. doi:10.1039/c5lc01104c
80. Shilton RJ, Travagliati M, Beltram F, Cecchini M (2014) Nanoliter-droplet acoustic streaming via ultra high frequency surface acoustic waves. Adv Mater 26(29):4941–4946. doi:10.1002/adma.201400091
81. Gracioso Martins AM, Glass NR, Harrison S, Rezk AR, Porter NA, Carpenter PD, Du Plessis J, Friend JR, Yeo LY (2014) Toward complete miniaturisation of flow injection analysis systems: microfluidic enhancement of chemiluminescent detection. Anal Chem 86 (21):10812–10819. doi:10.1021/ac502878p
82. Lin SS, Mao X, Huang TJ (2012) Surface acoustic wave (SAW) acoustophoresis: now and beyond. Lab Chip 12(16):2766–2770. doi:10.1039/c2lc90076a

83. Petersson F, Nilsson A, Holm C, Jonsson H, Laurell T (2004) Separation of lipids from blood utilizing ultrasonic standing waves in microfluidic channels. Analyst 129(10):938–943. doi:10.1039/b409139f
84. Laurell T, Petersson F, Nilsson A (2007) Chip integrated strategies for acoustic separation and manipulation of cells and particles. Chem Soc Rev 36(3):492–506. doi:10.1039/B601326K
85. Lenshof A, Laurell T (2010) Continuous separation of cells and particles in microfluidic systems. Chem Soc Rev 39(3):1203–1217. doi:10.1039/b915999c
86. Nawaz AA, Chen Y, Narna N, Nissly RH, Ren L, Ozcelik A, Wang L, McCoy JP, Levine SJ, Huang TJ (2015) Acoustofluidic fluorescence activated cell sorter. Anal Chem 87 (24):12051–12058. doi:10.1021/acs.analchem.5b02398
87. Ren L, Chen Y, Li P, Mao Z, Huang P, Rufo J, Guo F, Wang L, McCoy JP, Levine SJ, Huang TJ (2015) A high-throughput acoustic cell sorter. Lab Chip 15(19):3870–3879. doi:10.1039/c5lc00706b
88. Li S, Ding X, Mao Z, Chen Y, Nama N, Guo F, Li P, Wang L, Cameron CE, Huang TJ (2015) Standing surface acoustic wave (SSAW)-based cell washing. Lab Chip 15(1):331–338. doi:10.1039/c4lc00903g
89. Li P, Mao Z, Peng Z, Zhou L, Chen Y, Huang P, Truica CI, Drabick JJ, El-Deiry WS, Dao M, Suresh S, Huang TJ (2015) Acoustic separation of circulating tumor cells. Proc Natl Acad Sci U S A 112(16):4970–4975. doi:10.1073/pnas.1504484112
90. Chen Y, Wu M, Ren L, Liu J, Whitley PH, Wang L, Huang TJ (2016) High-throughput acoustic separation of platelets from whole blood. Lab Chip 16:3466–3472. doi:10.1039/c6lc00682e
91. Lin X, Sun X, Luo S, Liu B, Yang C (2016) Development of DNA-based signal amplification and microfluidic technology for protein assay: A review. TrAC. Trends Anal Chem 80:132–148. doi:10.1016/j.trac.2016.02.020
92. Zhang Y, Jiang H (2016) A review on continuous-flow microfluidic PCR in droplets: Advances, challenges and future. Anal Chim Acta 914:7–16. doi:10.1016/j.aca.2016.02.006
93. Dalerba P, Kalisky T, Sahoo D, Rajendran PS, Rothenberg ME, Leyrat AA, Sim S, Okamoto J, Johnston DM, Qian D, Zabala M, Bueno J, Neff NF, Wang J, Shelton AA, Visser B, Hisamori S, Shimono Y, van de Wetering M, Clevers H, Clarke MF, Quake SR (2011) Single-cell dissection of transcriptional heterogeneity in human colon tumors. Nat Biotechnol 29(12):1120–1127. doi:10.1038/nbt.2038
94. White AK, VanInsberghe M, Petriv OI, Hamidi M, Sikorski D, Marra MA, Piret J, Aparicio S, Hansen CL (2011) High-throughput microfluidic single-cell RT-qPCR. Proc Natl Acad Sci U S A 108(34):13999–14004. doi:10.1073/pnas.1019446108
95. Lin X, Wu J, Li H, Wang Z, Lin J-M (2013) Determination of mini-short tandem repeat (miniSTR) loci by using the combination of polymerase chain reaction (PCR) and microchip electrophoresis. Talanta 114:131–137. doi:10.1016/j.talanta.2013.04.012
96. Lin X, Wu J, Liu W, Li H, Wang Z, Lin J-M (2013) Detection of BCR-ABL using one step reverse transcriptase-polymerase chain reaction and microchip electrophoresis. Methods 64 (3):250–254. doi:10.1016/j.ymeth.2013.05.010
97. Zhu Z, Jenkins G, Zhang W, Zhang M, Guan Z, Yang CJ (2012) Single-molecule emulsion PCR in microfluidic droplets. Anal Bioanal Chem 403(8):2127–2143. doi:10.1007/s00216-012-5914-x
98. Margulies M, Egholm M, Altman WE, Attiya S, Bader JS, Bemben LA, Berka J, Braverman MS, Chen Y, Chen Z, Dewell SB, Du L, Fierro JM, Gomes XV, Godwin BC, He W, Helgesen S, Ho CH, Irzyk GP, Jando SC, Alenquer MLI, Jarvie TP, Jirage KB, Kim J, Knight JR, Lanza JR, Leamon JH, Lefkowitz SM, Lei M, Li J, Lohman KL, Lu H, Makhijani VB, McDade KE, McKenna MP, Myers EW, Nickerson E, Nobile JR, Plant R, Puc BP, Ronan MT, Roth GT, Sarkis GJ, Simons JF, Simpson JW, Srinivasan M, Tartaro KR, Tomasz A, Vogt KA, Volkmer GA, Wang SH, Wang Y, Weiner MP, Yu P, Begley RF, Rothberg JM (2005) Genome sequencing in microfabricated high-density picolitre reactors. Nature 437(7057):376–380. doi:10.1038/nature03959

99. Tanaka H, Yamamoto S, Nakamura A, Nakashoji Y, Okura N, Nakamoto N, Tsukagoshi K, Hashimoto M (2015) Hands-off preparation of monodisperse emulsion droplets using a poly (dimethylsiloxane) microfluidic chip for droplet digital PCR. Anal Chem 87(8):4134–4143. doi:10.1021/ac503169h
100. Wang P, Jing F, Li G, Wu Z, Cheng Z, Zhang J, Zhang H, Jia C, Jin Q, Mao H, Zhao J (2015) Absolute quantification of lung cancer related microRNA by droplet digital PCR. Biosens Bioelectron 74:836–842. doi:10.1016/j.bios.2015.07.048
101. Zhu H, Wang G, Xie D, Cai B, Liu Y, Zhao X (2014) Au nanoparticles enhanced fluorescence detection of DNA hybridization in picoliter microfluidic droplets. Biomed Microdevices 16(3):479–485. doi:10.1007/s10544-014-9850-8
102. Wang J, Aki M, Onoshima D, Arinaga K, Kaji N, Tokeshi M, Fujita S, Yokoyama N, Baba Y (2014) Microfluidic biosensor for the detection of DNA by fluorescence enhancement and the following streptavidin detection by fluorescence quenching. Biosens Bioelectron 51:280–285. doi:10.1016/j.bios.2013.07.058
103. Li H, Fang X, Cao H, Kong J (2016) Paper-based fluorescence resonance energy transfer assay for directly detecting nucleic acids and proteins. Biosens Bioelectron 80:79–83. doi:10.1016/j.bios.2015.12.065
104. Lubin AA, Plaxco KW (2010) Folding-based electrochemical biosensors: the case for responsive nucleic acid architectures. Accounts Chem Res 43(4):496–505. doi:10.1021/ar900165x
105. Fan CH, Plaxco KW, Heeger AJ (2003) Electrochemical interrogation of conformational changes as a reagentless method for the sequence-specific detection of DNA. Proc Natl Acad Sci U S A 100(16):9134–9137. doi:10.1073/pnas.1633515100
106. Xiao Y, Lubin AA, Baker BR, Plaxco KW, Heeger AJ (2006) Single-step electronic detection of femtomolar DNA by target-induced strand displacement in an electrode-bound duplex. Proc Natl Acad Sci U S A 103(45):16677–16680. doi:10.1073/pnas.0607693103
107. Ben-Yoav H, Dykstra PH, Bentley WE, Ghodssi R (2015) A controlled microfluidic electrochemical lab-on-a-chip for label-free diffusion-restricted DNA hybridization analysis. Biosens Bioelectron 64:579–585. doi:10.1016/j.bios.2014.09.069
108. Maerkl SJ (2011) Next generation microfluidic platforms for high-throughput protein biochemistry. Curr Opin Biotech 22(1):59–65. doi:10.1016/j.copbio.2010.08.010
109. Wang C, Ouyang J, Wang Y, Ye D, Xia X (2014) Sensitive assay of protease activity on a micro/nanofluidics preconcentrator fused with the fluorescence resonance energy transfer detection technique. Anal Chem 86(6):3216–3221. doi:10.1021/ac500196s
110. Wang C, Shi Y, Wang J, Pang J, Xia X (2015) Ultrasensitive protein concentration detection on a micro/nanofluidic enrichment chip using fluorescence quenching. ACS Appl Mater Inter 7(12):6835–6841. doi:10.1021/acsami.5b00383
111. Jiang L, Zeng Y, Sun Q, Sun Y, Guo Z, Qu JY, Yao S (2015) Microsecond protein folding events revealed by time-resolved fluorescence resonance energy transfer in a microfluidic mixer. Anal Chem 87(11):5589–5595. doi:10.1021/acs.analchem.5b00366
112. Benz C, Retzbach H, Nagl S, Belder D (2013) Protein-protein interaction analysis in single microfluidic droplets using FRET and fluorescence lifetime detection. Lab Chip 13 (14):2808–2814. doi:10.1039/c3lc00057e
113. Homola J (2008) Surface plasmon resonance sensors for detection of chemical and biological species. Chem Rev 108(2):462–493. doi:10.1021/cr068107d
114. Im H, Sutherland JN, Maynard JA, Oh S (2012) Nanohole-based surface plasmon resonance instruments with improved spectral resolution quantify a broad range of antibody-ligand binding kinetics. Anal Chem 84(4):1941–1947. doi:10.1021/ac300070t
115. Guo T, Liu F, Liang X, Qiu X, Huang Y, Xie C, Xu P, Mao W, Guan B, Albert J (2016) Highly sensitive detection of urinary protein variations using tilted fiber grating sensors with plasmonic nanocoatings. Biosens Bioelectron 78:221–228. doi:10.1016/j.bios.2015.11.047
116. Xiao B, Pradhan SK, Santiago KC, Rutherford GN, Pradhan AK (2016) Topographically engineered large scale nanostructures for plasmonic biosensing. Sci Rep 6:24385. doi:10.1038/srep24385

117. Tassa C, Liong M, Hilderbrand S, Sandler JE, Reiner T, Keliher EJ, Weissleder R, Shaw SY (2012) On-chip bioorthogonal chemistry enables immobilization of in situ modified nanoparticles and small molecules for label-free monitoring of protein binding and reaction kinetics. Lab Chip 12(17):3103–3110. doi:10.1039/c2lc40337d
118. Krivitsky V, Hsiung L, Lichtenstein A, Brudnik B, Kantaev R, Elnathan R, Pevzner A, Khatchtourints A, Patolsky F (2012) Si nanowires forest-based on-chip biomolecular filtering, separation and preconcentration devices: nanowires do it all. Nano Lett 12 (9):4748–4756. doi:10.1021/nl3021889
119. Lin X, Leung K, Lin L, Lin L, Lin S, Leung C, Ma D, Lin J (2016) Determination of cell metabolite VEGF165 and dynamic analysis of protein-DNA interactions by combination of microfluidic technique and luminescent switch-on probe. Biosens Bioelectron 79:41–47. doi:10.1016/j.bios.2015.11.089
120. Lin X, Chen Q, Liu W, Zhang J, Wang S, Lin Z, Lin J-M (2015) Oxygen-induced cell migration and on-line monitoring biomarkers modulation of cervical cancers on a microfluidic system. Sci Rep 5:9643. doi:10.1038/srep09643
121. Lin X, Chen Q, Liu W, Yi L, Li H, Wang Z, Lin J-M (2015) Assay of multiplex proteins from cell metabolism based on tunable aptamer and microchip electrophoresis. Biosens Bioelectron 63:105–111. doi:10.1016/j.bios.2014.07.013
122. Lin X, Chen Q, Liu W, Li H, Lin J-M (2014) A portable microchip for ultrasensitive and high-throughput assay of thrombin by rolling circle amplification and hemin/G-quadruplex system. Biosens Bioelectron 56:71–76. doi:10.1016/j.bios.2013.12.061
123. Liu W, Chen Q, Lin X, Lin J-M (2015) Online multi-channel microfluidic chip-mass spectrometry and its application for quantifying noncovalent protein-protein interactions. Analyst 140(5):1551–1554. doi:10.1039/c4an02370f

Chapter 11
Microfluidic Cell Culture Systems for Drug Research

Mingsha Jie and Jin-Ming Lin

Abstract The identification of new drugs and predicting drug responses to cancer patients are unprecedented challenges in pharmaceutical industry. Developing biologically relevant models of human tissues and organs is of vital importance for disease modeling and drug discovery. Microfluidic devices, with controllable chambers and channels in the micrometer range (10–1000 μm), are suitable for miniaturization analysis and they showed great promise as new and influential players in drug discovery. In this chapter, microfluidic technologies, which improve the efficiency of drug research at the microscale level, were summarized. To begin with, the attractive properties of lab-on-a-chip technology were described. Then, the development of three mainly microfluidic technologies were introduced and described in detail. Some practical applications of microfluidic-based, cell-culture systems in drug research particularly those pertaining to drug toxicity testing and those with a high-throughput capability are subsequently reviewed and highlighted. Finally, the challenges and opportunities in regard to numerous aspects of microfluidic system were discussed. It is believed that its excellent properties of miniaturization, high-throughput analysis, and highly integrated function, endows lab-on-a-chip device with great potential academic and commercial value in discovery and development of drug.

Keywords Drug research · Microfluidics · Cell culture · Organs on a chip
Pharmacokinetics system

M. Jie
School of Science, Beijing University of Chemical Technology, Beijing 100029, China

M. Jie · J.-M. Lin (✉)
Department of Chemistry, Tsinghua University, Beijing 100084, China
e-mail: jmlin@mail.tsinghua.edu.cn

J.-M. Lin (ed.), *Cell Analysis on Microfluidics*, Integrated Analytical Systems,
https://doi.org/10.1007/978-981-10-5394-8_11

11.1 Introduction

The pharmaceutical industry is facing unprecedented challenges due to rising costs and the declining efficiency of drug discovery [1, 2]. Since 1950, the number of new drugs approved per billion US dollars spent on drug research and development has halved roughly every 9 years [3]. The process of modern drug discovery continues to face increasing challenges, some of which include the biological hypothesis needing ultimately be tested in humans and the drug candidate analysis needing in a more rapid and accurate manner [4]. To solve these problems, microtechnologies, such as microfluidics are showing promise as new and influential players [5]. Microfluidic devices integrate various components, such as pump, valves, mixers, heaters, microchannels and can carry out experiments in the manner of higher throughput, less reagent consumption, faster reaction time. Therefore, there seems to be an emerging opportunity to develop new tools that may aid in the drug discovery and development process [6].

During the past few decades, there has been tremendous development in the field of microfluidic chip [7]. Unique characteristics of microfluidics make them excellent candidates for drug research and biological applications [8]. Microfluidic drug screening platforms offer greater utility in predicting subcellular and cellular physiological responses in humans in vivo and enable to access to physical phenomena and mechanisms that are not accessible on the macroscopic scale. The results reduce the incidence of costly late-stage failures during pharmaceutical clinical trials. Compared to animal experiments in drug testing, microfluidic systems reduce the sacrifice of animals and avoid ethical problems. Furthermore, microfluidic devices allow for reproducible screening at a significantly low cost and high-throughput and reduce the time consuming by performing operations quicker than conventional macroscopic device. Thus, microfluidic technology is already being used in many different stages of drug research and development [9]. In this chapter, the latest advances of microfluidic devices and technologies in drug research were summed. We initially discuss the applications of microfluidic systems for the drug discovery process by identifying emerging areas in two-dimensional (2D) cell culture, three-dimensional (3D) cell culture, organs chip and whole organisms chip and subsequently discuss the application of these technologies for the drug development process.

11.2 Microfluidic Chip

Microfluidics was initiated in the late 1970s, but received little attention until 1990, when Manz et al. [10] published the conceptual work on miniaturized total analysis systems, today known as 'lab-on-a-chip'. This initial work resulted in a tremendous increase in the research and development of chip-based microfluidic model systems and it has now grown to an interdisciplinary field engaging hundreds of research groups along with several companies. Microfluidics, the science and technology of fluid flow in microscale, integrate multiple components into a single microdevice and manipulate small (10^{-9}–10^{-18} l) amounts of fluids using pump, mixers, actuators, heaters, and channels with dimensions of tens to hundreds of micrometres [11].

11.2.1 Cell-Based Microfluidic Model Systems

Animal testing is different from in vivo microenvironments of human beings and difficult to determine the internal processes of human' body due to difference from in vivo microenvironments of human beings. Discarding the time consumption and resources waste, animal tests often also cause ethical. Discarding the time consumption and resources waste, animal tests often also cause ethical problems [12]. To overcome these limitations, in vitro cell-based microfluidic model systems are emerging as the tools to better understand of human biology and disease. For the research of disease pathogenesis and drug action mechanism, the direct observation of the human body is difficult and the in vitro biological models are essential. In vitro biological models are categorized into different levels to analyze biological phenomenon according to the pathophysiological relevance from cell lines to primary cells, and up to model organisms and human tissue explants, and eventually to rodents or nonhuman primates. However, labor consumption and time cost increase with the increasing of the biological relevance. Thus, the simplest model, cell-based in vitro culture system, is often used as the early stages of drug discovery, diagnostics and personalized medicine [13].

Traditional cell culture system usually relies on undefined biological materials such as polylysine, fetal bovine serum, fibronection and simple containers such as dishes, bottles, and flasks. As common material of fabrication chips, PDMS has various unique characteristics that make it suitable for cell culture in vitro and the advantages are as listed below [14]. Firstly, owing to the high gas permeability of PDMS, sufficient oxygen is supplied to cells in the microchannels and separate oxygenators aren't needed, which are crucial for drug discovery and toxicology studies. Additionally, due to the high flexibility of PDMS, various tissues and organs models could be fabricated by precise automated controlling of fluid flow and the shape and size of the microchannels. Further, this excellent optical transparency of PDMS, make the microdevices carry out real-time fluorescence analysis and optical monitoring of cell morphology and motility precisely [15, 16]. Based on above, microfluidic chips are emerging as a new kind of cell culture vessel and tremendously enhance human ability to control the local cellular microenvironment [17]. Microfluidic chips simultaneous manipulate and analyze from a single cell level, to cell populations and up to organs by controlling fluid flow in the micrometer and nanoliter scale in precisely defined geometries. On the microdevices, cells suspension is infused into microchannels, attached and grew in the basal (such as glass, PDMS). Cell numbers and density are controlled precisely in given complex geometries. The medium or other solutions go through the surface of cells by the mean of static or dynamic to provide nutrients for the cells [18, 19]. Simultaneously Cells images are monitored by ultra-high spatial and temporal resolution. The surface modification of chips is essential, general is carried out by some protein such as polylysine, serum, collagen, etc. [20]. Cell culture in microfluidic chips has the advantage of simple operation, easily injection the culture medium and convenience for observation, etc., thus there are a lot of cell culture carried out in 2D microfluidic devices [21, 22].

Mazutis et al. [23] reported a novel droplet-based microfluidic device for high-throughput analysis and sorting of single cells in 50-pl droplets by co-compartmentalizing for only 15 min. The single cell-droplets device can analyze protein release or secrete from cells and adapted for screening intracellular, cell-surface or secreted proteins and overcoming one of the major challenges of traditional flow cytometry and fluorescence-activated cell sorting. Two types of cells (orange and gray ellipses) were introduced together as a cell suspension into the microfluidic system by a bead suspension. Cells labelled gray produced mouse antibodies and cells labelled orange didn't that. After incubation for 15 min, droplets containing an antibody-producing cell were collected and the cells, which does not secrete antibody, are discarded (shown in Fig. 11.1).

An important application of cell-based microfluidic model systems is cell capture from blood and other biological samples. Stott et al. introduced a high-throughput microfluidic device, the herringbone-chip (Fig. 11.2), which provides an enhanced platform for isolating circulating tumor cells (CTCs) and investigating the role of epithelial-mesenchymal transition in human cancer [24]. The herringbone design disturbed the laminar flow, increased turbulence of the

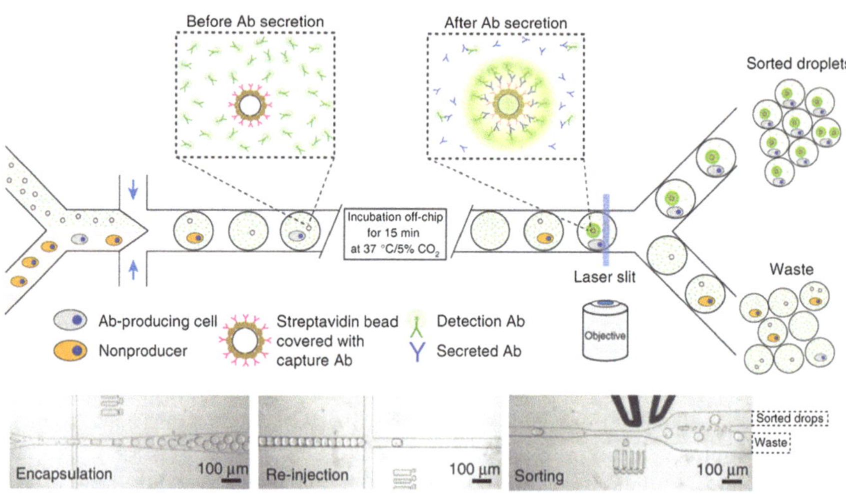

Fig. 11.1 Principle of single cell-droplets device design. Two types of cells (*orange* and *gray ellipses*) are introduced together as a cell suspension into a microfluidic device together with a bead suspension. One cell type produces mouse antibodies (*gray*) and the other does not (*orange*), with the latter being in 10-fold excess. Reproduced with permission from Ref. [23]

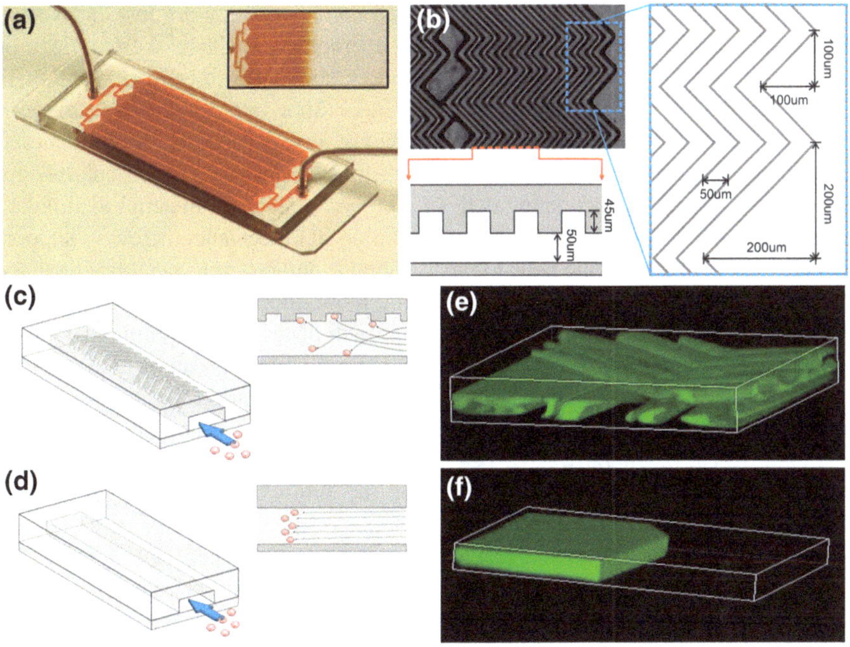

Fig. 11.2 The herringbone-chip. **a** The HB-chip consists of a microfluidic array of channels with a single inlet and exit. Inset illustrates the uniform blood flow through the device. **b** A micrograph of the grooved surface illustrates the asymmetry and periodicity of the herringbone grooves. Cartoon illustrating the cell-surface interactions in (**c**) the HB-Chip, and **d** a traditional flat-walled microfluidic device. Flow visualization studies using two paired streams of the same viscosity (one stream is *green*, the other clear) demonstrate **e** the chaotic microvortices generated by the herringbone grooves, and the lack of mixing in (**f**) traditional flat-walled devices. Reproduced with permission from Ref. [25]

chip, and caused the liquid mix in the chip, thus, it increased the cells contact with the chip's surface, and improved the efficiency of the CTCs capture [25].

11.2.2 3D Cell Culture in Microfluidic Chip

Conventional 2D cell cultures were developed almost a century. The advantages of 2D cell cultures are that the cells are cultured in a patterned way, the operation is simple and easy to realize [26, 27]. However, the disadvantages are that the cell growth situation in 2D state is different from the actual situation in living body, that is the cell is in 3D growth state, and cell experiments may not be able to accurately reflect the conditions in vivo. In addition, in the 2D culture state the cells will be affected by fluid shear stress that also will have a certain impact of the cell's physiological activity and function [28, 29]. In fact, tissue-specific architecture, mechanical and biochemical cues and cell-cell communication are lost under such simplified and highly biased conditions. To overcome the above shortcomings of 2D cell cultures, 3D cell culture in microdevices attracted widespread attention.

Moving from cell monolayers to 3D cultures is motivated by the need to work with cellular models that mimic the functions of living tissues [30]. Biomaterial scaffolds can provide a 3D microenvironment for cell growth. 3D cell cultures are typically performed by incubating cells in a biocompatible gel material to simulate the extracellular matrix (ECM) in vivo, providing 3D mechanical stress to support the cell growth and inducing cells to polarize and to interact with neighboring cells. Thus, cells cultured in 3D condition make the more effect response to the external environment, 3D cultures simulate the cells growth of the microenvironment closer to the real situation in vivo, and more realistic and reliable data can be obtained [31, 32]. Materials commonly used for 3D cell culture include various hydrogels such as agarose, collagen, matrigel, polyethylene glycol (PEG) b, sodium alginate, and so on [33–35]. They generally have better biocompatibility, and can provide better structural support and basic conditions for the biological behavior of 3D cell growth, proliferation, and movement. There are many methods for co-culture of cells on microfluidic chips, such as (a) the use of gel-wrapped cells for 3D cell co-culture to form cell spheres, (b) the formation of 3D structure by physical accumulation, and (c) induction multiple cells to form tissue to achieve the goal of 3D cell co-culture [36–38].

In addition to the use of gels, 3D cell co-culture can also be achieved by other methods. Choi et al. [39] described a circular microchannel with the orthogonal micro-grooves perpendicular to the axis of the microchannel by a strain responsive wrinkling method and PDMS replica molding process (Fig. 11.3). The microdevice holds promise as a utility platform for in-depth study of human aortic smooth muscle cells (HASMC) biology and vascular function. Yuan et al. [40] designed a 3D non-gel cells co-culture device (Fig. 11.4) by culturing different types of cells in different regions using a microchannel-defined method on a stretched PDMS membrane. The membrane is rolled up to form a 3D co-culture model of multi-layered cells that can be used to model many in vivo tissues composed of different cell types with tubular structures, such as blood vessels. Using a rolling membrane technology, tubes with controllable sizes were obtained and 2D membranes structure is transformed into 3D tubes structure. The results showed that 3D cells co-culture devices offer great possibilities to systematically study vascular cell biology and may be further used for tissue engineering.

Hsiao et al. [41] designed various high-throughput microcavities in a microfluidic chip to co-culture prostate cancer cells, osteoblasts, and endothelial cells to form 3D cell spheres by microfluidic spheroid formation technology (Fig. 11.5). The survival situation of the cells in the body was simulated successfully, and the metastasis mechanism of prostate cancer cells were studied. This device has great potential as an efficient platform for recapitulating the in vivo growth behavior of malignant cancer cells to understand human pathophysiology.

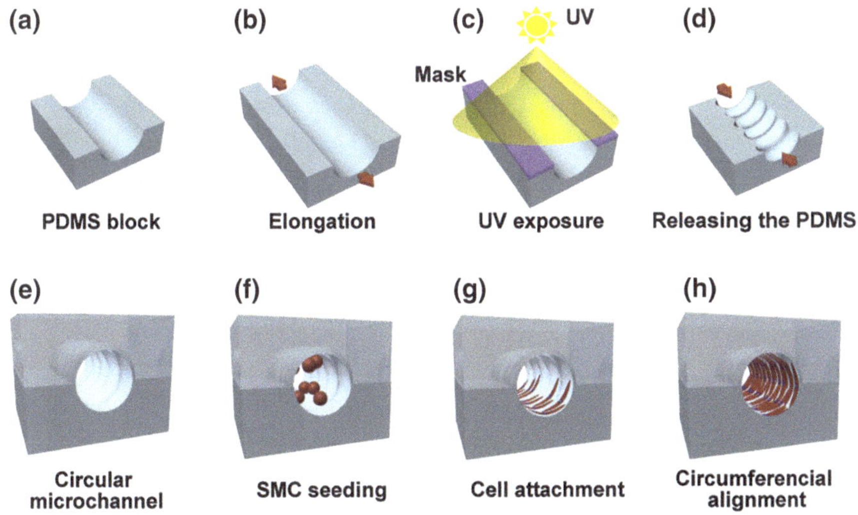

Fig. 11.3 Schematic illustrations to create an orthogonal microwrinkle structure inside the circular microchannel which guided the circumferential HASMC alignment. Reproduced with permission from Ref. [39]

11.2.3 Organs on a Chip

With the development of tissue engineering research, extracorporeal organs and even in vitro biological system has attracted more and more attention [42]. By simulating biological microenvironment in vitro and constructing biological system, it can provide a better platform for the research of life sciences, medicine and pharmacology, and advance the research progress in related fields [4]. Organ-on-a-chip systems as functionalization techniques for human physiology and pathology combine the advantages of microfluidics with cell culture [43, 44]. Microfluidic chip platform can precisely control cell culture conditions, spatial distribution, substance concentration gradient, biochemical stimulating conditions and other factors [45]. Owing to its excellent properties of miniaturization and high-throughput analysis, it can greatly reduce costs and consumption (cells, culture medium, and drugs), easily realize automation and commercialization [46]. Due to highly integrated structure and function, it is suitable for in vitro organ reconstruction and simulation, and ultimately establish physiology pathological model which close to the body situation [47]. The new in vitro model of organ-on-a-chip plays an important role in the clinical treatment and basic drug research [48]. It is particularly suitable for the research of biological phenomena that depend on tissue microarchitecture and perfusion and precisely monitoring of short-term pathophysiological processes [49].

Cell-based microfluidic model system, a simplest system, contains one type of cell cultured in a perfused microfluidic channel for one type of tissue function

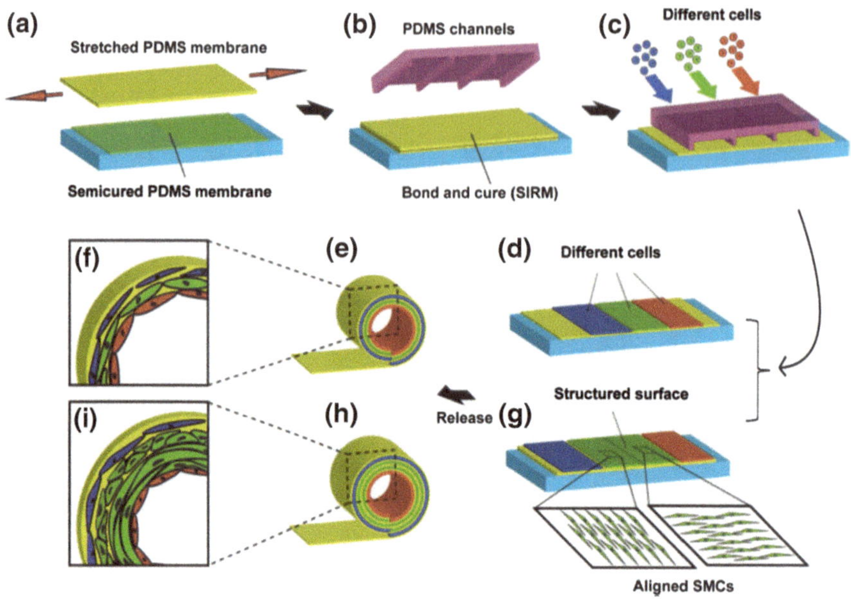

Fig. 11.4 Schematic illustration of a stress-induced rolling membrane (SIRM) and tubes with multiple types of cells as the walls. **a** A thin PDMS membrane is stretched as the top layer of the SIRM and covers a semicured PDMS membrane. After curing the two layers to cause adhesion (**b**), a SIRM is obtained that rolled up when the ends were released. Microfluidic channels cover the surface of the SIRM, and the direction of the channels is vertical to the stretching direction (**c**). Different cells are delivered via microfluidic channels to the surface of the SIRM. Once the cells attach to the SIRM surface, the microchannels are removed (**d**). One end of the SIRM is released by cutting its edge. The SIRM rolls up into a tube and each type of cell is delivered to a designated position as the tubular wall (**e**). The structure of the tube is similar to that of tubular tissues, such as small-diameter vessels (**f**). Cell alignment could be controlled by contact guidance on structured surfaces (microgrooves, **g**). After release, cell orientation inside the tubes is either circumferential or longitudinal (**h**), which is similar to the medium-sized vein (**i**). Reproduced with permission from Ref. [40]

simulation [50]. It can give large amount of inexpensive data, however, this model didn't represent complex pathophysiology in patients [51]. 3D cell co-culture in microfluidic chip applies two or more types of cells and usually relies on hydrogels to recreate interfaces between different tissues [52]. While as to organ-on-a-chip, the aim is not to build a whole organism on a chip but rather to synthesize integrated functional units that recapitulate tissue- and organ-level functions [53, 54]. The complex interactions of between functional units are mediated by various soluble and insoluble factors such as cytokines, nutrients, growth factors, hormones, the extracellular matrix and intercellular junctions. Additionally, fluid motions could reproduce various types of mechanical cues induced by physiological flow (such as, blood flow and interstitial flow) and tissue deformation (such as, breathing, peristalsis and heartbeat), including fluid shear stress, tension, compression and torque [30, 55]. Furthermore, the advantage of optical transparency

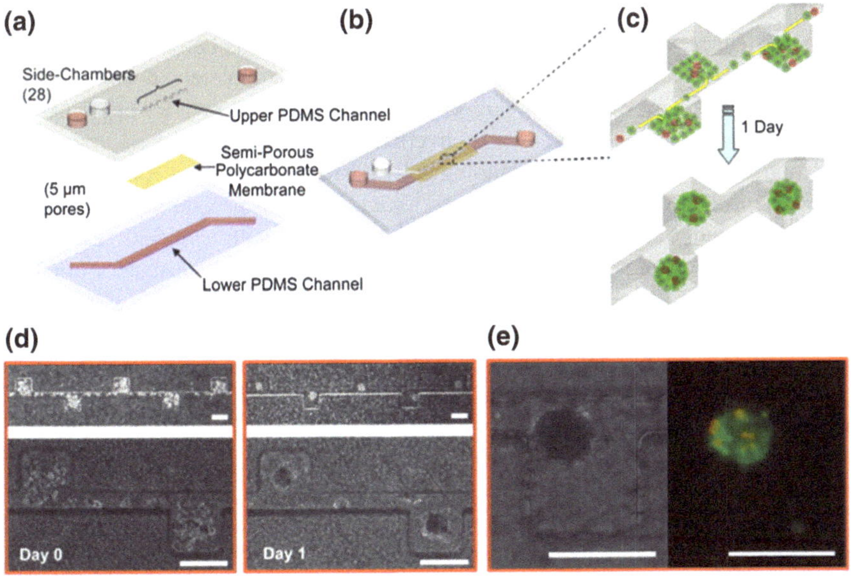

Fig. 11.5 3D cell co-culture on microfluidic chip. Schematic illustrations of the microfluidic spheroid formation device design (**a–b**) and PC-3^{DsRed} co-culture spheroid formation process (**c**). **d** Actual time-lapse images of PC-3^{DsRed} co-culture spheroid formation within microchannel (side-chambers: 200 × 200 × 200 μm, central microchannel: 50 μm width, 200 μm height). Optical images were taken immediately after seeding and 1 day after introducing the cells. **e** Optical and fluorescent images of a PC-3^{DsRed} co-culture spheroid after 1 day of cultlure in the microfluidic device. *Red* PC-3^{DsRed} cells, *Green* MC3T3 and HUVEC (support cells). *Scale bar* is 200 μm. Reproduced with permission from Ref. [41]

allows direct real-time visualization and quantitative high-resolution analysis in the chip, that is not possible in animals [11, 56, 57].

Cell sources of microfluidic organ simulation can be divided into cell lines, primary cells, tissue sections and pluripotent stem cells. The tissue sections are close to the body state, but it is difficult to achieve long-term culture and maintenance in vitro. Pluripotent stem cell technology is not yet mature, and the use of primary cells not only can achieve the aim of in vitro culture for a long time, but also can better simulate the state of the body, however, it is difficult to assemble different cells according to the situation of the human body. Recent development of organ-on-a-chip modeling includes vasculature, heart, brain, lung, liver, kidney, intestine and other organs, and a brief overview of some work is listed below [58–60].

11.2.3.1 Liver

The mainly function of liver is metabolism, and in the body liver plays a role in oxidation, storage of glycogen, synthesis of secretory proteins. The liver has biotransformation effects on many non-nutritive substances, such as various drugs, toxic substances and certain metabolites in the body, from the body and in the body. By the means of metabolism, these substances are completely decomposed or excreted in the prototype, and this effect is also known as the "detoxification function". Some toxicants can be converted into non-toxic or less toxic, easy excretion of substances through biological transformation; but some of the substances happens to the contrary, increased toxicity (such as pseudo-neurotransmitter formation), and decreased solubility (such as certain sulfonamides). Liver biotransformation contains many ways: the general water-soluble substances often are excreted in the prototype from the urine and bile; fat-soluble substances are easy to accumulate in the body and affect cell metabolism, thus, they must be inactivated by a series of enzyme system of the liver, or transformed to water-soluble substances, and then discharged. Due to its significance in drug metabolism, the liver is a major focus of drug research.

Most of in vitro metabolic systems were established on the basis of the liver, including microsomes, S9 (supernatant of 9000 × g centrifugated liver homogenates), liver slices, primary cultured hepatocytes, hepatocyte lines, mitochondria and cytoplasmic fluids. Kang et al. [61] described a microfluidic chip in which the primary liver cells and endothelial cells were co-cultured to in vitro simulate the physiological function of hepatic sinusoids based on polyethyleneterephthalate membrane and PDMS materials. As shown in Fig. 11.6, under the condition of co-culture, cell viability and maintenance of hepatocyte functions were monitored and observed by morphological observation. The result of viral replication analysis for the hepatotropic hepatitis B virus showed that the morphology and function of the cells were close to those in vivo.

11.2.3.2 Intestine

Oral administration is the most common and most convenient route of administration, and oral drugs are selectively absorbed by the small intestine into the liver and are metabolized by more than 90 metabolic enzymes. They are then transported to the target organ or distributed via the kidneys and bile [62]. The study and simulation of intestinal absorption and transport properties play an important role in drug development. Culturing Caco-2 cell monolayer in a plug-in filter is a relatively simple and most versatile standard method for predicting the intestinal epithelial absorption ratio of drugs [63, 64], which is the most common application in drug discovery. Caco-2 cells could differentiate into dense cell monolayers with microvilli structure, which is similar to the structure of small intestinal epithelial cells expressing a variety of protein carriers and enzymes, and also provide information on drug molecules at the cellular level absorbed, metabolized and transported through intestinal mucosa in vitro. Caco-2 cells contain enzymes associated with brush border cells such as glutathione transferase, alkaline phosphatase, glutathione S-transferase, sulfotransferase and P450 cytochrome isoenzymes [65]. 12 kinds of transport systems associated with intestinal active transport,

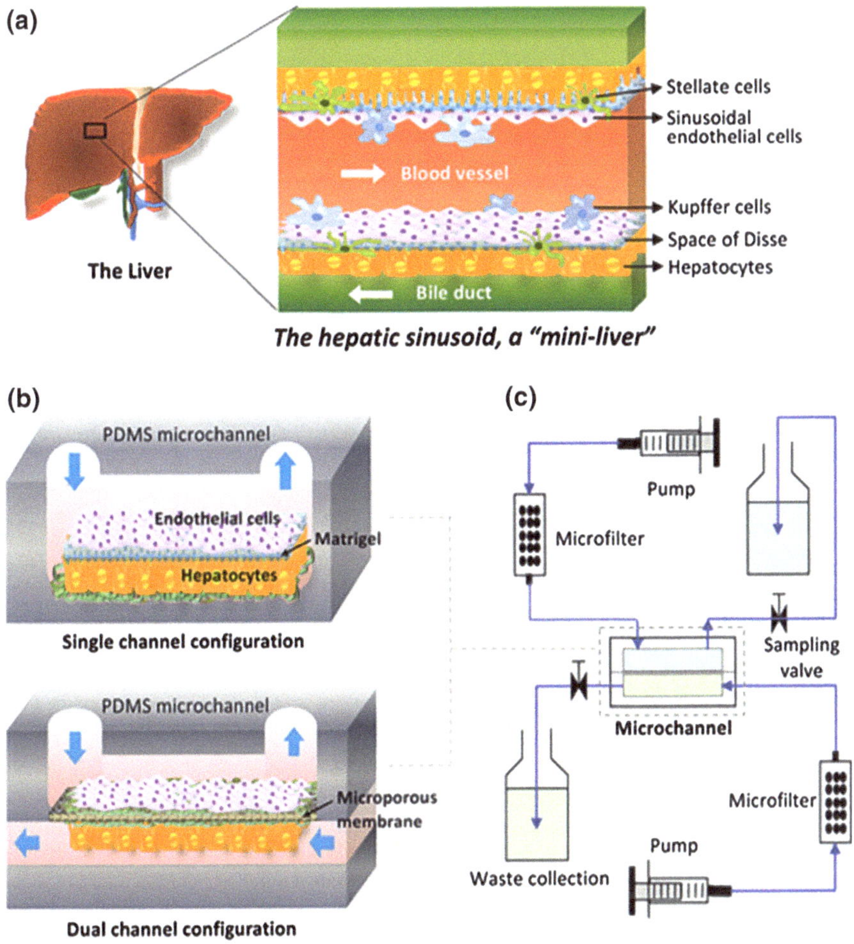

Fig. 11.6 The chip of hepatic sinusoids. **a** The liver sinusoid functional unit. **b** The single and dual channel configuration. **c** The cell culture platforms to mimic the liver sinusoid. Reproduced with permission from Ref. [61]

such as sugars, amino acids, dipeptides, cholic acid and vitamin B12, are also expressed in Caco-2 cells. Since Caco-2 cells have a tendency to differentiate into intestinal epithelial cells, the cellular model has a good correlation with integral absorption, and is a good tool to study drug absorption mechanisms, and to predict in vivo absorption and drug interactions, to design new drug and evaluate drugs security [66–68].

Kim et al. [69] developed a biomimetic 'human gut-on-a-chip' microdevice to mimic the mechanical, structural, absorptive, transport and pathophysiological properties of the intestine along with its crucial microbial symbionts. The device contains two PDMS channels sandwiched by a flexible PDMS membrane coated

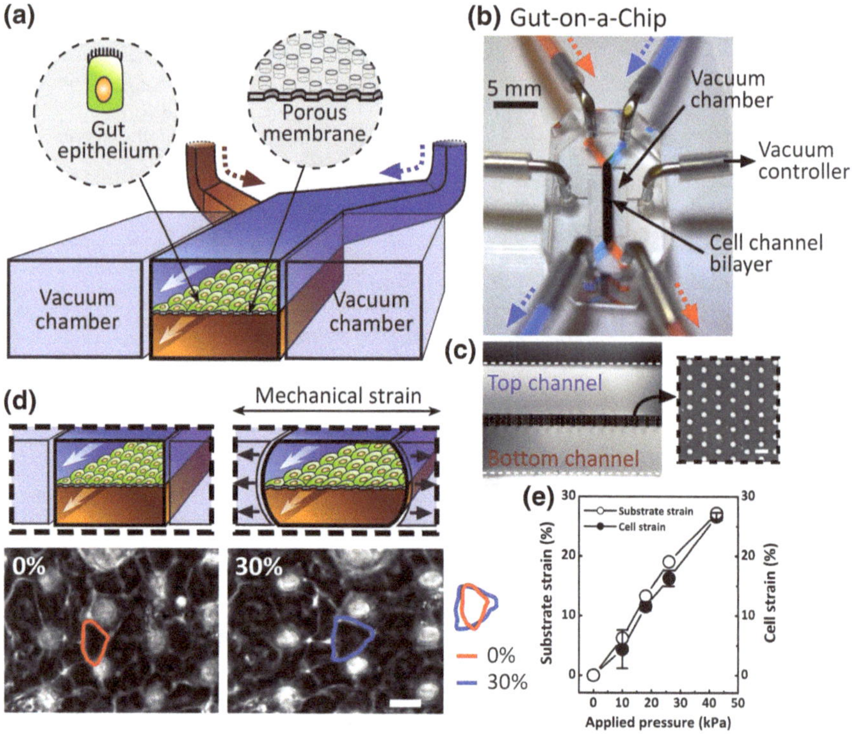

Fig. 11.7 The human gut-on-a-chip. **a** A schematic of the gut-on-a-chip device showing the flexible porous ECM-coated membrane lined by gut epithelial cells cross horizontally through the middle of the central microchannel, and full height vacuum chambers on both sides. **b** A photographic image of the gut-on-a-chip device composed of clear PDMS elastomer. A syringe pump was used to perfuse (direction indicated by *arrows*) *blue* and *red dyes* through tubing to the upper and lower microchannels, respectively, to visualize these channels. **c** A cross-sectional view of the top and bottom channels (both 150 mm high) of the gut-on-a-chip; square inset shows a top view of a portion of the porous membrane (10 mm pores; bar, 20 mm). **d** Schematics (*top*) and phase contrast images (*bottom*) of intestinal monolayers cultured within the gut-on-a-chip in the absence (*left*) or presence (*right*) of mechanical strain (30%; *arrow* indicated direction) exerted by applying suction to the vacuum chambers. *Red* and *blue* outlines indicate the shape of a single Caco-2 cell before (*red*) and after (*blue*) mechanical strain application (bar, 20 mm). Note that the cell distorts in the direction of the applied tension. The regular array of small *white circles* are pores visible beneath the epithelial monolayer. **e** Quantitation of the mechanical strain produced in the ECM-coated, flexible, porous PDMS membrane (*open circles*) and in the adherent gut epithelial cells (*closed circles*) as a function of pressure applied by the vacuum controller. Reproduced with permission from Ref. [69]

with ECM, and Caco-2 cells co-culturing a normal intestinal microbe (Fig. 11.7). A low shear stress and exerting cyclic strain for mimicking physiological peristaltic motions were successful applied to simulate whole intestine. Additionally, the normal intestinal microbe (*Lactobacillus rhamnosus GG*) co-culturing Caco-2 cells model improved barrier function as previously observed in humans.

Gao et al. [70] introduced a microfluidic platform for drug permeability assays of intestine coupled mass spectrometer to online detect curcumin directly. Cell culture and sample desalting and extraction were integrated in a chip by operating two independent channels separated by a semipermeable polycarbonate (PC) membrane, and micro solid-phase extraction columns. Permeation of curcumin on microfluidic platform was characterized by transport study and ESI-Q-TOF mass spectrum (Fig. 11.8).

11.2.3.3 Lung

Huh et al. [71] developed a microfluidic chip that simulates the motor function at the alveolar-tracheal interface of the lungs, integrating gas-liquid two-phase cells on the upper and lower sides of the membrane and cultured alveolar epithelial cells and vascular endothelial cells, respectively. The movement process of the alveolar expansion and contraction was simulated through the regulation of both sides of the gas chamber. The simulated alveolar organ on this chip can make responses to imported bacteria or immune factors. It was observed that the alveolar structure has more absorption to the nanoparticles in the state of motion in the device. Simultaneously, the inflammatory factors have a stronger response that consistent with the body. This study increased the awareness of alveolar cell movement and function. They adapted this approach to micro-engineer a human lung small airway-on-a-chip that supports full differentiation of a columnar, pseudostratified, mucociliary bronchiolar epithelium composed of cells isolated from healthy individuals or people with chronic obstructive pulmonary disease (COPD) and underlined by a functional microvascular endothelium [72].

11.2.3.4 Vasculature

Zheng et al. [73] developed a microfluidic vascular network using collagen in cultured endothelial cells in a microchannel. By this long-term on-chip vascular culture the morphology and function similar to in vivo endothelial cells were gradually produced. They used this model to study angiogenesis, the interaction of endothelial cells with pericytes, and the interaction of endothelial cells with blood components and flow, and validate that the model has an important feature that is consistent with the blood vessels in the body. The related tumor formation, inflammation, thrombosis and other pathological phenomena were studied.

Chung et al. [74] described a cell co-culture chip to mimic and evaluate the regeneration of capillaries and the migration of endothelial cells under different conditions, as shown in Fig. 11.9, in which the blue and green represent the flow channels, the brown represent the 3D collagen scaffold part, the endothelial cells are cultured in the middle channel. The stimuli are added or other cells are cultured in the lateral channel to study the proliferation of the endothelial cells.

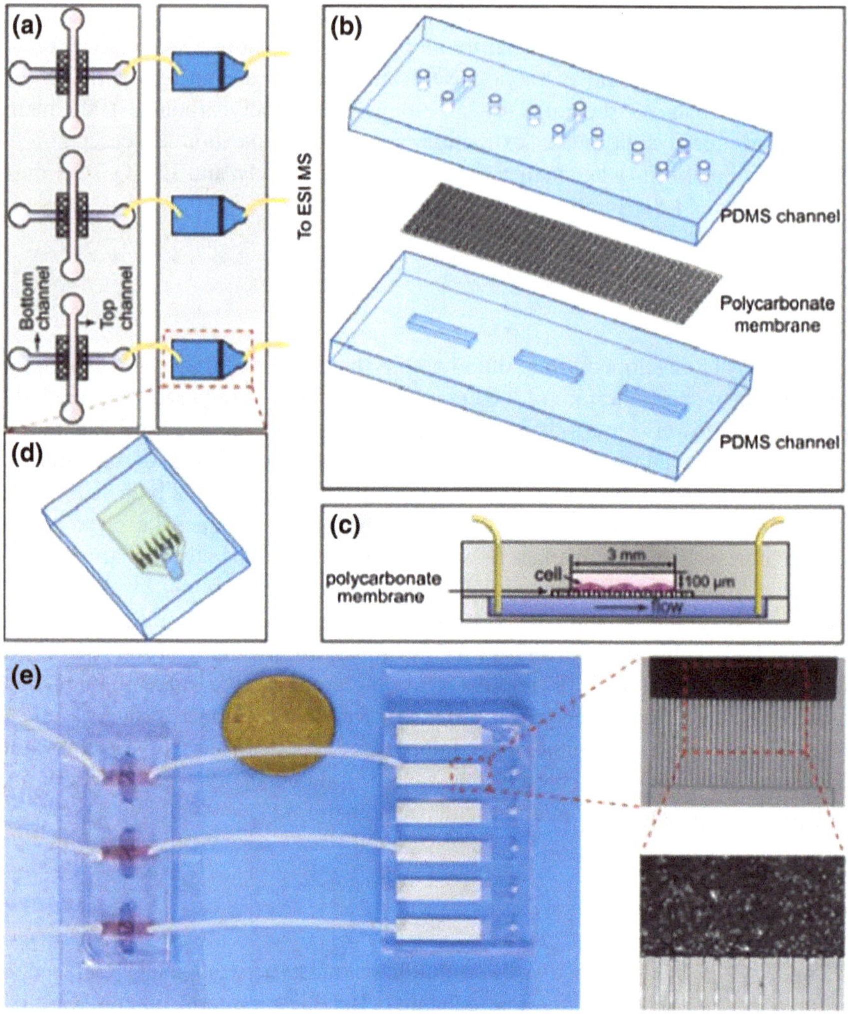

Fig. 11.8 Schematic illustration of the microfluidic device. **a** The overall schematic representation of the microfluidic chip with membrane-based cell culture and pretreatment zone. **b** A porous PC membrane sandwiched between two PDMS substrates. The *inlet* and *outlet holes* were punched on the upper PDMS layer. **c** Microfluidic membrane-based device design viewed from the side, not to scale. The top layer contained the culture well and the bottom layer involved the microfluidic flow line. **d** Scheme of the pretreatment zone. **e** An image of the microfluidic device filled with *blue* and *red dyes* separately in the upper and bottom channels of the membrane-based cell culture section. Reproduced with permission from Ref. [70]

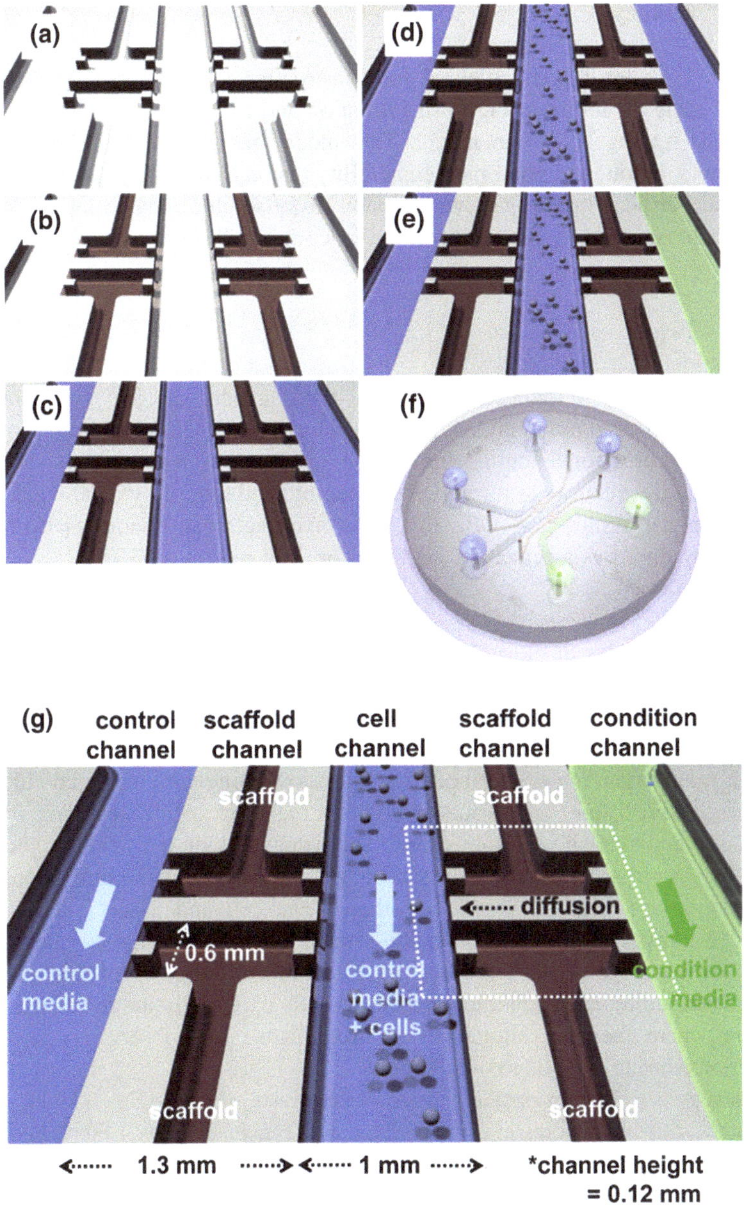

Fig. 11.9 Experimental protocol for the microfluidic assay development. **a** Prepared PDMS device made by soft lithography and surface treatment, **b** filled gel scaffold (indicated in *brown*) in the scaffold channel between the channels, **c** media (*blue*) filling both channels, **d** cell seeding (*spheres*) in the central cell channel, **e** chemical factors (*green*) applied in the condition channel. **f** Microfluidic device after filling of medium and chemical factors. Droplets are placed on all inlet ports to avoid evaporation of medium from the channels. Medium can be replaced with capillary forces generated by simply aspirating the existing droplets and adding new ones. **g** Schematic for microfluidic cell migration assay enabling direct comparison of cell migration behavior between the condition and control sides. Reproduced with permission from Ref. [74]

11.2.3.5 Kidney

Mu et al. [75] developed a method for constructing a microfluidic channel in a hydrogel material and by using fibril formation and fluid model of the collagen to form a 3D hydrogel nephron model. This model has good mechanical stability, permeable solution, and good biocompatibility that can make cells attached. In the functional verification, the passive diffusion of fat-soluble organic molecules was successfully achieved, it indicated that the function of renal cell was simulated well.

11.2.3.6 Brain

Fan et al. [76] introduced a novel brain cancer device composed of photo-polymerizable poly(ethylene) glycol diacrylate (PEGDA) hydrogel for drug screening. Glioblastoma multiforme (GBM) cells were cultured in the PEGDA hydrogel microwell array to form 3D cell spheroid and treated by combine drugs of pitavastatin and irinotecan (CPT-11). This hydrogel microfluidic platform was successfully tested as a screening platform for drug testing of GBM.

11.2.3.7 Multi-Organ-on-a-Chip

Recently, it is the focus of biomedical research to seek new ways of cell culture and multicellular synergistic drug metabolism. Wagner et al. introduced a microfluidic chip for multi-organ integrated modeling that allows long-term stable co-culture of primary cells, cell lines and tissue sections. In the experiment, they used artificial liver tissue and skin slices for co-culture and achieved long-term cell maintenance [77]. The system supports two different culture modes, one is the direct contact of cells and tissues with the flow, the other is the cell and tissue culture in the integrated transwell wells, and the culture medium and other solutions flow below it, as shown in Fig. 11.10. The usage of this platform not only enables long-term stable tissue culture, but also allows the growth of skin tissue at the gas-liquid junction where they are stimulated by troglitazone and observed tissue-tissue interactions during a 14-day co-culture.

Imura et al. [78] proposed a micro-bioanalysis system (Fig. 11.11), which integrated with micro-intestine, micro-liver and cancer cells was fabricated. Two channels were formed by the separation of the collagen-coated membrane to simulate the digestive tract and blood vessels, respectively. Dense colon cancer cells were cultured on the collagen-coated membrane, and the material exchange through absorption and diffusion of the intestinal cell membrane. This system is used to simulate the oral drug's absorption, metabolism, the role of a series of biological behavior, and forms a more complete drug analysis and detection system.

Jie et al. [79] developed a multi-type cell microfluidic integrated device for simulation oral drug absorption, metabolism, and anticancer effect. Caco-2, HepG2, and U251 cells were co-cultured to simulate the partly functions of the intestine,

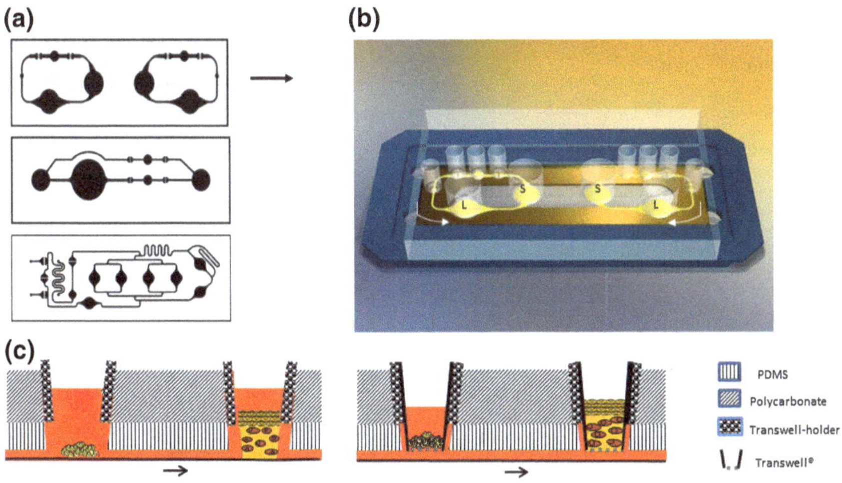

Fig. 11.10 The microfluidic MOC device at a glance. **a** Bird's-eye view of three different MOC designs. **b** 3D drawing of an assembled MOC with two microfluidic circuits in a PDMS-glass chip attached to a cover-plate and placed into a MOC holder (*blue*). *Arrows* indicate fluid flow direction of each circuit. S-skin culture compartments, L-liver culture compartments. **c** Schematic sections through the tissue culture compartments supporting both submersed tissue cultures in the fluid flow (*left image*) or cultures in Transwell1 (*right image*). Reproduced with permission from Ref. [77]

liver, and glioblastoma, respectively (Fig. 11.12). The microfluidic device was designed including three compartments: channels on the top and bottom layers of PDMS were vertically aligned and separated by a PC membrane and Caco-2 cells were cultured on PC membrane coated with fibronectin. The narrow channel array interconnects two separate channels on the bottom layer of chip which are cultured for HepG2 cells and U251 cells, respectively. Prodrug CPT-11 cytotoxicity was monitored by fluorescence visualization of drug cytotoxicity. CPT-11 and its active metabolite of 7-ethyl-10-hydroxycamptothecin were qualitatively and quantitatively characterized by liquid chromatography-tandem mass spectrometry.

11.2.4 Whole Organisms on a Chip

The ultimate goal of these works of organotypic simulation, tissue engineering and biological microenvironment simulation on a microfluidic chip is to build a more complete human body model, that is, to form whole-organisms-on-chip [14], as shown in Fig. 11.13. The platform has this ability to integrate multi-organ functional simulation, composed of gut, liver, lung, kidney, bone, and heart etc. It could carry out different functions, for example, absorption, delivery, transport,

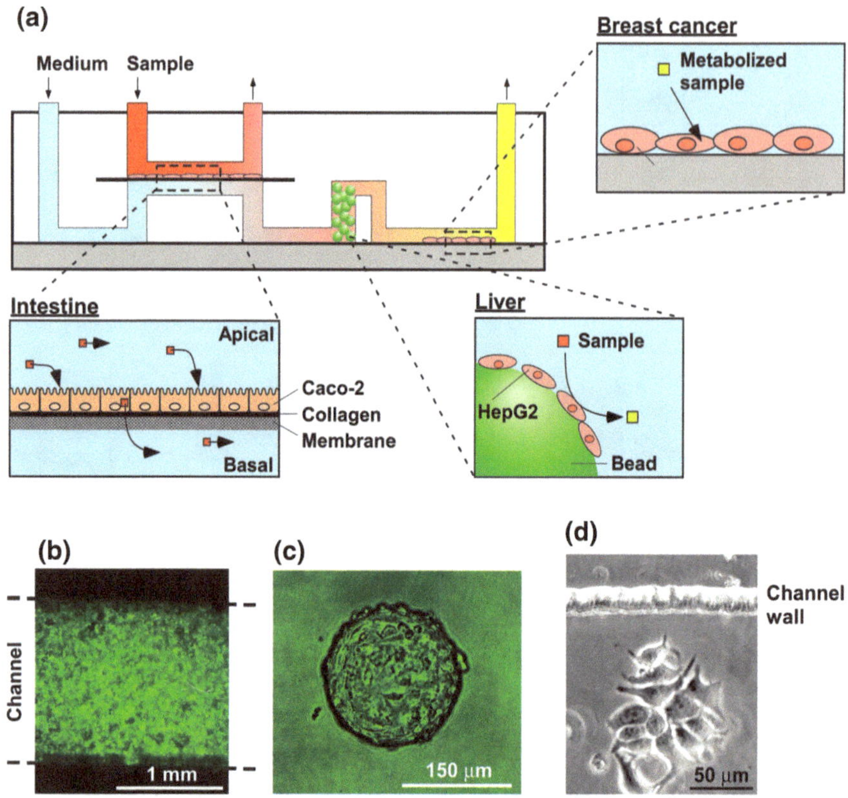

Fig. 11.11 A micro-bioanalysis system integrating micro-intestine, micro-liver and cancer cells. **a** Cross-sectional illustration of the micro total bioassay chip. Samples exhibit their bioactivity in the target cell component (breast cancer cells, in this case) after intestinal absorption and hepatic metabolism. Each solution was introduced with a microsyringe pump at a flow rate of 0.4 μL/min. No figure is to scale. **b** Fluorescent micrograph of Caco-2 cells cultured on a membrane support in a microchamber. Living cells were fluorescently stained with calcein-AM. **c** Micrograph of a Cytodex bead with HepG2 cells on its surface. **d** Micrograph of MCF-7 cells cultured on the fibronectin-modified bottom surface of the microchamber. Reproduced with permission from Ref. [78]

metabolism, toxicity and clearance of oral drugs, lung absorption and potential cardiotoxicity of aerosol drugs, contractility and conduction of heart, and immune response of bone. This platform will be an effective and reliable model for simulation of the real situation of physiology, pathology, and application of clinical diagnosis, disease mechanism, disease treatment, drug screening, metabolomics, and life exploration.

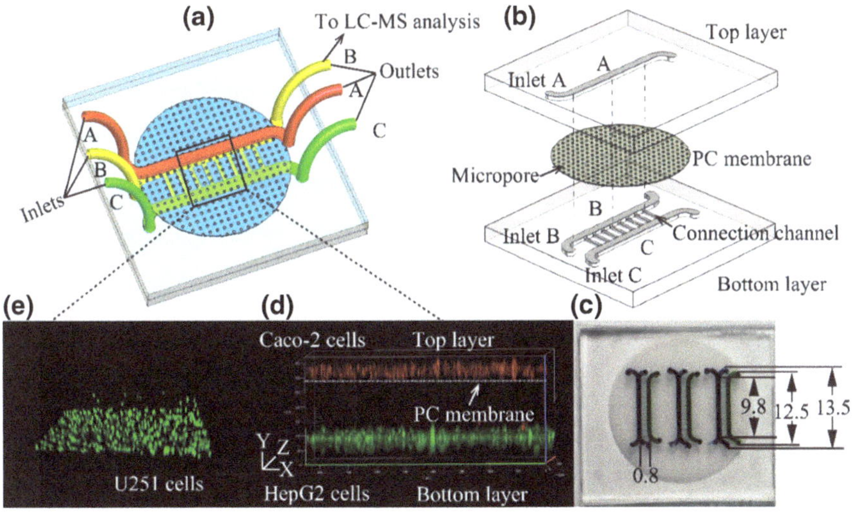

Fig. 11.12 Schematic diagram of the 3D microfluidic integrator. **a** Overall schematic illustration of the microfluidic chip. **b** Layer-sectional assembly illustration. **c** Photograph of the fabricated microfluidic chip (*scale bar* 5 mm). **d** Confocal morphology showing the co-culture of Caco-2 cells, HepG2 cells and U251 cells: Caco-2 cells labeled by red fluorescent carbocyanine DiI probes, HepG2 cells labeled by green fluorescent carbocyanine Dio probes, and **e** U251 cells stained with calcein AM fluorescent probes. Reproduced with permission from Ref. [79]

11.3 Chip-MS Platform

Mass spectrometry has the advantages of high sensitivity, good selectivity and powerful ability to analyze and treat complex samples. It is especially suitable for the detection and identification of molecular structures such as organic molecules, nucleic acids and proteins. The combination of mass spectrometry and microfluidic chips is of great interest to drug research. After several years of continuous efforts, the author's research group has made great progress in the application of microfluidic chip mass spectrometry, and put forward Chip-MS concept and development idea in the world [80, 81]. There are four main combinations of microfluidic chip and mass spectrometry:

1. Ionization on a microfluidic chip. The microfluidic chip is ionized by the design of the architecture [82, 83]. Ramsey's group pioneered the microfluidic chip-mass spectrometric analysis to generate a liquid spray the terminus of an etched channel and detect the ionized substances with a mass spectrometer, that is an off-line non-contact chip mass spectrometry techniques [84]. The advantage of this method is high integration, but there are many shortcomings, such as low efficiency of ionization, small scope of application, and so on. This method can realize the ionization detection of certain substances, but because of its own limitations, applications became less.

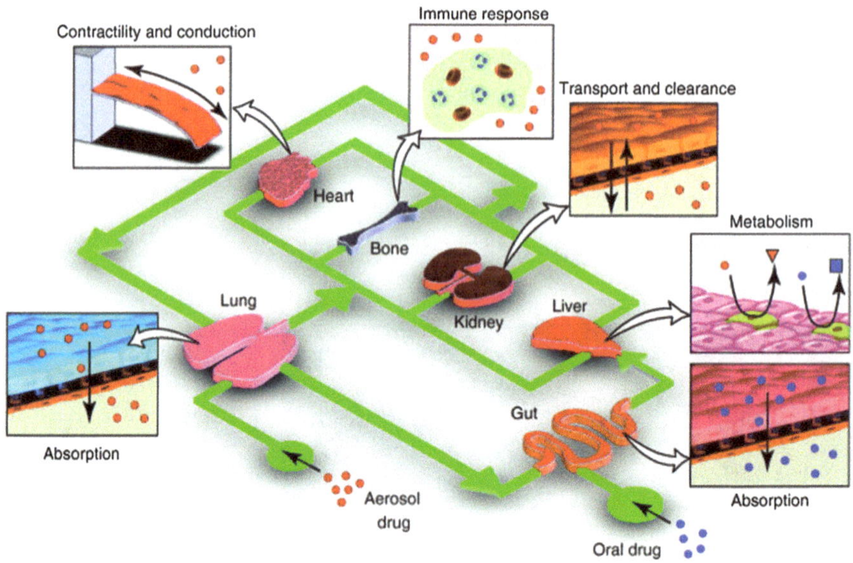

Fig. 11.13 The human-on-a-chip concept. Biomimetic microsystems representing different organs can be integrated into a single microdevice and linked by a microfluidic circulatory system in a physiologically relevant manner to model a complex, dynamic process of drug absorption, distribution, metabolism and excretion, and to more reliably evaluate drug efficacy and toxicity. As shown in this example, an integrated system of microengineered organ mimics (lung, heart, gut, liver, kidney and bone) can be used to study the absorption of inhaled aerosol drugs (*red*) from the lung to microcirculation, as well as to measure their cardiotoxicity (e.g. changes in heart contractility or conduction), transport and clearance in the kidney, metabolism in the liver, and immune-cell contributions to these responses. Drug substances (*blue*) also can be introduced into the gut compartment to investigate interplay between orally administered drugs and molecular transporters and metabolizing enzymes expressed in the various organs. Reproduced with permission from Ref. [14]

2. Integrating the ionization nozzle on the microfluidic chip. Harrison et al integrated the nozzle into the chip, and this method provides higher ionization efficiency and wider applicability, compared to the direct ionization on the microfluidic chip [85].
3. Connection to mass spectrometry by capillary. This method is simple and easy, and mainly operating steps are as follows: the outlet of microfluidic channel is connected to the inlet of the mass spectrometer through the capillary, and the sample containing the analyte on the microfluidic chip is pushed into the mass spectrometer by a syringe pump. The integration of this method is not high, but sample loss of that is less. It is now widely used in drug screening and cell metabolite detection [86, 87]. In 1997, Harrison et al. combined capillary and mass spectrometry for direct mass spectrometric detection [85]. This device can be further connected to a wide variety of external detectors with minimal dead

volume. Liu and Lin integrated microfluidic chip-mass spectrometry to monitor lactate metabolism and efflux of normal cells and tumor cells for rapid drug screening [88]. In this work, they introduced a cell culture unit in a microfluidic chip as a platform for cell metabolism and used cell lactate efflux as a metabolic model to prove the validity of the established method. Our group directly connected the channels of microfluidic chip to the MS ion source interface using a capillary, and integrated a solid-phase microextraction column for sample desalting and purification in the chip channel, eventuallly achieved the goal of continuous detection of analytes [89].

4. Paper-based ionization mass spectrometry. Cooks et al. proposed the concept of paper-based ionization in 2010 [90]. Analyte of dried blood spots or whole blood spiked with imatinib was transported and deposited in a porous triangular chromatographic paper with a macroscopically sharp point, and a high electric field is used to perform ionization. The sample mainly was deposited on a small triangular chromatographic paper, by adding voltage to achieve ionization on paper and alignment MS detection interface position to achieve the information of substances. Since its inception, many research groups have contributed to the introduction and further development of the combined techniques of microfluidic chip and mass spectrometry [91, 92]. An integrated platform for microdialysis-paper spray ionization-mass spectrometry was constructed and applied to online chemical monitoring of cell culture by Lin et al. [93]. Monitoring glucose concentration in hepatocytes demonstrates the potential for broader applications in drug screening, cell signaling transduction, and metabolites monitoring.

This MS detection method requires less sample matrix, provides good ability of soft ionization, low detection limit, and easy to couple microfluidic chip. In the cellular metabolites detection, the composition of metabolite solution is complex, while this method with its low requirements on the substrate has a great advantage compared to the electrospray ionization mass spectrometry. This platform is expected to play a major role in the detection of cells metabolites. Microfluidic chip and mass spectrometry have been developed extensively since then, and have been further developed in interface technology, multifunctionalization, automation and extension applications [94], and are used to implement microfluidics continuous monitoring of droplets on a flow control chip [95].

11.4 Application in Drug Research

Approximately 10 million people suffer from different kinds of cancer per year and many of them unfortunately die due to lack of better treatment strategies. Apparently, microfluidics devices have the potential to serve different scientific needs of healthcare and biomedical sectors and as we discussed earlier, their several successful applications have already been reported.

The major advantages associated with miniaturized systems are faster/more accurate diagnoses, better epidemiological data for disease modeling, vaccine introduction, and utilization of minimally trained healthcare workers and better use of existing therapeutics but still many hurdles are there in broader applications of microfluidics systems. However, there is always a silver lining and due to vastly increased interest in global health issues, the current funding climate for the development of diagnostics kits is significantly good.

11.4.1 Drug Delivery

Drug delivery is a field of vital importance to medicine and healthcare. Controlled drug delivery improves bioavailability by preventing premature degradation and enhancing uptake, maintains drug concentration within the therapeutic window by controlling the drug release rate, and reduces side effects by targeting to disease site and target cells [96]. Bottaro et al. proposed an "off-the-shelf" microfluidic chip approach, which was used to produce the liposomes and verified the effect of ivermectin delivery of liposomes. The result demonstrated that this microdevice allowed the rapid production of narrowly dispersed and very reproducible unilamellar liposomes for drug delivery [97].

11.4.2 Drug Metabolism and Toxicity

Wu et al. [98] introduced a novel multi-layer microfluidic device for characterization of drug metabolism in human liver microsomes (HLMs) and their cytotoxicity on tumor cells composed of PDMS sheet, PC membrane, and solid phase extraction (SPE) micro-column. The metabolites could be directly detected online with an ESI-Q-TOF MS after micro-SPE pretreatment. Wei et al. established an integrated microfluidic system in a controlled coculture device, in which the secreted proteins were qualitatively and semiquantitatively determined by a directly coupled mass spectrometer (shown in Fig. 11.14). PC12 cells and GH3 cells were cocultured under various conditions as a model of the regulation of the organism by the nervous system. A SPE column was integrated in order to remove salts from the cells secretion prior to mass spectrometry detection. A three layer PDMS microfluidic device was fabricated to integrate valves for avoiding contamination between the cells coculture zone and the pretreatment zone. ESI-Q-TOF-mass spectrometry was employed to realize highly sensitive qualitative analysis and to implement semiquantitative analysis [21].

Chen et al. [81] reported a microfluidic chip integrated ESI-MS platform based on a stable isotope labeling technology for qualitative and quantitative analysis of cell metabolism treated by antitumor drug genistein. The device contained three components: one part composed of drug and culture medium injection, the second

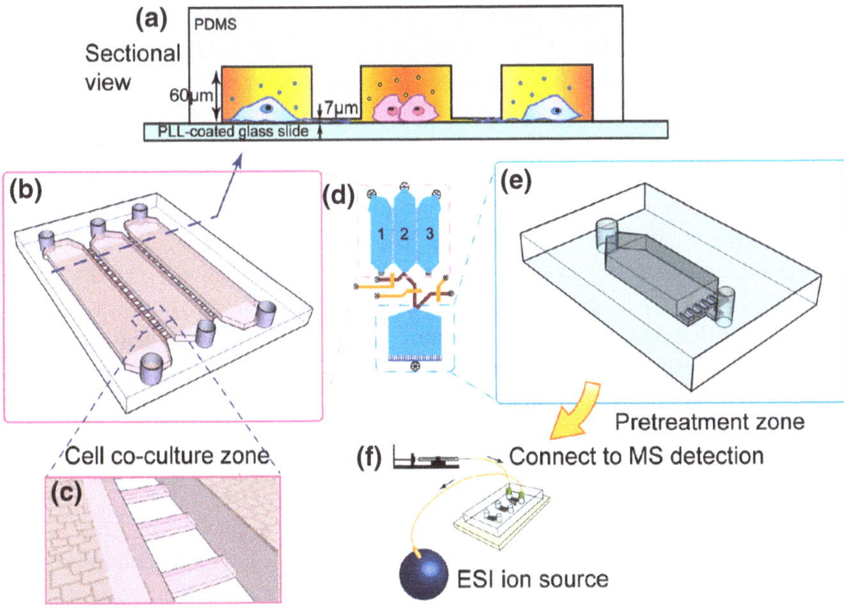

Fig. 11.14 Schematic illustration of the microfluidic device. **a** The sectional view of the cell coculture zone cut from the *dash line* in (**b**). Different types of cells could be cocultured, with the signal factor communicating by diffusion. **b** Scheme of the cell coculture zone. **c** A magnified illustration of the connected small channels between the neighboring cell culture channels. **d** The overall schematic representation of the microfluidic chip with cell coculture zone, control valves, and pretreatment zone. **e** Scheme of the pretreatment zone. **f** Scheme for the coupling of the microfluidic device to the mass spectrometry for detection. Reproduced with permission from Ref. [21]

part was used as cell culture and drug-induced cell apoptosis analysis, the third part was composed of array of SPE microcolumns for sample desalination.

11.4.3 Oral Drug Pharmacokinetics System

Drug research and development often exists this problem that in vitro assays is poor in prediction drug toxicity. One important reason for this phenomenon is that it is difficult to reproduce the pharmacokinetics (PK) of drugs in vitro. Many researchers have focused on in vitro modeling of pharmacokinetic processes using microfluidic chips. Gut-on-a-chip models integrated with physiologically based pharmacokinetic computational models have developed to predict the process of absorption and excretion, which determine the bioavailability of drug compounds as in vitro models. Kim et al. introduced a microfluidic platform composed of two overlapping cell culture channels separated by an ECM-coated elastomeric membrane lined with

Caco-2 cells to mimic the physiology of intestine. This device enabled prolonged culture of normal intestinal microbe on the epithelial surface using fluid at a low rate and cyclic strain to mimic peristaltic motions to recapitulate the physiological peristaltic motions of the intestine. Importantly, this model recapitulated absorptive features and the barrier function of intestine, which have great value for development of novel intestinal disease models to be applied to drug absorption studies [69, 99].

Advances in microfluidics technology have also made it possible to develop liver-on-a-chip models that govern drug bioavailability and model critical pharmacokinetic processes, for example, drug metabolism. Chao et al. [100] developed a microfluidic chip with primary human hepatocytes for the prediction of hepatic clearance of six marketed drugs (carbamazepine, caffeine, timolol, sildenafil, imipramine, and buspirone). Baudoin et al. [101] established a novel microfluidic liver-on-a-chip platform to evaluate metabolisms and clearances of 7 kinds of drugs by cultivating cryopreserved human primary hepatocytes in 12 parallel microfluidic chambers. This research illustrated the parallelized microfluidic device enable to reproduce hepatic tissue functions.

Though these individual organ model studies have a conspicuous advantage, more and more people have realized that multi-organ-on-a-chip technology has great potential for pre-clinical testing, especially for imitating the pharmacokinetic, and it can realize the human body physiology reconstruction by integrating these models. Based on this consideration, researchers have begun to focus on the research of multiple organ model which enable to predict the ADME process of drugs and replicate the function and interaction of multiple organ. Recently, there are a lot of biomedical research to seek new ways of cell culture and multicellular synergistic effect.

Imura and colleagues [78] established a network of interconnected cell culture chambers and introduced a combined model integrated with micro-intestine, micro-liver and breast cancer cells to demonstrate the sequential absorption, metabolism, and efficacy of anticancer drugs. A new feature of gastrointestinal digestion was added to the previous system by incorporating a gastric acid stomach chamber before intestinal absorption, and it demonstrated the integrated platform can form a more complete drug analysis system and determine bioavailability and efficacy of orally administered drugs [102]. An extension of this system consisting of a microengineered peristaltic pump, a dialysis part and a chamber containing breast cancer cells to replicate the process of the circulation and excretion [103].

The fabrication and operation of these multi-organ models are often challenging and require careful consideration of allometric scaling principles [104]. To address these scaling considerations, Maschmeyer et al. [105] established a microphysiological system integrating the functionality of four organs into standard cell culture inserts at a size 100,000-fold smaller than their human counterpart organs over 28 days in co-culture. In this model, the residence times of medium in each organ compartment were near to physiological fluid-to-tissue ratios and a second circuit assured drainage of the fluid excreted through the kidney epithelial cell layer. Further, this system enabled to repeated dose systemic toxicity testing of drug candidates over 28 days that is reported for the first time. These studies illustrate that microfluidics technology enable to be used to functional tools for mimicking and predicting key physiological responses for drug screening.

11.4.4 Drug-Droplet Technology

The extensive understanding of the drug-protein interactions is essential for the absorption, distribution, metabolism, and excretion (ADME) properties of the drug. The conventional investigation of the interactions typically immobilizes a drug or protein in the static well of a microplate. Recently, increasing attention has been paid to use droplet microfluidics to study the drug-protein interaction. Lombardi et al. [106] implemented a technique of integrated droplet, magnetic beads and microfluidic chip to study between human serum albumin (HSA) and the anticoagulant drug warfarin. First, magnetic beads were coated with HSA and warfarin were injected into a stream of oil to form microdroplets in a microfluidic device. Daughter droplets are formed at the T-junction exposed asymmetrically to a magnetic field and the droplets containing magnetic beads (including HAS and warfarin complexes) were separated from the solution containing the free. The free and total (bound + free) warfarin concentrations in the outlets of microfluidic chip were measured and finally determined the association constant (Fig. 11.15). The platform provided a new tool to investigate the interactions of drugs with specific targets including enzymes and cells. Furthermore, the continuous method could be employed for pharmacokinetic (ADME) studies.

One of the most important application of droplet technology in drug analysis may be single-cell encapsulation. Single-cell droplets are compatible with high-throughput screening and sort-ing. Brouzes et al. developed a fully integrated droplet-based microfluidic chip for conducting a mammalian cell cytotoxicity analysis with high-throughput screening and sort-ing [107]. This device used intact

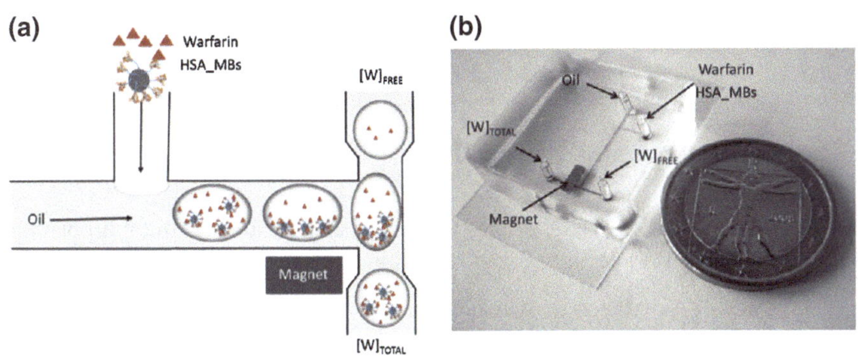

Fig. 11.15 Scheme of droplet microfluidics with magnetic beads. **a** Scheme of the method for measuring the affinity constant of warfarin and human serum albumin (HSA). **b** Photograph of the fabricated microfluidic chip (not to scale). Reproduced with permission from Ref. [106]

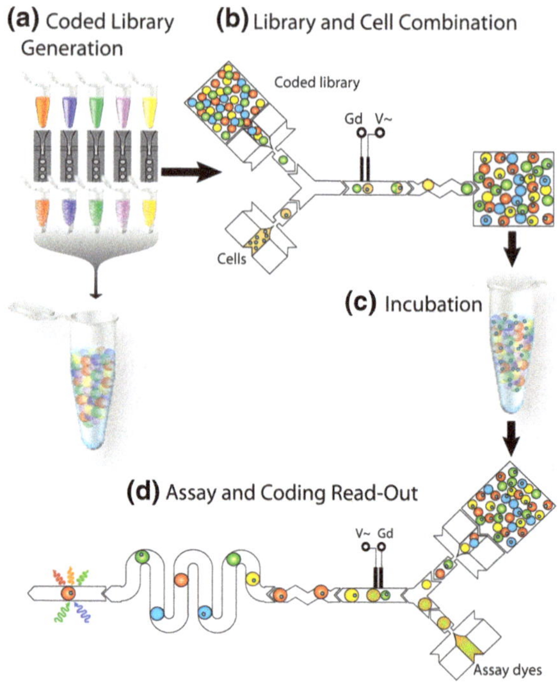

Fig. 11.16 Droplet screening workflow. The droplet screen has 4 steps. **a** A reformatting step to emulsify compound-code pairs and pool them into a droplet library. **b** Merging each library member with one of the cell containing droplets that are continuously generated. Hence, each cell droplet has a specific composition defined by the compound droplet it has merged with. **c** Incubation. **d** Merged droplets are reinjected into an assay chip to identify each compound via their code and assess their specific effect on cells. Reproduced with permission from Ref. [107]

droplets to carry out the quantitative scoring of cell viability, and enable to screen entire drug library for cytotoxic effects against U937 cells. The device combined a cell viability assay and the automated generation of coded droplet libraries containing vari-ous concentrations of the drug mitomycin C in a simple microfluidic chip (Fig. 11.16). By incubating them for approximately 15 min, the droplet containing cells enabled to characterized cell viability by fluorescence imaging. Further, this droplet-microfluidc platform enable to be used with different types of libraries (e.g., DNA, siRNA, drug), fluorescent assay read-outs, or additional droplet manipulation modules.

11.5 Challenges and Prospects

Despite its demonstrated value in biomedical research, achieving the full potential of microfluidic chip, especially organ-on-a-chip, still faces many challenges:

1. The physical properties of these materials such as PDMS, PC and polyester, are not suitable for mimicking the extracellular matrix, and new materials need to be developed [108]. Due to its absorption small hydrophobic molecules, 20,88 PDMS will cause the effective concentration of drugs and efficacy decline, further,

drug quantitative analysis is not accurate. This can be improved by surface modification. In addition, new alternative materials have emerged in succession.

2. Current cell sources on microfluidic chips are mostly differentiated cells and more reliable and sustainable sources human cells are needed, which can represent the corresponding organs in vivo. Human embryonic stem cells and induced pluripotent cells can serve as cellular sources and can be engineered to meet the specific needs of organ-on-a-chip development [109].
3. Miniaturization of the chips reduces the sensitivity of the detection, and high-throughput analysis can make up for this deficiency.
4. The balance between complexity and practicality is needed to consider carefully. This challenge can be solved by precisely controlling cell culture parameters, making use of simple models, developing automated instruments, and collecting and analyzing samples in real time.
5. Although organ chips provide a meaningful biological model, alternative tools are still needed. Organ chip can be used as an important auxiliary tools of the basic drug development stage, and it is still a long way to go that it applied to clinical practice as a standardized method.

Despite their challenges, microfluidic chips have the potential to play a transformative role in streamlining the clinical trial process and they could present exciting new avenues for drug discovery and development, that determines they will remain in the drug industry for a long time, and continue to grow mature and grow.

References

1. Esch EW, Bahinski A, Huh D (2015) Organs-on-chips at the frontiers of drug discovery. Nat Rev Drug Discov 14(4):248–260. doi:10.1038/nrd4539
2. Paul SM, Mytelka DS, Dunwiddie CT, Persinger CC, Munos BH, Lindborg SR, Schacht AL (2010) How to improve R&D productivity: the pharmaceutical industry's grand challenge. Nat Rev Drug Discov 9(3):203–214. doi:10.1038/nrd3078
3. Scannell JW, Blanckley A, Boldon H, Warrington B (2012) Diagnosing the decline in pharmaceutical R&D efficiency. Nat Rev Drug Discov 11(3):191–200. doi:10.1038/nrd3681
4. Polini A, Prodanov L, Bhise NS, Manoharan V, Dokmeci MR, Khademhosseini A (2014) Organs-on-a-chip: a new tool for drug discovery. Expert Opin Drug Discov 9(4):335–352
5. Editorial (2007) Five years on…and four challenges for the pharmaceutical industry. Nat Rev Drug Discov 6(1):3–3
6. Kang L, Chung BG, Langer R, Khademhosseini A (2008) Microfluidics for drug discovery and development: from target selection to product lifecycle management. Drug Discov Today 13(1–2):1–13
7. Pihl J, Karlsson M, Chiu DT (2005) Microfluidic technologies in drug discovery. Drug Discov Today 10(20):1377–1383
8. Tsui JH, Lee W, Pun SH, Kim J, Kim DH (2013) Microfluidics-assisted in vitro drug screening and carrier production. Adv Drug Deliv Rev 65(11):1575–1588
9. Duncombe TA, Tentori AM, Herr AE (2015) Microfluidics: reframing biological enquiry. Nat Rev Mol Cell Biol 16(9):554–567
10. Manz A, Graber N, Widmer HM (1990) Miniaturized total chemical analysis systems: a novel concept for chemical sensing. Sens Actuators B Chem 1(1–6):244–248

11. Whitesides GM (2006) The origins and the future of microfluidics. Nature 442(7101):368–373
12. Squires TM, Quake SR (2004) Microfluidics: fluid physics at the nanoliter scale. Rev Mod Phys 77(3):977–1026
13. Mehling M, Tay S (2014) Microfluidic cell culture. Curr Opin Biotechnol 25(2):95–102
14. Huh D, Hamilton GA, Ingber DE (2011) From 3D cell culture to organs-on-chips. Trends Cell Biol 21(12):745–754
15. Ning R, Wang F, Lin L (2015) Biomaterial-based microfluidics for cell culture and analysis. TrAC Trends Anal Chem 80:255–265
16. Dittrich PS, Manz A (2006) Lab-on-a-chip: microfluidics in drug discovery. Nat Rev Drug Discov 5(3):210–218
17. Sackmann EK, Fulton AL, Beebe DJ (2014) The present and future role of microfluidics in biomedical research. Nature 507(7491):181–189
18. Kaji H, Yokoi T, Kawashima T, Nishizawa M (2010) Directing the flow of medium in controlled cocultures of HeLa cells and human umbilical vein endothelial cells with a microfluidic device. Lab Chip 10(10):2374–2379
19. Lamb BM, Barrett DG, Westcott NP, Yousaf MN (2008) Microfluidic lithography of SAMs on gold to create dynamic surfaces for directed cell migration and contiguous cell cocultures. Langmuir 24(16):8885–8889
20. Chen Z, Chen W, Yuan B, Xiao L, Liu D, Jin Y, Quan B, Wang JO, Ibrahim K, Wang Z (2010) In vitro model on glass surfaces for complex interactions between different types of cells. Langmuir 26(23):17790–17794
21. Wei H, Li H, Mao S, Lin JM (2011) Cell signaling analysis by mass spectrometry under coculture conditions on an integrated microfluidic device. Anal Chem 83(24):9306–9313
22. Xu T, Yue W, Li CW, Yao X, Cai G, Yang M (2010) Real-time monitoring of suspension cell-cell communication using an integrated microfluidics. Lab Chip 10(17):2271–2278
23. Mazutis L, Gilbert J, Ung WL, Weitz DA, Griffiths AD, Heyman JA (2013) Single-cell analysis and sorting using droplet-based microfluidics. Nat Protoc 8(5):870–891. doi:10.1038/nprot.2013.046
24. Yu M, Bardia A, Wittner BS, Stott SL, Smas ME, Ting DT, Isakoff SJ, Ciciliano JC, Wells MN, Shah AM (2013) Circulating breast tumor cells exhibit dynamic changes in epithelial and mesenchymal composition. Science 24(3):225–226
25. Stott SL, Hsu CH, Tsukrov DI, Yu M, Miyamoto DT, Waltman BA, Rothenberg SM, Shah AM, Smas ME, Korir GK, Floyd FP Jr, Gilman AJ, Lord JB, Winokur D, Springer S, Irimia D, Nagrath S, Sequist LV, Lee RJ, Isselbacher KJ, Maheswaran S, Haber DA, Toner M (2010) Isolation of circulating tumor cells using a microvortex-generating herringbone-chip. Proc Natl Acad Sci U S A 107(43):18392–18397. doi:10.1073/pnas.1012539107
26. Wu J, Chen Q, Liu W, He Z, Lin JM (2016) Recent Advances in microfluidic 3D cellular scaffolds for drug assays. TrAC Trends Anal Chem 87(10):19–31
27. Pampaloni F, Reynaud EG, Stelzer EH (2007) The third dimension bridges the gap between cell culture and live tissue. Nat Rev Mol Cell Biol 8(10):839–845
28. Kim D, Lin YS, Haynes CL (2011) On-chip evaluation of shear stress effect on cytotoxicity of mesoporous silica nanoparticles. Anal Chem 83(22):8377–8382
29. Park ES, Brown AC, Difeo MA, Barker TH, Lu H (2010) Continuously perfused, non-cross-contaminating microfluidic chamber array for studying cellular responses to orthogonal combinations of matrix and soluble signals. Lab Chip 10(5):571–580
30. Shamir ER, Ewald AJ (2014) Three-dimensional organotypic culture: experimental models of mammalian biology and disease. Nat Rev Mol Cell Biol 15(10):647–664
31. Tan JL, Tien J, Pirone DM, Gray DS, Bhadriraju K, Chen CS (2003) From the Cover: Cells lying on a bed of microneedles: an approach to isolate mechanical force. Proc Natl Acad Sci U S A 100(4):1484–1489
32. Kunze A, Giugliano M, Valero A, Renaud P (2011) Micropatterning neural cell cultures in 3D with a multi-layered scaffold. Biomaterials 32(8):2088–2098

33. Lii J, Hsu WJ, Parsa H, Das A, Rouse R, Sia SK (2008) Real-time microfluidic system for studying mammalian cells in 3D microenvironments. Anal Chem 80(10):3640–3647
34. Tumarkin E, Tzadu L, Csaszar E, Seo M, Zhang H, Lee A, Peerani R, Purpura K, Zandstra PW, Kumacheva E (2011) High-throughput combinatorial cell co-culture using microfluidics. Integr Biol (Camb) 3(6):653–662
35. Liu J, Gao D, Li HF, Lin JM (2009) Controlled photopolymerization of hydrogel microstructures inside microchannels for bioassays. Lab Chip 9(9):1301–1305
36. Wu J, Chen Q, Liu W, Zhang Y, Lin JM (2012) Cytotoxicity of quantum dots assay on a microfluidic 3D-culture device based on modeling diffusion process between blood vessels and tissues. Lab Chip 12(18):3474–3480
37. Ong S, Zhang C, Toh Y, Sh Foo H, Tan C, Van-Noort D, Park S, Yu H (2008) A gel-free 3D microfluidic cell culture system. Biomaterials 29(22):3237–3244
38. Toh YC, Zhang C, Zhang J, Khong YM, Chang S, Samper VD, Van ND, Hutmacher DW, Yu H (2007) A novel 3D mammalian cell perfusion-culture system in microfluidic channels. Lab Chip 7(3):302–309
39. Choi JS, Piao Y, Seo TS (2014) Circumferential alignment of vascular smooth muscle cells in a circular microfluidic channel. Biomaterials 35(1):63–70
40. Yuan B, Jin Y, Sun Y, Wang D, Sun J, Wang Z, Zhang W, Jiang X (2012) Artificial vessels: a strategy for depositing different types of cells in three dimensions to mimic tubular structures in tissues (Adv. Mater. 7/2012). Adv Mater 24(7):890–896
41. Hsiao AY, Torisawa YS, Tung YC, Sud S, Taichman RS, Pienta KJ, Takayama S (2009) Microfluidic system for formation of PC-3 prostate cancer co-culture spheroids. Biomaterials 30(16):3020–3027
42. Selimović Š, Dokmeci MR, Khademhosseini A (2013) Organs-on-a-chip for drug discovery. Curr Opin Pharmacol 13(5):829–833
43. Lee JB, Sung JH (2013) Organ-on-a-chip technology and microfluidic whole-body models for pharmacokinetic drug toxicity screening. Biotechnol J 8(11):1258–1266
44. Caplin JD, Granados NG, James MR, Montazami R, Hashemi N (2015) Microfluidic organ-on-a-chip technology for advancement of drug development and toxicology. Adv Healthc Mater 4(10):1426–1450
45. Zhang YS, Khademhosseini A (2015) Seeking the right context for evaluating nanomedicine: from tissue models in petri dishes to microfluidic organs-on-a-chip. Nanomedicine 10(5):685–688
46. Zhang B, Montgomery M, Chamberlain MD, Ogawa S, Korolj A, Pahnke A, Wells L, Massé S, Kim J, Reis L (2016) Biodegradable scaffold with built-in vasculature for organ-on-a-chip engineering and direct surgical anastomosis. Nat Mater 15(6):669–678
47. Kim S, Takayama S (2015) Organ-on-a-chip and the kidney. Kidney Res Clin Pract 34 (3):165–169
48. Luni C, Serena E, Elvassore N (2014) Human-on-chip for therapy development and fundamental science. Curr Opin Biotechnol 25(1):45–50
49. Duinen VV, Trietsch SJ, Joore J, Vulto P, Hankemeier T (2015) Microfluidic 3D cell culture: from tools to tissue models. Curr Opin Biotechnol 35:118–126
50. Wu MH, Huang SB, Lee GB (2010) Microfluidic cell culture systems for drug research. Lab Chip 10(10):939–956
51. Elali J, Sorger PK, Jensen KF (2006) Cells on chips. Nature 442(7101):403–411
52. Xu Z, Gao Y, Hao Y, Li E, Wang Y, Zhang J, Wang W, Gao Z, Wang Q (2013) Application of a microfluidic chip-based 3D co-culture to test drug sensitivity for individualized treatment of lung cancer. Biomaterials 34(16):4109–4117
53. Bhatia SN, Ingber DE (2014) Microfluidic organs-on-chips. Nat Biotechnol 32(8):760–772
54. Bowes J, Brown AJ, Hamon J, Jarolimek W, Sridhar A, Waldron G, Whitebread S (2012) Reducing safety-related drug attrition: the use of in vitro pharmacological profiling. Nat Rev Drug Discov 11(12):909–922
55. Ghaemmaghami AM, Hancock MJ, Harrington H, Kaji H, Khademhosseini A (2012) Biomimetic tissues on a chip for drug discovery. Drug Discov Today 17(3–4):173–181

56. Huh D, Torisawa YS, Hamilton GA, Kim HJ, Ingber DE (2012) Microengineered physiological biomimicry: organs-on-chips. Lab Chip 12(12):2156–2164
57. Inamdar NK, Borenstein JT (2011) Microfluidic cell culture models for tissue engineering. Curr Opin Biotechnol 22(5):681–689
58. Yeon JH, Ryu HR, Chung M, Hu QP, Jeon NL (2012) In vitro formation and characterization of a perfusable three-dimensional tubular capillary network in microfluidic devices. Lab Chip 12(16):2815–2822
59. Forouzan O, Burns JM, Robichaux JL, Murfee WL, Shevkoplyas SS (2011) Passive recruitment of circulating leukocytes into capillary sprouts from existing capillaries in a microfluidic system. Lab Chip 11(11):1924–1932
60. Huh D, Leslie DC, Matthews BD, Fraser JP, Jurek S, Hamilton GA, Thorneloe KS, Mcalexander MA, Ingber DE (2012) A human disease model of drug toxicity-induced pulmonary edema in a lung-on-a-chip microdevice. Sci Transl Med 4(159):3068–3072
61. Kang YB, Rawat S, Cirillo J, Bouchard M, Noh HM (2013) Layered long-term co-culture of hepatocytes and endothelial cells on a transwell membrane: toward engineering the liver sinusoid. Biofabrication 5(4):380–387
62. Artursson P, Palm K, Luthman K (1996) Caco-2 monolayers in experimental and theoretical predictions of drug transport. Adv Drug Deliv Rev 22(1–2):67–84
63. Artursson P (1990) Epithelial transport of drugs in cell culture. I: a model for studying the passive diffusion of drugs over intestinal absorbtive (Caco-2) cells. J Pharm Sci 79(6):476–482
64. Hilgers AR, Conradi RA, Burton PS (1990) Caco-2 cell monolayers as a model for drug transport across the intestinal mucosa. Pharm Res 7(9):902–910
65. Boulenc X, Bourrie M, Fabre I, Roque C, Joyeux H, Berger Y, Fabre G (1992) Regulation of cytochrome P450IA1 gene expression in a human intestinal cell line, Caco-2. J Pharmacol Exp Ther 263(3):1471–1478
66. Oulianova N, Cheng D, Huebert N, Chen Y (2007) Human oral drugs absorption is correlated to their in vitro uptake by brush border membrane vesicles. Int J Pharm 336 (1):115–121
67. Volpe DA (2008) Variability in Caco-2 and MDCK cell-based intestinal permeability assays. J Pharm Sci 97(2):712–725
68. Shah P, Jogani V, Bagchi T, Misra A (2006) Role of Caco-2 cell monolayers in prediction of intestinal drug absorption. Biotechnol Prog 22(1):186–198
69. Kim HJ, Huh D, Hamilton G, Ingber DE (2012) Human gut-on-a-chip inhabited by microbial flora that experiences intestinal peristalsis-like motions and flow. Lab Chip 12 (12):2165–2174. doi:10.1039/c2lc40074j
70. Gao D, Liu H, Lin JM, Wang Y, Jiang Y (2013) Characterization of drug permeability in Caco-2 monolayers by mass spectrometry on a membrane-based microfluidic device. Lab Chip 13(5):978–985
71. Huh D, Matthews BD, Mammoto A, Montoyazavala M, Hong YH, Ingber DE (2010) Reconstituting organ-level lung functions on a chip. Science 328(5986):1662–1668
72. Benam KH, Villenave R, Lucchesi C, Varone A, Hubeau C, Lee HH, Alves SE, Salmon M, Ferrante TC, Weaver JC (2015) Small airway-on-a-chip enables analysis of human lung inflammation and drug responses in vitro. Nat Methods 13(2):151–157
73. Zheng Y, Chen J, Craven M, Choi NW, Totorica S, Diaz-Santana A, Kermani P, Hempstead B, Fischbach-Teschl C, López JA (2012) In vitro microvessels for the study of angiogenesis and thrombosis. Proc Natl Acad Sci U S A 109(109):9342–9347
74. Chung Sudo S, Mack R, Peter J (2009) Cell migration into scaffolds under co-culture conditions in a microfluidic platform. Lab Chip 9(2):269–275
75. Mu X, Zheng W, Xiao L, Zhang W, Jiang X (2013) Engineering a 3D vascular network in hydrogel for mimicking a nephron. Lab Chip 13(8):1612–1618
76. Fan Y, Nguyen DT, Akay Y, Xu F, Akay M (2016) Engineering a brain cancer chip for high-throughput drug screening. Sci Rep 6:25062. doi:10.1038/srep25062

77. Wagner I, Materne EM, Brincker S, Sussbier U, Fradrich C, Busek M, Sonntag F, Sakharov DA, Trushkin EV, Tonevitsky AG, Lauster R, Marx U (2013) A dynamic multi-organ-chip for long-term cultivation and substance testing proven by 3D human liver and skin tissue co-culture. Lab Chip 13(18):3538–3547. doi:10.1039/c3lc50234a
78. Imura Y, Sato K, Yoshimura E (2010) Micro total bioassay system for ingested substances: assessment of intestinal absorption, hepatic metabolism, and bioactivity. Anal Chem 82 (24):9983–9988
79. Jie M, Li HF, Lin L, Zhang J, Lin JM (2016) Integrated microfluidic system for cell co-culture and simulation of drug metabolism. RSC Adv 6(59):54564–54572
80. Kim SH, Weitz DA (2011) One-Step emulsification of multiple concentric shells with capillary microfluidic devices. Angew Chem Int Ed Engl 50(37):8731–8734
81. Chen Q, Wu J, Zhang Y, Lin JM (2012) Qualitative and quantitative analysis of tumor cell metabolism via stable isotope labeling assisted microfluidic chip electrospray ionization mass spectrometry. Anal Chem 84(3):1695–1701
82. Kim JS, Knapp DR (2001) Miniaturized multichannel electrospray ionization emitters on poly(dimethylsiloxane) microfluidic devices. Electrophoresis 22(18):3993–3999
83. Xue Q, Foret F, Dunayevskiy YM, Zavracky PM, Mcgruer NE, Karger BL (1997) Multichannel microchip electrospray mass spectrometry. Anal Chem 69(69):426–430
84. And RSR, Ramsey JM (1997) Generating electrospray from microchip devices using electroosmotic pumping. Anal Chem 69(13):1174–1178
85. Bings NH, Wang C, Skinner CD, Colyer CL, Thibault P, Harrison DJ (1999) Microfluidic devices connected to fused-silica capillaries with minimal dead volume. Anal Chem 71 (15):3292–3296
86. Gao D, Wei H, Guo GS, Lin JM (2010) Microfluidic cell culture and metabolism detection with electrospray ionization quadrupole time-of-flight mass spectrometer. Anal Chem 82 (13):5679–5685
87. Benetton S, Kameoka J, Tan A, Wachs T, Craighead AH, † JDH (2003) Chip-based P450 drug metabolism coupled to electrospray ionization-mass spectrometry detection. Anal Chem 75(23):6430–6436
88. Liu W, Lin JM (2016) Online monitoring of lactate efflux by multi-channel microfluidic chip-mass spectrometry for rapid drug evaluation. ACS Sen 1(4):344–347
89. Wei H, Li H, Lin JM (2009) Analysis of herbicides on a single C(30) bead via a microfluidic device combined with electrospray ionization quadrupole time-of-flight mass spectrometer. J Chromatogr A 1216(52):9134–9142
90. Wang H, Liu J, Cooks RG, Ouyang Z (2010) Paper spray for direct analysis of complex mixtures using mass spectrometry. Angew Chem Int Ed Engl 49(5):877–880. doi:10.1002/anie.200906314
91. Li A, Wang H, Ouyang Z, Cooks RG (2011) Paper spray ionization of polar analytes using non-polar solvents. Chem Commun 47(10):2811–2813
92. Yin H, Killeen K, Brennen R, Sobek D, Werlich M, Van de GT (2005) Microfluidic Chip for peptide analysis with an integrated HPLC column, sample enrichment column, and nanoelectrospray tip. Anal Chem 77(2):527–533
93. Liu W, Wang N, Lin X, Ma Y, Lin JM (2014) Interfacing microsampling droplets and mass spectrometry by paper spray ionization for online chemical monitoring of cell culture. Anal Chem 86(14):7128–7134
94. Fortier M, Bonneil E, §, Goodley AP, Thibault P, ‡, § (2005) Integrated microfluidic device for mass spectrometry-based proteomics and its application to biomarker discovery programs. Anal Chem 77(6):1631–1640
95. Dr RTK, Dr JSP, Dr IM, Dr KT, Dr RDS (2009) Dilution-free analysis from picoliter droplets by nano-electrospray ionization mass spectrometry. Angew Chem Int Ed Engl 48 (48):6832–6835
96. Ying Z, Chan HF, Leong KW (2013) Advanced materials and processing for drug delivery: the past and the future. Adv Drug Deliv Rev 65(1):104–120

97. Bottaro E, Nastruzzi C (2016) "Off-the-shelf" microfluidic devices for the production of liposomes for drug delivery. Mater Sci Eng C Mater Biol Appl 64:29–33
98. Wu Q, Gao D, Wei J, Jin F, Xie W, Jiang Y, Liu H (2014) Development of a novel multi-layer microfluidic device towards characterization of drug metabolism and cytotoxicity for drug screening. Chem Commun 50(21):2762–2764
99. Kim HJ, Ingber DE (2013) Gut-on-a-Chip microenvironment induces human intestinal cells to undergo villus differentiation. Integr Biol (Camb) 5(9):1130–1140. doi:10.1039/c3ib40126j
100. Chao P, Maguire T, Novik E, Cheng KC, Yarmush ML (2009) Evaluation of a microfluidic based cell culture platform with primary human hepatocytes for the prediction of hepatic clearance in human. Biochem Pharmacol 78(6):625–632. doi:10.1016/j.bcp.2009.05.013
101. Baudoin R, Prot JM, Nicolas G, Brocheton J, Brochot C, Legallais C, Benech H, Leclerc E (2012) Evaluation of seven drug metabolisms and clearances by cryopreserved human primary hepatocytes cultivated in microfluidic biochips. Xenobiotica 43(2):140–152
102. Imura Y, Yoshimura E, Sato K (2012) Micro total bioassay system for oral drugs: evaluation of gastrointestinal degradation, intestinal absorption, hepatic metabolism, and bioactivity. Anal Sci 28(3):197–199
103. Imura Y, Yoshimura E, Sato K (2012) Microcirculation system with a dialysis part for bioassays evaluating anticancer activity and retention. Anal Chem 85(3):1683–1688
104. Wikswo JP, Curtis EL, Eagleton ZE, Evans BC, Kole A, Hofmeister LH, Matloff WJ (2013) Scaling and systems biology for integrating multiple organs-on-a-chip. Lab Chip 13 (18):3496–3511
105. Maschmeyer I, Lorenz AK, Schimek K, Hasenberg T, Ramme AP, Hubner J, Lindner M, Drewell C, Bauer S, Thomas A, Sambo NS, Sonntag F, Lauster R, Marx U (2015) A four-organ-chip for interconnected long-term co-culture of human intestine, liver, skin and kidney equivalents. Lab Chip 15(12):2688–2699. doi:10.1039/c5lc00392j
106. Lombardi D, Dittrich PS (2011) Droplet microfluidics with magnetic beads: a new tool to investigate drug–protein interactions. Anal Bioanal Chem 399(1):347–352
107. Brouzes E, Medkova M, Savenelli N, Marran D, Twardowski M, Hutchison JB, Rothberg JM, Link DR, Perrimon N, Samuels ML (2009) Droplet microfluidic technology for single-cell high-throughput screening. Proc Natl Acad Sci U S A 106(34):14195–14200. doi:10.1073/pnas.0903542106
108. Flaim CJ, Chien S, Bhatia SN (2005) An extracellular matrix microarray for probing cellular differentiation. Nat Methods 2(2):119–125
109. Mathur A, Loskill P, Hong SG, Lee JY, Marcus SG, Dumont L, Conklin BR, Willenbring H, Lee LP, Healy KE (2013) Human induced pluripotent stem cell-based microphysiological tissue models of myocardium and liver for drug development. Stem Cell Res Ther 4(Suppl 1):1–5

Chapter 12
Cell Metabolite Analysis on Microfluidic Platform

Xuexia Lin and Jin-Ming Lin

Abstract Cell metabolite is related to every aspect of biology and biochemistry. Cell metabolite analysis can help us to understand disease like tumor. It has been reported that abnormal metabolic states impact on cellular function and tissue dysfunction leading to many human diseases. Microfluidic platform is a powerful and promising analytical tool for cell metabolite analysis with the advantages of high-throughput, convenience, reduced reagent and time consumption, good separation ability and portability. This chapter summarizes recent progresses in microfluidic platform of cell metabolite analysis, the models of cell culture and how to control cell microenvironment, the strategy for cell metabolite analysis and the platform application. It considers the prospects of developing new metabolic approaches to disease treatment.

Keywords Microfluidic · Cell metabolite · Cell microenvironment
Methods · Application

Abbreviations

2D	Two-dimensional
3D	Three-dimensional
GC	Gas
HPLC	High-pressure
LC	Liquid
μTAS	Micro
ECM	Extracellular
PDMS	Polydimethylsiloxane

X. Lin
College of Chemical Engineering, Huaqiao University,
Xianmen 361021, People's Republic of China
e-mail: linxuexia@hqu.edu.cn

J.-M. Lin (✉)
Department of Chemistry, Tsinghua University,
Beijing 100084, People's Republic of China
e-mail: jmlin@mail.tsinghua.edu.cn

J.-M. Lin (ed.), *Cell Analysis on Microfluidics*, Integrated Analytical Systems,
https://doi.org/10.1007/978-981-10-5394-8_12

12.1 Introduction

Metabolism of living organisms is the sum of life-sustaining chemical transformations [1]. Core metabolism contains metabolic pathways involving carbohydrates, amino acids, and fatty acids, macromolecular synthesis and energy homeostasisin humans [2]. The scientific study of initial metabolism can be traced back to several centuries ago. Subsequently, animals testing are studied in early studies, followed by individual metabolic reactions has been developed. In 1614, Santorio Santorio is the first people who determined human metabolism by controlling experiments in his book. He found that most of food consumed through insensible perspiration. However, metabolism research development is accompanied with human diseases. About 2000 years ago, found Celsus knew found gout was greatly related to rich foods and drink precipitated [2]. The research has revealed that ants can be attracted by the urine of diabetic patients, but normal urine can not. The period of metabolic activities in biological research has blossomed from 1920s to 1960s, which was named "the golden age". But, newer areas of metabolic research are gradually displayed the momentum in biological research in the latter half of the 20th century. The search for the metabolic detection is benefit for clinic diagnoses and for study on the mechanisms of various diseases such as cancer, not only study on metabolic states in these diseases. More and more attention of researcher is focused on diseases in terms of inherited or somatic mutations such as signal transduction, cellular differentiation, not on bioenergetic or metabolic terms.

Before the year 2009, the definition of metabolism is not as clearly defined. In the year 2009, Kraly et al. [3] defined metabolomics related to metabolomics, metabonomics, metabolic fingerprinting, metabolic profiling, and targeted metabolic profiling. Metabolomics are similar to genomics and proteomics, which is quantitative analysis of all metabolites in an organism, study of relation between the metabolites and physiological and pathological changes. Metabonomics are aimed to monitor the time-depending concentration of metabolic markers. The goal of making up metabolomics and metabonomics together is to search biomarkers of disease. It also aim to illustrate pathways related with metabolite changes. Based on pattern recognition, the goal of metabolic fingerprinting is the determination of a whole profile of metabolites, which can be used to identify specific biomarker's profile. The field of metabolomics is targeted metabolic profiling. The history of this research is long and it often analyzes one or two targets tracked with time.

Metabolism analysis is vital consequences for disease detection and therapeutic strategy development which can provide the evidence for cancer pathophysiology. In order to predict the physiological response by applying new drug or new chemical substance in vivo, the metabolism analysis in vitro screening systems has been developed with long history [4]. Moreover, to reduce the cost and the risk of failure during clinical trials, in vivo animal studies are applied. Although the environment of study in animal testing is similar to that in human body, animal studies are expensive and labor-intensive. In vitro screening systems are widely used to investigate drug development, screening, and toxicology [5]. Conventional in vitro cell-based assays is limited by the lack of the complexity of the physiological environment and thus leads to deviation of results. In addition, because the functions of cells can be easily lost during long-time culturing, they are limited in analysis metabolites for periods of time. With micrometer chambers and channels, microfluidics has been widely used to analyze cell metabolite [6]. Comparison to conventional in vitro cell-based assays, microfluidics technology offers various advantages. It allows simulation in vivo micro-environment, integration of functional components, high-throughput, ability to accurately control fluid flow, and automatization [7]. Importantly, with low volume and high sensitivity, microfluidics is beneficial for analyzing cellar secretions as well as suiting for single-cell analysis [3].

Looking into the past, microfluidic-based cell metabolite analysis involved various techniques, such as hydrodynamics, cell culture, separation method, and analytical method. As development of fluid propulsion in the nanolitre (nL) and picolitre (pL) range, the first microfluidic technology in biological research was developed in the early 1950s [8]. By microfluidic technology, Terry et al. [9] has realized a miniaturized gas chromatograph (GC) in the year 1979. Based on Si-Pyrex technology, the first microfluidic device was fabricated including high-pressure liquid chromatography (HPLC) column microfluidic device in 1990 [10]. During the beginning of the 1990s Manz et al. first proposed "lab-on-a-chip" or "micro total analysis systems" (μTAS), which provided controllable automation of liquid handling by microfluidic integration. Microfluidic technology for biosensor has blossomed in the early 2000s. The microdevices coupled with fluorescent imaging are feasible and highly sensitive. Meanwhile, microfluidic platform combined with highly sensitive methods can be applied to dynamic study, which is benefical to investigate the procedure of cell metabolism. Therefore, recent work of cell metabolite analysis on microfluidic platform has been focused on the following aspects: (1) study numerous mechanisms of disease and identify the regulatory pathway. (2) Investigation of dynamic mechanisms. (3) Metabolism alters cell signaling such as enzyme activity.

The microfluidic-based cell metabolites analysis is major to mimic cells in vitro. Thus, the technologies of cell culture are of significance. The original two-dimensional (2D) cell culture on microfluidic platforms is like commercial culture plates. Although simple 2D culture has been well developed, some features of the cell are eventually lost. Consequently, the cell culture on-chip are developed three-dimensional (3D) culture that are used to mimic microenvironment in vivo, finally complex 3D cultures are developed to organ on chip.

This chapter reviews the microfluidic technology are employed for cell metabolism. The incorporation of cell culture models in microfluidic systems, different microenvironment on chip, and separation and detection system is reviewed. In addition, the applications of clinical diagnostics, drug research and development, and toxicology study are discussed.

12.2 Models of Cell Culture on Microfluidic Platform

12.2.1 Cell Culture Models in Microfluidic Systems

Cell culture has played a very important role in biochemical study including clinic diagnose, drug development and screening. Different cell culture method will affect cell metabolites, and metabolites can feed back cell culture methods [11]. People have to rack their brains to develop the cells culture method to simulate the native "in vivo" microenvironments [12]. Even though some kinds of cell culture techniques are reliability and facility in amount of research, they are limited for more advanced and complex research. The interactions of cell and cell, cell and the extracellular matrix (ECM), even systemic factors are determined the differentiation and phenotype of cells in vivo [13]. The most common technologies of cell culture are the usage of dishes and flasks. These strategies cannot control the spatial distribution of the cells and others in vivo. Microfluidic has received more and more attention because it can manipulate cells such as single cell analysis [14], enhance the spatial and temporal resolution, and improve the microenvironments of cultured cells [15, 16]. As a result, a major revolution of microfluidic cell cultures is "organs-on-chips" that are highly complex models with organ-level functions. The increased spatiotemporal control makes more information achieved. Hence, the thermodynamics, kinetics as well as mechanism can be better understood.

12.2.1.1 2D Cell Culture

2D cell culture platform is the most widely used models in vitro experiments [17]. In 2D platform, cells cannot grow directly in the microchannels or on the substrate. It often requires pre-coating of ECM to facilitate cell adhesion as well as to maintain cellular function [18]. The cell cultured in microchip can simplifies the manipulation not only of single cell but also of large quantities of cells (Fig. 12.1a) because researchers can change the channel geometries and designed microchannel structures to get more functionality. Likewise, to control of spatiotemporal factors in microchip, 2D monoculture is lack of simulating the cellular microenvironment in vivo, cell coculture models are employed to simulate tissue microenvironments in vivo by co-culturing multiple relevant cell types.

Cell growth, division, survival, movement etc. have effect on the interactions of cell–cell. Cell–cell communication includes autocrine, paracrine, and juxtacrine are also altered signaling pathways. The application of microfluidic technologies can realize cell coculture, and precise control spatial and temporal by designing microchip. With these advantages, physiologically relevant distances and concentrations can be achieved, the investigation of dynamic diffusion as well as manipulation cell can be realized. Huibin Wei and co-workers designed an integrated microfluidic device that can coculture PC12 cells and GH3 cells, as a model of the nervous tissue (Fig. 12.1b) [19]. Microfluidic device combined with mass

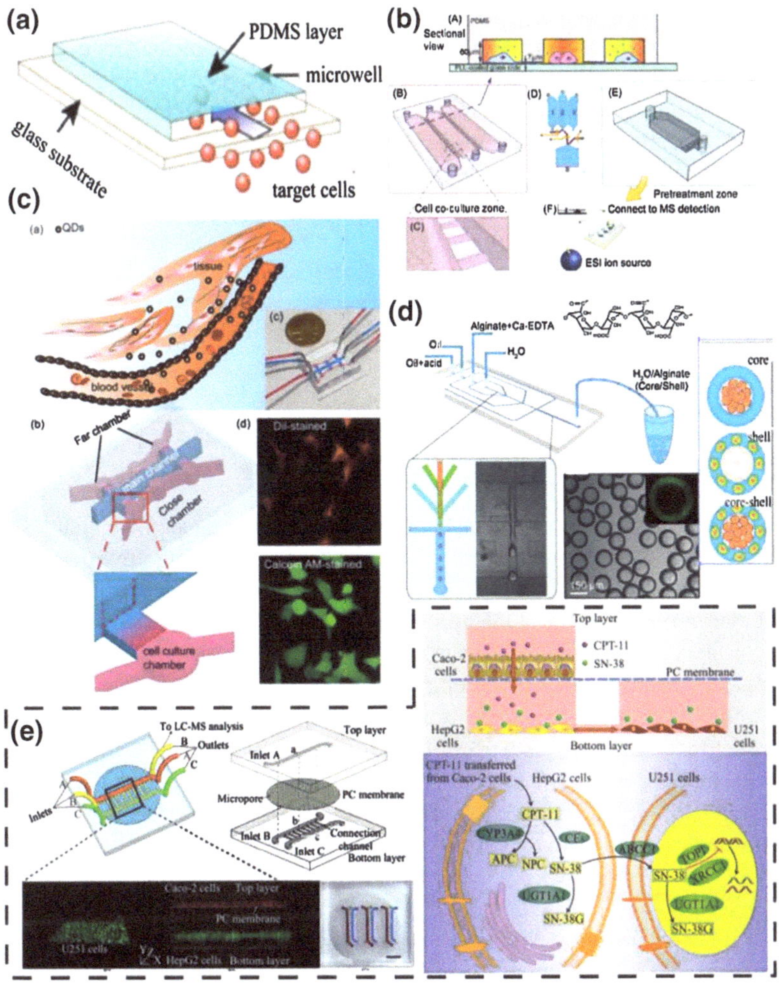

Fig. 12.1 Overview of microfluidic cell culture. **a** Monolayer 2D cell culture [30]; **b** 2D cell coculture [19]; **c** 3D cell culture [25]; **d** 3D spheroid cell culture [28]; **e** organ-on-chip system [32]. (Reprinted with permission from Refs. [19, 25, 28, 30, 32])

spectrometer make semiquantitative analysis of secreted proteins. Microfluidic cell culture is also used to study cell migration processes and their mechanism, which is different from macroscopic methods. Taking the advantages of design flexibility, real-time on-chip analysis of metabolism and low reagent consumption, Lin and co-workers developed a two-layer microfluidic cell culture device which resembles oxygen concentration gradient generator to study cell migration [20]. The microfluidic platform allowed investigation of various oxygen and distances effect on cell-to-cell communication. The cellular movement mechanisms and cell morphology are studied, and cell secretions are also on-chip detected.

Microfluidic cell culture method has ability to incorporate analytical instrumentations for the determination of living cells and cellular metabolites. One of challenges in metabolites analysis is reproduction quality. With high automation and integration, microfluidic combined with aforementioned analysis chemistry enhance the analytic precision, thereby increasing reproducibility. For instance, a two-layer microdevice was developed. This device can realize two kinds of cell coculture, and can real-time analyze the secretion proteins on the same chip. Sifeng et al. [21] developed a microfluidic system for co-culture 293 cells and L-02 cells. Cells were stimulated with different concentrations of epinephrine, and the metabolites including glucose and epinephrine were treated by on-chip solid-phase extraction and were detected directly by an electrospray ionization-quadrupole-time of-flight-mass spectrometer (ESI-Q-TOF-MS).

12.2.1.2 3D Cell Culture

Although 2D models are convenient, cells cultured by these methods often lost their functions such as differentiation due to the lack of surrounding environments. 3D culture is beneficial to understanding cellular behavior and function in their native microenvironment [22, 23]. One approach is using specific ECM [24], which can fabricate layers of extracellular matrix to achieve cell coculture [4]. Importantly, this approach reduces the lose of cell functions during the process of cell culture as well as improves tissue organization [22]. More, the secretions, nutrients and gas can be transported via the networks of 3D ECM and microchannel. In microfluidic platform, mechanical strain such as shear stress, can be controlled in the physiological range. Researchers have found cancer cells differently respond to drug when they cultured in a 3D ECM platform and in 2D platform. Jing Wu and co-workers has proposed an integrated 3D culture microdevice that can analyze quantum dot (QD) via imitating the diffusion process between blood vessels and tissues (Fig. 12.1c) [25]. The novel microdevice has been utilized with multiple constructs of 3D hydrogels and allows for a real-time monitoring cell behavior and cellular morphology exposed to both autocrine and paracrine signaling. The research showed that cell apoptosis were decreased by giving the drug 3-methyladenine (3-MA) at low concentrations of QDs, which proves that QD cytotoxicity effect on cell autophagy.

Nowadays, the most applied 3D cellular model is the spheroid [26, 27]. With cell growing, one kind of cells aggregates to form spheroids. The inner core of spheroids contains another kind of cells. The availability of oxygen and nutrients are gradually decreased from the medium to the inner core move away, which is a common tumor microenvironment. The standard in vitro model is 3D spheroid culture in the application of drug discovery and screening. Based on droplet-based microfluidics, Chen et al. [28] developed 3D core-shell scaffold by assembling hepatocytes cells and fibroblasts cells, forming an artificial liver (Fig. 12.1d). This in vitro model has potential for drug assays. Although 3D cellular models in the microfluidic technology are ECM platform and spheroid, the most applied model is mimic vessels. Based on a blood vessel and adjacent tissue model, Wu Jing developed a microfluidic 3D-culture device that is composed of a main channel and two sets of cell culture chambers [25]. In this device, HepG2 cells were cultured in an agarose matrix on the microfluidic device. In this model, cell apoptosis, intracellular reactive oxygen species (ROS) and glutathione (GSH) were detected and QD cytotoxicity was evaluated.

12.2.1.3 Organ-on-Chip System

Building in vitro model of organ level on chip holds a promising future for drug discovery and pathological study. In vitro culture cell is an alternative for animal model. With great capacity of displaying, cell culture technique has advanced many breakthroughs in molecular biology and pathology. But surely it is retarded by not taking into consideration the complex interactions between cells and extracellular matrix and different cells. Recently, 3D cell culture incorporated with extracellular matrix has appeared as a new approach to give cell culture in a way closer to in vivo situation. Succeeding in rehabilitate cell morphology, still it fails in constructing organ-level model, and thus limits the application in pathology and drug discovery. Organ is a highly organized and functional unit consisting of different cell types and unique structures [29]. Cell co-culture methods incorporate 2 or even more cell types to study cell communication and enable partly or even fully simulation of in vivo organ. However, there are challenged about organ model building in chip, first of which lay is in the culture of primary body cell. Still, it is impossible for modern cell culture technique to launch a long term primary cell culture. Most of the existing works about organ on chip circumvented the bottleneck by using immortal cell line instead, and thus reduced fidelity to in vivo human body. A few crucial problems need to be tackled before constructing organ on chip model. Interfaces between different tissues make the barrier to separate functional parts of organ and limit flow of fluids like blood as well as myriads of signal molecular. In vitro mimicking biological interfaces can be a determining step to arrange cells or tissues into organ. One current solution is the employment of thin porous film with pores in micro scale which could stop cell but permit free exchange of cellular factors and nutrient, and herein implement cell co-culture in opposite sides [30, 31]. Jie and co-workers have developed a multi-type cell microfluidic integrator for co-cultivation and simulation of drug absorption, metabolism, and anticancer activity (Fig. 12.1e) [32]. Caco-2, HepG2, and U251 cells were co-cultured as mimics of the intestine, liver, and glioblastoma, respectively. The usage of a polycarbonate semipermeable membrane, HepG2 cell channel was separated from the Caco-2 cell channel. Then, U251 cell were cultured in anther channels through the connection with a narrow channel array. In this model, irinotecan (CPT-11) metabolism and cytotoxic analysis were performed simultaneously. These studies have revealed that 3D cell cultures have a great advantage in dynamic metabolism study, which has a potential application for high throughput drug screening and personalized cancer therapy. The appropriate tissue structure, complex biochemical environment and dynamic mechanical microenvironment such as fluid shear are important for fabricating an organ-on-chip system [33]. Fortunately, microfluidic coculture systems can integrate to study the interaction of different cells for simulation physiology, which have the potential for fabrication human-relevant disease models. Based on 3D and organ-on chip technologies development, they have potential to study not only the absorption of drugs from the lung, metabolism in the liver and others, but also prediction clinical responses.

12.2.2 Controlling the Microenvironment In Vitro by Microfluidic Technology

Metabolites assay is aimed to quantify the level of all intermediates of metabolism. It is well known that changes of cell metabolites greatly depending on cells reside in a milieu. Chemical and mechanical parameters are important for the microenvironment (Fig. 12.2a). The surrounding of cells often consist of soluble factors, cell–matrix and its interactions, and cell–cell contacts, and so on. The cells often live in a specific physicochemical environment including pH, oxygen gradient, temperature change, and osmolality. These make up a physiological environment and effect on cell metabolism. Cell microenvironment contains biochemical, physical, and physicochemical factors. Growth factors, cytokines, amino acids and other biomolecules make up of biochemical microenvironment. Combination of these elements form complex signaling pathways, and determine the fate of the cell. Studying cell microenvironments helps examining cell metabolites for understanding the progress of disease and diagnosis. Although the construction of ideal comprehensive cell microenvironment is still difficult, multiple environmental parameters can be get using soft-lithography. The major interests of microfluidics in cell microenvironmental is: (1) controlled the scale; (2) changing chip design to control the cell microenvironment, which is more close to the physiological conditions in vivo [34].

12.2.2.1 Controlled Chemical Microenvironment

Because the richness of issues, the challenge is to develop a cellular model in vivo. Chemical environment around the cell is the compositions of cell culture medium. Therefore, researchers often firstly control the cell medium in microfluidics. Chemical gradients have been to mimic natural stimuli. These gradients in human body are mainly in the role of diffusion, and effect biological processes such as cell migration, differentiation, or development. Because concentration gradients of local region are higher 2% of its diameter, cells are able to response [35]. Thus, it is required to fine control local concentration in spatial resolution. The design of microfluidic devices have good ability with controlled spatiotemporal distribution

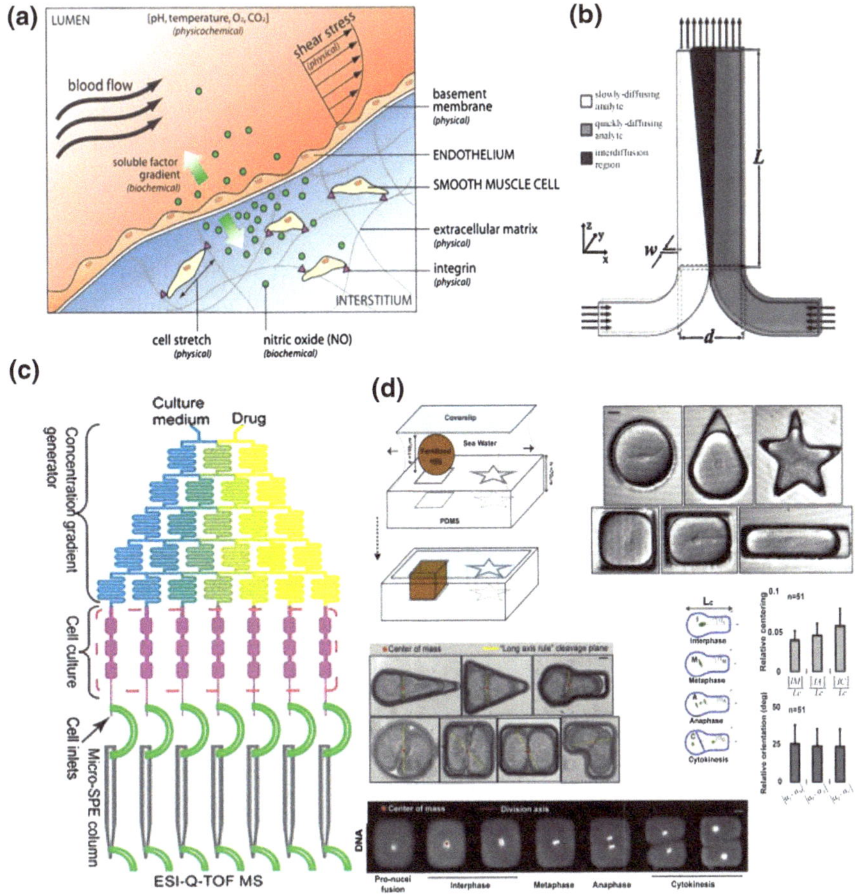

Fig. 12.2 Microfluidic plarform for fabrication cell microenvironment. **a** The cell microenvironment consists of physical, biochemical, and physicochemical factors [34]; **b** T gradient generator [36]; **c** premixer gradient generators [61]; **d** PDMS patch-clamp chip for study mechanical influence [53]. (Reprinted with permission from Refs. [34, 36, 53, 61])

and have fast response ability, which would be benefit for study the response of immune system. Researchers have successfully carried out a time invariant gradients and dynamic gradients on a microfluidic chip as well as continuous or discrete gradients based on laminar flows. Molecular gradients has been made by diffusive mixing among sever laminar streams of different composition. These gradient generators mainly depend on the flow rate and the time the streams. T-sensor is the simplest gradient generator. Time resolution of T-sensor can be achieved 10^{-2} s (Fig. 12.2b) [36], and this simple T-sensor devices are enabled to realize long-time cell culture. The simple T-sensor can also applied to study bacterial chemotaxis, endothelial cell migration. But, T-sensor device is reagent-consuming because it often requires continuous flow. The autocrine/paracrine or cell metabolites are removed, and cells subject to shear stresses [34, 37].

One of improvement of the T-sensor is premixer gradient generators. These generators splits and recombines fluids to generate more complex gradients before enter to cell culture area (Fig. 12.2c) [15]. Dertinger et al. [38] have developed this device for generation overlapping gradients by adjusting the inlet flows. This type of device has applied to study neutrophil chemotaxis [39], neural stem cell differentiation [40]. Another kind of gradient generators including a lot of channels or chambers split the solution to generate many profiles of concentrations. A generator has been developed by Cooksey et al. [41] is composed of 16 inlets, and can combine 64 of chambers. By integrating bypass valves, it enables to format of complex biomolecules gradients at the same time with subsecond temporal resolution. Although laminar flow based on a gradient generator has been well developed, it required control of the flow rate precisely. Furthermore, Shear stresses from flow can change cells behavior and metabolism. The undesired metabolites can be found and some metabolites undergo changes. To overcome this limitation, flow resistive generators were established to eliminate shear stress around the cells. Take advantage of generation steady-state gradients, decreasing convection, and preserving metabolites by cells, the device often assembled hydrogel [42], nanopore membrane [43], or microchannel, which can promote diffusion of biomolecules to generate gradients [44].

The high permeability of PDMS has been used to control gas composition in cell cultures [30, 45]. PDMS device has been developed for generation gas gradient of the medium by controlling flowing gas. Oxygen is important for maintain growth, proliferation, and controlled cell death. Tumor cells are often incubated at 37 °C under 20% O_2 as a control, while at 37 °C under 1–5% O_2 as hypoxia because the inside tumor is found in hypoxic condition [46]. Hypoxia can increase radiation and drug resistance, and enhance metastasis of tumor cells, which is a critical factor of the tumor microenvironment [47]. Oxygen tensions on cell behavior have been commonly studied by gas gradient generator. A gas gradient generator is often formed by channels, a media chamber and a gas-permeable membrane, and then oxygen gradient microenvironment is generated by mixing carbon dioxide, nitrogen and oxygen in a predefined ratio. Furthermore, in order to obtain the desired oxygen gradients in microfluidic devices, Oppegard et al. [48] developed a microfluidic device to control oxygen tension temporally and spatially by designing multiple microchannels into a PDMS membrane to divide mixed gas. 3D cell culture in microfluidic device often need the multiple-layer microfluidic systems, and accurate control gas flow rates is a limitation. Therefore, a new microfluidic device of generating oxygen gradients was developed based on on-line chemical reaction to generate oxygen and gas diffusion. Chen et al. [49] used spatially confined chemical reactions to generate oxygen gradient in the microchip. This design applied two chemical reactions involved NaClO and H_2O_2 to generate oxygen and pyrogallol and NaOH to scavenge oxygen. Oxygen gradient was generated by the distance of generate oxygen area and scavenging area with gas diffusion in the channel. This design simplified the structure of device and fluid flow systems, which have potential in the application of biological labs for study cellular responses although the oxygen gradients created were nonlinear.

12.2.2.2 Controlled Mechanical Microenvironment

Mechanical environment can directly affect metabolites of cells, and different metabolites can effect on mechanical environment. The common mechanical environment fabrication is to change ECM [50]. It has been reported that tumor cells show different mechanical properties to healthy cells [51]. Tumor cells are more tortuosity. This tortuosity has effect on metastasis spreading over tissues and cancer cell autophagy. Chen et al. [52] has developed a micro-device to investigate the relation between cell metabolite and mechanical environment by changing the cell culture chamber geometry. Minc et al. [53] fabricated different geometrical channels and different parameters of channels to study spatial response of cells organization (Fig. 12.2d). Chaw et al. [54] used a PDMS-matrigel device to study mechanical interactions between cells and their underlying substrates. They found that cell adhesion and mechanics were regulated by biochemical and mechanical signals. The mechanical force detection is a difficult for fabrication mechanical microenvironment. Besides of ECM, PDMS pillars or walls are applied to fabricate mechanical microenvironment of cells, which are also used to measure forces of adherent cells and substrate [55]. In most of these studies, the pillars were in the range of 2–3 μm wide, 6–10 μm high and 4–9 μm spacing apart. This method has low precision and deformable membranes have been exploited. The side force can be measured by monitoring the membrane deformation when the stress was imposed. Hohne et al. [56] developed a microfluidic device with PDMS membrane that enabled to characterize the pressure in the range of 0.1–100 kPa. Atomic force microscope, magnetic beads and so on are also applied to measure the mechanical forces, and then investigate the relativity between mechanical environment and cells.

Continuous flow microfluidic devices can control perturbation of the cellular environment in temporal and spatial resolution. Fluid flow not only transports nutrient substance, but also shear stress. Cells in a microchannel often suffer flow-stream laminar shear and splitting. Some works have showed that the migration of cells was affected fast flow of liquid caused mechanical force. Compared to fast flow, diffusion is too slow for mixing solutions, but it creates different liquid-phase environment on the changes in distance. Takayama et al. [57] applied laminar flow to study the movement of subcellular organ (mitochondria), and find that the the laminar flow can change the cytoskeletal structure of cells. In addition, magnetic field gradient are also studied by magnetic aggregates, which is used to produce cell surface stresses to investigate cell mechanobiology [58].

12.3 Strategy for Cell Metabolite Analysis on Microfluidic Platform

Although microfluidics are still developing, microfluidic systems are already used in many different aspects of cell metabolism and development. Microfluidic technology has the ability to integrate biosensor with separation system [59]. In combination with separation techniques possessing faster separation with smaller volumes, microfluidic chips are powerful tools for highly parallel, fast, easy and multiplexed metabolites assays. Researchers have developed sample separation technologies such as liquid chromatogram (LC), droplet technologies on microfluidic platform for cell metabolite analysis. Membrane techniques and electrokinetic separation techniques are often used for cell metabolite assay for sample pretreatment. The majoy technology of membrane technique is microdialysis sampling (MD). The frequently used electrokinetic separation techniques contained capillary electrophoresis (CE), gel electrophoresis, electrochromatography or isoelectric focusing (IEF).

12.3.1 Sample Separation on Microfluidic Platform for Cell Metabolite Analysis

12.3.1.1 LC Technology

The combination of micro separation or enrichment columns in microchip for cell metabolites treatment, and mass spectrometry or other detector can be directly detected cell metabolites. For instance, Chen et al. developed a microfluidic system for drug metabolite assay. The device consists of a microfluidic network for reagent supply, cell culture chambers and on-chip solid-phase extraction micro-columns for sample pretreatment. The quantitative analysis of drug metabolite was based on the carbonation of sample pretreatment system and stable isotopelabeling-assisted electrospray ionization mass spectrometry (Fig. 12.3a) [60]. Furthermore, this system was used to study drug-induced cell apoptosis. Gao et al. [61] also established a similar microfluidic system cell culture system to investigate on the HepG2 and Caco-2 cells absorption of methotrexate and the drug's effect. Cooksey et al. [62] applied ribosome activity inhibitor cycloheximide to investigate used green fluorescent protein decay in microfluidic cell culture system and tradition cell culture system. They found that the comparable results were obtained from different cell culture system, and microfluidic cell culture system showed higher levels of confidence.

Fig. 12.3 Strategy for cell metabolite analysis on microfluidic platform. **a** On-chip micro-SPE ▶ columns for drug metabolite analysis [60]; **b** a microdialysis-paper spray ionization-mass spectrometry (MS) system for cell glucose study [71]; **c** microchip electrophoresis for cell metabolite's analysis [73]. (Reprinted with permission from Refs. [60, 71, 73])

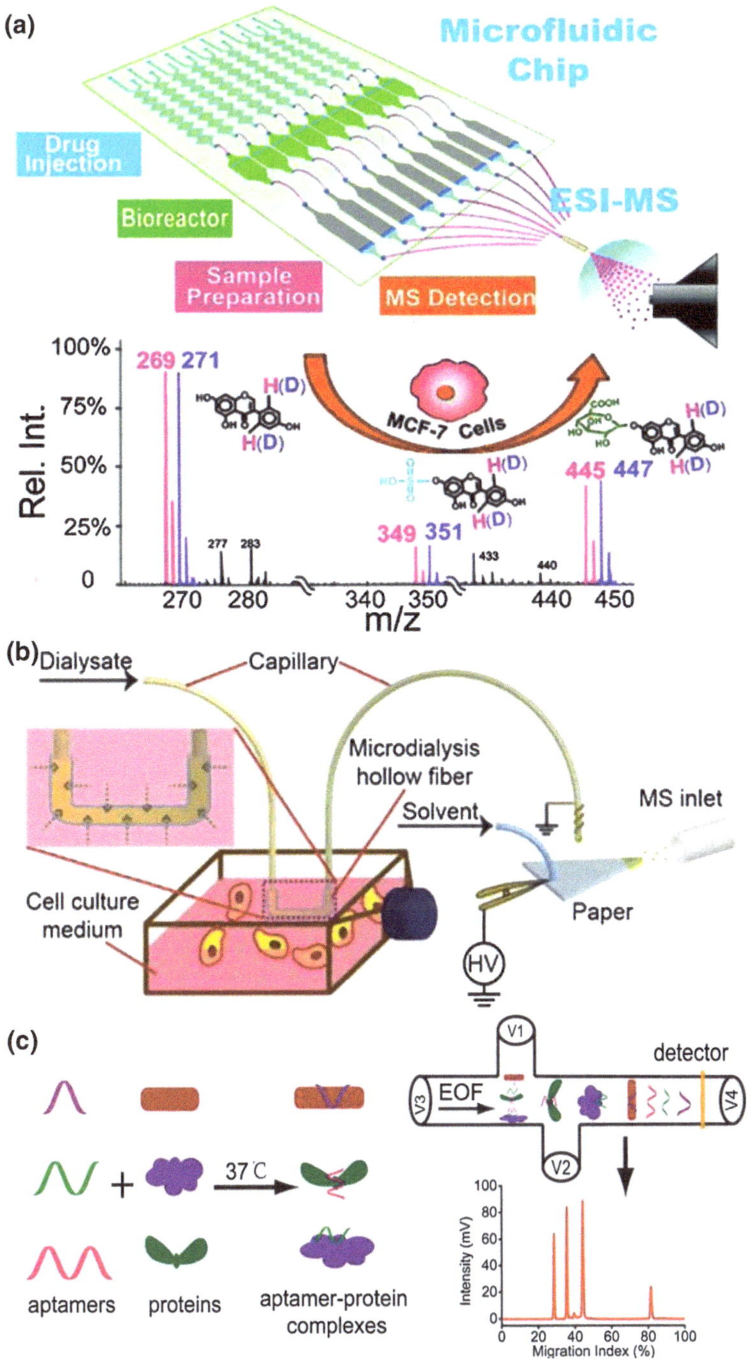
(a)
Microfluidic Chip
Drug Injection
Bioreactor
Sample Preparation
MS Detection
ESI-MS
MCF-7 Cells
Rel. Int.
m/z
(b)
Dialysate
Capillary
Microdialysis hollow fiber
Solvent
MS inlet
Paper
Cell culture medium
HV
(c)
aptamers
proteins
37℃
aptamer-protein complexes
EOF
detector
Intensity (mV)
Migration Index (%)

12.3.1.2 Droplet Technology

Among many microfluidic techniques used for cell analysis, droplet microfluidic (including digital microfluidics) has attracted increasingly attention in biochemical assay because the low volume [14]. The volume often is pL to nL [63]. Based T-sensor devices and the developed technologies, droplets can be generated by the application of kinds of immiscible medium [64], such as a water and oil, water or gas [65]. In order to achieve uniform size, shape, and monodispersity of the droplets, hydrodynamic methods have been developed by using channel structure and properties of fluid flow because controlling droplets generation are depended on shear forces in fluids. Electrohydrodynamic technologies, electrowetting and other are also been developed to produce fine droplet in molecular biology such as gene mutation, DNA assays and molecular evolution. Chaoyong James Yang produced water-in-oil droplets for cancer biomarker EpCAM and small toxin molecule aflatoxin B1 analysis [66]. One of application is to encapsulate single-cells into individual droplets that made of hydrogels such as agarose. Then, single cell physiology and genomic content can be investigated. More, the cell-secreted molecules can be remain within the droplet, and thus can rapidly analyze. Another application is for the high-precision timing of reaction kinetics because mix pL volumes of reagents, which can be used to identifying the productivity and activity of enzymes [67]. On the other hand, with the advantages of identify such super-producing cells, droplet microfluidic technology allows for identification of cells metabolite that can be as biomarker or be used to distinguish cell type and identify genetic elements that are responsible for phenotype study. Nowadays, Chen et al. [28] used droplet-based microfluidics to reconstitute “organ in a droplet”. Heterotypic cells were assembled in a 3D core-shell hydrogel to imitate liver tissue (Fig. 12.1d). These droplets can arranged multiple types of cells to create an artificial liver. Then, the time-depended albumin secretion and urea synthesis were investigated [68].

12.3.1.3 Microdialysis

Microdialysis technique is a powerful tool for analysis of cell metabolism and drug metabolism [69]. With the advantage of collective molecule of interest from the extracellular fluid, microdialysis membrane was integrated on microchip to do on-line microdialysis [70]. Furthermore, microdialysis technique can be combined with almost any detector which is very important for cell metabolites analysis with low volume requirements. The most common system is microdialysis coupled with microchip electrophoresis and laser-induced fluorescence detection. The usage of microdialysis, Scott and co-workers has developed an on-animal separation-based sensor for monitoring drug metabolism in freely roaming sheep. By the application of microdialysis, Liu et al. [71] have developed an online microdialysis-paper spray ionization-mass spectrometry (MS) platform for cell metabolism studies

(Fig. 12.3b). In this system, cell metabolite glucose rates for cells were detected, and hormone regulation of the glucose concentration was investigated.

12.3.1.4 Microchip Electrophoresis

A high-resolution separation technique is used for separation targets from complex mixtures of cell metabolites. Using microfluidics techniques, electrophoresis has been successfully integrated to microchip, and has been widely applied in various fields. Microchip electrophoresis (MCE) has the advantages of portability, simplify, rapid analysis, flexibility and low cost. The ability of this technique is to integrate sample preparation, derivatization, separation, detection and other functional units in a single device. By applying voltages to integrated electrodes, samples can be analyzed throughout channels. Analytes are separated in microchannels based on the different mobility of substances, and the length of separation channels often range from 10 to 75 mm. Electroosmotic flow is an important factor for the function of separation. Surface modification of the microchannels and changing composition of the buffer can alter electroosmotic flow, resulting in changing the separation behavior. Electrochemical methods are more portable due to it do not need derivatization, which is the most frequently analysis method coupled with MCE for the determination of cell metabolites. Holcomb et al. [3] coupled electrical conductivity detector with MCE device for monitoring metabolites such as glucose, catechins, and so on. Compared with electrochemical detection, fluorescence analysis is high sensitivity with several orders of magnitude. As an interesting development, MCE coupled with detector are employed to monitoring protein levels by aptamer-based microfluidics [72]. In addition, MCE can be used for real-time sample analysis due to its short separation time. The affinity probe for fluorescence analysis can be labeled and has been developed for multiple proteins assay containing $VEGF_{165}$, PDGF-BB, and thrombin (Fig. 12.3c) [73]. It uses non-specific DNA as an internal standard to achieve quantitative assay and reduce the interference. Finally, the established method was successfully applied to cell metabolite assay.

12.3.2 Cell Metabolite Detection Systems

Cells remain in a homeostatic state through the release of ions, small molecules, peptides, and proteins. Analysis of cellular metabolite faces analytical challenges. One is the frequent sampling collection for the dynamic nature of the release of bimolecular [74]. Otherwise, the dynamic nature would be altered average and transient events [75]. It also requires high sample collection rate, which makes low level of analysts, analytical methods with high sensitivities. In addition, cells often release a lot of substances at any given time. Thus, the assays required high selectivity for the analysts of interest. The final challenge is that cells are often in high-salt media, which is problematic for many analytical techniques. To overcome these problems, many creative approaches have been developed. Microfluidic devices have become a powful tool for investigation of cellular secretions. Nowadays, kinds of biosensors have been designed and are applied to various fields such as diagnosing disease, food safety, and metabolites assay. A biosensor is often contained three parts: (a) a sample preparation system that ensures the capture substance is the target analysts, (b) a detection system can transform the biological phenomenon into a detectable signal, and (c) a data processing system [76]. Cell metabolites assay aims to quantify the level of all intermediates of metabolism. The mostly applied method of cellular metabolites is enzymatic assays. These assays are based on antibody. Although many antibodies are available for common peptides and proteins, it is well known antibodies are difficult to analyze small molecules or toxic compounds. To develop assays without antibodies, other methods have been established such as nucleic acid-based cell metabolite analysis, or enzymes-based assay.

12.3.2.1 Antibody-Based Immunoassays for Cell Metabolite

Antibody-based immunoassays in microsystems are the most common type in diagnostics, medical research, and disease mechanisms research and so on. Because of antibody specificity, these technologies that involved with the rapid analysis of biomolecules have been rapidly developed. In these methods, the most widely used antibody-based assay is the enzyme-linked immunosorbent assay (ELISA) [77]. A typical ELISA involves the surface subtract coated with antibodies to capture the targets, and the secondary antibodies are labeled with enzymes, fluorophores or others to detect the target [78]. Mangru and Harrison developed antibody-based chemiluminescence detection for IgG immunoassay after MCE separation. ELISA was also employed to simultaneously analyze multiple cell secretions by a fluorescence based immunosensor [79]. Recognition elements are fixed on the surface of microchannel to capture the analysts. The detection system employs a small diode laser, CCD camera, and data processing system. The signal can be correlated with image analysis software. Another method performed on chip is competitive immunoassays. This method often involved an immunoreactor with MCE separation and fluorescence detection, which has been applied in drug monitoring in human serum. Immunoaffinity chromatography assays are also established with various kinds of patterns [80]. Arash Dodge and his co-workers immobilized protein A (PA) on the surface of glass and Cy5 labled rabbit immunoglobulin G (rIgG) as sample. The behavior of immunoaffinity chromatographic in the chip was studied by using electrokinetical method [81]. Although these methods have been developed and applied, they limit sensitivity of cell metabolites assay. In the past decades, many efforts have been made to improve the

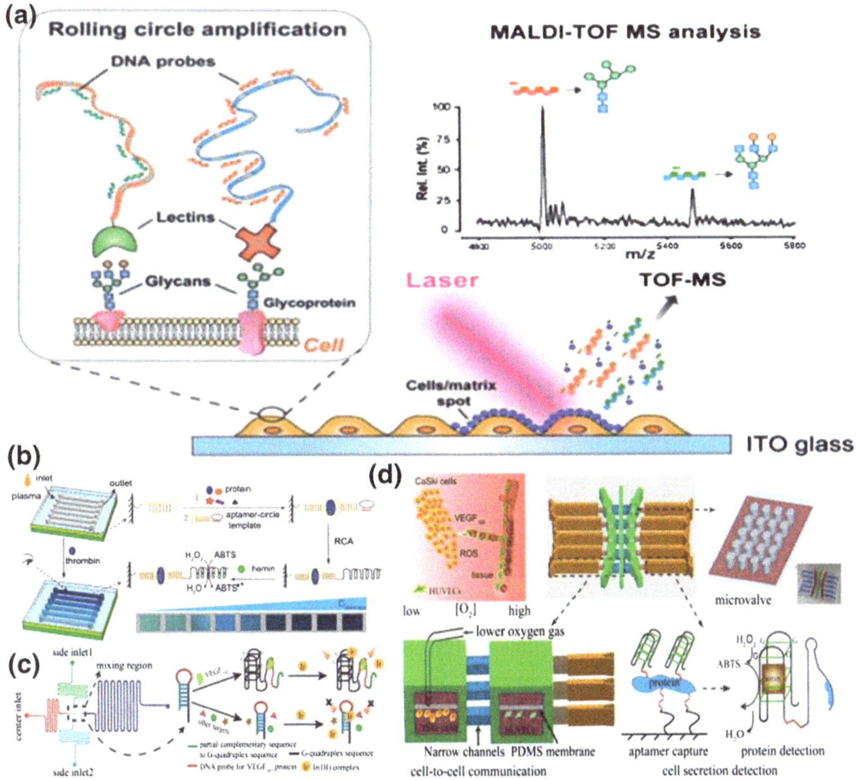

Fig. 12.4 Cell metabolite detection systems. **a** Antibody-based immunoassays for cell metabolite [82]; **b** aptamer-based biosensors coupled with RCA technology [84]; **c** real time monitoring cell metabolites with mixing device [73]; **d** microfluidic device for determination of cell metabolite with cell co-culture [20]. (Reprinted with permission from Refs. [20, 73, 82, 84])

analytic sensitivity. Heziyi et al. [82] have also developed this method for analysis of glycoproteins on cell surface (Fig. 12.4a). In this work, oligonucleotides are employed covalently linked to antibodies that recognize cell surface glycans. Upon antibody binding to lectins, the oligonucleotide probes are used as a primer, which can be utilized in the subsequent rolling circle amplification (RCA). After RCA, long single-strand DNA (ss-DNA) with repetitive sequence units is produced. After hybridizing with complementary short DNA probes, DNA duplexes are broken under laser irradiation, which makes the short DNA probes detached from the cell surface and detected by TOF-MS. This method converts the analysis of glycans to the detection of DNA probes. A benefit of it in situ proximity ligation assay is that the signal arises from specific sites in the cell, thus providing new information and improves the sensitivity and low ionization efficiency of glycans.

12.3.2.2 Nucleic Acid-Based Biosensors for Cell Metabolite Analysis

In recent years, there has been increasing interest in finding new molecules that enable to replace antibodies. Nucleic acids carry important biological information. DNA fragments store and impart genetic instructions, while RNA segments are significant for gene expression. Because nucleic acids can be rationally influenced by the base sequence, and have exciting physicochemical properties. Thus, nucleic acids have become important in cell metabolite detection because they can be used as recognition molecules, and as a material for building machine like nanodevices. The detection of DNA or RNA can be hybridized with its complementary strand with microfluidic technologies [83]. In order to improve the sensitivity, researchers often applied PCR to obtain repeated sequences of DNA fragements, and then used electrophoresis coupled with fluorescence spectrometer to separate and detect the DNA fragments. Tachibana et al. [1] have developed a self-propelled continuous-flow polymerase chain reaction in microfluidic device for amplification 232 bp of AH1pdm influenza virus and 95 bp of 16S rDNA of Escherichia coli genomic DNA. Nucleic acid-based biosensors have been widely used to detect proteins, DNA or RNA. Molecular beacon is a particular format that attracted lots of interest. Molecular beacon is often designed to various kinds of hairpin structure. This structure contains a loop domain, and a complementary sequence of analytes. A luminescence probe and a quench probe placed in start site and end site of a molecular beacon. When the target is hybrid with beacon molecules, the hairpin structure was broken, and the luminescence signal will turns on. Furthermore, as recognition factors, these probes are also applied as intracellular sensors for various molecules such as ions, proteins, DNA and RNA *in* cell *vivo*, thus enabled to do cell imaging analysis.

One of nucleic acid-based biosensors is aptamer. After the discovery of aptamers, nucleic acid-based biosensors are applied to proteins, virus, drug and even whole cells. Aptamers are single-stranded DNA or RNA molecules that are able to specific bind to targets with high affinity comparable to those of antibodies. Various aptamer-based methods have been reported in recent years, and the most common detection methods are luminescence or electrochemical methods. Lin group described immobilizing aptamer molecules on microchannel, labeling these DNA molecules with G-quadruple fragment, and replicating G-quadruple fragments, using hemin/G-quadruplex system to determinate thrombin in blood serum (Fig. 12.4b) [84]. Moreover, this method are more suitable than other methods for study the kinetics of cell metabolism because the signal can be directly obtained from analyte bind and it can decrease analytical process such as washing or labeling steps. Lin et al. have also been integrated aptasensors into microfluidic devices that could be used for on-chip aptasensor for monitoring secrets ($VEGF_{165}$, PDGF-BB, and thrombin) between cells (Fig. 12.4c, d) [20, 73]. This on-chip aptamer-based sensor is important because it extended the lifetime of the affinity biosensor, and enabled to realize the long-time monitor cell metabolism. In this case, the dynamic changes of cell metabolites are investigated.

12.3.2.3 Enzyme-Based Biosensors for Cell Metabolite Analysis

Enzymes, as catalytic biosensors, are widely applied for cell metabolite analysis. Enzyme-based biosensors are used to record analyte changes depending on time, which are the earliest biosensors in the metabolite assay. Because enzymes are not consumed, and enabled to reuse between catalyzing reaction cycles, Enzyme-based biosensors have lower detection limits over other affinity-based biosensors. Although some commercial biosensor such as glucose meter is in use, great efforts still are made. However, one of the challenges in making enzyme-based microfluidic platform is immobilized enzyme molecules and can remain functional over time. Another difficulty is to obtain the information hidden inside the cell. This requires no disruption of the cell membrane and maintain cell viability for a long time. Microfluidic technologies have their unique properties with controllable fluid transport, enzyme immobilization, and feasible manipulation, showing great potential in analysis of intracellular molecules and cell metabolites, even the whole cell. I-Jane Chen and his worker designed an interdigitated array microelectrode with enzyme-linked DNA hybridization assay into a microfluidic channel [85]. This strategy improves the limit of detection compared to conventional electrochemical biosensors. The detection of 4-aminophenol (PAP) can low to sub-nM. Batalla et al. [80] has fabricated an electrochemical microfluidic device for determination of enantiomer-biomarkers (D-methionine and D-leucine). This microfluidic strategy allowed for enantiomeric separation and covalent immobilization of the enzyme into the wall channels. In addition, hybrid polymer/grapheme based electrodes coupled with the microfluidic system enable to improve D-methionine and D-leucine analysis.

12.4 Application

12.4.1 Clinical Diagnostics

Cell metabolite assay have considerable potential in the field of clinical diagnostics. Numerous sensor technologies in the industry have potentially been used for clinical applications. The usage of microfluidic systems in metabolite assay have attracted more and more attention in the investigation of disease mechanisms and enhance understanding of pathology. The results would be potential for the monitoring disease processes, but these techniques are not mature and reliable for routine clinical practice. The challenges are the detection of low-abundance proteins or low concentration of other molecular. Great efforts have been made over the past decade and fruitful results were obtained. The most common class of diagnostic tests for metabolites analysis is the lateral flow test based on microfluidic platforms. Biomarkers are immobilized on a membrane or paper strip to analyze the targets. The most advantage is the induction of sample by capillary action without using intervention. When samples flow across the membrane, targets are captured in the membrane with the labeling reagents embedded. Then the targets are detected by eye to form a visible band or by specific instrument. The most application is pregnancy test. The streptococcus infections are also well tested in developed countries. In developing countries, this technology is widely used for diagnose human immunodeficiency virus. Although it is simple, the test is significant in a clinic diagnostic market. During the last few decades, many people tried to improve the method, but so far there is no significant progress [86]. Another critical application of membrane and lateral flow immunoassays is blood glucose test. The glucose test has the key effect on the entire diagnostics industry, which improve diabetic patients healthy. With the development of diagnostic technologies, microfluidic device are required much more functionalization. Indeed, iSTAT system is the first commercially successful microfluidic device in the current concept based on the combination microfluidic system and electrochemical detection [86]. The clinical medicine is greatly required new microfluidic devices, especially for high sensitivity assay with an ability to quantify.

12.4.2 Drug Research and Development

The drug development is challenging high costs and decreasing efficiency of drug. The poor advanced research of preclinical models brings the unsatisfactory clinical outcomes, resulting in drug failures in clinical trial [87]. Moreover, drugs often weakly selectivity but distribute to the whole of body, causing side effects. Therefore, great efforts have been made for drug development including drug discovery and delivery, drug sensitivity and toxicity, and drug preparation. Furthermore, metabolites are the key indicators for drug research. Microfluidic device is extensively applied to drug discovery, drug delivery, drug toxicity and so on. In order to decrease the cost and time, drug discovery and development greatly requires the rapid and high-throughput drug screening platform and metabolite analysis platform. Lin group developed a microfluidic device performed high-throughput drug screening based on integrated drug gradient generator, cell culture, and sample pretreatment system. This microchip can also rapidly analyze metabolites coupled with ESI-Q-TOF MS [61]. Although traditional culture models in microfluidic device can provide lots of relatively inexpensive data, they are lack of complex physiological conditions to predict drug responses in vivo. Microengineered cell culture models provide a solution to this requirement. Xu et al. [88] have developed 3D co-culture microchip for coculturing lung cancer cells and stromal cells, which enables to imitate lung tissue microenvironment in vivo. Then, different anticancer drugs (gefitinib, paclitaxel, and gemcitabine) with different concentration were given by a gradient concentration generator inside the chips. The cell apopotosis was used to access the drug sensitivity and toxicity. Battig et al. [89] fabricated drug delivery system based on microfluidic technologies, and evaluated the system by the metabolites (VEGF and PDGF-BB). Au and co-workers fabricated a three-dimensional hydrogel-based microfluidic system to perform drug screening. On this plarfom, HepG2 and NIH-3T3 cells were co-cultured to simulate live microtissues and as a tissue model for drug metabolism [90]. The albumin secretion profiles and cytochrome P450 3A4 are evaluated for the delivery of acetaminophen. Recently, microfluidics is applied to synthesize drug carriers. Microfluidic based on droplet and photolithography technologies has been successfully applied to make colloids of drug. Combination with aptamer and microfluidic technologies, the selectivity of the drug carriers was improved by the application of aptamer and the release time would be controlled to the desired place. Oh et al. [91] employed an efficient "click chemistry" technique to synthesize aptamer-polymer hybrids. On recognizing the targeted cell-surface marker, tumor cells were targeted and killed by doxorubicin. Furthermore, Tan group proved that aptamer-functionalized micelles were able to go through the surface of immobilized tumor cells, modeled drug delivery in the blood system, and demonstrated the dynamic specificity of aptamer–micelles [92].

12.4.3 Toxicology Study

Cell metabolite analysis has been increasingly used in toxicological studies mostly because the analytical results have the ability to provide more detailed information to elucidate mechanism of toxicity. During a toxicological study, it can be monitored time-depending on metabolites changes by drug dose or new chemicals treatment. In vitro models and analytical methods are desirable for predicting drug or other substance effects in humans. These technologies would also provide the mechanism and manifestation of toxicity based on patterns of metabolite changes. Metabolites have different kinds of properties with polarities, volatilities, chemical reactivities, and concentrations. These characteristics of metabolites and variations are greatly challenging tasks. Therefore, the selectivity, sensitivity, and continuous monitoring are the most important factors for metabolite analysis. An ideal analytical platform for metabolite analysis should have an ability to separate metabolites as much as possible and to quantify their dynamic concentration ranges. The advantages of microfluidic systems provide opportunities to realize these requirements by the combination with various detectors. Microfluidic cell culture technology provides opportunities to fabricate bioartificial organs. Organ-on-chip can perform not only cellular microenvironment but also extracellular variety with high spatiotemporal resolution. Furthermore, the metabolite can be analyzed at any time with the desired location. These properties are critical for toxicology study especially in drug discovery process, environmental pollutants and food toxicity screening. Bergström and his co-workers cultured clusters of human induced pluripotent stem cell derived cardiomyocytes in the integrated microfluidic device to simulate cardiomyocyte clusters beating. The application of this platform, the effects of drugs doxorubicin, verapamil and quinidine were assessed by the combination of video imaging system [94]. Shintu et al. [93] designed an organ-on-chip system to imitate liver and kidney tissue. This design system is used to evaluate the toxicity of environmental pollutant such as *N*-acetyl-para-aminophenol and ammonia based NMR technology.

12.5 Conclusions and Perspectives

Microfluidic platforms with simplify, portability, miniaturization and integration have become basic research tools for cell metabolite analysis. Great progresses have been made in analysis of metabolites over the past decades especially in the drug discovery and development, clinical diagnose and environmental contaminant studies. Although many strategies have been greatly used to analyze metabolites, challenges remain and the application facilitate to more development. If these challenges are resolved and scientific concept are reliable, cell metabolite analysis on microfluidic platform will benefit an amount of domains on drug screening, contamination assessment, disease diagnosis, and personalized healthcare, altering many fields of our life. Much remains to be done, and so there are a lot of opportunities to take part in what we hope innovations and the potential of cell metabolite analysis on microfluidic platform holds.

References

1. Tachibana H, Saito M, Tsuji K, Yamanaka K, Le Quynh H, Tamiya E (2015) Self-propelled continuous-flow PCR in capillary-driven microfluidic device: microfluidic behavior and DNA amplification. Sensors Actuat B Chem 206:303–310
2. DeBerardinis RJ, Thompson CB (2012) Cellular metabolism and disease: what do metabolic outliers teach us? Cell 148:1132–1144
3. Kraly JR, Holcomb RE, Guan Q, Henry CS (2009) Review: Microfluidic applications in metabolomics and metabolic profiling. Anal Chim Acta 653:23–35
4. Novik E, Maguire TJ, Chao P, Cheng K, Yarmush ML (2010) A microfluidic hepatic coculture platform for cell-based drug metabolism studies. Biochem Pharmacol 79:1036–1044
5. van Midwoud PM, Verpoorte E, Groothuis GM (2011) Microfluidic devices for in vitro studies on liver drug metabolism and toxicity. Integr Bioly 3:509–521
6. Ning R, Wang F, Lin L (2016) Biomaterial-based microfluidics for cell culture and analysis. TrAC Trends Anal Chem 80:255–265
7. Derkus B (2016) Applying the miniaturization technologies for biosensor design. Biosens Bioelectron 79:901–913
8. Le HP (1998) Progress and trends in ink-jet printing technology. J Imaging Sci Technol 42:49–62
9. Terry SC, Jerman JH, Angell JB (1979) A gas chromatographic air analyzer fabricated on a silicon wafer. IEEE Trans Electron Dev 26:1880–1886
10. Manz A, Graber N, Widmer H (1990) Miniaturized total chemical analysis systems: a novel concept for chemical sensing. Sensors Actuat B Chem 1:244–248
11. Halldorsson S, Lucumi E, Gómez-Sjöberg R, Fleming RM (2015) Advantages and challenges of microfluidic cell culture in polydimethylsiloxane devices. Biosens Bioelectron 63:218–231

12. Ziolkowska K, Kwapiszewski R, Brzozka Z (2011) Microfluidic devices as tools for mimicking the in vivo environment. New J Chem 35:979–990
13. Meyvantsson I, Beebe DJ (2008) Cell culture models in microfluidic systems. Annu Rev Anal Chem 1:423–449
14. Vasdekis AE, Stephanopoulos G (2015) Review of methods to probe single cell metabolism and bioenergetics. Metab Eng 27:115–135
15. Hung PJ, Lee PJ, Sabounchi P, Lin R, Lee LP (2005) Continuous perfusion microfluidic cell culture array for high-throughput cell-based assays. Biotechnol Bioeng 89:1–8
16. Zare RN, Kim S (2010) Microfluidic platforms for single-cell analysis. Annu Rev Biomed Eng 12:187–201
17. Kafi MA, El-Said WA, Kim T-H, Choi J-W (2012) Cell adhesion, spreading, and proliferation on surface functionalized with RGD nanopillar arrays. Biomaterials 33:731–739
18. Mabry KM, Payne SZ, Anseth KS (2016) Microarray analyses to quantify advantages of 2D and 3D hydrogel culture systems in maintaining the native valvular interstitial cell phenotype. Biomaterials 74:31–41
19. Wei H, Li H, Mao S, Lin J-M (2011) Cell signaling analysis by mass spectrometry under coculture conditions on an integrated microfluidic device. Anal Chem 83:9306–9313
20. Lin X, Chen Q, Liu W, Zhang J, Wang S, Lin Z, Lin J-M (2015) Oxygen-induced cell migration and on-line monitoring biomarkers modulation of cervical cancers on a microfluidic system. Sci Rep 5:09643
21. Mao S, Zhang J, Li H, Lin J-M (2012) Strategy for signaling molecule detection by using an integrated microfluidic device coupled with mass spectrometry to study cell-to-cell communication. Anal Chem 85:868–876
22. Huh D, Hamilton GA, Ingber DE (2011) From 3D cell culture to organs-on-chips. Trends Cell Biol 21:745–754
23. Haycock JW (2011) 3D cell culture: a review of current approaches and techniques. Methods Mol Biol 695:1–15
24. Tibbitt MW, Anseth KS (2009) Hydrogels as extracellular matrix mimics for 3D cell culture. Biotechnol Bioeng 103:655–663
25. Wu J, Chen Q, Liu W, Zhang Y, Lin J-M (2012) Cytotoxicity of quantum dots assay on a microfluidic 3D-culture device based on modeling diffusion process between blood vessels and tissues. Lab Chip 12:3474–3480
26. Montanez-Sauri SI, Beebe DJ, Sung KE (2015) Microscale screening systems for 3D cellular microenvironments: platforms, advances, and challenges. CMLS Cell Mol Life Sci 72:237–249
27. Fennema E, Rivron N, Rouwkema J, van Blitterswijk C, de Boer J (2013) Spheroid culture as a tool for creating 3D complex tissues. Trends Biotechnol 31:108–115
28. Chen Q, Utech S, Chen D, Prodanovic R, Lin J-M, Weitz DA (2016) Controlled assembly of heterotypic cells in a core–shell scaffold: organ in a droplet. Lab Chip 16:1346–1349
29. Bhatia SN, Ingber DE (2014) Microfluidic organs-on-chips. Nature 201:4
30. Chen Q, Wu J, Zhang Y, Lin Z, Lin J-M (2012) Targeted isolation and analysis of single tumor cells with aptamer-encoded microwell array on microfluidic device. Lab Chip 12:5180–5185
31. Huh D, Y-s T, Hamilton GA, Kim HJ, Ingber DE (2012) Microengineered physiological biomimicry: organs-on-chips. Lab Chip 12:2156–2164
32. Jie M, Li H-F, Lin L, Zhang J, Lin J-M (2016) Integrated microfluidic system for cell co-culture and simulation of drug metabolism. RSC Adv 6:54564–54572
33. Huh D, Matthews BD, Mammoto A, Montoya-Zavala M, Hsin HY, Ingber DE (2010) Reconstituting organ-level lung functions on a chip. Science 328:1662–1668
34. Young EWK, Beebe DJ (2010) Fundamentals of microfluidic cell culture in controlled microenvironments. Chem Soc Rev 39:1036–1048
35. Parent CA, Devreotes PN (1999) A cell's sense of direction. Science 284:765–770
36. Kamholz AE, Weigl BH, Finlayson BA, Yager P (1999) Quantitative analysis of molecular interaction in a microfluidic channel: the T-sensor. Anal Chem 71:5340–5347

37. Velve-Casquillas G, Le Berre M, Piel M, Tran PT (2010) Microfluidic tools for cell biological research. Nano Today 5:28–47
38. Dertinger SKW, Chiu DT, Jeon NL, Whitesides GM (2001) Generation of gradients having complex shapes using microfluidic networks. Anal Chem 73:1240–1246
39. Jeon NL, Baskaran H, Dertinger SKW, Whitesides GM, Van de Water L, Toner M (2002) Neutrophil chemotaxis in linear and complex gradients of interleukin-8 formed in a microfabricated device. Nat Biotechnol 20:826–830
40. Sip CG, Bhattacharjee N, Folch A (2014) Microfluidic transwell inserts for generation of tissue culture-friendly gradients in well plates. Lab Chip 14:302–314
41. Cooksey GA, Sip CG, Folch A (2009) A multi-purpose microfluidic perfusion system with combinatorial choice of inputs, mixtures, gradient patterns, and flow rates. Lab Chip 9:417–426
42. Saadi W, Rhee SW, Lin F, Vahidi B, Chung BG, Jeon NL (2007) Generation of stable concentration gradients in 2D and 3D environments using a microfluidic ladder chamber. Biomed Microdevices 9:627–635
43. Luo XL, Vo T, Jambi F, Pham P, Choy JS (2016) Microfluidic partition with in situ biofabricated semipermeable biopolymer membranes for static gradient generation. Lab Chip 16:3815–3823
44. Somaweera H, Ibraguimov A, Pappas D (2016) A review of chemical gradient systems for cell analysis. Anal Chim Acta 907:7–17
45. Polinkovsky M, Gutierrez E, Levchenko A, Groisman A (2009) Fine temporal control of the medium gas content and acidity and on-chip generation of series of oxygen concentrations for cell cultures. Lab Chip 9:1073–1084
46. Lo JF, Sinkala E, Eddington DT (2010) Oxygen gradients for open well cellular cultures via microfluidic substrates. Lab Chip 10:2394–2401
47. Funamoto K, Zervantonakis IK, Liu YC, Ochs CJ, Kim C, Kamm RD (2012) A novel microfluidic platform for high-resolution imaging of a three-dimensional cell culture under a controlled hypoxic environment. Lab Chip 12:4855–4863
48. Oppegard SC, Blake AJ, Williams JC, Eddington DT (2010) Precise control over the oxygen conditions within the Boyden chamber using a microfabricated insert. Lab Chip 10:2366–2373
49. Chen YA, King AD, Shih HC, Peng CC, Wu CY, Liao WH, Tung YC (2011) Generation of oxygen gradients in microfluidic devices for cell culture using spatially confined chemical reactions. Lab Chip 11:3626–3633
50. Haynl C, Hofmann E, Pawar K, Forster S, Scheibel T (2016) Microfluidics-produced collagen fibers show extraordinary mechanical properties. Nano Lett 16:5917–5922
51. Guck J, Schinkinger S, Lincoln B, Wottawah F, Ebert S, Romeyke M, Lenz D, Erickson HM, Ananthakrishnan R, Mitchell D et al (2005) Optical deformability as an inherent cell marker for testing malignant transformation and metastatic competence. Biophys J 88:3689–3698
52. Chen CC, Folch A (2006) A high-performance elastomeric patch clamp chip. Lab Chip 6:1338–1345
53. Minc N, Burgess D, Chang F (2011) Influence of cell geometry on division-plane positioning. Cell 144:414–426
54. Chaw KC, Manimaran M, Tay EH, Swaminathan S (2007) Multi-step microfluidic device for studying cancer metastasis. Lab Chip 7:1041–1047
55. Tan JL, Tien J, Pirone DM, Gray DS, Bhadriraju K, Chen CS (2003) Cells lying on a bed of microneedles: an approach to isolate mechanical force. Proc Natl Acad Sci USA 100:1484–1489
56. Hohne DN, Younger JG, Solomon MJ (2009) Flexible microfluidic device for mechanical property characterization of soft viscoelastic solids such as bacterial biofilms. Langmuir 25:7743–7751
57. Takayama S, Ostuni E, LeDuc P, Naruse K, Ingber DE, Whitesides GM (2001) Laminar flows: subcellular positioning of small molecules. Nature 411:1016

58. Gupta M, Kocgozlu L, Sarangi BR, Margadant F, Ashraf M, Ladoux B (2015) Micropillar substrates: a tool for studying cell mechanobiology. Methods Cell Biol 125:289–308
59. Neuzil P, Giselbrecht S, Lange K, Huang TJ, Manz A (2012) Revisiting lab-on-a-chip technology for drug discovery. Nat Rev Drug Discov 11:620–632
60. Chen QS, Wu J, Zhang YD, Lin J-M (2012) Qualitative and quantitative analysis of tumor cell metabolism via stable isotope labeling assisted microfluidic chip electrospray ionization mass spectrometry. Anal Chem 84:1695–1701
61. Gao D, Li HF, Wang NJ, Lin JM (2012) Evaluation of the absorption of methotrexate on cells and its cytotoxicity assay by using an integrated microfluidic device coupled to a mass spectrometer. Anal Chem 84:9230–9237
62. Cooksey GA, Elliott JT, Plant AL (2011) Reproducibility and robustness of a real-time microfluidic cell toxicity assay. Anal Chem 83:3890–3896
63. Anna SL (2016) Droplets and bubbles in microfluidic devices. Annu Rev Fluid Mech 48:285–309
64. Martino C, deMello AJ (2016) Droplet-based microfluidics for artificial cell generation: a brief review. Interface Focus 6:20160011
65. Raj A, Halder R, Sajeesh P, Sen AK (2016) Droplet generation in a microchannel with a controllable deformable wall. Microfluid Nanofluid 20:102
66. Zhu Z, Song YL, Li C, Zou Y, Zhu L, An Y, Yang CJ (2014) Monoclonal surface display SELEX for simple, rapid, efficient, and cost-effective aptamer enrichment and identification. Anal Chem 86:5881–5888
67. Hassan SU, Nightingale AM, Niu XZ (2016) Continuous measurement of enzymatic kinetics in droplet flow for point-of-care monitoring. Analyst 141:3266–3273
68. Liu YW, Wei JJ, Lu JF, Lei DM, Yan SL, Li XH (2016) Micropatterned coculture of hepatocytes on electrospun fibers as a potential in vitro model for predictive drug metabolism. Mat Sci Eng C Mater 63:475–484
69. Lu CM, Hou ML, Lin LC, Tsai TH (2014) Development of a microdialysis system to monitor lamivudine in blood and liver for the pharmacokinetic application in herbal drug interaction and the gene expression in rats. Mater Sci Eng C Mater 96:231–240
70. Scott DE, Willis SD, Gabbert S, Johnson D, Naylor E, Janle EM, Krichevsky JE, Lunte CE, Lunte SM (2015) Development of an on-animal separation-based sensor for monitoring drug metabolism in freely roaming sheep. Analyst 140:3820–3829
71. Liu W, Wang N, Lin X, Ma Y, Lin J-M (2014) Interfacing microsampling droplets and mass spectrometry by paper spray ionization for online chemical monitoring of cell culture. Anal Chem 86:7128–7134
72. de Jong S, Krylov SN (2011) Protein labeling enhances aptamer selection by methods of kinetic capillary electrophoresis. Anal Chem 83:6330–6335
73. Lin XX, Chen QS, Liu W, Yi LL, Li HF, Wang ZH, Lin J-M (2015) Assay of multiplex proteins from cell metabolism based on tunable aptamer and microchip electrophoresis. Biosens Bioelectron 63:105–111
74. Schrell AM, Mukhitov N, Yi L, Wang X, Roper MG (2016) Microfluidic devices for the measurement of cellular secretion. Annu Rev Anal Chem 9:249–269
75. Lin XX, Leung KH, Lin L, Lin LY, Lin S, Leung CH, Ma DL, Lin JM (2016) Determination of cell metabolite VEGF(165) and dynamic analysis of protein–DNA interactions by combination of microfluidic technique and luminescent switch-on probe. Biosens Bioelectron 79:41–47
76. Zhou Q, Son K, Liu Y, Revzin A (2015) Biosensors for cell analysis. Annu Rev Biomed Eng 17:165–190
77. Zheng CH, Wang JW, Pang YH, Wang JB, Li WB, Ge ZG, Huang YY (2012) High-throughput immunoassay through in-channel microfluidic patterning. Lab Chip 12:2487–2490
78. Ng JH, Ilag LL (2002) Biomedical applications of protein chips. J Cell Mol Med 6:329–340

79. Mangru SD, Harrison DJ (1998) Chemiluminescence detection in integrated post-separation reactors for microchip-based capillary electrophoresis and affinity electrophoresis. Electrophoresis 19:2301–2307
80. Batalla P, Martin A, Lopez MA, Gonzalez MC, Escarpa A (2015) Enzyme-based microfluidic chip coupled to graphene electrodes for the detection of D-amino acid enantiomer-biomarkers. Anal Chem 87:5074–5078
81. Dodge A, Fluri K, Verpoorte E, de Rooij NF (2001) Electrokinetically driven microfluidic chips with surface-modified chambers for heterogeneous immunoassays. Anal Chem 73:3400–3409
82. He ZY, Chen QS, Chen FM, Zhang J, Li HF, Lin JM (2016) DNA-mediated cell surface engineering for multiplexed glycan profiling using MALDI-TOF mass spectrometry. Chem Sci 7:5448–5452
83. Lin XX, Sun XY, Luo SD, Liu B, Yang CX (2016) Development of DNA-based signal amplification and microfluidic technology for protein assay: a review. Trac Trends Anal Chem 80:132–148
84. Lin XX, Chen QS, Liu W, Li HF, Lin JM (2014) A portable microchip for ultrasensitive and high-throughput assay of thrombin by rolling circle amplification and hemin/G-quadruplex system. Biosens Bioelectron 56:71–76
85. Chen IJ, White IM (2011) High-sensitivity electrochemical enzyme-linked assay on a microfluidic interdigitated microelectrode. Biosens Bioelectron 26:4375–4381
86. Chin CD, Linder V, Sia SK (2012) Commercialization of microfluidic point-of-care diagnostic devices. Lab Chip 12:2118–2134
87. Khan IU, Serra CA, Anton N, Vandamme T (2013) Microfluidics: a focus on improved cancer targeted drug delivery systems. J Control Release 172:1065–1074
88. Xu ZY, Gao YH, Hao YY, Li EC, Wang Y, Zhang JN, Wang WX, Gao ZC, Wang Q (2013) Application of a microfluidic chip-based 3D co-culture to test drug sensitivity for individualized treatment of lung cancer. Biomaterials 34:4109–4117
89. Battig MR, Soontornworajit B, Wang Y (2012) Programmable Release of Multiple Protein Drugs from Aptamer-Functionalized Hydrogels via Nucleic Acid Hybridization. J Am Chem Soc 134:12410–12413
90. Au SH, Chamberlain MD, Mahesh S, Sefton MV, Wheeler AR (2014) Hepatic organoids for microfluidic drug screening. Lab Chip 14:3290–3299
91. Parveen S, Misra R, Sahoo SK (2012) Nanoparticles: a boon to drug delivery, therapeutics, diagnostics and imaging. Nanomedicine 8:147–166
92. Zhu HJ, Li J, Zhang XB, Ye M, Tan WH (2015) Nucleic acid aptamer-mediated drug delivery for targeted cancer therapy. ChemMedChem 10:39–45
93. Shintu L, Baudoin R, Navratil V, Prot JM, Pontoizeau C, Defernez M, Blaise BJ, Domange C, Pery AR, Toulhoat P et al (2012) Metabolomics-on-a-chip and predictive systems toxicology in microfluidic bioartificial organs. Anal Chem 84:1840–1848
94. Bergstrom G, Christoffersson J, Schwanke K, Zweigerdt R, Mandenius CF (2015) Stem cell derived in vivo-like human cardiac bodies in a microfluidic device for toxicity testing by beating frequency imaging. Lab Chip 15:3242–3249

Chapter 13
Microfluidic Platforms for Microbial

Lin Zhou and Jin-Ming Lin

Abstract The combination of microfluidic chip systems and microbial science is a reciprocal innovation. Microfluidic microchips provide convenient and stable experimental conditions for microbiological research. In return, microbes provide assembly of fixed-point components for microfluidic systems. In the latest twenty years, with the miniaturization of microfluidic chip system and high-throughput optimization, research on the microbial science is becoming deeper, and relatively abundant knowledge has been accumulated. Microbiologists continue to innovate to use this technology as much as possible to explore more results. In this chapter, the application of microfluidics in microbiology will be described in detail from the most common field of life sciences to multidisciplinary crossover applications, such as microbial genetic diagnosis, portable processing device, rare mutant capture, bacterial sensitivity test, bacterial chemotaxis, tracking individual bacteria dynamic changes.

13.1 Introduction

As a special group in nature, microorganisms play an important role in the research of life sciences. From the sedimentary soil of volcanic to the human gastrointestinal tract [1, 2], various spatial structures can be the habitat of microbial populations [3, 4]. Compared with animal cells and plant tissues, microorganisms have a relatively short period of proliferation cycle and simplicity of growth, especially suitable for the operation as genetic tools and large-scale industrial culture. To date, the basic theory of life science and industrial application has shown great potential by using microorganism as the main materials, which rely on genetic engineering

L. Zhou
College of Biological Science and Technology, Beijing Forestry University, Beijing 100083, China

L. Zhou · J.-M. Lin (✉)
Department of Chemistry, Tsinghua University, Beijing 100084, China
e-mail: jmlin@mail.tsinghua.edu.cn

J.-M. Lin (ed.), *Cell Analysis on Microfluidics*, Integrated Analytical Systems,
https://doi.org/10.1007/978-981-10-5394-8_13

technology. Ranging from delivering drugs for cancer treatment to producing biofuels for vehicles and many other aspects have formed a relatively mature research means. With the deep research on the diversity of microbial species, the 3D model have been used to study the complex microbial communities in the environment. At even in smaller spaces, the normal gut microbes play an important role in human digestion, bacteria can be squeezed through the contraction, attached to the gap in the intestinal wall. Yet this ecosystem remains incompletely characterized and its diversity poorly defined. New methods need to be established and used to solve the dilemma of the present study. Research on microbial resources can help to solve the complex environmental problems and social problems in the long term [5], such as the development of green energy and energy [6–8], the production of medical [9] and health care products and the treatment of environmental pollution [10].

Microfluidic technology is a controlled volume of fluid in the micro-flow channel system, which can be sample preparation, reaction, separation, detection and other operations integrated in a small chip to form a network of microchannel [11, 12]. Controllable fluid throughout the system, thereby replacing conventional biochemical laboratory functions. The biggest advantage of multi-cell technology is on a tiny platform and flexible combination of large-scale integration [13–17], a controlled microenvironment can be created to accurately drive and control the progress of the experiment, increasing the efficiency of interaction and detection sensitivity. The combined use of microdevices and microfluidic chips has contributed to the development of single cell analysis [18, 19], gene engineering [20] and proteomics [21–23]. Compared with the traditional experimental methods, it greatly reduces the consumption of biological samples and chemical reagents. Another advantage is that the research object is a single cell, relying on real-time in situ observation and a more in-depth study of the multi-level parallel tests to study the cell behavior and some biochemical processes [11, 24–26]. This research on microbiology can be regarded as a historic progress, it provides a new method and means for the study of microorganisms.

The combination of microfluidic and microbial engineering is of great significance to study microorganisms. The development of microfluidic provides convenience for microbial research. In return, microorganisms as the important components serve the microsystem. The combination is mutually beneficial. With its unique advantages, microfluidic has solved a series of complicated problems in traditional research. Research has been applied in many aspects of microorganisms using microfluidic chip technology, such as microbial genetic diagnosis, portable processing device, rare mutant capture, bacterial sensitivity test [27], bacterial chemotaxis [28], tracking individual bacteria dynamic changes [29].

In this chapter, we will discuss the advantages of microfluidic platforms, both the most basic research methods and the widest range of applications. We hope that such an overview will facilitate more and more microbiologists to find a suitable method for their applications, and promote the further development of microbiology and microfluidic technology. In the next sections we will describe the advantages of

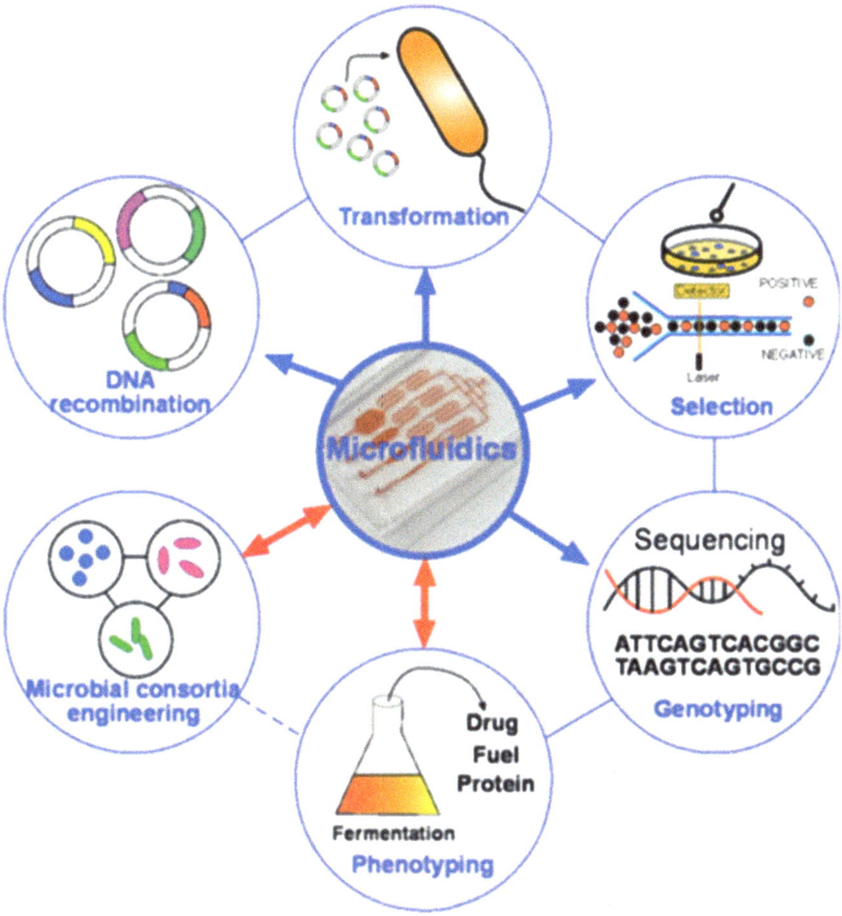

Fig. 13.1 Application of microfluidic in microbiology. (Reprinted with permission from Ref. [31])

micro fluidic technology in microbial research [30, 31], the most basic research methods, examples of successful applications and future development direction (Fig. 13.1).

13.2 Microbial Characteristics

For microorganisms, an essential attribute that distinguishes it from other large organisms is that it can form a unique system with small volume and large area. The formation of five common microorganisms: small volume, (diameter less than 0.1 mm, large area of bacterial community); rapid the absorption of nutrients and metabolism; vigorous growth and rapid propagation (generally under suitable conditions some bacteria can breed 20 min generation); Strong adaptability and easy variation; widely distributed and a lot of species (many kinds of microbial species have been identified, it has reached more than 100 thousand). Same with other organisms, microbes also have precise material metabolism and signal transduction pathways. Due to the special characteristics of the microorganism, and the culture condition is simple, especially suitable for gene engineering operation. It has become one of the most interested research objects in the research of basic theory. The use of microfluidic chip technology for scientific research has created a new method to solve the problems in the past. We will discuss in the following two sections.

13.2.1 Traditional Methods of Analysis and Research

A good microbiological device basic conditions are scientific designed according to the growth of microorganisms, keeping a continuous supply of the required nutrients and the microorganism to obtain suitable temperature and the required ventilation qualification. In addition, it should also be ensure good physical and chemical environment and strict anti-pollution the measures. At present, there are two main methods in the laboratory, which are solid culture and liquid culture. Liquid culture is usually inoculated with ring or needle tools vaccinate the bacteria in liquid medium for shaking culture. The solid medium is usually composed of liquid culture medium adding coagulants. The specific method is generally put the culture medium into the culture dish to cool, then coating or other methods to inoculation culture. Another method is mixed bacteria with the culture medium. Two kinds of training methods are required to operate the vessels, drugs, and operating tools for strict sterilization operation, in the experimental process, but also to ensure the strict aseptic experimental procedures.

Overall, the current research on microbial mainly focused on the screening of microorganisms and optimization of culture conditions, growth and metabolism analysis, genetic transformation, which involves the bacterial chemotaxis, operation resistance and sensitivity test and high-throughput gene engineering. In the traditional experimental methods, the need to manually configure the drug, the concentration of control, drug addition and other micro operations, the ability of the relevant personnel is a great challenge. By using the microfluidic chip technology, the specific design of the chip can solve most of the problems.

13.2.2 New Technologies for the Study of Microorganisms

Now, from the separation of bacteria to packaging and culture have a more mature means, a variety of functions of microfluidic devices have been developed.

Research methods are also diverse, through the development of different materials, special structures and other methods, has been able to use liquid droplets, bubbles, particles and other methods to achieve the encapsulation of bacteria [32–34]. Scientists have also developed PDMS or plastic micro chamber array through the formation of lithography, high throughput chip design and complex device structure are particularly suitable for the tenacious vitality of microorganisms. At the same time, with the combination of nano-particles and micro electro mechanical systems, the research of human beings to the bacteria is deep into the complex micro circuit, electromagnetic, optical, acoustic and other fields [32, 34–39]. Remarkably, not all commercial design chips include microfluidic circuits or a fully simulated laboratory environment. We will focus on some of the characteristics of the chip in the field of microbiology. Through the design of the microbial culture combining with tests on a chip, the automatic detection and analysis of bacteria reached the appropriate concentration, such as the extensive use of fluorescent protein labeling method, can achieve the purpose of real-time tracking detection.

For some specific flux requirements, we can change the shape ratio of device, adjusting the concentration of bacteria to optimize the use of these advantages. Make use of these beneficial condition further researches of cell mechanics, cell migration and subcellular organization behavior. A clear advantage of microfluidic chips in microbial research is the physical isolation, which is entirely important in the study of the isolation of a particular organism from a natural colony. Traditional research methods, there are a majority of micro-organisms cannot be in the general culture dish, even it can grow, in many cases will be because of competition and the advantages of bacteria and other reasons cannot be separated. Nichols et al. have designed and tested an isolation chip composed of several diffusion microchambers, each inoculated with a single cell [40]. A total of three layers, each layer with a porous poly carbonate film separated. The chip is placed in the mixed bacteria, microbes into the micro chamber device, by screw in the chamber, bacteria can rely on membrane permeability with environmental material free exchange, this method can effectively separate some of the traditional methods are difficult to obtain the strain. Scientist show that microbial recovery in the chip exceeds many folds that afforded by standard cultivation, and this device can retain nearly 50% of the species in the sample. Pivetal et al. demonstrates the selective isolation of bacterial cells from a mixture by combining magnetic probe-based cell with the use of microfluidics [41]. While non-target cells pass through the device, the magnetically labeled target cells can be efficiently captured thanks to the high magnetic field gradients generated by the micro-flux sources. QPCR sequencing analysis showed that the initial concentration of target cells from 0.04% to final concentration was 98.46%, with a recovery rate of 50%, with a throughput of 10^9 cells per hour. These methods provide great potential to solve the problem of bacterial diversity exploration in complex ecosystems, in which the labeling strategy is required for the application of non-cultured bacteria.

In the separation of bacteria, the magnetic nanoparticles and external magnetic field can be used to reduce the complex pretreatment process. The microfluidic chip through 3D printing technology to produce a spiral, the mixed solution of composite particles containing magnetic particles and bacteria through the screw channel, the sheath flow drag and lift under the action of bacteria composite particles were isolated by spectroscopy, the concentration of the *E. coli* bacteria was determined using a light absorption spectrometer, and the limit of detection was found to be 10 cfu/mL in buffer solution and 100 cfu/mL in milk [42]. Jung et al. have developed an approach to clearing bacteria and endotoxin from the bloodstream, using magnetic nanoparticles (MNPs) modified with bis-Zn-DPA, combination of gram-positive and gram-negative bacteria for the synthesis of ligands [43]. This device can be used to remove bacteria from patients with sepsis.

Another obvious feature of microfluidic chips is the high throughput test. The simple design of a plurality of groups of parallel experiments are integrated in a chip, and now owing to the development of fluorescent labeling technology and wide application in microbiology, more and more scientists began to research on a microfluidic platform. The idea of high flux chip has been widely used in the research of various types of micro fluidic chip, the simplest layout is to design a large array of chips on a chip. Taninguchi et al. designed 96 channels on a chip, it can simultaneously carry out 96 sets of research to avoid the interference between the bacteria [44]. Zou et al. have developed a polydimethylsiloxane (PDMS)/paper/glass hybrid microfluidic system integrated with aptamer-functionalized graphene oxide nano-biosensors for simple, using fluorescence resonance energy transfer (FRET) for rapid detection of multiplexed pathogen [45]. The chip is designed into 4 groups of reaction channels and detection of micro chamber, the adapter is connected with the fluorescent dye to be modified on the graphene oxide, through the chromatography paper installed in the micro chamber. After the combination of bacterial fluid and the appropriate ligand, the graphite is off to reproduce the original rough fluorescence. The method not only avoids the complex steps such as DNA extraction, but also can complete the high throughput detection in 10 min. Similar to cell research, it is not only possible to combine with fluorescent labeling technology in bacteria, but also to connect a variety of detection devices in parallel. In short, from the bacterial sorting to culture, to the downstream of the whole gene analysis can be achieved through a series of ingenious design.

13.3 Basic Research Methods

In the previous chapters, the design and application of microfluidic have been introduced in detail. In this section, we will focus on some of the points in the practical operation of microbiology. In the study of microbial experiment, according to the experimental requirements and different experiments, the chip model design of the actual operation of the general shape is different, not limited to a single complex design method, usually by a variety of simple model combination, some even adding a lot of other material mosaic. Next we will introduce several simple microfluidic chip culture of bacteria, it mainly included in the channel culture [21, 42, 46–48], micro chamber culture [22, 49, 50] and agar package [27, 51, 52].

13.3.1 Channel Culture

Channel is a relatively simple culture method and can be connected in parallel with a number of analytical devices for real-time monitoring of the reaction. The design of the channel is flexible, and the manufacturing process is simple and convenient, and it is the most widely used method. When culturing bacteria in the channel, the culture medium can be fully contacted with the bacteria, and it is worth to pay attention to control the flow rate at a lower level or at rest, to prevent the bacteria from being washed away. The experiments show that the bacterial culture in the channel, good growth state, such as the static culture of Escherichia coli in microfluidic channel 200 μm in depth, results show that the bacteria in the channel of growth than in shake flask and tablet [53]. Kim et al. propose a simple, novel analytical drug sensitive bacterial biofilm using a microfluidic system for the continuous production of an antibiotic concentration gradient [54]. In the channel, the flow rate of 0.1 L/min was continuously through the culture medium, and the biofilm was formed in the channel after 24 h. The results of minimal biofilm eradication concentration. It was clearly confirmed that the required concentration of Pseudomonas aeruginosa was higher than the minimum inhibitory concentration for the removal of biofilm growth.

13.3.1.1 Diffusion Culture

In the current study, the research of the channel is mainly concentrated in the following categories. The first is to establish a concentration gradient of the channel to study the growth state, chemotaxis and sensitivity of microorganisms. The establishment of a concentration gradient mainly has two kinds of diffusion and liquid flow. Diffusion method requires special treatment of the channel (Fig. 13.2a). The general method is to add a permeable medium in the production process of the chip, like gel and agarose [55]. The agarose layer was inserted between the PDMS and the glass sheets, and the bacteria were embedded in agarose. The culture medium was continuously spread to the agarose through the channel, and the bacteria had been growing for a long time. The effects of Iso-butyl alcohol on growth and morphological changes of bacteria were traced. And the growth curves and the dynamic changes of morphology of bacteria were obtained by bacterial count and length measurement [29]. Three parallel channels were designed in the agarose layer (Fig. 13.2c), and the two side channels were respectively connected to the high concentration solution and the blank solution [56]. The concentration gradient of the intermediate channel was formed by agarose diffusion (Fig. 13.2b). The agarose is heated and dissolved, and the droplets are in the middle of the channel as a semi permeable membrane [57]. The concentration gradient was formed on both sides of the bacterial solution and the target substance. A new technique (Fig. 13.2d) for the in situ formation of the hydrogel with controllable thickness was developed [58].

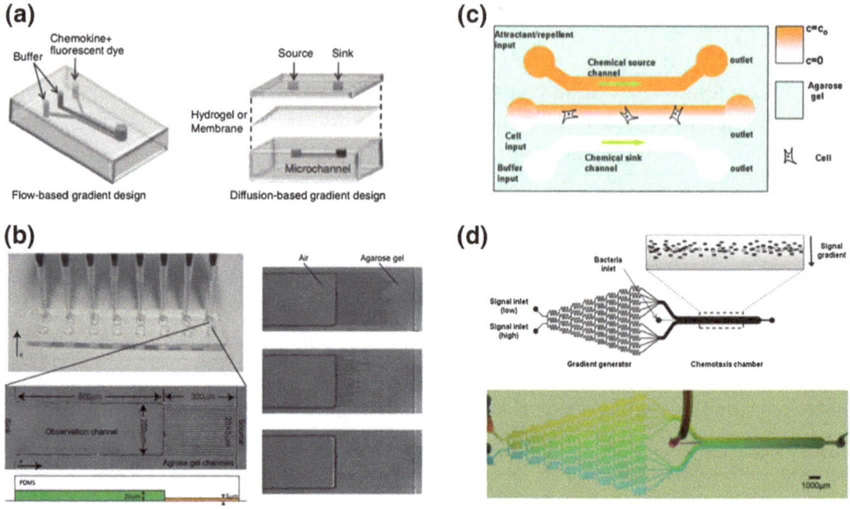

Fig. 13.2 Diffusion culture method. **a** In situ monitoring of antibiotic susceptibility of bacterial biofilms in a microfluidic device (Reprinted with permission from Ref. [55]). **b** A parallel diffusion-based microfluidic device for bacterial chemotaxis analysis (Reprinted with permission from Ref. [56]). **c** A hydrogel-based microfluidic device for the studies of directed cell migration (Reprinted with permission from Ref. [57]). **d** Flow-based microfluidic device for quantifying bacterial chemotaxis in stable, competing gradients (Reprinted with permission from Ref. [59])

The film thickness is controlled by adjusting the reaction time and the speed of the micro fluid. The effectiveness of chemical gradients was confirmed by monitoring the migration of *Salmonella*.

Based on the concentration gradient of the liquid flow, the liquid flow mixing in the microfluidic channel is mainly utilized. The classic design is that the two sampling holes are respectively connected to the high concentration solution and the blank liquid, and the concentration gradient is formed through the layer by layer streaming into the same channel. This design was used to investigate the chemotaxis of *Escherichia coli*, and to analyze the factors of colonization of intestinal bacteria [59]. In addition, there is injection pump speed control or design of multiple water inlet to form a concentration gradient.

13.3.1.2 Liquid Flow Culture

The second kind is to construct a simple parallel multichannel for high flux experiment. In general, this method is used to study the behavior characteristics of cells efficiently. The optimal culture conditions of the bacteria were cultured in the chip channel, and a variety of culture conditions were realized on one chip, and the optimal pH, the concentration of the culture medium and the temperature of the bacteria were obtained [60]. 96 channels are integrated on a single chip to achieve a single cell level detection of E.coli protein and changes in the copy number of messenger RNA [45]. Through the design of trapezoidal channel, the single bacteria control in the ladder structure, on both sides of the culture medium used to study the growth kinetics of bacteria [61].

13.3.2 Chamber Culture

The design of the micro chamber can provide a stable environment for the growth of bacteria. There are a lot of similarities with the channel culture, the micro chamber is connected to the main channel through a very narrow passage, the target or medium through the main channel into the chamber continuously [62]. The initial concentration of the bacteria in the micro chamber was changed (Fig. 13.3c), the bacteria could be found in the micro chamber, and the antibiotic sensitivity test was carried out [63]. Luo et al. present a simple and high-throughput microfluidic system (Fig. 13.3a) for diffusion based monolayer cell culture monitoring [64]. Yeast cell monolayer growth by controlling the height of the micro chamber. The cell is patterned into a suitable high-level microcavity array, which is fixed in the membrane. Different sizes of holes and different times of repeated injections were tested in order to achieve as many single cells/holes as possible. Single-cell/cavity has been achieved in about 40% of 100 parallel cavities. Design of different shapes of the micro chamber (Fig. 13.3b), can only contain a bacterial cell to explore the effect of protein metabolism on cell shape [65]. The design of the micro chamber into a maze topology is used to simulate the quorum sensing of bacteria in complex environments.

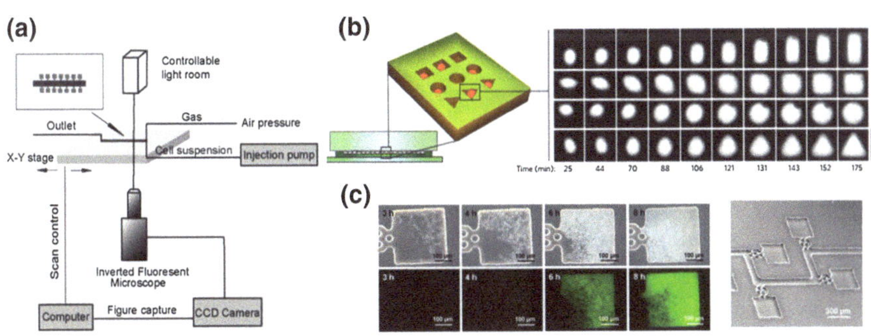

Fig. 13.3 Chamber culture method. **a** High-through put microfluidic system for monitoring diffusion-based monolayer yeast cell culture over long time periods (Reprinted with permission from Ref. [64]). **b** Symmetry and scale orient Min protein patterns in shaped bacterial sculptures (Reprinted with permission from Ref. [65]). **c** High-throughput microfluidic system for long-term bacterial colony monitoring and antibiotic testing in zero-flow environments (Reprinted with permission from Ref. [63])

13.3.3 Agar Package

The bacteria in the channel or in the micro chamber are in a free state. Although the precision of the design can be fixed, the individual bacteria is small and difficult to operate, thus making the manufacturing process of the chip complicated. These problems can be effectively solved by the method of gel package or diffusion. General gels were prepared by agarose and alginate two methods, the alginate gel loose structure of microorganism and easy leakage, so the agarose gel become the main research methods. About the preparation of the gel, please refer to the previous chapters. It is worth noting that the microbial growth and reproduction speed, individual smaller than animal cells ten times, so when preparing bacteria to adjust the concentration according to the need to ensure the complete package of individual bacterial cells. In addition to the gel package, some laboratories have also developed some new methods. Gel has become more and more common, especially mentioned in the passage to the study of the operation, to provide a new way for the study of microbiology.

The above research shows that using the microfluidic chip to study bacteria not only avoids the complex steps in the traditional experiment, but also greatly improves the efficiency and accuracy of the experiment.

13.4 Single Cell Analysis

All along, the human study of microorganisms, are based on the function and nature of groups. The traditional research methods are followed by a central idea, that is, from the nature of the separation of pure bred in the artificial environment, the study of genetic characteristics, macro culture parameters, physical and chemical characteristics. However, an unavoidable fact is that these studies are based on the assumption that the microbial community is an ideal model. In fact, bacterial independence and individual differences are common even in populations of the same genotype.

The most characteristic of micro fluidic chip is to reduce the research object from a population to a single cell level. The use of micro fluidic chip technology to control microorganisms to study individual differences, physical and chemical interactions between the individual has set off a revolution in microbial research. In this regard, scientists have made a lot of achievements [66]. Moffitt et al. microfluidic dissection platform features an optically prealigned single focal plane and an integrated array of soft elastomer-based micropads [51]. Integrated arrays are used together to allow for mother cell capture, daughter cell removal, monitoring of aging, and an unprecedented microscopic imaging of the entire aging process. Lee et al. follow the growth dynamics of individual cells with minute-resolution using a single-cell assay in a microfluidic device during this shift (Fig. 13.4a), while also monitoring lac expression [67]. Gel coating is widely used in the research of single cell, Escherichia coli wrapped in gel ball diameter of 30 μm, using fluorescence activated cell separation method (FACS) can realize the separation of cells, this technique has important significance for the study of cell metabolism [68]. Probst et al. design a chip (Fig. 13.4b), single bacteria was inoculated into the dot matrix laser tweezers chip, chip array is designed on the micron size [69]. Single cell elongation, morphological changes and a large number of amplification processes can be observed. Single cell control in the micro level of the trap of the new cells will be quickly swept away,-growth rates and speciation analysis can be achieved. The design of the channel was radioactive (Fig. 13.4d), single bacteria were encapsulated in the central, where single bacteria are enclosed in the center. Single bacterial movement and responses to different concentrations of antibiotics can be observed [27]. The sorting device and the chip are combined, and the liquid drop of the yeast mutant is directed to collect, and the reaction efficiency is improved by 1000 times.

The production of single cell channels by agarose, creating a chemostat culture, because of the characteristics of soft agarose, the different forms of single bacteria clip in the channel, so the growth can also be observed in many bacteria [51]. Single bacterial growth in the channel can be calculated by the elongation of individual bacteria in the growth process. Different shapes of nano structures were designed, and the single bacterial control was observed in the micro chamber [69]. The distribution of the protein in the cells changed obviously, and the bacteria showed the adaptability to the environment [64]. And this is similar to the agarose channel design into a different curve, the study of *Escherichia coli* single cells in different environmental resistance movement and morphological changes [70]. Combining

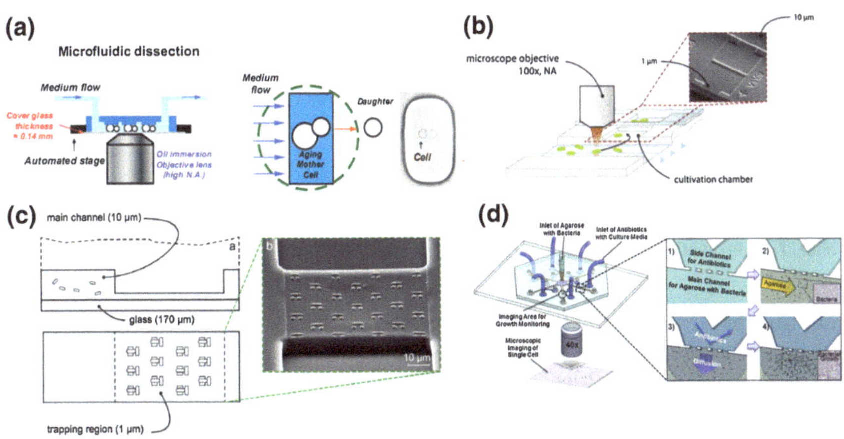

Fig. 13.4 **a** Whole lifespan microscopic observation of budding yeast aging through a microfluidic dissection platform (Reprinted with permission from Ref. [66]). **b** Microfluidic growth chambers with optical tweezers for full spatial single-cell control and analysis of evolving microbes (Reprinted with permission from Ref. [69]). **c** Polydimethylsiloxane (PDMS) sub-micron traps for single-cell analysis of bacteria (Reprinted with permission from Ref. [124]). **d** Rapid antibiotic susceptibility testing by tracking single cell growth in a microfluidic agarose channel system (Reprinted with permission from Ref. [27])

chip with microelectrode and optical detection, using electrophoresis capture *Escherichia coli* single cell, the study of another innovation is that the same channels alternately pass into the culture medium and antibiotics to screen single bacterial resistance through the optical detection and tracking the growth state of single cells [71].

Through these studies, we can see that the analysis of microbial diversity and personalized depth, will meet the needs of various fields of science in the future. It might help our understanding of the interplay between the extracellular environment and cellular processes in natural ecosystems, mixed cultures and isogenic populations in technical ecosystems (reactors), as used in environmental applications, medical research and industrial processes. It will also be the most important and widely applicable method for the analysis of the single cell parameters of the basic microorganisms. This is particularly applicable to the quantification of intracellular biosynthesis, material absorption, and metabolite excretion. Even the most basic cell properties, like the specific growth rate of a cell, usually described biomass or cell volume increase, there is not a universally applicable research method. These parameters are the key to describe a single cell system. They will ultimately determine the role of individual cells in the process of microbial populations and even in the course of natural ecosystems.

13.5 Applied Technology

In the field of microbial research, microfluidic control technology is not only limited to the field of life science research, it has also penetrated into the pharmaceutical production, environmental monitoring, food safety, and many other aspects. In the following sections we will list some of the applications of microfluidic chips in daily lives and production.

13.5.1 Basic Sciences

Microfluidic technology has opened up a new world for the application of microbiology. It completely liberated the limitation of the traditional method in the research object, simplified the complex research method, and took the research step further. Today, microfluidic chip technology has been able to replace almost all of the basic experimental microbial operation, and even in the high-throughput study of gene flow can also show its unique advantages. From the isolation and cultivation of fragile strains of environmental requirements, the study of the behavior of single bacteria, the introduction of PCR technology, the micro world of microorganisms is more and more clearly present in the human eyes. Many of the applications and research are based on the basic research as the main body. In this we want to focus on the study of the relative maturity of bacteria chemotaxis research. Bacteria in the water soluble medium to survive, the migration is vulnerable to the impact of fluid force. *Escherichia coli* and *Salmonella* have been used as the model system for the study of bacterial chemotaxis [47, 72–75]. The linear and nonlinear fluid gradient was established by using the microfluidic chip, which was used to study the chemotaxis of bacteria [28]. The results were consistent with the conclusion of the mathematical model. A one dimensional fluid model was simulated on the chip to simulate the human gastrointestinal tract environment [59]. The results showed that the induction of AI-2 in the group of induction molecules preferentially induced the chemotaxis of E. coli, which is a significant signal of bacterial colonization in the gastrointestinal tract. Kalinin uses a three channel microarray to study the sensing properties of Escherichia coli under the logarithm concentration gradient [76]. Besides *E. coli* and *Salmonella*, the study of other bacteria such as *Pseudomonas aeruginosa* has become more widely (Fig. 13.5b) [77, 78]. In addition to chemotaxis research, the study of the physiological and biochemical characteristics of the bacterial base is also more comprehensive. With the use of complex design and high technology, the chip is also becoming more and more automated. A pressure driven valve is added to the chip to study the deformation of the bacteria in the face to environmental stress (Fig. 13.5a), and the use of high resolution microscopy to observe the single cell and subcellular components [79].

Kalsi et al. integrated 16,800 electrodes on a single chip, the chip can be carried out independently of the simultaneous operation of multiple single droplets [80]. Impedance sensor is arranged on the chip can detect the droplet size and position. The chip also integrates a thermistor and a heater for real-time detection and control of temperature, the researchers use this chip was successfully amplified and DNA gene analysis. Czilwik et al. developed the labdisk chip system (Fig. 13.5c), the extraction of DNA, the whole PCR amplification and specific strains of real-time PCR integration on a chip [81]. Only about 4 h to complete the rapid detection of bacterial pathogens, almost no manual operation and complex laboratory equipment

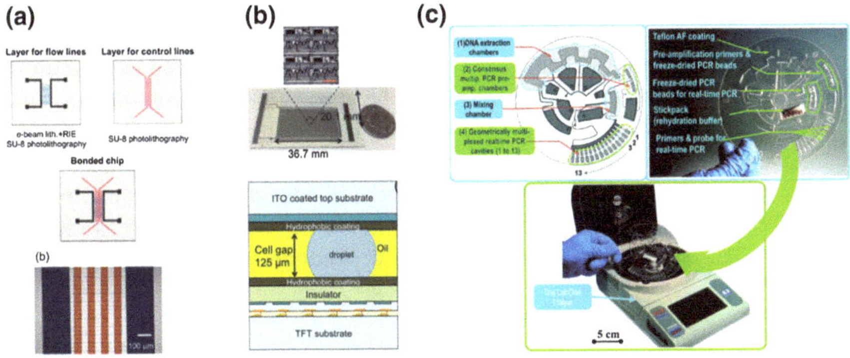

Fig. 13.5 **a** Characterization of small microfluidic valves for studies of mechanical properties of bacteria (Reprinted with permission from Ref. [79]). **b** Rapid and sensitive detection of antibiotic resistance on a programmable digital microfluidic platform (Reprinted with permission from Ref. [80]). **c** Rapid and fully automated bacterial pathogen detection on a centrifugal-microfluidic labdisk using highly sensitive nested PCR with integrated sample preparation (Reprinted with permission from Ref. [81])

can be molecular diagnostics. Mohan et al. design multiple parallel sampling channels, two different bacteria can be passed, respectively [82]. This device was used to study the interaction between different strains and their sensitivity to antibiotics in the environment of interaction. In another study, a pair of static arrays is designed, the single cell of *Escherichia coli* was captured and the concentration of tryptophan in the cells was analyzed by fluorescence protein [83]. It was proved that the intracellular tryptophan concentration could be controlled by the regulation of the RNA and the high yield strains were screened out.

13.5.2 Resistance Detection

The abuse of antibiotics has become a major problem in the field of public health and drug sensitivity test of the slow detection rate also caused the treatment is not timely caused the unnecessary death.

In view of this series of questions, the researcher introduces the microfluidic chip technology in the detection of drug resistance and drug sensitivity test in clinical medicine, which greatly improves the detection efficiency. Because the previous studies have shown that bacteria in the microfluidic chip growth state is suitable for bacterial resistance test. Studies show that the method can greatly shorten the test time, and quickly get the minimum inhibitory concentration (MIC) value, so it plays a significant role in clinical treatment. Mohan et al. designed a ladder shaped channel, by testing the expression of fluorescent protein to carry out the antibiotic sensitivity test [84]. The experiment can be detected only 4 h, in addition, the experiment showed that not a variety of antibiotics used in conjunction with the best results. Mixed bacteria and different concentrations of antibiotics were added to the radial channels, minimal inhibitory concentrations of antibiotics were available only within 3–4 h. This method can achieve the same effect as the standard method, the time is more than 5 times shorter than the standard method [58]. Tang et al. also cultured bacteria in microfluidic channels, but the sensor was used to test the growth of bacteria in the channel [85]. The minimal inhibitory concentration of antibiotics was also obtained in 2 h. Encapsulated bacteria on a chip, the response of cells in the gel to antibiotics was observed by using time delayed imaging technique [86]. Finally, the growth curve and pharmacodynamics curve were obtained by software process. The design of microfluidic chip self-priming, rely on the vacuum, bacteria liquid flows into the small holes of different concentrations of antibiotics. To determine the growth of strains depending on the color changes [87].

13.5.3 Toxicity Test

With the development of microfluidic chip technology, biosensors made by this method have also been developed, which are widely used in environmental monitoring and detection of toxic substances. Generally, the structure of electrochemical instrument is complex, detection of heavy metal ions (HMIs) using conventional methods needs skilled operation and methods requiring expensive instruments such as atomic absorption spectroscopy, Such as atomic absorption spectrometry and coupled plasma atomic emission spectrometry/mass spectrometry (Fig. 13.6c) [88–90]. The biological sensor based on micro fluidic technology shows great advantage. Long et al. developed a portable microfluidic device of dimethyl sulfoxide and ursodeoxycholic acid osmium bezoar [61]. The chip is designed to be in the form of micro pits, controlled by a characteristic magnetic device. First, they were transformed into bacteria that can express the ATP dependent luciferase by genetic engineering technique, and the transformation of D-fluorescein dependent luciferase requires ATP energy supply. So the growth state and metabolism of bacteria can be detected through the light reaction. The experiments show that the diluted DMSO and TCDCA can be effectively detected. It is a kind of widely used sensor model is the gradient dilution model. Roda et al. use of biocompatible charged nanoparticles in the treatment of yeast cells, the yeast cells with fluorescent labeling are arranged in the chamber of the downstream of the chip (Fig. 13.6b) [91]. The toxicity of methyl methanesulfonate (MMS) on yeast cells was detected by the detection of the fluorescence expression of GFP. On this basis, researchers constantly optimize the device to develop a rapid detection of heavy metal ions (arsenic and cadmium) microbial sensor (Fig. 13.6c) [92]. In the detection of arsenic, more chips are designed to detect. The labeled *Escherichia coli* was encapsulated into microspheres by agarose, these microspheres are encapsulated in a 500 * 500 micron cage (Fig. 13.6a) [93]. By detecting the expression of EGFP can detect arsenic levels in drinking water; the accuracy of this chip can reach 10 micrograms per liter. And the chip performance is stable, which can be

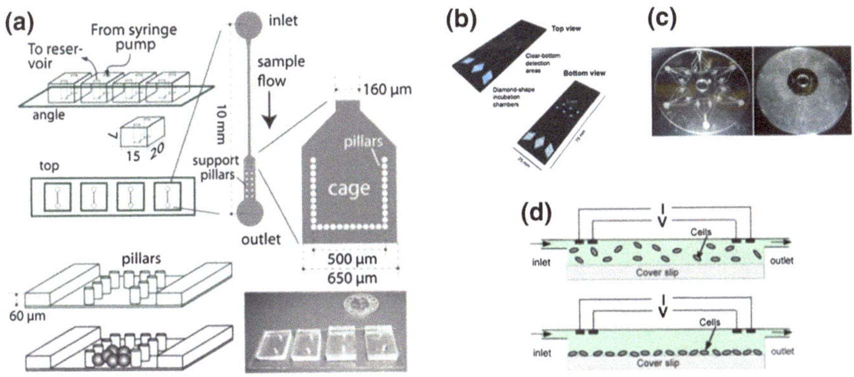

Fig. 13.6 **a** Development of a microfluidics biosensor for agarose-bead immobilized Escherichia coli bioreporter cells for arsenite detection in aqueous samples (Reprinted with permission from Ref. [93]). **b** Bioengineered bioluminescent magnetotactic bacteria as a powerful tool for chip-based whole-cell biosensors (Reprinted with permission from Ref. [91]). **c** Integration of spore-based genetically engineered whole-cell sensing systems into portable centrifugal microfluidic platforms (Reprinted with permission from Ref. [92]). **d** On-chip microfluidic biosensor for bacterial detection and identification (Reprinted with permission from Ref. [94])

stored in the −20° of the refrigerator for a month, especially suitable for the actual production and life applications.

It is not only the detection of toxic substances, the use of micro flow control chip for the same can quickly detect harmful microorganisms identification. Boehm et al. developed a method to detect bacteria by using electrical impedance, by immobilized monoclonal antibodies on the chip to identify bacteria, the concentration of 10^4 cfu/ml in the water of E. coli can be detected (Fig. 13.6d) [94]. Sakamoto et al. developed a kind of micro fluidic chip which can detect the number of bacteria in drinking water quickly and quantitatively [95]. After the samples were stained with fluorescence, the chip can get the number of bacteria in the 15 min.

13.5.4 Cancer Surveillance and Treatment

Small malignant lesions are found to be effective in the treatment of cancer and greatly reduce mortality. Like secondary tumors, it is often found that until severe pathological symptoms or significant changes in the body's organs. With the development of modern medicine, increasingly advanced imaging technology, such as positron emission tomography (PET), nuclear magnetic resonance imaging (MRI), electron computed tomography (CT) can only identify a relatively large number of tumor tissues, and its ability to identify small lesions is limited. In recent years, bacteria as a target to detect lesions were developed. Bacteria have great potential in cancer treatment, and there is increasing evidence that certain bacterial species preferentially replicate and accumulate in tumor cells. For example, *Salmonella* has a specific tropism for the malignant organization (Fig. 13.7b) [96]. The expression of fluorescent protein was detected by gene engineering modification of Salmonella attenuated strain, and the protein was detected in the microfluidic chip to detect the tumor tissue. This strategy was able to find a 0.12 mm tumor with an accuracy of more than a scanning electron microscope.

Based on the characteristics of bacteria, in addition to the detection of tumor, but also can be used for tumor targeted therapy. Park et al. developed a bacterial micro robot, the robot was attached to streptavidin–perCP–Cy5.5 coated polystyrene microbeads via biotinylated outer membrane proteins (Fig. 13.7a). The researchers conducted a migration experiment on a microfluidic chip [97]. The results show that this device is rapidly moving to the tumor cell lysates or spheroids. This chemotaxis can be used for targeted drug delivery, and provides a way for cancer diagnostics/therapeutics with emphasis on targeting and treating tumor mass quantification.

13.5.5 Microfluidic for Fuel Cell

Microbial fuel cell is a small neutral electrode device, which uses microbial degradation of organic matter and waste water and other substances to renewable

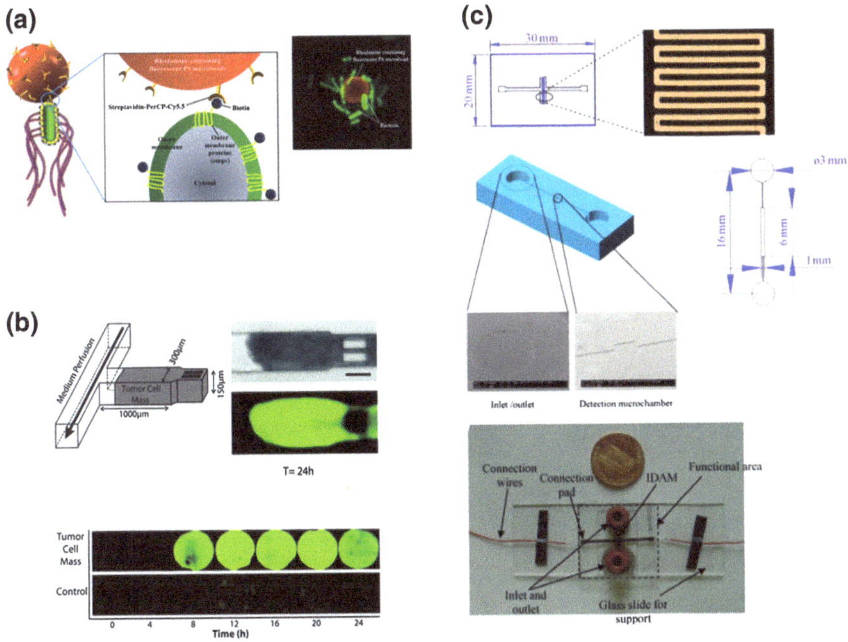

Fig. 13.7 **a** New paradigm for tumor theranostic methodology using bacteria-based microrobot (Reprinted with permission from Ref. [97]). **b** Genetically modified bacteria as a tool to detect microscopic solid tumor masses with triggered release of a recombinant biomarker (Reprinted with permission from Ref. [96]). **c** A label-free, microfluidics and interdigitated array microelectrode-based impedance biosensor in combination with nanoparticles immunoseparation for detection of *Escherichia coli* O157: H7 in food samples (Reprinted with permission from Ref. [111])

energy. The bacteria are inoculated into the anode compartment. The catalytic species produce electrons, protons and other metabolites. In the reduction reaction, protons diffuse into the cathode chamber and electrons flow through an external circuit. Traditional microbial fuel cell has a great disadvantage of the low voltage generated by the need to consume high cost. The main reason for this disadvantage is the limited number of bacteria that can be colonized on a small size electrode, resulting in an excessively high internal resistance. The use of microfluidic chip laminar flow technology can be a good solution to this problem [98–100].

Choi et al. designed a chip model, the height of micro-chamber were limited to 20, 55, 155 μm (Fig. 13.8a) [101]. The results showed that after inoculation *Geobacter sulfurreducens* the current and power are limited when micro-chamber height is less than 55 microns. The maximum power generation capacity of 95 $\mu w/cm^2$, which the highest level in the field of microbial fuel cells has reached, even comparable to the traditional power generation capacity of the device. Hou et al. design of the air as the cathode chip, 24 independent micro-chambers can be integrated in a chip on a parallel test and screening high-yield microbial strains (Fig. 13.8b) [102]. The anode and cathode were designed to have a capacity of

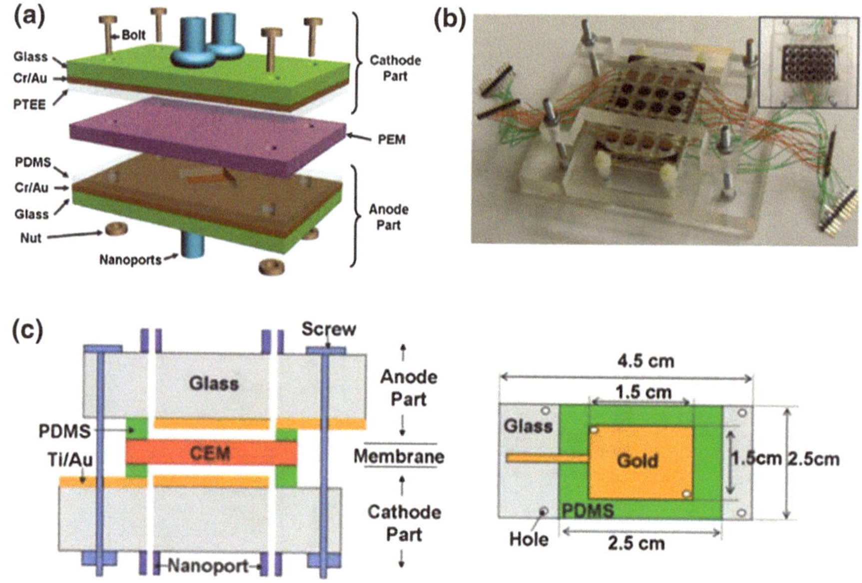

Fig. 13.8 **a** A microfluidic microbial fuel cell array that supports long-term multiplexed analyses of electricigens (Reprinted with permission from Ref. [101]). **b** Optimal biofilm formation and power generation in a micro-sized microbial fuel cell (MFC) (Reprinted with permission from Ref. [102]). **c** Air-cathode microbial fuel cell array: a device for identifying and characterizing electrochemically active microbes (Reprinted with permission from Ref. [103])

4.5 μl, efficient enrichment culture strain of anode chamber (Fig. 13.8c) [103]. To minimize the effect of oxygen intrusion into the anode chamber, the anode chamber specific surface area leading to lower internal resistance, which greatly enhance the production capacity.

13.5.6 Food-Brone Bacteria Detection

Microfluidic chip technology can also be used in food-borne bacteria detection and quantification and so on. Conventional detection methods are time-consuming and laborious, and require intensive and complicated operations from cell enrichment to isolation and identification. Microfluidic chip detection technology has attracted wide attention in the industry because of its fast, sensitive and low cost [104, 105]. The application of nano materials and magnetic particles further speed up the recognition and signal transduction process of microfluidic chip for detection of pathogens. In recent years, more and more researchers are devoted to the optimization of this technology, develop different identification principle and enlarge the signal method, greatly improve the detection speed and sensitivity [106–109].

Fluorescence-labeled antibodies were used to bind to bacteria. Droplets were formed by shunting, and the fluorescence of the droplets was detected to determine the content of bacteria in the food [110]. A microelectrode system was embedded in the microfluidic chip. The magnetic nanoparticles were combined with the impedance biosensor technology (Fig. 13.7c) [111]. The sample was injected into the chip after a simple pretreatment. The impedance sensor could be used to detect *E. coli* in beef and mutton samples effectively. The chip was designed as eight nanotube channels. Channel surface is covered with a layer of antibodies against Staphylococcus aureus enterotoxin B, enzyme-linked immunosorbent assay (ELISA) was used to determine the levels of enterotoxins in food samples. This method improves the detection sensitivity of 6 times [112].

In addition, the use of microbial properties and metabolic processes to solve a variety of social problems is necessary, such as environmental remediation, sewage treatment, regulating gastrointestinal function and so on. Microfluidics plays an important role in the study of ecological diversity in addressing microbes and their potential synergistic effects. So far, this type of research is still in its infancy.

13.6 Opportunities and Challenges

With the combination of many technologies, the history of microfluidic technology is not long but the rapid development and broad prospects.

Microfluidic technology provides a complete quantitative system for microbial research, which not only solves a series of problems such as microbial growth, chemotaxis and cell cycle research, but also develops some very valuable achievement in scientific research. And simulate the real environment to explore the parameter for microbe under ambient pressure, population dynamics and genetic evolution of the phenomenon. In this paper and many other studies of scientists, we can see that microfluidic technology has a lot of mature models and common methods, such as micro-chamber arrays, linear channels and gel encapsulation, etc. Wait. This emerging technology is showing great potential to explore the complex microbial world.

Since then, scientists have developed more and more complex structure with a variety of functions of microfluidic devices, which have outstanding performance in exploring the physical properties of bacteria. The systematic study of the behavior of microorganisms is more mature, the time and space of the drug control gene expression, the group induced and biofilm formation can be controlled accurately [113, 114].

Despite many advantages and reliable techniques, this study still has many imperfections in practice, especially to individual smaller microbe cells. For example, the current use of micro flow control technology for cell sorting over reliance on fluorescence technology, is only applicable to physical separation of cells, new methods need to be developed according to cell size, electrode, hydrophilic ability, etc. At the same time, the universal applicability of microfluidic chip technology has been queried, and the scientific research is facing new challenges.

The integration of a new technology into the microbiology experiment is the need to have two advantages, first is there are far superior to the daily method, the second is whether there are cheap costs and labor consumption. As integration of single cell sequencing technology to micro-chip is an ideal model, is still unable to meet the chip manufacturing use the existing technology, but widely adopted flow cytometry technology has been able to encapsulate and cell classification. In the field of microbial analysis of single cells to go a long way, for the study of cells in non-equilibrium state, like transmembrane transport, metabolic kinetics of intracellular research has established a few methods, but is still an extension of the classical research methods. The development of new technologies and methods has become the focus of the research of microorganisms using microfluidic technologies.

The rise of nanotechnology has greatly stimulated the development of cheap materials and the emerging manufacturing technology, the use of nanotechnology to produce microbial fuel cells can reduce the impact of internal resistance, with the matching imaging technology has been able to achieve subcellular resolution levels [115, 116]. At current level, the integration of pretreatment and chip samples may be somewhat complicated, but the use of this technology to obtain high resolution information of bacteria can basically achieved, surface chemical structure on bacteria and some other phenotypes can be easily obtained in the future. Another emerging trend is to establish a microbial community for the study of microbial social behavior. By tracking the survival of a large number of single cells in a complex environment, assessment of the interaction between microorganisms and the environment can predict the development trend of the corresponding ecological environment, and clarify the social behavior of microorganisms [51, 116–118]. In addition, to improve the flux is also the future direction of development, the high flux of the electrode printing technology, paper microfluidic technology, 3D printing is also beginning to be applied in the manufacture of the chip. Using 3D printing technology, *Pseudomonas aeruginosa* and *Staphylococcus aureus* were encapsulated in the gel, which was defined in a specific space and maintained the normal exchange of substances [119–123]. The integration of new technologies will become the focus of future research. More scientists are committed to the microfluidic chip in series with other detection analysis technology, to achieve the integration of the whole process from sample preparation to the results analysis.

The use of microfluidic technology to study bacteria has been nearly 10 years of histories, the development of this powerful tool to promote the exploration of human life science. However, many problems remain unresolved, and sparking by microbiology and microfluidic collisions has led researchers to explore a new world, and the integration of different disciplines will also promote a deeper understanding of microbes.

References

1. Vos M, Wolf AB, Jennings SJ, Kowalchuk GA (2013) Micro-scale determinants of bacterial diversity in soil. FEMS Microbiol Rev 37(6):936–954. doi:10.1111/1574-6976.12023
2. Raynaud X, Nunan N (2014) Spatial ecology of bacteria at the microscale in soil. PLoS ONE 9(1):e87217. doi:10.1371/journal.pone.0087217
3. Li Y, Liu B, Zhang X, Gao M, Wang J (2015) Effects of Cu exposure on enzyme activities and selection for microbial tolerances during swine-manure composting. J Hazard Mater 283:512–518. doi:10.1016/j.jhazmat.2014.09.061
4. Kamada N, Núñez G (2014) Regulation of the immune system by the resident intestinal bacteria. Gastroenterology 146(6):1477–1488. doi:10.1053/j.gastro.2014.01.060
5. Srivastava S, Agrawal SB, Mondal MK (2015) A review on progress of heavy metal removal using adsorbents of microbial and plant origin. Environ Sci Pollut R 22(20):15386–15415. doi:10.1007/s11356-015-5278-9
6. Liang JH, Gao ZL, Yu JS (2010) A review of the substrates used in microbial fuel cells for sustainable energy production. Amino Acids & Biotic Resour 32(3):20–25. doi:10.1016/j.biortech.2009.10.017
7. Pant D, Bogaert GV, Diels L, Diels L (2010) A review of the substrates used in microbial fuel cells (MFCs) for sustainable energy production. Bioresour Technol 101(6):1533–1543. doi:10.1016/j.biortech.2009.10.017
8. Rabaey K, Verstraete W (2005) Microbial fuel cells: novel biotechnology for energy generation. Trends Biothchnol 23(6):291–298. doi:10.1016/j.tibtech.2005.04.008
9. Ross CL, Hansel TT (2014) New drug therapies for COPD. Clin Chest Med 35(1):219–239. doi:10.1016/j.ccm.2013.10.003
10. Ribes M, Coma R, Gili J (2015) A Review on synthesis, characterization and applications of nano-Zero Valent Iron (nZVI) for environmental remediation. Dev Comp Immunol 35(3):360–365. doi:10.1080/10643389.2015.1103832
11. Mcdonald JC, Duffy DC, Anderson JR, Chiu DT (2000) Review—fabrication of microfluidic systems in poly (dimethylsiloxane). Electrophoresis 21(1):27–40. doi:10.1002/(SICI)1522-2683(20000101)21:1<27:AID-ELPS27>3.0.CO;2-C
12. Abgrall P, Gué A (2007) Lab-on-chip technologies: making a microfluidic network and coupling it into a complete microsystem—a review. J Micromech Microeng 17(5):R15–R49 (35). doi:10.1088/0960-1317/17/5/R01
13. Fox MB, Esveld DC, Valero A, Luttge R, Mastwijk HC, Bartels PV, Berg AVD, Boom RM (2006) Electroporation of cells in microfluidic devices: a review. Analytical & BioAnal Chem 385(3):474–485. doi:10.1007/s00216-006-0327-3
14. De JJ, Lammertink RG, Wessling M (2006) Membranes and microfluidics: a review. Lab Chip 6(9):1125–1139. doi:10.1039/b603275c
15. Kjeang E, Djilali N, Sinton D (2009) Microfluidic fuel cells: a review. J Power Sour 186 (2):353–369. doi:10.1016/j.jpowsour.2008.10.011
16. Lee CY, Chang CL, Wang YN, Fu LM (2010) Microfluidic mixing: a review. Int J Mol Sci 12(5):3263–3287. doi:10.3390/ijms12053263
17. Kuswandi B, Nuriman Huskens J, Verboom W (2007) Optical sensing systems for microfluidic devices: A review. Anal Chim Acta 601(2):141–155. doi:10.1016/j.aca.2007.08.046
18. Tzu-Chiao Chao AR (2008) Microfluidic single-cell analysis of intracellular compounds. J R Soc Interface 5(Suppl 2):S139–S150. doi:10.1098/rsif.2008.0233.focus
19. Emma E, Kristin S, Fredrik L, Martin S, Jan S, Dag H, Mattias G, Annette G (2009) A microfluidic device for reversible environmental changes around single cells using optical tweezers for cell selection and positioning. Lab Chip 10(5):617–625. doi:10.1039/b913587a
20. Benedetto A, Accetta G, Fujita Y, Charras G (2014) Spatiotemporal control of gene expression using microfluidics. Lab Chip 14(7):1336–1347. doi:10.1039/c3lc51281a

21. Chiu DT, Jeon NL, Huang S, Kane RS, Wargo CJ, Choi IS, Ingber DE, Whitesides GM (2000) Patterned deposition of cells and proteins onto surfaces by using three-dimensional microfluidic systems. P Natl Acad Sci USA 97(6):2408–2413. doi:10.1073/pnas.040562297
22. Huber DL, Manginell RP, Samara MA, Kim BI, Bunker BC (2003) Programmed adsorption and release of proteins in a microfluidic device. Science 301 (5631):352–354. Accessed 4 June 2003
23. Zheng B, Roach LS, Ismagilov RF (2003) Screening of protein crystallization conditions on a microfluidic chip using nanoliter-size droplets. J Am Chem Soc 125(37):11170–11171. doi:10.1021/ja037166v
24. Nilsson J, Evander M, Hammarström B, Laurell T (2009) Review of cell and particle trapping in microfluidic systems. Anal Chim Acta 649(2):141–157. doi:10.1016/j.aca.2009.07.017
25. Khandurina J, Guttman A (2002) Review: bioanalysis in microfluidic devices. J Chromatogr A 943(2):159–183. doi:10.1016/S0021-9673(01)01451-0
26. Sun T, Morgan H (2010) Single-cell microfluidic impedance cytometry: a review. Microfluid Nanofluid 8(4):423–443. doi:10.1007/s10404-010-0580-9
27. Choi J, Jung YG, Kim J, Kim S, Jung Y, Na H, Kwon S (2013) Rapid antibiotic susceptibility testing by tracking single cell growth in a microfluidic agarose channel system. Lab Chip 13(2):280–287. doi:10.1039/c2lc41055a
28. Ahmed T, Shimizu TS, Stocker R (2010) Bacterial chemotaxis in linear and nonlinear steady microfluidic gradients. Nano Lett 10(9):3379–3385. doi:10.1021/nl101204e
29. Wong I, Atsumi S, Huang WC, Wu TY, Hanai T, Lam ML, Tang P, Yang J, Liao JC, Ho CM (2010) An agar gel membrane-PDMS hybrid microfluidic device for long term single cell dynamic study. Lab Chip 10(20):2710–2719. doi:10.1039/c004719h
30. Sackmann EK, Fulton AL, Beebe DJ (2014) The present and future role of microfluidics in biomedical research. Nature 507(7491):181–189. doi:10.1038/nature13118
31. Kou S, Cheng D, Sun F, Hsing IM (2016) Microfluidics and microbial engineering. Lab Chip 16(3):432–446. doi:10.1039/C5LC01039J
32. Whitesides GM (2006) The origins and the future of microfluidics. Nature 442:368–373. Nature 442(7101):368–373. doi:10.1038/nature05058
33. Eun YJ, Utada AS, Copeland MF, Takeuchi S, Weibel DB (2010) Encapsulating bacteria in agarose microparticles using microfluidics for high-throughput cell analysis and isolation—ACS Chem Biol 6:260–266. doi:10.1021/cb100336p
34. Leung K, Hansen CL (2012) A programmable droplet-based microfluidic device applied to multiparameter analysis of single microbes and microbial communities. P Natl Acad Sci Usa 109(20):7665–7670. doi:10.1073/pnas.1106752109
35. Thorsen T, Maerkl SJ, Quake SR (2002) Microfluidic large-scale integration. Science 298 (5593):580–584
36. Balagaddé FK, You L, Hansen CL, Arnold FH, Quake SR (2005) Long-term monitoring of bacteria undergoing programmed population control in a microchemostat. Science 309 (309):137–140
37. Kamei K, Guo S, Yu ZT, Takahashi H, Gschweng E, Suh C, Wang X, Tang J, Mclaughlin J, Witte ON (2009) An integrated microfluidic culture device for quantitative analysis of human embryonic stem cells. Lab Chip 9(4):555–563. doi:10.1039/b809105f
38. Malvankar NS, Vargas M, Nevin KP, Franks AE, Leang C, Kim BC, Inoue K, Mester T, Covalla SF, Johnson JP (2011) Tunable metallic-like conductivity in microbial nanowire networks. Nat Nanotechnol 6(9):573–579. doi:10.1038/NNANO.2011.119
39. Hol FJ, Dekker C (2014) Zooming into see the bigger picture: microfluidic and nanofabrication tools to study bacteria. Science 346(6208):1251821. doi:10.1126/science.1251821
40. Nichols D, Cahoon N, Trakhtenberg EM, Pham L, Mehta A, Belanger A, Kanigan T, Lewis K, Epstein SS (2010) Use of Ichip for high-throughput in situ cultivation of "uncultivable" microbial species. Appl Environ Microb 76(8):2445–2450. doi:10.1128/AEM.01754-09

41. Osman O, Pivetal J, Frenea-Robin M, Haddour N, Zanini LF, Reyne G, Dempsey N, Dumas-Bouchiat F, Buret F (2013) On-chip cell sorting using an autonomous microfluidic magnetic device.
42. Lee W, Kwon D, Choi W, Jung GY, Jeon S (2015) 3D-printed microfluidic device for the detection of pathogenic bacteria using size-based separation in helical channel with trapezoid cross-section. Sci Rep 5(2):7717. doi:10.1038/srep07717
43. Lee JJ, Jeong KJ, Hashimoto M, Kwon AH, Rwei A, Shankarappa SA, Tsui JH, Kohane DS (2013) Synthetic ligand-coated magnetic nanoparticles for microfluidic bacterial separation from blood. Nano Lett 14(1):1–5. doi:10.1021/nl3047305
44. Taniguchi Y, Xie XS (2010) Quantifying E. coli proteome and transcriptome with single-molecule sensitivity in single cells. Science 329(5991):533–538. doi:10.1126/science.1188308
45. Zuo P, Li X, Dominguez DC, Ye BC (2013) A PDMS/paper/glass hybrid microfluidic biochip integrated with aptamer-functionalized graphene oxide nano-biosensors for one-step multiplexed pathogen detection. Lab Chip 13(19):3921–3928. doi:10.1039/c3lc50654a
46. Matsumoto Y, Hayama K, Sakakihara S, Nishino K, Noji H, Iino R, Yamaguchi A (2011) Evaluation of multidrug efflux pump inhibitors by a new method using microfluidic channels. Plos One 6(4). doi:10.1371/journal.pone.0018547
47. Lanning LM, Ford RM, Tao L (2008) Bacterial chemotaxis transverse to axial flow in a microfluidic channel. Biotechnol Bioeng 100(4):653–663. doi:10.1002/bit.21814
48. Balasubramanian AK, Beskok A, Pillai SD (2007) In situ analysis of bacterial capture in a microfluidic channel. J Micromech Microeng 17(8):1467–1478. doi:10.1088/0960-1317/17/8/008
49. Janakiraman V, Englert D, Jayaraman A, Baskaran H (2009) Modeling growth and quorum sensing in biofilms grown in microfluidic chambers. Ann Biomed Eng 37(6):1206–1216. doi:10.1007/s10439-009-9671-8
50. De LFL, Montanes E, Meng Y, Li Y, Burr TJ, Hoch HC, Wu M (2007) Assessing adhesion forces of type I and type IV pili of Xylella fastidiosa bacteria by use of a microfluidic flow chamber. Appl Environ Microb 73(8):2690–2696. doi:10.1128/AEM.02649-06
51. Moffitt JR, Lee JB, Cluzel P (2012) The single-cell chemostat: an agarose-based, microfluidic device for high-throughput, single-cell studies of bacteria and bacterial communities. Lab Chip 12(8):1487–1494. doi:10.1039/b000000x
52. Churski K, Kaminski TS, Jakiela S, Kamysz W, Baranskarybak W, Weibel DB, Garstecki P (2012) Rapid screening of antibiotic toxicity in an automated microdroplet system. Lab Chip 12(9):1629–1637. doi:10.1039/c2lc21284f
53. Chen CH, Lu Y, Sin MLY, Mach KE, Zhang DD, Gau V, Liao JC, Wong PK (2010) Rapid antimicrobial susceptibility testing using high surface-to-volume ratio microchannels. Anal Chem 82(82):1012–1019. doi:10.1021/ac9022764
54. Kim KP, Kim YG, Choi CH, Kim HE, Lee SH, Chang WS, Lee CS (2010) In situ monitoring of antibiotic susceptibility of bacterial biofilms in a microfluidic device. Lab Chip 10(23):3296–3299. doi:10.1039/c0lc00154f
55. Chung BG, Choo J (2010) Microfluidic gradient platforms for controlling cellular behavior. Electrophoresis 31(18):3014–3027. doi:10.1002/elps.201000137
56. Cheng SY, Heilman S, Wasserman M, Archer S, Shuler ML, Wu M (2007) A hydrogel-based microfluidic device for the studies of directed cell migration. Lab Chip 7:763–769. doi:10.1039/b618463d
57. Si G, Yang W, Bi S, Luo C, Ouyang Q (2012) A parallel diffusion-based microfluidic device for bacterial chemotaxis analysis. Lab Chip 12(7):1389–1394. doi:10.1039/c2lc21219f
58. Choi E, Jun I, Chang HK, Park KM, Shin H, Park KD, Park J (2012) Quantitatively controlled in situ formation of hydrogel membranes in microchannels for generation of stable chemical gradients. Lab Chip 12(2):302–308. doi:10.1039/c1lc20777f
59. Derek L, Englert MDM, Jayaraman Arul (2009) Flow-based microfluidic device for quantifying bacterial chemotaxis in stable. Competing Gradients. Appl Environ Microbiol 75 (13):4557–4564. doi:10.1128/AEM.02952-08

60. Tang Y, Gan M, Xie Y, Li X, Chen L (2014) Fast screening of bacterial suspension culture conditions on chips. Lab Chip 14(6)1162–1167. doi:10.1039/c3lc51332g
61. Long Z, Nugent E, Javer A, Cicuta P, Sclavi B, Cosentino LM, Dorfman KD (2013) Microfluidic chemostat for measuring single cell dynamics in bacteria. Lab Chip 13(5):947–954. doi:10.1039/c2lc41196b
62. Renner LD, Weibel DB (2011) Physicochemical regulation of biofilm formation. MRS Bull 36(5):347–355. doi:10.1557/mrs.2011.65
63. Sun P, Liu Y, Sha J, Zhang Z, Tu Q, Chen P, Wang J (2011) High-throughput microfluidic system for long-term bacterial colony monitoring and antibiotic testing in zero-flow environments. Biosens Bioelectron 26(5):1993–1999. doi:10.1016/j.bios.2010.08.062
64. Luo C, Jiang L, Liang S, Ouyang Q, Ji H, Chen Y (2009) High-throughput microfluidic system for monitoring diffusion-based monolayer yeast cell culture over long time periods. Biomed Microdevices 11(5):981–986. doi:10.1007/s10544-009-9315-7
65. Wu F, van Schie BG, Keymer JE, Dekker C (2015) Symmetry and scale orient min protein patterns in shaped bacterial sculptures. Nat. Nanotech. 10(8):719–726. doi:10.1038/NNANO.2015.126
66. Boulineau S, Tostevin F, Kiviet DJ, ten Wolde PR, Nghe P, Tans SJ (2013) Single-cell dynamics reveals sustained growth during diauxic shifts. PLoS ONE 8(4):7377–7382. doi:10.1371/journal.pone.0061686
67. Lee SS, Vizcarra IA, Huberts DHEW, Lee LP, Heinemann M (2012) Whole lifespan microscopic observation of budding yeast aging through a microfluidic dissection platform. P Natl Acad Sci USA 109(13):4916–4920. doi:10.1073/pnas.1113505109
68. Eun YJ, Utada A, Copeland MF, Takeuchi S, Weibel DB (2011) Encapsulating bacteria in agarose microparticles using microfluidics for high-throughput cell analysis and isolation. ACS Chem Biol 6(6):260–266. doi:10.1021/cb100336p
69. Probst C, Grünberger A, Wiechert W, Kohlheyer D (2013) Microfluidic growth chambers with optical tweezers for full spatial single-cell control and analysis of evolving microbes. J Microbiol Meth 95(3):470–476. doi:10.1016/j.mimet.2013.09.002
70. Shoji Takeuchi, ‡, † DL, And DBW, †, † GMW (2005) Controlling the shape of filamentous cells of Escherichia coli. Nano Lett 5(9):1819–1823
71. Peitz I, Van LR (2010) Single-cell bacteria growth monitoring by automated DEP-facilitated image analysis. Lab Chip 10(21):2944–2951. doi:10.1039/c004691d
72. Wu M (2010) Microfluidics for bacterial chemotaxis. Integr Biol 2(11–12):604–629. doi:10.1039/c0ib00049c
73. Mao H, Cremer PS, Manson MD (2003) A sensitive, versatile microfluidic assay for bacterial chemotaxis. P Natl Acad Sci USA 100(9):5449–5454. doi:10.1073/pnas.0931258100
74. Lin H, Zhang W, Jia S, Guan Z, Yang CJ, Zhu Z (2014) Microfluidic approaches to rapid and efficient aptamer selection. Biomicrofluidics 8(4):041501. doi:10.1063/1.4890542
75. Jun Kim M, Breuer‡ KS (2007) Controlled mixing in microfluidic systems using bacterial chemotaxis. Anal Chem 79(3):955–959. doi:10.1021/ac0614691
76. Smart AG (2010) Two experiments, two takes on electric bacteria. Phys Today 63(12):18–20. doi:10.1063/1.3529396
77. Park S, Wolanin PM, Yuzbashyan EA, Silberzan P, Stock JB, Austin RH (2003) Motion to form a quorum. Science 301(5630):188–188. doi:10.1126/science.1079805
78. Park S, Wolanin PM, Yuzbashyan EA, Lin H, Darnton NC, Stock JB, Silberzan P, Austin R (2003) Influence of topology on bacterial social interaction. P Natl Acad Sci USA 100(24):13910–13915. doi:10.1073/pnas.1935975100
79. Yang D, Greer CM, Jones BP, Jennings AD, Retterer ST, Männik J (2015) Characterization of small microfluidic valves for studies of mechanical properties of bacteria. J. Vac. Sci. Technol., B: Nanotechnol. Microelectron.: Mater., Process., Meas., Phenom. 33(6). doi:10.1116/1.4929883
80. Kalsi S, Valiadi M, Tsaloglou MN, Parry-Jones L, Jacobs A, Watson R, Turner C, Amos R, Hadwen B, Buse J (2015) Rapid and sensitive detection of antibiotic resistance on a

programmable digital microfluidic platform. Lab Chip 15(14):22–28. doi:10.1039/c5lc00462d
81. Czilwik G, Messinger T, Strohmeier O, Wadle S, Von SF, Paust N, Roth G, Zengerle R, Saarinen P, Niittymäki J (2015) Rapid and fully automated bacterial pathogen detection on a centrifugal-microfluidic labdisk using highly sensitive nested PCR with integrated sample preparation. Lab Chip 15(18):3749–3759. doi:10.1039/c5lc00591d
82. Mohan R, Sanpitakseree C, Desai AV, Sevgen SE, Schroeder CM, Kenis PJA (2015) A microfluidic approach to study the effect of bacterial interactions on antimicrobial susceptibility in polymicrobial cultures. Rsc Adv 5(44):35211–35223. doi:10.1039/c5ra04092b
83. Jang S, Lee B, Jeong HH, Jin SH, Jang S, Kim SG, Jung GY, Lee CS (2016) On-chip analysis, indexing and screening for chemical producing bacteria in a microfluidic static droplet array. Lab Chip 16(10):1909–1916. doi:10.1039/c6lc00118a
84. Mohan R, Mukherjee A, Sevgen SE, Sanpitakseree C, Lee J, Schroeder CM, Kenis PJA (2014) A multiplexed microfluidic platform for rapid antibiotic susceptibility testing. Biosens Bioelectron 49(35):118–125. doi:10.1016/j.bios.2013.04.046
85. Tang Y, Zhen L, Liu J, Wu J (2013) Rapid antibiotic susceptibility testing in microfluidic pH sensor. Anal Chem 85(5):2787–2794. doi:10.1021/ac303282j
86. Hou Z, An Y, Hjort K, Hjort K, Sandegren L, Wu Z (2014) Time lapse investigation of antibiotic susceptibility using a microfluidic linear gradient 3D culture device. Lab Chip 14 (17):3409–3418. doi:10.1039/c4lc00451e
87. Cira NJ, Ho JY, Dueck ME, Weibel DB (2011) A self-loading microfluidic device for determining the minimum inhibitory concentration of antibiotics. Lab Chip 12(6):1052–1059. doi:10.1039/c2lc20887c
88. And KWJ, Chen G (1994) Atomic absorption, atomic emission, and flame emission spectrometry. Anal Chem 66(66):169R–184R. doi:10.1021/ac00163a017
89. Burlingame AL, Boyd RK, Gaskell SJ (2010) Chem inform abstract: mass spectrometry. Chem Inform 27(39)
90. Ryan MD, Bowden EF, Chambers JQ (1994) Dynamic electrochemistry: methodology and application. Anal Chem 66(12):4497–4520. doi:10.1021/ac00084a015
91. Roda A, Cevenini L, Borg S, Michelini E, Calabretta MM, Schüler D (2013) Bioengineered bioluminescent magnetotactic bacteria as a powerful tool for chip-based whole-cell biosensors. Lab Chip 13(24):4881–4889. doi:10.1039/c3lc50868d
92. Date A, Pasini P, Daunert S (2010) Integration of spore-based genetically engineered whole-cell sensing systems into portable centrifugal microfluidic platforms. Anal Bioanal Chem 398(1):349–356. doi:10.1007/s00216-010-3930-2
93. Buffi N, Merulla D, Beutier J, Barbaud F, Beggah S, Van LH, Renaud P, Jr VDM (2011) Development of a microfluidics biosensor for agarose-bead immobilized Escherichia coli bioreporter cells for arsenite detection in aqueous samples. Lab Chip 11(14):2369–2377. doi:10.1039/c1lc20274j
94. Boehm DA, Gottlieb PA, Hua SZ (2007) On-chip microfluidic biosensor for bacterial detection and identification. Sensor Actuat 126(2):508–514. doi:10.1016/j.snb.2007.03.043
95. Sakamoto C, Yamaguchi N, Nasu M (2005) Rapid and simple quantification of bacterial cells by using a microfluidic device. Appl Environ Microb 71(2):1117–1121. doi:10.1128/AEM.71.2.1117-1121.2005
96. Panteli JT, Forkus BA, Van DN, Forbes NS (2015) Genetically modified bacteria as a tool to detect microscopic solid tumor masses with triggered release of a recombinant biomarker. Integr Biol-UK 7(4):423–434. doi:10.1039/c5ib00047e
97. Park SJ, Park SH, Cho S, Kim DM, Lee Y, Ko SY, Hong Y, Choy HE, Min JJ, Park JO (2013) New paradigm for tumor theranostic methodology using bacteria-based microrobot. Sci Rep 3(12):3394–3394. doi:10.1038/srep03394
98. Chung D, Jeong D, Koo D, Yoshida H, Song KM, Chang MH, Kang HG, Yun SH, Cho S, Jung KJ, Chung H (2011) Fusion fuel gas recovery and delivery characteristics on a

tray-type ZrCo bed. Fusion Eng Des 86(9–11):2233–2236. doi:10.1016/j.fusengdes.2010.11.026
99. Chang MH, Cho S, Kang HG, Yun SH, Song KM, Kim D, Chung H (2011) Process simulation for fuel delivery from storage and delivery system in fusion power plant. Fusion Eng Des 86(9–11):2200–2203. doi:10.1016/j.fusengdes.2011.01.130
100. Hou H, Li L, Ceylan CÜ, Haynes A, Cope J, Wilkinson HH, Erbay C, De FP, Han A (2012) A microfluidic microbial fuel cell array that supports long-term multiplexed analyses of electricigens. Lab Chip 12(20):4151–4159. doi:10.1039/c2lc40405b
101. Choi S, Chae J (2013) Optimal biofilm formation and power generation in a micro-sized microbial fuel cell (MFC). Sens Actuators, A 195(6):206–212. doi:10.1016/j.sna.2012.07.015
102. Hou H, Li L, Figueiredo PD, Han A (2011) Air-cathode microbial fuel cell array: a device for identifying and characterizing electrochemically active microbes. Biosens Bioelectron 26(5):2680–2684. doi:10.1016/j.bios.2010.06.037
103. Choi S, Lee HS, Yang Y, Parameswaran P, Torres CI, Rittmann BE, Chae J (2011) A μL-scale micromachined microbial fuel cell having high power density. Lab Chip 11(6):1110–1117. doi:10.1039/c0lc00494d
104. Boehm DA, Gottlieb P, Hua SZ (2007) Surface functionalization of a microfluidic biosensor for bacteria detection and identification. Proceedings of SPIE—The International Society for Optical Engineering 6529. doi:10.1117/12.714441
105. Lonigro SL, Valerio F, Angelis MD, Bellis PD, Lavermicocca P (2009) Microfluidic technology applied to cell-wall protein analysis of olive related lactic acid bacteria. Int J Food Microbiol 130(1):6–11. doi:10.1016/j.ijfoodmicro.2008.12.018
106. Ikeda M, Yamaguchi N, Tani K, Nasu M (2006) Rapid and simple detection of food poisoning bacteria by bead assay with a microfluidic chip-based system. J Microbiol Meth 67(2):241–247. doi:10.1016/j.mimet.2006.03.014
107. Chen J, Li-Hong TU, Chen YX, Qin SY (2014) Lab-on-chip for the detection of food borne pathogenic bacteria. Shanghai J Preven Med
108. Oh SJ, Park BH, Jung JH, Seo TS (2014) Centrifugal isothermal amplification microdevice for rapid, multiplex and colorimetric food poisoning bacteria detection
109. Zhang B, Luo Y, Zhou B, Wang Q, Millner PD (2015) A novel microfluidic mixer-based approach for determining inactivation kinetics of Escherichia coli O157: H7 in chlorine solutions. Food Microbiol 49:152–160. doi:10.1016/j.fm.2015.01.013
110. Schemberg J, Grodrian A, Römer R, Gastrock G, Lemke K (2010) Application of segmented flow for quality control of food using microfluidic tools. Physica Status Solidi 207(4):904–912. doi:10.1002/pssa.200983315
111. Varshney M, Li Y, Srinivasan B, Tung S (2007) A label-free, microfluidics and interdigitated array microelectrode-based impedance biosensor in combination with nanoparticles immunoseparation for detection of Escherichia coli O157: H7 in food samples. Sensor Actuat, B 128(1):99–107. doi:10.1016/j.snb.2007.03.045
112. Yang M, Sun S, Kostov Y, Rasooly A (2010) Lab-On-a-Chip for carbon nanotubes based immunoassay detection of Staphylococcal Enterotoxin B (SEB). Lab Chip 10(8):1011–1017. doi:10.1039/b923996k
113. Wilkens AS (2015) Microfluidic platforms for studying growth and motility of individual bacteria. Dissertations & Theses—Gradworks
114. Bj Q, Vm E, Ir F (2009) Microfluidic confinement of single cells of bacteria in small volumes initiates high-density behavior of quorum sensing and growth and reveals its variability. Angew Chem 48(32):5908–5911. doi:10.1002/anie.200901550
115. Wessel AK, Hmelo L, Parsek MR, Whiteley M (2013) Going local: technologies for exploring bacterial microenvironments. Nat Rev Microbiol 11(5):337–348. doi:10.1038/nrmicro3010
116. Moon HS (2009) Continuous microfluidic airborne bacteria separation using dielectrophoresis

117. Moon HS, Nam YW, Park JC, Jung HI (2009) Continuous microfluidic airborne bacteria separation using dielectrophoresis. In: Solid-State Sensors, Actuators and Microsystems Conference, 2009. Transducers 2009. International pp 2038–2041
118. Loens K, Heirstraeten L V, Malhotrakumar S (2009) Optimal sampling sites and methods for detection of pathogens possibly causing community-acquired lower respiratory tract infections. J Clin Microbiol 47(1):21-31. doi:10.1128/JCM.02037-08
119. Leineweber W, Williams M (2015) Assured bacterial detection towards paper-based microfluidic chip for resource-limited areas. Santa Clara Santa Clara University
120. Lee W, Kwon D, Choi W, Jung GY, Au AK, Folch A, Jeon S (2015) Erratum: corrigendum: 3D-printed microfluidic device for the detection of pathogenic bacteria using size-based separation in helical channel with trapezoid cross-section. Sci Rep 5:(7717). doi:10.1038/srep07717
121. Kim J, Hegde M, Kim H, Wood K, Jayaraman A (2012) Lab chip paper a microfluidic device for high throughput bacterial biofilm studies †. doi:10.1039/c2lc20800h
122. Fronczek CF, Park TS, Yoon JY (2013) Paper microfluidic extraction of bacterial and viral nucleic acid from field and clinical samples towards a direct microtas apparatus
123. Choi G, Choi S (2015) Bacterial cell transportation in paper-based microfluidics. Transducers 2015:21–25. doi:10.1109/TRANSDUCERS.2015.7181297
124. Siva PRM (2013) A microfluidic bacteria culturing device with MALDI mass spectrometry detection
125. Seymour JR, Ahmed T, Marcos Stocker R (2008) A microfluidic chemotaxis assay to study microbial behavior in diffusing nutrient patches. Limnol Oceanogr-Meth 6(5):477–488. doi:10.4319/lom.2008.6.477
126. Drescher K, Nadell C, Stone H, Wingreen N, Bassler B (2013) Solutions to the public goods dilemma in bacterial biofilms. Curr Biol 24(1):50–55

Index

J.-M. Lin (ed.), *Cell Analysis on Microfluidics*, Integrated Analytical Systems,
https://doi.org/10.1007/978-981-10-5394-8

D

E

F

FSC
www.fsc.org
MIX
Papier | Fördert gute Waldnutzung
FSC® C083411

Kolibri 360 GmbH
Ferdinand-Jühlke-Straße 7
99095 Erfurt, Deutschland
produktsicherheit@kolibri360.de